Toward Global Planning of Sustainable Use of the Earth

Development of Global Eco-Engineering

Toward Global Planning of Sustainable Use of the Earth

Development of Global Eco-Engineering

A Collection of Contributions Based on Lectures Presented at the Eighth TOYOTA Conference, Mikkabi, Shizuoka, Japan, November 8 to 11, 1994

Edited by

SHUNJI MURAI

School of Advanced Technologies
Asian Institute of Technology
Bangkok, Thailand

ELSEVIER

Amsterdam – London – New York – Tokyo 1995

ELSEVIER SCIENCE B.V.
Sara Burgerhartstraat 25
P.O. Box 211, 1000 AE Amsterdam, The Netherlands

Library of Congress Cataloging-in-Publication Data

Toyota Conference (8th : 1993 : Shizuoka-shi, Japan)
Toward global planing of sustainable use of the Earth : development of global eco-engineering : a collection of contributions based on lectures presented at the Eighth Toyota Conference, Mikkabi, Shizuoka, Japan, November 8 to 11, 1994 / edited by Shunji Murai.
p. cm.
Includes bibliographical references and index.
ISBN 0-444-81904-5 (alk. paper)
1. Environmental policy--Congresses. 2. Environmental engineering--Congresses. 3. Human ecology Congresses. 4. Sustainable development--Congresses. I. Murai, Shunji. II. Title.
GE170.T72 1993
363.7--dc20 95-18909
CIP

ISBN: 0 444 81904 5

This book is printed on acid-free paper.

Printed in The Netherlands

Preface

Shunji Murai, Chairman of the Organizing Committee
December, 1994

Backgrounds

Rapid expansion of human activities results in serious impacts on the global environmental system, which threatens the long-term habitability of the Earth. Better understanding of global environmental systems and the magnitude of human impacts is the most fundamental research task for developing an ecologically sound basis for the continuous human habitation and sustainable development of the earth's limited resources. Although many research projects are already underway to begin addressing these issues, using global data mainly obtained from remote sensing technologies, our knowledge is far from sufficient.

To secure ecologically sound basis for the continuous human habitation, strategies for the sustainable uses and conservation of global environmental resources are urgently needed.

Through intensive discussion and exchange of information among experts in a wide variety of fields, such as global monitoring of environmental resources, modeling of environmental system dynamics and interactions with human activities, the conference will contribute to formulation of a framework for further research. A break through is expected toward the development of global eco-engineering which seeks continuous human habitation and improvement of human welfare based on the sustainable utilization of global environmental resources and preservation of global eco-systems.

The scope of the conference

The scope of the conference includes:

1. Global monitoring and evaluation of human impacts on global environmental systems
   Global monitoring of land cover and vegetation changes
   Evaluation of global environmental resources
   Large-scale changes of land cover and climate changes
   Impacts on global environment of Mega-cities or overpopulated cities

2. Analysis and modeling of global environmental systems and the interactions with human activities
   Modeling of terrestrial and oceanic ecosystem / Modeling of biogeochemical cycles
   Analysis of impacts of land cover changes on climate systems
   Analysis on impacts of land use and land cover changes on biogeochemical cycles
   Long term growth model of population and economy under resources and environmental constraints

3. Establishing global environmental databases and information networks
   Global Geo-Information System
   Integrated models of physical environment and socio-economic activities
   Visualization of global environmental information

4. Environmental ethics and concept of eco-soundness in the global age
   - Environmental ethics, philosophical views on the environment
   - Social norms and life styles for eco-sound society
   - Environmental education in the global age

5. International cooperation in the global age and decision making under large scale of uncertainties and risks
   - Strategies for promoting consensus for international cooperation in a global information network environment
   - Decision making under large scale of uncertainties and risks

6. Development of global eco-engineering
   - Establishing global eco-planning
     - Underlying philosophy and planning objectives
     - Planning techniques (Accounting system of environmental resources, resources management, optimum control, risk management etc.)
     - Proposals for global eco-technology development
     - Underlying philosophy of global eco-technology
     - Technologies for sustainable use of global environmental resources
     - Technologies for substance circulation
     - Controlling strategies of socio-economic system and implementation of global eco-technology and eco-planning

Summaries

1. The 8th TOYOTA Conference was remarkably valuable with the following findings: a) it was quite interdisciplinary with a variety of disciplines covering remote sensing, geography, meteorology, biology, biochemistry, ecology, marine science, hydrology, agriculture, environmental engineering, urban planning, social science, economy, ethics and philosophy, b) top rank experts in the field of global change research gathered with high quality presentation, c) communication among the experts present was fostered with potential interdisciplinary cooperation in future, d) current status and future prospect about global change research and the applications were demonstrated with expertized experience and observations, and e) a new concept of "Global Eco-engineering" which bridges between human and physical dimensions has been proposed at the Conference and accepted by the participants.

2. All participants expressed a great appreciation to TOYOTA MOTOR CORPORATION and TOYOTA Central Research and Development Laboratory for their sponsorship and administrative arrangements to make the Conference productive and successful particularly with respect to the following items; a) provision of modern conference facilities, b) the proceedings to be published in a book by Elsevier Publisher after the Conference, and c) dissemination of CD-ROM with global dataset in cooperation of Institute of Industrial Science, University of Tokyo under TOYOTA crown "Global Eco-engineering" project.

Recommendations

1. Realizing the advantages of "Global Eco-engineering" approach, "Global Eco-engineering" should be promoted for decision support to construct an eco-sound society.

2. Global dataset with socio-economical and physical items to support "Global Eco-engineering" should be generated and disseminated with specific reliability.

3. Experts in the field of global change research should exchange information and available data among the experts to accelerate their research.

4. Such conference with a focus on "Global Eco-engineering" should be periodically organized.

Organizing Committee of the Eighth TOYOTA Conference

Chairman: Professor Shunji Murai
Asian Institute of Technology
G. P. O. Box 2754, Bangkok 10501, Thailand

Members: Mr. Tokuta Inoue
TOYOTA Motor Corporation
1200 Mishuku, Susono-shi, Sizuoka-ken 410-11
Japan

Dr. Akira Miyawaki
Japanese Center for International Studies in Ecology
(JISE)
1560-39 Kamiyamaguchi, Hayama-cho, Miura-gun
Kanagawa 240--1, Japan

Dr. Masaaki Naito
National Institute for Environmental Studies
16-2 Onogawa, Tsukuba 305, Japan

Professor Mikio Takagi
Institute of Industrial Science
University of Tokyo
7-22-1 Roppongi, Minato-ku, Tokyo 106, Japan

Professor Ryosuke Shibasaki
Institute of Industrial Science
University of Tokyo
7-22-1 Roppongi, Minato-ku, Tokyo 106, Japan

Professor Yoshiaki Honda
Chiba University
1-33 Yayoi-cho, Inage-ku, Chiba 263, Japan

Dr. Jun Matsumoto
University of Tokyo
7-3-1 Hongo, Bunkyo-ku, Tokyo 113, Japan

Professor Ryutaro Tateishi
Chiba University
1-33 Yayoi-cho, Inage-ku, Chiba 263, Japan

Dr. Yoshifumi Yasuoka
National Institute for Environmental Studies
16-2 Onogawa, Tsukuba 305, Japan

Auditor: Mr. Noboru Nakagawa
Director
Toyota Central R&D Labs., Inc.
41-1 Yokomichi, Nagakute, Nagakute-cho,
Aichi-gun, Aichi 480-11, Japan

Secretary General: Mr. Masato Kimura
Toyota Central R&D Labs., Inc.
41-1 Yokomichi, Nagakute, Nagakute-cho,
Aichi-gun, Aichi 480-11, Japan

TOYOTA Conferences
Sponsored by **TOYOTA MOTOR CORPORATION**

The First TOYOTA Conference
Molecular Conformation and Dynamics of Macromolecules in Condensed Systems, September 28 - October 1, 1987 (Aichi, Japan), edited by M. Nagasawa

The Second TOYOTA Conference
Organization of Engineering Knowledge for Product Modelling in Computer Integrated Manufacturing, October 2 - 5, 1988 (Aichi, Japan), edited by T. Sata

The Third TOYOTA Conference
Integrated Micro Motion Systems - Micromachining, Control and Applications, October 22 - 25, 1989 (Aichi, Japan), edited by F. Harashima

The Fourth TOYOTA Conference
Automation in Biotechnology, October 21 - 24, 1990 (Aichi, Japan), edited by I. Karube

The Fifth TOYOTA Conference
Nonlinear Optics - Fundamentals, Materials and Devices, October 6 - 9, 1991 (Aichi, Japan), edited by S. Miyata

The Sixth TOYOTA Conference
Turbulence and Molecular Processes in Combustion, October 11 - 14, 1992 (Shizuoka, Japan), edited by T. Takeno

The Seventh TOYOTA Conference
Towards the Harnessing of Chaos, October 31 - November 3, 1993 (Shizuoka, Japan), edited by M. Yamaguti

The Eighth TOYOTA Conference
Toward Global Planning of Sustainable Use of the Earth - Development of Global Eco-Engineering, November 8 - 11, 1994 (Shizuoka, Japan), edited by S. Murai

The Ninth TOYOTA Conference
Brain & Mind - for better understanding of the dynamic function of mind and its supporting brain mechanism, December 5 - 8, 1995 (Shizuoka, Japan), edited by M. Ito (to be published)

The Eighth TOYOTA Conference Nov. 8-11, 1994

CONTENTS

Cover photopraph: World Map of Potential Natural Vegetation
(Dark to Bright: Desert, Semi-desert, Glassland, Forest, Tropical forest) -by Y. Honda

TOWARD GLOBAL PLANNING OF SUSTAINABLE USE OF THE EARTH
Development of Global Eco-Engineering
S. Murai (Editor)

Development of Global Eco-engineering using Remote Sensing and Geo-information Systems

Shunji Murai

School of Advanced Technologies
Asian Institute of Technology (AIT)
P.O.Box 2754, Bangkok 10501, Thailand

Institute of Industrial Science
University of Tokyo
7-22-1 Roppongi, Minatoku, Tokyo 106, Japan

Abstract

Human impacts due to over consumption of natural resources cause drastic global changes on the earth eco-systems.

Satellite remote sensing and geo-information system (GIS) with global data set are now powerful tools for mapping the resource inventory and the land use change as well as for scientific analysis of the mechanism of global change. This is a sort of global earth science. However we need an engineering view point in order to propose the countermeasures for eco-sound human response in a term of global "earth use planning" rather than regional or national planning.

To support the global eco-engineering concept, remote sensing and GIS on global scale will provide with the following useful information.

a. Actual Vegetation
b. Potential Vegetation
c. Human Impacts on Loss of Carbon Dioxide Stock due to Deforestation
d. Macro Balance of Carbon Dioxide
e. Habitability of Human Population with respect to Crop Production
f. Suitability Study for Forest Conservation with Global GIS

Establishment of global eco-planning is a major target in global eco-engineering.

1. INTRODUCTION

The author has been engaged in global study since 1986, mainly at Murai Laboratory, Institute of Industrial Science (IIS), University of Tokyo (UOT) , Japan where computer processing of global data set with remote sensing and GIS has been implemented by the following collaborators.

Dr. R. Shibasaki, Associate Professor, IIS
Dr. M. Takagi, Research Associate, IIS
Prof. E.O. Box, Visiting Professor, IIS

Dr. Denis Dye, Visiting Associate Professor, IIS
Dr. S. Gotoh, Associate Professor of Kanazawa Institute of Technology
Dr. Y. Honda, Lecturer, Yokohama National University
Dr. Bai Xue Mei, Environmental Research Center Yokohama

The budget sources of this study were as follows;
1989 FY ~ 1991 FY Three Years
Better Understanding of Earth Environment
Ministry of Education; 72M Japanese Yen
1991 FY ~ 1994 FY Three Years
Global Engineering Project
TOYOTA Corporation; 200M Japanese Yen

The author has served for the above projects as Coordinator.

Under these projects, all possible global data sets including NOAA AVHRR GVI data, weather data, soil data, population data, agricultural census, coastal lines and country boundaries, DEM (digital elevation model) etc. were collected.

Concerning raster base data set of temperature and rainfall have been rasterized from monthly contour maps of temperature and rainfall which were manually drawn from point wise weather data available at the existing stations in cooperation with meteorological specialists.

Dr. Honda, Dr. Gotoh and Dr. Bai took doctor degrees (Doctor of Engineering, University of Tokyo) during the above mentioned periods as the results of global eco-engineering issues.

Prof. Box and Dr. Dye have been invited as visiting faculty at IIS, University of Tokyo from Georgia University and Maryland University, USA respectively.

2. CONCEPT OF GLOBAL ECO-ENGINEERING UNDER GLOBAL INFORMATION SYSTEM

As proposed by ICSU/IGBP, two issues of human and physical dimension should be in a linkage each other as shown in Figure 1 which is added some details by the author.

Physical dimensions of human impacts and global environmental change will be analysed by "Earth System Science", which will be monitored by remote sensing.

On the other hand, human dimensions of human driving forces and human response will be studied by "Social Science", which will be supported by geo-information system (GIS).

Human driving forces will be implemented in a form of human impacts on the environment by the existing science, technology or engineering.

However we need a new descipline, that is called "Global Eco-engineering" by the author to feed back information about global environmental change by remote sensing to human response with respect to policy making, planning and management with an aid of GIS.

Therefore global eco-engineering is to bridge between physical and human dimensions for better strategies for building an eco-sound society.

3. FRAMEWORK OF GLOBAL ECO-ENGINEERING

Global eco-engineering includes the following framework as shown in Figure 2.

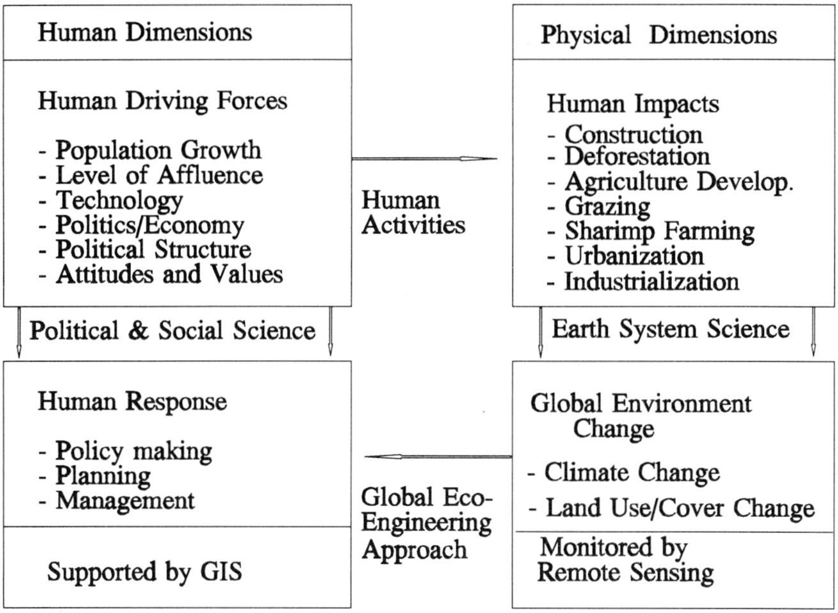

Figure 1 Concept of Global Information System

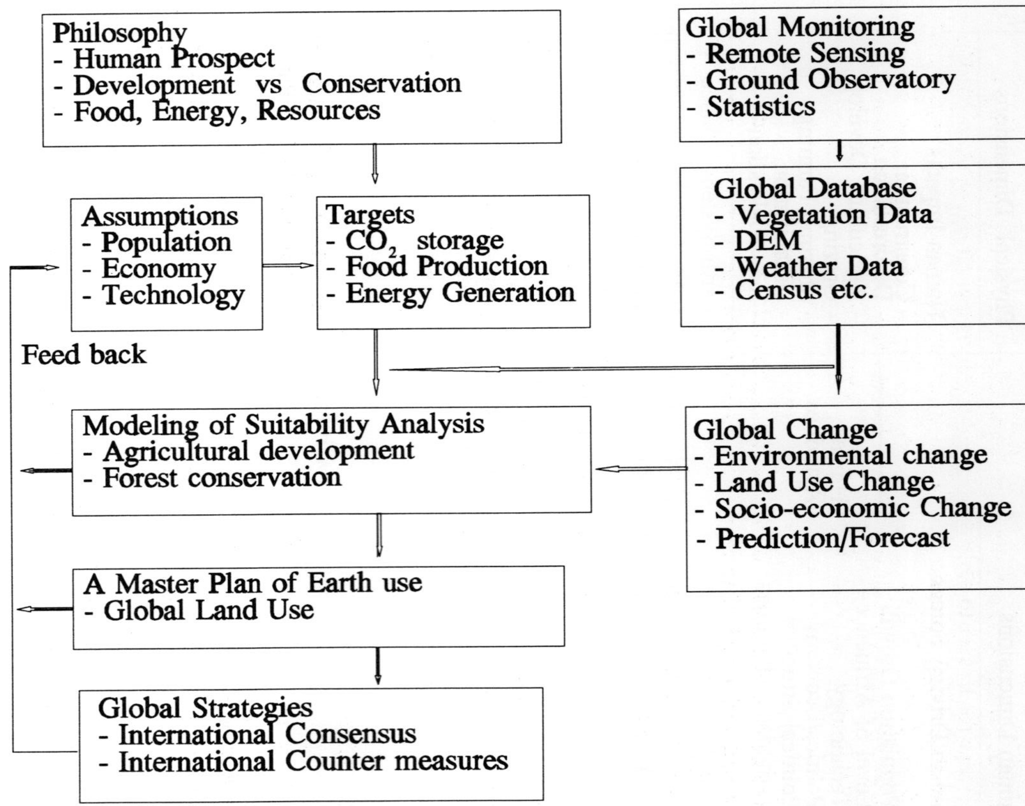

Figure 2 Framework of Global Eco-engineering

3.1 Philosophy

First of all, we need to establish a new philosophy for sustainable development. For example, in order for human to survive, can we cut all forests on the earth? Certainly not. Then how we should develop agricultural lands to support potentially increasing population. How should we control industrial development to keep fixation of carbon dioxide by vegetation to avoid green house effect? We need an international consensus about philosophy on what should be human prospect or sustainable development.

3.2 Targets

According to a philosophy as well as assumptions about population increase, energy consumption and technological development, targets should be launched for a model of suitability study.

Targets will include the followings for example;

a. How much CO_2 storage should be kept?
b. How much food production should we obtain?
c. How much energy should we generate?

3.3 Global Monitoring/Global Database

Global scale monitoring should be operationalized with satellite remote sensing ground base observations, global scale statistical survey etc. Then global database with vegetation, soil, rivers, topography, landuse, coastal lines, political boundaries, weather, agricultural/industrial census etc. should be established.

3.4 Global Change Study

Depending on global monitoring and global database, global change study will be conducted periodically or at rather short interval, which can result in prediction or forecasting.

3.5 Modeling of Suitability Analysis

A model or models can be built for locating suitable lands for agricultural development, forest conservation or forestation with use of the targets, global database and/or global change information.

3.6 A Master Plan of Earth Use

A master plan of global land use will be output from the above model (s).

3.7 Global Strategies

All countries, possibly through an agreement at United Nations should follow and implement "Earth Use Planning" based on international consensus.

4. ACTUAL AND POTENTIAL VEGETATION MAPS

An actual vegetation map at global scale was produced by Honda and Murai with Global Vegetation Index (GVI) of NOAA AVHRR using 1985, 1986 and 1987 years of very and normal vegetation conditions.

On the other hand, an potential vegetation map was also produced by Honda and Murai with an improved Martone's model as shown in Table 1.

Table 2 shows the results of actual and potential vegetation classification for five vegetations types; tropical forest, other forest, grassland (including agricultural land), semi-desert and desert.

According to the table, it was found that actually tropical forest and other forest have been deforested in the past 7.5% and 7.8% respectively, while grassland, semi-desert and desert have been increased 5.2%, 4.3% and 5.8% respectively due to human impacts.

5. HUMAN IMPACTS ON LOSS OF CO_2 STOCK

One can estimate the loss of CO_2 stock in term of Net Primary Production (NPP) or unit of carbon. Average NPP for the five vegetation types are given in Table 3 in unit of gDW/sq.m/year of NPP, which can be converted to unit of carbon/year for CO_2 stock.

Table 4 shows the CO_2 stock for actual and potential vegetation types as well as the loss from the potential vegetation's. Due to deforestation, 15.8 G ton carbon/year has been lost while the total loss including some gains from the increase of grassland is 12.4 G ton/year.

6. MACRO BALANCE OF CARBON DIOXIDE

According to Table 4, carbon stock by actual vegetation was estimated 49.4 G ton carbon/year by Honda and Murai.

Gotoh and Murai produced a chlorophyll map with NIM BUS satellite data on ocean at global scale for 1978-1981.

An improved model to estimate NPP in unit of gC/sq.m/day has been developed by Gotoh and Murai. The total NPP for all oceans at global scale was estimated 31.05 G ton carbon/year.

Carbon stock on both of land and sea areas at global scale has been estimated with satellite data for the first time in the world, while the existing estimates have been calculated by summing up point-wise measurement values.

Table 1: Zoning Criteria for Potential Vegetation

Type	
Tropical Forest	AI > 40 (T ≥ 24°C)
Other Forest	20 < AI ≥ 40
Grassland	10 < AI ≥ 20
Semi-desert	5 < AI ≤ 10
Desert	AI ≤ 5

Remarks: AI (Arid Index) = P/(T + 10)
P: annual rainfall (mm)
T: annual average temperature (°C)

Table 2: Change of Vegetation Type *Unit: %*

	Tropical Forest	Other Forest	Grassland	Semi-Desert	Desert	Total
Actual Vegetation	5.9	28.1	33.9	14.9	17.2	100.0
Potential Vegetation	13.4	35.9	28.7	10.6	11.4	100.0
Human Impact	-7.5	-7.8	+5.2	+4.3	+5.8	0.0

Remark: - decrease + increase

Table 3: Net Primary Production *Unit: gDW/sq.m/year*

Types	Net Primary Productivity
Tropical Forest	2,200
Other Forest	1,250
Grassland	900
Semi-desert	90
Desert	90

Table 4: Loss of Carbon Dioxide Stock by Vegetation Cover Change *Unit: Gton carbon/year*

	Tropical Forest	Other Forest	Grassland	Semi-Desert	Desert	Total
Actual Vegetation	7.9	21.3	18.3	0.8	0.9	49.4
Potential Vegetation	17.8	27.3	15.6	0.6	0.6	61.8
Loss	-9.9	-5.9	+2.9	+0.2	+0.3	-12.4

There is no way to verify the truth of the values, but it is important to demonstrate that satellite image data can provide such global information with area coverage.

7. HABITABILITY OF HUMAN POPULATION

There would be a limit for human population with a constraint of crop production. a question arises how much population can be supported potentially with respect to crop production. In order to answer the question, a potential crop land should be estimated. According to the author's philosophy, we should develop new crop land only in the area of the existing grassland but not in the forest land, because further deforestation will affect the drastic decrease of carbon stock which accelerates green house effect.

Table 5 shows an estimate of potential crop land that has been computed by Gotoh and Murai for five continents as well as the existing agricultural land and forest area.

Table 5: Potential Crop Land, Existing Crop Land and Forest Area

1,000 km^2

	Africa	North America	South America	Asia Europe	Oceania	World
Potention Crop Land (Gotoh and Murai)	4,681	4,604	1,446	8,832	1,812	21,375
Existing Crop Land (FAO 187)	1,688	2,671	1,163	7,752	478	13,732
Forest Area (Honda and Murai)	8,383	8,038	12,221	15,221	1,586	45,774

FAO has announced annually the statistics of agricultural productivity particularly main crops important for survival.

From those statistics, Gotoh and Murai calculated the existing and potential agricultural productivity in unit of k cal as shown in Table 6.

Table 6: Existing and Potential Agricultural Productivity

10^{12} k cal

	Africa	North America	South America	Asia Europe	Oceania	World
A. Existing Productivity	293	1,369	361	3,643	68	5,734
B. Potential Productivity	857	2,533	702	4,407	424	8,923
B/A	2.93	1.85	1.95	1.21	6.23	1.56

As seen in Table 6, there will be a potential to increase the agricultural production by 1.56 times of the existing agricultural production in the world.

Potential habitability will be estimated with the information about how much energy in unit of kcal is necessary to support a person. In this study, the three cases of energy consumption as shown in Table 7 were to be considered to estimate the potential habitability.

Table 7: Potential Habitability for Three Cases

Case	Energy Consumption	Potential Habitability (billion people)
1	as same at present in each country	7.34
2	world average	8.97
3	American standard	4.01

The current population as of 1993 is about 5.3 billion in the world. It implies that we cannot survive if we eat as American people eat. If we east as we eat at present, only 7.34 billions people would be a limit, which shows that the earth will be saturated with human population in comming about twenty years.

8. SUITABILITY STUDY FOR FOREST CONSERVATION WITH GLOBAL GIS

A model for forest conservation and reforestation has been proposed by Shibasaki and Murai with global GIS dataset as shown in Figure 3. The concept of the model is basically based on two aspects.

a. Contribution of fulfillment of a target to keep CO_2 stock at a certain level to protect from green house effect

b. Ease of forest conservation or reforestation in consideration of potential natural conditions such as climate and soil as well as existing land cover condition.

Figure 2 shows a procedure of suitability evaluation for forest conservation and reforestation.

In order to keep the current CO_2 level, that is 350 ppm of CO_2 density by reforestation, we need further reforestation shown in Table 8.

Table 8: Required Reforestation

10^3 km^2

	Africa	North America	South America	Asia Europe	Oceania	World
Required Reforestation (Gotoh and Murai)	389	276	263	705	110	1,734
Existing Forest Area (Honda and Murai)	8,383	8,038	12,546	15,221	1,586	45,774

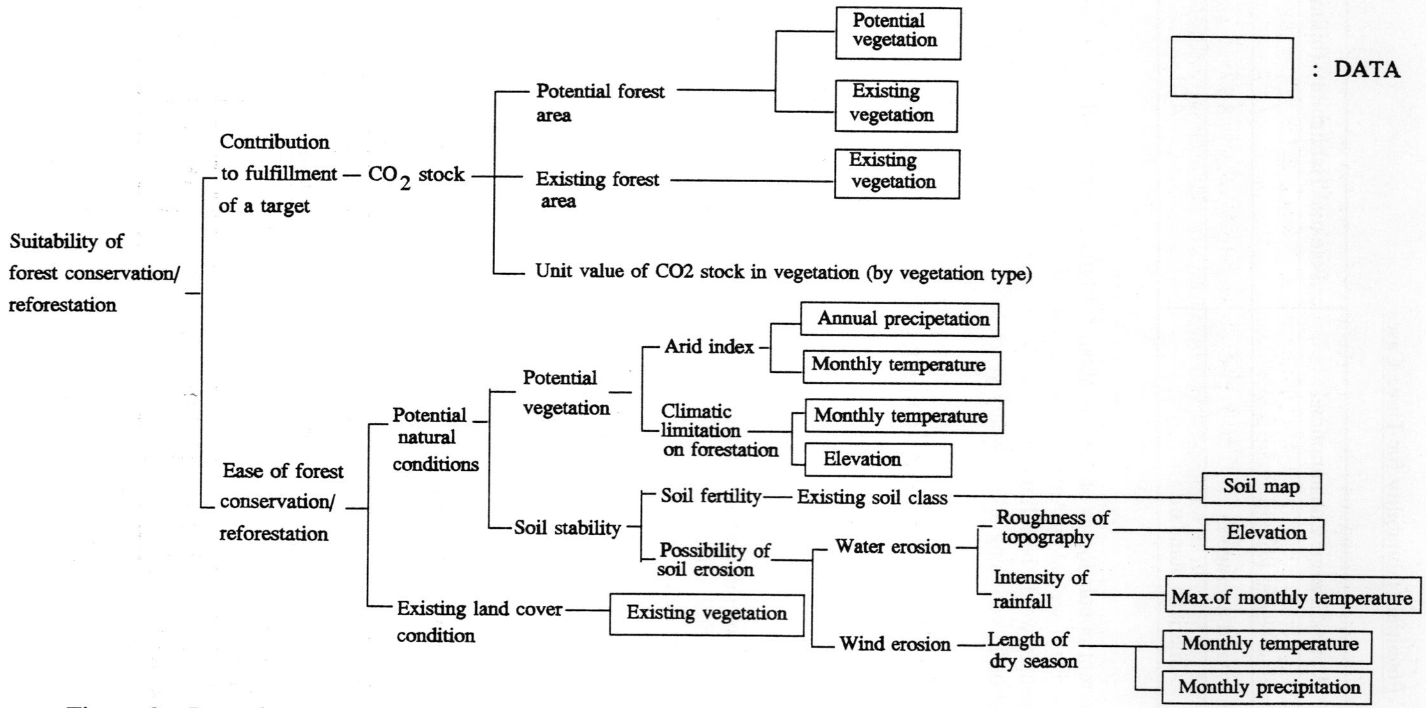

Figure 3 Procedure of Suitability evaluation for forest conservation / reforestation

9. CONCLUSIONS

1) Satellite remote sensing technologies together with GIS techniques are very powerful to assess the current global environmental condition and also predict the future trend or potential about human impacts, CO_2 stock, habitability, suitability for forest conservation or reforestation etc.
2) The results showed a lot of capability to develop global change analysis even though the results have not yet verified because there has been few reference data.

TOWARD GLOBAL PLANNING OF SUSTAINABLE USE OF THE EARTH
Development of Global Eco-Engineering
S. Murai (Editor)

Satellite Remote Sensing Studies of Desertification and Deforestation: Actual Findings vs. Assumed Situations

Compton J. Tucker

Laboratory for Terrestrial Physics, Code 923
National Aeronautics and Space Administration
Goddard Space Flight Center, Greenbelt, Maryland 20771 USA.

Abstract: Satellite data, coupled with selected ground observations, have been used to study two continental-scale environmental problems: Desert expansion and contraction in arid and semi-arid Africa on the south side of the of Sahara; and Tropical deforestation in the Amazon Basin of Brazil. In both situations, actual conditions differed from assumed conditions.

In Sahelian Africa, contrary to reports in the popular press, substantial recovery from the 1984 drought was found. Since 1984 there has been a marked improvement in rainfall, and hence vegetation production in the Sahel zone, with substantial inter-annual variation. These results cast doubts on the assumption of Sahelian desertification, where the recovery from the combined effect of drought and over grazing is considered to be a slow process.

Deforestation in the Brazilian Amazon was found to be lower than most current estimates, both in total extent as well as the 1978 to 1988 rate of deforestation. While tropical deforestation is less than assumed in the Brazilian Amazon, the pattern of deforestation was shown to be the principal determinant for the effect of deforestation upon biological diversity. This results from the combined effect of destruction of habitat (i.e., deforestation), "edge effects" where remaining tropical forest is in contact with deforestation, and the isolation of forest fragments by deforestation. The combined effects of tropical deforestation (destruction of tropical forest, edge effects, and isolation of forest fragments) upon biological diversity is two to three times greater than the tropical deforestation *per se*.

Introduction

Large-scale environmental problems, such as desertification and tropical deforestation, have been widely discussed and debated the past 20 years. It has proven difficult to quantitatively determine and specify exact conditions or situations with respect to both desertification and tropical deforestation. Recently, the use of satellite remote sensing has permitted both of these large-scale environmental problems to be quantified. I will review some recent work which my coworkers and I have completed regarding aspects of these two important environmental problems.

Desertification in Saharan-Sahelian Africa

Desertification has been defined as the process by which more productive arid and semi-arid lands become less productive or more "desert like". This can result from many causes, but generally is attributed to over-use by man and/or by climatic tendencies toward dryer climates.

In the 1980's international interest was focused on the south side of the Sahara Desert, upon the Sahel Zone, the semi-arid grassland or steppe area immediately to the south of the Sahara. One of the prevalent assumptions at that time was the continuous expansion of the Sahara Desert to the south. I will now review some of our NOAA satellite work on this topic from (Tucker et al., 1991 and Tucker and Newcomb, 1995).

The largest desert of our planet, the Sahara of Africa, is between ~7,000,000 to ~10,000,000 km^2 in size. The northern boundary has been arbitrarily set to follow the south side of the Atlas Mountains to Biskra (Algeria) where it dips southward through Tunisia to the Gulf of Gabes and from there continues along the Mediterranean to the Suez Canal (McGinnies et al., 1968; Cloudsley-Thompson, 1984; Swift, 1975). The southern boundary stretches east to west at ~ 16° to 17° N for 6,000 km from the Atlantic Ocean off Mauritania to the Red Sea off Sudan and is a vegetation-zone boundary. At this boundary there is an almost insensible gradient of plants, animals, and physiographic characteristics into steppe vegetation of the Sahel zone (Le Houerou, 1980; Monod, 1986) resulting from a mean annual precipitation gradient of ~ 1 mm yr^{-1} km^{-1} from north to south.

Uncertainty in and variation of the location of the boundary between the Sahara and the Sahel zone has prevented precise estimates of the extent of the Sahara (note: We include the Nubian Desert of Sudan, the Eastern and Western Deserts of Egypt, the Libyan Desert, and all other subdeserts of North Africa contiguous to the Sahara in our figures for the Sahara).

Although deserts have a number of common features of climate, weather, and a low density of vegetation, most workers have defined "desert" according to their discipline (Goodall and Perry, 1979; McGinnies, 1979; Shmida, 1985; Evenari, 1985). We use "desert" to be synonymous with "arid". "Semi-arid" represents the gradation of desert into "steppe" and denotes areas of less aridity, higher and more evenly distributed rainfall, and higher levels of primary production. Steppe is also imprecise and its use varies widely (Evenari, 1985).

Largely anecdotal information has accumulated from early this century to the present suggesting that the Sahara has expanded toward the south (Cana, 1915; Bovill, 1921; Stebbing, 1935; Rodd, 1938; Stebbing, 1938; Aubreville et. al., 1973; Eckholm and Brown, 1977; Smolowe, 1987; Norman, 1987; Ellis, 1987; Lamprey, 1975). The alleged expansion was attributed in part to climate variation (droughts)(Smith, 1986) and to land mismanagement such as overgrazing, increased cultivation, and firewood cutting (Eckholm and Brown, 1977; Ellis, 1987).

This process of land degradation is called "desertification" by some (Kassas, 1970) and "desertization" by others (Le Houerou, 1968). The occurence of a period of wet years (1950 to 1968) followed by dry years (1969 to 1990) has also contributed to the controversy in the location of the southern Sahara boundary (Fig. 1) (Nicholson, 1989; Lamb et al., 1986; Toupet, 1972).

Lamprey (1975) compared the location of the southern boundary of desert vegetation in western Sudan in 1958 with its location in 1975. He estimated that the boundary had shifted southward by 90 to 100 km during the 17-year period, a desert expansion of ~5.5 km yr^{-1} and this value has been repeated often in the popular press (Smolowe, 1987; Norman, 1987; Ellis, 1987). However, a 1984 field study by Hellden (1984) in the same area found no evidence to support such an expansion.

In order to measure directly the changes in vegetation at his boundary, we used a satellite-derived vegetation index to map interannual changes in vegetative cover and, by inference, rainfall, along the Saharan-Sahelian boundary from the Atlantic Ocean to the Red Sea for the period 1980 to 1993. We hypothesized that annual variations in rainfall would bring corresponding changes in the density of vegetative cover. Malo and Nicholson (1990) have shown that the satellite-derived vegetation index we used is linearly related to precipitation in the area of our study (Fig. 2).

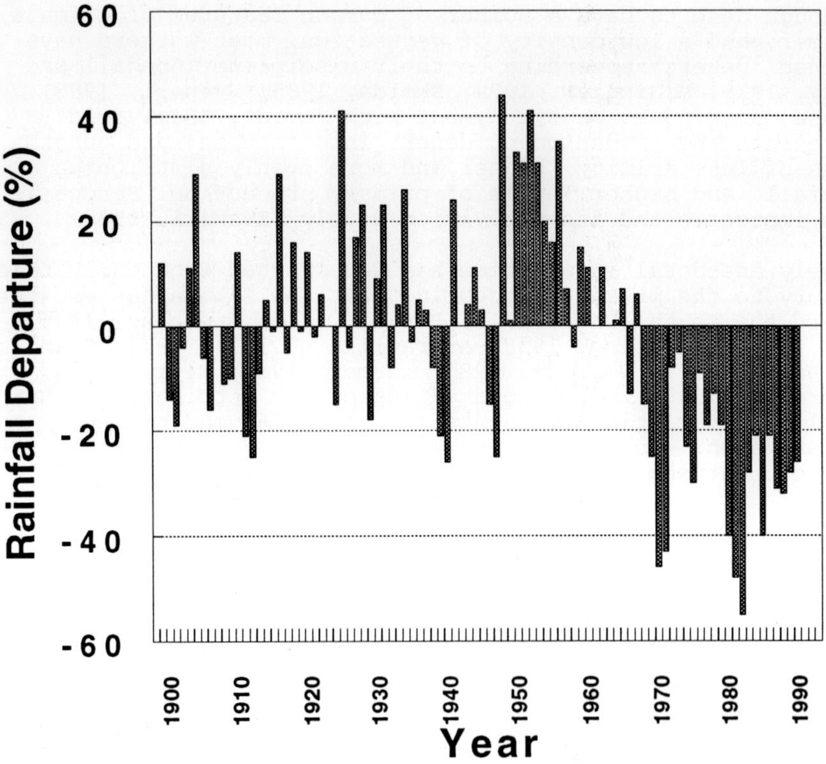

Figure. 1. Rainfall departures for the Sahel zone of Africa from 1900 to 1992 from Nicholson (1989) with additional data to 1992. Note the run of wet years (circa 1950-1965) followed by a run of dry year (circa 1970 to the present).

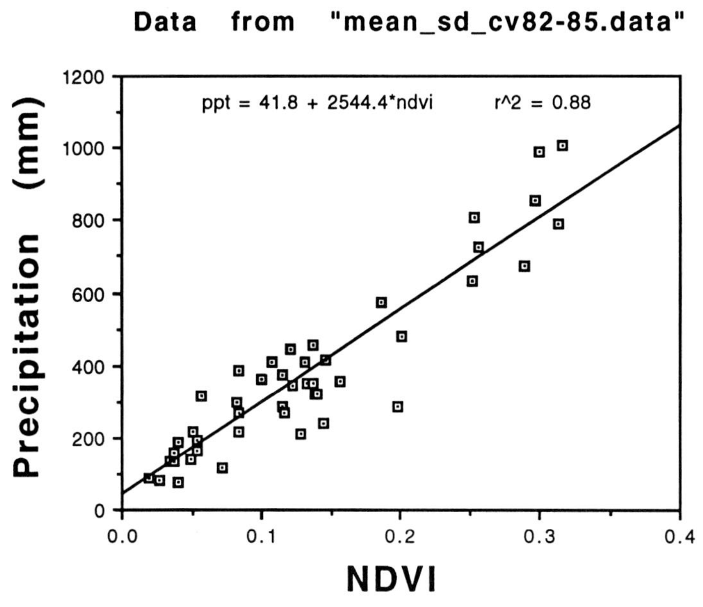

Figure. 2. Relation between the mean normalized difference vegetation index and mean annual precipitation for 1982-1985. The linear correlation coefficient between these variables is 0.93.

Vegetation indices are commonly used spectral measures derived from remotely-sensed data in the red and near-infrared spectral regions. The red spectral response is inversely related to the chlorophyll density, and the near-infrared spectral response is directly related to scattering in individual leaves and between leaves in the canopy. Combinations of these two adjacent spectral regions provides compensation for differences in irradiance and an estimate of the intercepted fraction of the photosynthetically active radiation or photosynthetic capacity (Tucker, 1979; Curran, 1983; Tucker and Sellers, 1986; Sellers, 1985; Monteith, 1977; Hatfield et al., 1984; Daughtry,et al., 1983; Asrar et al., 1986). These data have also been shown to be highly related to total primary production when summed or averaged over the growing season (Tucker et al., 1981; Asrar,et al., 1985; Tucker et al., 1985; Gosse *et al.*, 1986; Prince, 1991). We used a satellite-derived vegetation index calculated from meteorological satellite data and expressed this index in terms of estimated annual precipitation (Fig. 2).

The area we studied lies between 16° W and 39° E longitude and 10° and 25° N latitude. This area includes a sizeable part of the Sahara Desert as well as the Sahel. Precipitation varies from <100 mm/yr in the Sahara Desert to ~400 mm/yr at the southern boundary of the Sahel zone; a precipitation gradient of ~1 mm yr^{-1} km^{-1} occurs from ~18° N to ~10° N (Le Houerou, 1980; White, 1983). The Sahara Desert, Sahel zone, and their transition zone have been defined as: 0 to 100 mm yr^{-1}, Sahara Desert; 100 to 200 mm yr^{-1}, Saharan-Sahelian transition zone; and 200 to 400 mm yr^{-1}, Sahel proper. Rainfall in this area is unimodal and occurs from July to October; the rest of the year is dry (Le Houerou, 1980; Monod, 1986; White, 1983; Breman and De Wit, 1983). We will use the 200 mm/yr precipitation isoline as the boundary between the Sahara Desert and the Sahel zone realizing that this is actually the boundary between the Sahel proper and the Saharan-Sahelian transition zone. This is unavoidable as the utility of our satellite vegetation index approach is limited below the ~150-200 mm/yr precipitation isoline because of limited green vegetation.

Satellite data from the U.S. National Oceanic and Atmospheric Administration's (NOAA) polar-orbiting meteorological satellites were used from 1980 to 1993 (1994 is presently being analyzed). Data from the advanced very high resolution radiometer (AVHRR) channel 1 (0.55-0.68 μm), channel 2 (0.73-1.1 μm) and channel 5 (11.5-12.5 μm) were used to calculate the normalized difference vegetation index. Data from NOAA-6 (1980), NOAA-7 (1981-1984), NOAA-9 (1985-1988), and NOAA-11 (1989-1990) were used from July 1 to October 31 of 1980 to 1993. Daily AVHRR 4-km data were acquired for Africa, the normalized difference vegetation index was formed from channel 1 and channel 2 as (2-1)/(2+1), a cloud mask using channel 5 was applied labeling everything colder than 12°C as cloud, and the data mapped to an equal-area projection with a ~7.6-km grid cell size. The daily mapped data were formed into 10-day images and averaged for each year. Formation of composite images minimizes atmospheric effects, scan angle effects, cloud contamination, and solar zenith angle effects (Holben, 1986; Shibayama et al.,1986; Ranson and Daughtry, 1987). Calibration corrections after Kaufman and Holben (1993) were applied. A total of 7,000 orbits of AVHRR data were used.

Normalized difference vegetation index data were computed for the Sahel zone as well as a 1,000,000 km^2 area of the central Sahara and plotted against year (Fig. 3). Rainfall and hence primary production decreased progressively from 1980 to 1984 in the Sahel zone (Figs. 1 and 3). Nicholson has reported that 1980 had a rainfall departure of about -13% from the long-term mean precipitation for the Sahel zone, and 1984 was one of the driest years this century (Nicholson, 1989; Lamb et al., 1986). Higher mean normalized difference vegetation index values for the Sahel zone were found for 1985 to 1993 than 1984; these values indicate that conditions improved in 1985 to 1993 versus 1984. Significant variation of the normalized difference vegetation index was observed in the

Sahel zone over the 14 years of observation; 1980 was the year of the highest normalized difference vegetation index and 1984 the lowest. This observation is in agreement with the meteorological data of Nicholson (1989) and Lamb *et al.*(1986). Annual precipitation for all 12 years of our study was below the long term mean.

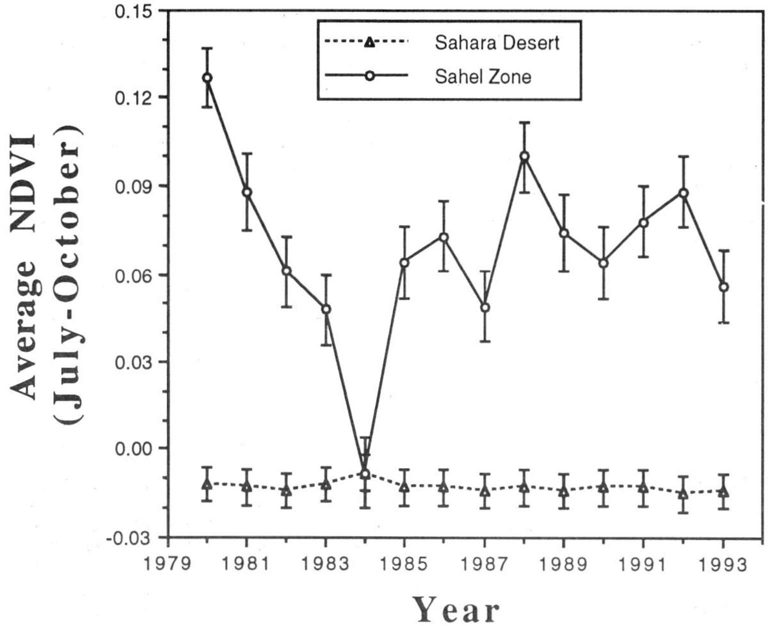

Figure. 3. Average vegetation index values for the approximate 200 to 400 mm/yr long-term mean precipitation zone and the Sahara Desert from 1980 to 1993. 1984 was an exceptionally dry year (see also Fig. 1). There is a direct relation between primary production and the vegetation index (Tucker et al., 1981; Asrar et al., 1985; Tucker et al., 1985; Gosse *et al.*, 1986; Prince, 1991), precipitation and primary production (Houerou and Hoste, 1977; Rutherford, 1980; Seely, 1978), and hence precipitation and the vegetation index (Fig. 2).

Our data document a progressively southward movement of the Saharan-Sahelian boundary (the 200 mm/yr precipitation isoline) from 1980 to 1984 of 240 km (~60 km/yr). This rate was computed by averaging the boundary location at each half degree of longitude from 15.5°W to 38.5°E (Table 1). We estimate that the error associated with our boundary determination is ± 2 pixels or 15 km.

Table 1. Area of the Sahara Desert and Saharan-Sahelian transition zone from 1980 to 1993 using the estimated 200 mm/yr precipitation isoline as the southern boundary. In the central Sahara, the Adrar des Iforas, Air, and Tibesti mountain areas were included as were the desert areas for the Eastern Desert of Egypt and the Nubian Desert of Sudan. Changes are relative to 1980. Nicholson's (1989) Sahelian mean annual precipitation departures from the long-term mean are given from 1980-1987. The northern boundary of the Sahara was the line running from Wadi Draa (29°N 10.5°W) on the Atlantic coast along the Saharan Fault to Figuig (32°N 2°W), Morocco, from there to Biskra (35°N 6°E), Algeria, and from there southward through Tunisia to the Gulf of Gabes (34°N 10°E). Areas of cultivation on the Mediterranean coast were excluded as was the Nile delta and river valley. We estimate the mean latitude position error to be ± 15 km. The error associated with the total Saharan area is estimated to be ± 90,000 km^2 and is the product of the north-south error (±15 km or ±0.14° of latitude) and the length of the southern boundary (6,000 km). The standard deviation appears in parenthesis (Table 1 fron Tucker and Newcomb, 1994).

year	Saharan area (km^2)	Change relative to 1980 (km^2)	Annual precipitation departure(%)	mean latitude and range of the 200-mm/yr isoline (degrees)	
1980	8,633,000	0	-13%	16.3 (0.9)	14.3 to 17.9
1981	8,942,000	308,000	-19%	15.8 (1.1)	14.1 to 17.6
1982	9,260,000	627,000	-40%	15.1 (0.9)	12.9 to 16.7
1983	9,422,000	789,000	-48%	15.0 (0.9)	13.3 to 16.7
1984	9,982,000	1,349,000	-55%	14.1 (1.0)	12.1 to 17.0
1985	9,258,000	625,000	-28%	15.1 (0.9)	13.3 to 17.3
1986	9,093,000	460,000	-21%	15.4 (0.8)	14.0 to 17.1
1987	9,411,000	778,000	-40%	14.9 (0.9)	13.3 to 16.8
1988	8,882,000	248,000	no data	15.8 (0.8)	14.2 to 17.3
1989	9,134,000	501,000	no data	15.4 (1.3)	12.8 to 17.6
1990	9,269,000	635,000	no data	15.1 (1.1)	12.8 to 17.4
1991	9,072,000	445,000	no data	15.3 (0.8)	13.5 to 17.0
1992	8,987,000	360,000	no data	15.8 (0.9)	13.9 to 17.3
1993	9,593,000	966,000	no data	14.8 (1.1)	12.9 to 16.9

The average southward movement of the Saharan-Sahelian boundary (the estimated 200 mm/yr annual precipitation isoline) from 1980 to 1981 was 55 km, from 1981 to 1982 was 77 km, from 1982 to 1983 was 11 km, and from 1983 to 1984 was 99 km. From 1984 to 1985 there was south-to-north movement (retreat of the desert) of 110 km followed by a further northward movement from 1985 to 1986 of 33 km. In 1987 the mean position moved southward 55 km. In 1988, it moved northward 100 km; in 1989 and 1990 it again moved southward 44 and 33 km, respectively. The mean position of the southern boundary in 1991 was indistinguishable from the boundary in 1990. The mean position of the estimated 200 mm/yr precipitation boundary in 1992 moved northward ~55 km from 1991, followed by an ~110 km southward movement (expansion of the desert) in 1993 from 1992.

Clearly it is perhaps inappropriate to think about deserts expanding and contracting annually. Climates, especially arid and semi-arid climates, are determined from mean climatic conditions. Our satellite data show a tremendous interannual variation in precipitation and hence vegetation development on the south side of the Sahara.

The estimated mean position of the 200 mm/yr boundary in 1993 was thus ~160-170 km to the south of its position in 1980. Desert expansion varied in the fourteen years of our analysis from 248,000 km^2 (1988) to 1,349,000 km^2 (1984), all relative to 1980. The 1984 expansion represents a 15% increase of the Sahara from its size in 1980. We estimated the area of the Sahara and the Sahara-Sahelian transition zone was 8,633,000 km^2 in 1980 (Table 1).

The north-south movement of the 200 mm/yr isoline varied greatly across the south side of the Sahara (Fig.4).

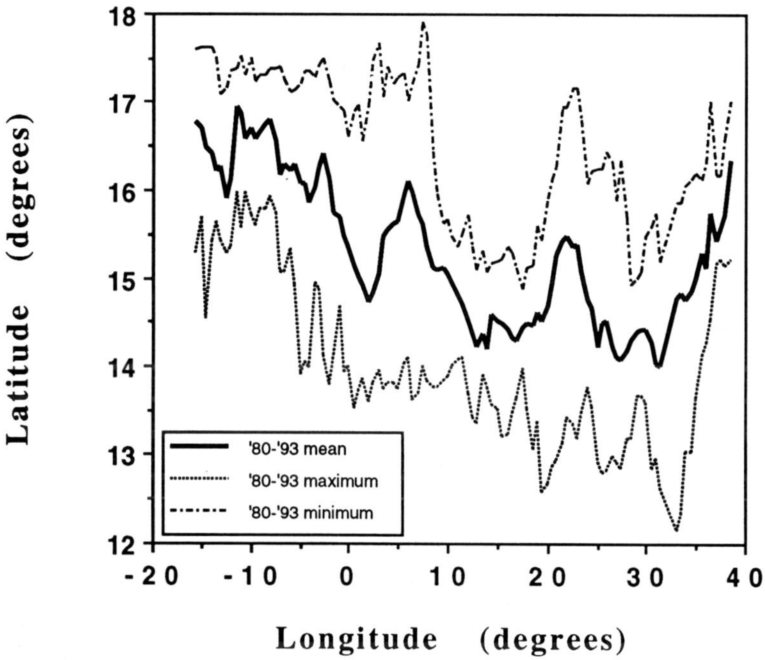

Figure. 4. Summary of the estimated mean, most northern, and most southern positions of the 200-mm/yr-boundary location from 1980 to 1993 at every 0.5^o of longitude from 15.5^o W to 38.5^o E. See also Table 1.

Variation was high from Mali through central Niger and from eastern Chad through western Sudan. Variation was low in southern central and southeastern Mauritania and from central Niger through western Chad. Rainfall, and hence vegetation distributions, vary in space as well as in time in this area.

The observed low variability in interannual movement of the estimated 200 mm/yr isoline in Mauritania and from central Niger through western Chad may have implications for detecting desertification. Detection of changes should be more effective in areas of lower interannual variability. Interannual variations mean that it would require a decades-long study to determine whether long-term expansion or contraction of the Sahara is occurring.

While our satellite data provide the measurement baseline against which to compare future data, it will be necessary to continue similar types of high-temporal frequency low-spatial resolution (i.e., 500-m, 1-km, 4-km) for the next several tens of years. It is only by a continued systematic series of measurements that global processes such as arid and semi-arid precipitation-primary production dynamics can be understood. The point should be clearly stated that highly detailed data for 2-3 years are much less useful than adequate but simple data obtained for tens of years for studying global processes.

Tropical Deforestation and Habitat Fragmentation in the Amazon of Brazil

The second large-scale environmental problem I wish to review involves the destruction of tropical forest through conversion to agriculture, pasture, logging, or other human activities. This is an environmental problem of tremendous importance due to the habitat destruction where half of our total number of plant and animal species reside. The last time a comparable species loss occurred was ~70 million years ago when the dinosaurs and many other species vanished forever.

Deforestation has been occurring in temperate and tropical regions throughout history (Tucker and Richards, 1983; Richards, 1984; Williams, 1989 and 1990). In recent years much attention has focused upon tropical forests, where as much as 50% of its original extent may have been lost to deforestation in the last two decades, primarily due to agricultural expansion (Myers, 1991).

Global estimates of tropical deforestation range from 69,000 km^2 yr^{-1} in 1980 (FAO, 1981) to 100,000 to 165,000 km^2 yr^{-1} in the late 1980's; 50% to 70% of the more recent estimates have been attributed to deforestation in the Brazilian Amazon, the largest continuous tropical forest region in the world (Table 2) (Myers, 1991; FAO, 1992; and World Resources

Table 2. Tropical forest area by country and reported tropical deforestation rates for the 1970's and the late 1980's (from Skole and Tucker, 1993).

Country	Total Forest Area (km^2)	% World Total	1970's FAO Deforestation Rate (km^2)	% World Total	{<----------Deforestation Rates, Late 1980's--------->} Myers(1991) (km^2)	% World Total	WRI(1990) (km^2)	% World Total
Brazil	3,562,800	30.7%	13,600	19.7%	50,000	36.1%	80,000	48.4%
Indonesia	1,135,750	9.8%	5,500	8.0%	12,000	8.7%	9,000	5.4%
Zaire	1,056,500	9.1%	1,700	2.5%	4,000	2.9%	1,820	1.1%
Peru	693,100	6.0%	2,450	3.6%	3,500	2.5%	2,700	1.6%
Columbia	464,000	4.0%	8,000	11.6%	6,500	4.7%	8,200	5.0%
India	460,440	4.0%	1,320	1.9%	4,000	2.9%	15,000	9.1%
Bolivia	440,100	3.8%	650	1.0%	1,500	1.1%	870	0.5%
Papua N.G.	337,100	2.9%	210	0.3%	3,500	2.5%	220	0.1%
Venezuela	318,700	2.7%	1,250	1.8%	1,500	1.1%	1,250	0.8%
Burma	311,930	2.7%	920	1.3%	8,000	5.8%	6,770	4.1%
Others*	2,829,930	24.4%	33,300	48.3%	44,100	31.8%	39,610	23.9%
Total	11,610,350	100.0%	68,900	100.0%	138,600	100.0%	165,440	100.0%

*63 other countries

Institute, 1990). The FAO Forest Assessment 1990 Project (FAO, 1992) has produced several reports, and estimates from them have varied considerably. The 1992 release of another report (P. Aldhous, Science **259**, p. 1390) provides slightly different estimates than those reported in 1992.

The area and rate of deforestation in Amazonia are not well known, nor are there quantitative measurements of the effect of deforestation on habitat degradation. Skole and Tucker (1993) used 1:500,000 scale photographic imagery from Landsat Thematic Mapper data and a geographic information system (GIS) to create a computerized map of deforestation and evaluate its influence on forest fragmentation and habitat degradation. Areas of deforestation were digitized into the GIS and forest fragments and edge effects which result from the spatial pattern of forest conversion were determined.

Background. Tropical deforestation is a major component of the carbon cycle and has profound implications for biological diversity. Deforestation increases atmospheric CO_2 and other trace gases, possibly affecting climate (Gash and Shuttleworth, 1991; Houghton et al., 1985; Houghton and Skole, 1990; Houghton et al., 1991; Keller et al., 1991; Salati, 1987; Salati and Nobre, 1991; Salati and Vose, 1983 and 1984; Shukla et al., 1990). Conversion of forests to cropland and pasture results in a net flux of carbon to the atmosphere, because carbon contained in forests is higher than that in the agricultural areas that replace them.

The paucity of data on tropical deforestation limits our understanding of the carbon cycle and possible climate change (IPCC, 1992). Furthermore, while occupying less than 7% of the terrestrial surface, tropical forests are the home to half or more of all plant and animal species (Wilson, 1988; Gentry, 1988). The primary adverse effect of tropical deforestation is massive extinction of species including, for the first time, large numbers of vascular plant species (Prance and Elias, 1982; Wilson, 1985 and 1986; Erwin, 1982 and 1988; Wilson and Peters, 1988; Reid and Miller, 1989; May, 1988; and Knoll, 1984).

The immense biological diversity of tropical forests is difficult to comprehend. For example, 10 selected 1-ha plots in Borneo contained 700 species of trees and one hectare of tropical Peru contained 300 tree species. By comparison, 700 tree species occur in all of North America. 'Species' is used to mean organisms that can breed freely with each other and conversely, cannot breed freely with other species.

Deforestation affects biological diversity in three ways: destruction of habitat, isolation of fragments of formerly contiguous habitat, and edge effects within a boundary zone between forest and deforested areas. This boundary zone extends some distance into the remaining forest. In this zone there is greater exposure to winds; dramatic micrometeorological differences occur over short distances; easier access is possible by livestock, other non-forest animals, and hunters; and a range of other biological and physical effects occurs. The cumulative result is loss of plant and animal species in the affected edge areas (Franklin and Forman, 1987; Harris, 1984 and 1988; Janzen, 1983; Lovejoy et al., 1984; Saunders et al., 1991; Terborgh, 1976; Wilcove et al., 1986; Wilcox, 1980; and Redford, 1992).

There is a wide range in current estimates of the area deforested and the rate of deforestation in Amazonia. Brazilian scientists at the Instituto Nacional de Pesquisas Espaciais (INPE) have estimated Amazonian deforestation (Tardin et al., 1979 and 1980; Fearnside et al., 1990; Tardin and da Cunha, 1990; and INPE, 1992). These unpublished studies report a total deforested area of 280,000 km^2 as of 1988, and that the average annual rate from 1978 to 1988 was 21,000 km^2 yr^{-1}.

Other studies have reported rates which range from 50,000 km^2 yr^{-1} to 80,000 km^2 yr^{-1} (Table 2) (Myers, 1991; FAO, 1992; and World Resources Institute, 1990). Additional deforestation estimates have been made for geographically limited study areas in the southern Amazon Basin of Brazil using Landsat or meteorological satellite data (Mahar, 1989; Fearnside, 1990; Cross et al., 1991; Malingreau and Tucker, 1988; Nelson and Holben, 1986; Nelson et al., 1987; Tucker et al., 1984; Woodwell et al., 1987; Setzer and Pereira, 1991).

The Amazon Basin of Brazil has been defined by law to include all of the states of Acre, Amapa, Amazonas, Para, Rondonia, Roraima, plus part of Mato Grosso, Maranhao and Tocantins and is referred to as the Legal Amazon (IBGE, 1991). It covers an area of ~5,000,000 km^2, of which ~4,090,000 km^2 is forested, ~850,000 km^2 is cerrado or tropical savanna, and ~90,000 km^2 is water (Table 3). Confusion has arisen among researchers regarding the stratification of the Brazilian Amazon into forest, cerrado, and water strata (Table 3), preventing comparison of different deforestation studies. The use of satellite data and GIS make it possible to explicitly stratify Amazonia on the basis of cover types (IBGE, 1988), thereby providing a means to compare deforestation results from other studies. This approach is also necessary for spatial analysis of habitat fragmentation and edge effects of deforestation. Finally, GIS provides a data management tool for managing large amounts of spatial data, and for the precise merging and geocoding of information from more than 200 satellite images use in this study.

Table 3. Predeforestation water, forest, and cerrado land cover for the Brazilian Amazon by state as used in this study. The values determined in this study were based upon the IBGE vegetation map and by interpretation of satellite data. Summary values are also given from other sources for comparison. Areas obscured by clouds were excluded from deforestation and affected forest habitat analyses. 97% of the cloud-affected data were over tropical forest. All units are in km^2. Table from Skole and Tucker (1993).

State	Water Area	Forest Area	Cloud Total	Cerrado Area	Total Area
Acre	393	152,394	0	0	152,787
Amapa	1,188	137,444	53,566	978	139,610
Amazonas	29,842	1,531,122	94,058	14,379	1,575,343
Maranhao	1,344	145,766	13,444	114,675	261,785
Mato Grosso	4,212	527,570	8,630	368,658	900,440
Para	49,522	1,183,571	56,807	28,637	1,261,730
Rondonia	1,462	212,214	474	24,604	238,280
Roraima.	1,817	172,425	15,232	51,464	225,706
Tocantins	2,914	30,325	0	244,005	277,244
Total	**92,694**	**4,092,831**	(242,211)	**847,400**	**5,032,925**
FAO/UNEP(1981)	(not given)	3,562,800	--	(not given)	(not given)
Fearnside(1990)	(not given)	4,195,660	--	793,279	(not given)
Fearnside et al'90	(not given)	(not given)	--	(not given)	4,906,784
IBGE(1988)	20,972	3,793,664	--	1,149,943	4,964,920

Remote Sensing. The large area of the Brazilian Amazon necessitates a straightforward and accurate method of measurement. Landsat thematic mapper photo products are inexpensive and of sufficient spatial and spectral resolution for determining deforestation. Their analysis using visual interpretation techniques produces similar quantitative results to digital processing of full-resolution, multi-spectral thematic mapper and SPOT data (Skole, 1992).

We acquired 210 Landsat thematic mapper channel 5 (1.55-1.75 µm) black and white photographic images at 1:500,000 scale for the entire Brazilian Amazon of Brazil from primarily 1988 (Skole and Tucker, 1993). 175 images were from 1988; 8 images were from 1987; 7 images were from 1989, and 20 were from 1986. All data were from the Brazilian Landsat receiving station. The exact boundary between intact forest and deforested land was digitized in the Universal Transverse Mercator projection, and then edited and error-checked using clear velum plots of the line-work overlaid on each photographic image. Each Landsat scene contained coordinate control points in decimal degree units, such that each scene could be geographically registered within precise tolerances and mosaicked together.

When digitizing, vertices were placed approximately every 50 m of ground position.

Tests of positional accuracy in digitizing followed those of R. Dunn et al. (1990) and indicated encoding and hence area-estimation errors were less that 3%. The variance associated with interpretation and delineation of boundaries between intact forest and deforested areas was less than 10% overall. Further accuracy assessment was made in test sites established in Rondonia, where fragmentation was very high. An explicit spatial comparison between our estimate of deforestation and the same derived from high resolution (20 m resolution) SPOT satellite imagery was highly correlated (r2 = 0.98, y = 1.11x - 57.358). Additional ground checking and verification was made in eastern Para state, north of Manaus, and along the Rio Negro, both in Amazonas state.

We digitized the deforested areas shown on individual Landsat scenes using visual deforestation interpretation and standard vector GIS techniques. The digitized scenes were projected into equal-area geographic coordinates (latitude, longitude), edge-matched and merged in the computer to form a single, seamless data set for the entire Brazilian Amazon.

Spatial analysis of the geometry of deforestation is critical to the estimation of forest fragmentation and the edge effect. If 100 km^2 of tropical deforestation occurs as a square 10 by 10 km area, and we assume that the edge effect is 1 km, the total area affected would be ~143 km^2. In contrast, if the 100 km^2 of deforestation is distributed as 10 strips of deforestation, each 10 x 1 km in dimension, the resulting area affected would be ~350 km^2.

We extracted forest fragments <100 km^2 that were isolated by deforestation and computed edge effects for a zone of 1 km along the boundaries. All areas of closed canopy tropical forest deforested by 1988 were delineated, including areas of secondary growth on abandoned fields and pastures where visible. Areas of long-term forest degradation along river margins in central Amazonia were also included, as were scattered small clearings associated with rubber tappers, mining operations, airfields, and other small disturbance features. All visible roads, power line right-of-ways, pipelines, and similar human-made features were also digitized into the GIS and treated as deforestation. 50 digital Landsat MSS scenes from 1986 and 15 digital thematic mapper images from 1988 were used for detailed examination from the states of Acre, Amazonas, Mato Grosso, Para, and Rondonia.

To determine the extent of deforestation in 1978, we used the GIS to digitize areas of deforestation shown on 1:500,000 scale maps from single-channel Landsat MSS data, produced jointly by the Instituto Brasiliero de Desenvolvimento Florestal (IBDF) and the Instituto de Pesquisas Espaciais (INPE) in the early 1980s (Tardin et al., 1980; Skole, 1992). These maps did not differentiate between forest and cerrado clearing. We compiled a forest-cerrado-water data set by combining a vegetation map with analysis of Landsat images and meteorological satellite data (Skole and Tucker, 1993). Our deforestation and affected habitat analyses for 1978 and 1988 were restricted to only closed-canopy forest of the Brazilian Amazon.

Deforestation and Forest Fragmentation. Distribution of deforestation and affected habitat in the Brazilian Amazon for 1978 and 1988 was concentrated in two areas: a crescent along the southern and eastern fringe of the Amazon, a spatial pattern similar to the distribution of fires observed from thermal anomalies in AVHRR data (Setzer and Pereira, 1991); and along major transportation corridors in the interior of the Amazon. Deforestation increased from 68,000 km^2 in 1978 to 230,000 km^2 in 1988, while the total affected habitat increased from 151,000 km^2 in 1978 to 588,000 km^2 in 1988 (Table 4). The total area deforested increased by a factor of two to three or more in every state except Amapa. It is likely that the deforested area in Amapa is higher than our assessment because excessive cloud cover in this region prevented complete analysis (Table 3). We found 6% of closed-canopy forest had been cleared as of 1988 and ~15% of the forested Amazon was affected by deforestation-caused habitat destruction, habitat isolation, and edge effects (Table 4).

Our analysis of the spatial pattern of deforestation found a strong tendency toward spatial concentration; areas of undisturbed tropical forest tended to be sizeable (Table 5). This is further pronounced than Table 5 indicates, because many of the large areas of undisturbed tropical forest are contiguous between and among states.

For the entire Brazilian Amazon, our deforestation estimate is close to, but lower than the estimates of Fearnside et al. (1990) and of the INPE (1992) of ~280,000 km^2 of Brazilian Amazonia deforested as of 1988. The difference is due to three factors: (i) different stratification of forest, cerrado, and water; (ii) slightly different estimates of secondary growth, which is spectrally similar to intact forest in channel 5; and (iii) positional accuracy, interpretation, and boundary generalization. We estimate that ~30,000 km^2 of the difference is due to different evaluation of the forest-cerrado boundaries in Mato Grosso and Tocantins states. By comparison, our analysis suggests that deforestation estimates based on coarse resolution meteorological satellite data in the southern Amazon of Brazil have over-estimated deforestation by up to ~50%.

Table 4. Tropical deforestation, forest isolated or cut off by deforestation, and the area of forest adversely affected by a 1 km "edge effect" from adjacent areas of deforestation. Tabular summaries are given by state for the Brazilian Amazon for 1978 and 1988. Areas of tropical forest which were obscured by clouds were omitted from this analysis and all units are in km^2. The numbers in parentheses following the 1-km "Edge Effects" entries are the ratio between a 500-m buffer and a 1,000-m buffer. Table from Skole and Tucker (1993).

State	Deforested	Isolated Forest	Edge Effects (1-km)	Total Affected Forest	Deforested Area	Isolated Forest	Edge Effects (1-km)	Total Area Affected
	{<----------- 1978			----------->}	{<------------- 1988			--------->}
Acre	2,612	18	4,511	7,141	6,369	405	23,686 (0.517)	30,460
Amapa	182	0	368	550	210	1	689 (0.537)	900
Amazonas	2,300	36	6,498	8,833	11,813	474	36,392 (0.582)	48,679
Maranhao	9,426	705	13,120	23,251	31,952	2,123	28,147 (0.626)	62,222
Mato Grosso	21,134	776	25,418	47,288	47,568	2,542	71,128 (0.580)	121,238
Para	30,449	2,248	49,794	82,488	95,075	6,837	116,669 (0.633)	218,581
Rondonia	6,281	991	17,744	25,015	23,998	2,408	52,345 (0.657)	78,751
Roraima.	196	4	812	1,011	1,908	1	5,236 (0.521)	7,145
Tocantins	5,688	337	6,584	12,609	11,431	1,437	6,760 (0.659)	19,628
Total	78,268	5,115	124,846	**208,229**	230,324	16,228	341,052 (0.610)	**587,604**

Table 5. Spatial characteristics of isolated forest and remaining intact tropical forest within the Legal Amazon of Brazil by state as determined by the analysis of 1988 Landsat thematic mapper satellite imagery. Isolated forest refers to areas of forest surrounded by deforestation. Remaining forest refers to tropical forest which has not been deforested and includes both isolated forest areas < 100 km^2 and larger areas of forest. Many of the largest remaining areas of tropical forest are contiguous among states. Refer also to tables 2 and 3. From Skole and Tucker (1993).

State	Isolated Forest Areas (< 100 km^2)	# of Isolated Forest Polygons	Undisturbed Remaining Forest (km^2)	Total # of Remaining Undisturbed Forest Polygons	Range of Total Isolated and Remaining Forest Areas (km^2) min.	max.
Acre	405	603	146,025	605	< 1	139,215
Amapa	1	2	83,676	3	< 1	83,675
Amazonas	474	464	1,425,253	465	< 1	1,424,779
Maranhao	2,123	1,035	100,554	1,042	< 1	70,057
Mato Grosso	2,542	2,016	478,619	2,027	< 1	471,792
Para	6,837	4,030	1,032,194	4,032	< 1	1,021,263
Rondonia	2,408	1,587	187,743	1,588	< 1	185,335
Roraima	1	2	155,326	6	< 1	152,414
Tocantins	1,437	493	18,894	508	< 1	6,982
Total	**16,228**	**10,232**	**3,628,284**	**10,276**		

Fundamental to our analysis was a specified representation for water, cerrado or savanna, and forest for the Brazilian Amazon. We used a vegetation map that was augmented by Landsat thematic mapper and meteorological satellite imagery for more accurate depiction of cerrado and water.

The average deforestation rate in the closed-canopy forests of the Brazilian Amazon from 1978 to 1988 was ~16,000 km^2 yr^{-1} (Table 4). This rate is higher than the rate from 1975 to 1978 (FAO, 1981) but considerably lower than recent estimates (Myers, 1991; FAO, 1992; World Resources Institute, 1990; Setzer and Pereira, 1991). Our estimates can be used in assessments of net flux of carbon from land clearing and biomass burning in the Brazilian Amazon. Current estimates of these fluxes have largely been based on model calculations with deforestation values much higher than we report (tables 2 and 4). In addition, many deforested areas are in stages of regrowth following abandonment (Buschbacher, 1986; Uhl et al., 1988; Buschbacher et al., 1988). If regrowth is widespread, estimates of the net flux of carbon should be further reduced, because carbon accumulates in regrowing biomass.

The preponderance of affected habitat results from proximity to areas of deforestation (~ 341,000 km^2 for a 1-km edge effect) and not from isolation of forest (~16,000 km^2) or deforestation per se (~230,000 km^2). While the rate of deforestation averaged ~16,000 km^2 yr^{-1} in the Brazilian Amazonia from 1978 to 1988, the rate of habitat fragmentation and degradation was ~44,000 km^2 yr^{-1}. Implications for biological diversity are not encouraging and provide added impetus to minimize tropical deforestation.

The work reported here for the Amazon of Brazil is being extended through the mid-1990's, and also extended to the Amazon Basin of Bolivia, Peru, Ecuador, Columbia, Venezuela, Guyana, Surinam, and French Guyana for the 1970's, 1980's and 1990's.

References

G. Asrar, E. T. Kanemasu, R. D. Jackson, P. J. Pinter, *Remote Sens. Environ.* **17**, 211 (1985).

G. Asrar, E. T. Kanemasu, G. P. Miller, R. L. Weiser, *IEEE Trans. Geosci. Remote Sens.* **GE24**, 76 (1986).

Aubreville, A. et. al.,*Bois et Forets des Tropiques 148*, 3 (1973).

E. W. Bovill, *J. Afr. Soc.* **20**, 175 and 259 (1921).

H. Breman and C. T. De Wit, *Science* **221**, 1341 (1983).

R. B. Buschbacher, *Bioscience* **36**, 22 (1986).

R. B. Buschbacher, C. Uhl, E. A. S. Serrao, *J. Ecol.* **76**, 682 (1988).

F. R. Cana, *Geograph. J.* **46**, 333 (1915).

J. L. Cloudsley-Thompson, *Sahara Desert.* (Pergamon Press ,New York, 1984).

A. M. Cross, J. J. Settle, N. A. Drake, and R. T. M. Paivinen, *Int. J. Remote Sensing,* **12**, 1119 (1991).

P. J. Curran, *Phil. Trans. R. Soc.* **A-309**, 257 (1983).

C. S. T. Daughtry, K. P. Gallo, M. E. Bauer, *Agron. J.,***75**, 527 (1983).

R. Dunn, R. Harrison, and J. C. White, *Int. J. Geograph. Inform. Syst.* **4**, 385 (1990).

E. Eckholm and L. R. Brown, *Worldwatch Pap. 13*, 1 (Worldwatch Institute, Washington, D.C., 1977).

W. S. Ellis, *National Geographic* **172**, 140 (1987).

P. R. Ehrlich and E. O. Wilson, *Science* **253**, 758 (1991).

T. L. Erwin, *Coleopt. Bull.* **36**, 74 (1982).

T. L. Erwin, *Bull. Ent. Soc. Am.* **30**, 14 (1983).

T. L. Erwin, in *Biodiversity*, E. O. Wilson and F. M. Peters, Eds. (National Academy Press, Washington, DC, 1988), pp. 123-129.

M. Evenari in *Ecosystems of the World: Hot Deserts and Arid Shrublands,* M. Evenari, I. Noy-Meir, and D. W. Goodall, Eds., (Elsevier, New York, 1985), pp. 1-22.

FAO/UNEP, *Los Recursos Forestales de la America Tropical*, 32/6.1301-78-04, Technical Report No. 1 (Food and Agriculture Organization of the United Nations, Rome, 1981).

FAO, *Forest Resources of Tropical Africa*, 32/6.1301-78-04, Technical Report No. 2 (Food and Agriculture Organization of the United Nations, Rome, 1981).

FAO, *Forest Resources of Tropical Asia*, 32/6.1301-78-04, Technical Report no. 3 (Food and Agriculture Organization of the United Nations, Rome, 1981).

FAO, *The Forest Resources of the Tropical Zone By Main Ecological Regions*, Report to the United Nations Conference on Environment and Development by the Forest Resource Assessment 1990 Project (Food and Agriculture Organization of the United Nations, Rome, 1992).

P. M. Fearnside, *Environ. Cons.* **17**, 213 (1990).

P. M. Fearnside, A. T. Tardin, L. G. M. Meira, *Deforestation Rate in Brazilian Amazonia* (National Secretariat of Science and Technology, Brasília, Brazil 1990).

J. F. Franklin and R. T. T. Forman, *Landsc. Ecol.* **1,** 5 (1987).

J. H. C. Gash and W. J. Shuttleworth, *Climatic Change* **19**, 123 (1991).

A. H. Gentry, *Proc. Natl. Acad. Sci.* **85**, 156 (1988).

D. W. Goodall, and R. A. Perry, *Arid-Land Ecosystems: Structure, Functioning and Management, 1* (Cambridge Univ. Press, Cambridge, 1979).

G. Gosse *et al., Agronomie* **6,** 47 (1986).

L. D. Harris, *The Fragmented Forest. Island Biogeographic Theory and the Preservation of Biotic Diversity.* (Univ. of Chicago Press, Chicago, 1984).

L. D. Harris, *Conserv. Biol.* **2**, 330 (1988).

J. L. Hatfield, G. Asrar, and E. T. Kanemasu, *Remote Sens. Environ.* **14**, 65 (1984).

U. Hellden, *Drought Impact Monitoring* (Lund Univ. Naturgeografiska Inst., Lund Sweden, 1984), 61.pp.

B. N. Holben, *Int. J. Remote Sens.* **7**, 1417 (1986).

R. A. Houghton, *Climatic Change* **19**, 99 (1991).

R. A. Houghton, R. D. Boone, J. M. Melillo, C. A. Palm, G. M. Woodwell, N. Myers, B. Moore, D. L. Skole, *Nature* **316**, 617 (1985).

R. A. Houghton and D. L. Skole, in *The Earth as Transformed by Human Action,* B. L. Turner II, W.C. Clark, R.W. Kates, J.F. Richards, J.T. Mathews, and W.B. Meyer, Eds. (Cambridge Univ. Press, Cambridge, 1990) pp. 393-408.

R. A. Houghton, D. L. Skole, D. S. Lefkowitz, *J. For. Ecol. Manag.* **38**, 173 (1991).

Instituto Nacional de Pesquisas Espaciais, 3 pp., São José dos Campos, Brazil (1992).

IBGE, *Anuario Estatistico do Brasil 1991.* Fundacao Instituto Brasileiro de Geografia e Estatística, Rio de Janeiro, Brazil, V. 51, p. 1-1024, 1991 (ISSN 0100-1299).

IBGE, *Mapa de Vegetacão do Brasil* (Fundacão Instituto Brasileiro de Geografia e Estatística, Rio de Janeiro, Brazil, 1988).

IPCC, *Climate Change 1992: the Supplementary Report to the IPCC Scientific Assessment.* J. T. Houghton, B. A. Callander, S. K. Varney, Eds. (Intergovernmental Panel on Climate Change, Cambridge Univ. Press, Cambridge, 1992).

D. H. Janzen, *Oikos* **41**, 402 (1983).

T. E. Lovejoy et al., in *Extinctions*, M. H. Nitecki, Ed. (University of Chicago Press, Chicago,1984), pp. 295-325.

M. Kassas, *Arid Lands in Transition.*, (American Association for the Advancement Science, Washington, D.C., 1970), pp. 123-142.

Y. J. Kaufman and B. N. Holben, *Int. J. Remote Sens.,* **14**, 21 (1993).

M. Keller, D. J. Jacob, S. C. Wofsy, R. C. Harriss, *Climatic Change* **19**: 139 (1991).

A. H. Knoll, in *Extinctions,* M. H. Nitecki, Ed. (Univ. of Chicago Press, Chicago, 1984), pp. 21-68.

P. J. Lamb, R. A. Peppler, S. Hastenrath, *Nature* **322**, 238 (1986).

H. F. Lamprey, Report on the desert encroachment reconnaissance in Northern Sudan, United Nations Educational, Scientific, and Cultural Organization/United Nations Environment Program (Nairobi, 1975), pp. 1-16 [*Desertification Control Bulletin*, **17**, 1 (1988)].

H. N. Le Houerou, *Ann. Alger. Geogr.* **6**, 2 (1968).

H. N. Le Houerou, *J. Range Manage.* **33**, 41 (1980).

H. N. Le Houerou and C. H. Hoste, *J. Range Manag.* **30**, 181 (1977).

T. E. Lovejoy et al. in *Conservation Biology. The Science of Scarcity and Diversity*, M. E. Soule, Ed. (Sinauer Associates, Sunderland, MA 1988), pp. 257-285.

D. J. Mahar, *Government Policies and Deforestation in Brazil's Amazon Region* (World Bank, Washington, D.C., 1989).
A. R. Malo, and S.N. Nicholson, *J. Arid Environ.* **19**, *24,* (1990).

J. P. Malingreau and C. J. Tucker, *Ambio* **17**, 49 (1988).

R. M. May, *Science* **241**, 1441 (1988).

W. G. McGinnies, B. J. Goldman, P. Paylore, *Deserts of the World,* (Univ. of Arizona Press, Tucson, 1968).

W. G. McGinnies in *Arid-Land Ecosystems: Structure, Functioning and Management, 1,* D. W. Goodall. and R. A. Perry, Eds. (Cambridge Univ. Press, Cambridge, 1979) pp. 299-314.

T. Monod, in *Ecosystems of the World: Hot Deserts and Arid Shrublands,* M. Evenari, I. Noy-Meir, D. W. Goodall, Eds. (Elsevier, New York, 1986), pp. 203-243.

J. L. Monteith, *Phil. Trans. R. Soc.* **B-281**, 277 (1977).

N. Myers, *Climatic Change* **19**, 3 (1991).

R. F. Nelson and B. N. Holben, *Int. J. Remote Sensing* **7**, 429 (1986).

R. F. Nelson, N. Horning, and T. A. Stone, *Int. J. Remote Sensing* **8**, 1767 (1987).

S. E. Nicholson, *Weather* **44**, 47 (1989).

C. Norman, *Science* **235**, 963 (1987).

G. T. Prance and T. S. Elias, Eds., *Extinction is Forever* (New York Botanical Garden, New York, 1982).

S. D. Prince, *Int. J. Remote Sens.* **12**, 1301 (1991).

K. J. Ranson and C. S. T. Daughtry, *IEEE Trans. Geosci. Remote Sens.* **GE-25**, 502 (1987).

K. H. Redford, *Bioscience.* **42**, 412 (1992).

W. V. Reid and K. R. Miller, *Keeping Options Alive: The Scientific Basis for Conserving Biodiversity* (World Resources Institute, Washington, DC, 1989).

J. F. Richards, *Environment* **26**, 6 (1984).

F. Rodd, *Geograph. J.* **91**, 354,(1938).

M. C. Rutherford, *S. Afr. J. Sci.* **76**, 53 (1980).

E. Salati, in *The Geophysiology of Amazonia: Vegetation and Climate Interaction,* R. E. Dickinson, Ed. (Wiley Interscience, New York, 1987), pp. 273-296.

E. Salati and C. A. Nobre, *Climatic Change* **19**, 177 (1991).

E. Salati and P. B. Vose, *Ambio* **12**, 67 (1983).

E. Salati and P. B. Vose, *Science* **225**, 129 (1984).

D. A. Saunders, R. J. Hobbs, C. R. Margules, *Cons. Biol.* **5**, 18 (1991).

M. K. Seely, *S. Afr. J. Sci.* **74**, 295 (1978).

P. J. Sellers, *Int. J. Remote Sens.* **6**, 1335 (1985).

A. W. Setzer and M. C. Pereira, *Ambio* **20**, 19 (1991).

A. W. Setzer, Relatorio INPE-4534-RPE/565, Instituto Nacional de Pesquisas Espaciais, São José dos Campos, Brazil (1988).

M. Shibayama, C. L. Wiegand, A. J. Richardson, *Int. J. Remote Sens.* **7**, 233 (1986).

A. Shmida in *Ecosystems of the World: Hot Deserts and Arid Shrublands,* M. Evenari, I. Noy-Meir, and D. W. Goodall, Eds. (Elsevier, New York, 1985), pp. 23-77.

J. Shukla, C. Nobre, P. Sellers, *Science* **247**, 1322 (1990).

D. L. Skole, Ph.D. Dissertation, Univ. of New Hampshire (1992).

D. L. Skole and C. J. Tucker, *Science* **260**, 1905 (1993).

S. E. Smith, *J. Soil Water Cons.* **41,** 297 (1986).

J. Smolowe, *Time* **127** (33), 36 (1987).

E. P. Stebbing, *Geograph. J.* **85,** 506 (1935).

E. P. Stebbing, *Geograph. J.* **91,** 356 (1938).

J. Swift, *The Sahara,* (Time-Life Books, New York, 1975).

A. T. Tardin, et al., Rep. 411-NTE/142 (Instituto Nacional de Pesquisas Espaciais, São Jose dos Campos, Brazil, 1979).

A. T. Tardin, et al., *Subprojeto Desmatamento.* IBDF/CNPq-INPE (Inst. de Pesquisas Espaciais, São Jose dos Campos, Brasil, 1980).

A. T. Tardin and R. P. da Cunha, Rep. INPE-5015-RPE/609 (Instituto de Pesquisas Espaciais, São José dos Campos, Brazil 1990).

J. Terborgh, *Science* **193**, 1029 (1976).

C. Toupet, *C. R. Somm. Soc. Biogeogr.* **48,** 39 (1972).

C. J. Tucker, *Remote Sens. Environ.* **8**, 127 (1979).

C. J. Tucker and P. J. Sellers, *Int. J. Remote Sens.* **7**, 1395 (1986).

C. J. Tucker, B. N. Holben, T. E. Goff, *Remote Sensing Environ.* **15**, 255 (1984).

C. J. Tucker, B. N. Holben, J. H. Elgin, J. E. McMurtrey, *Remote Sens. Environ.* **11**, 171 (1981).

C.J. Tucker, H.E. Dregne, and W.W. Newcomb, *Science* **253,** 299 (1991).

C. J. Tucker, C. L. Vanpraet, M. J. Sharman, and G. Van Ittersum, *Remote Sens. Environ.* **14**, 233 (1985).

R. P. Tucker and J. F. Richards, *Global Deforestation and the Nineteenth Century World Economy.* (Duke Univ. Press, Durham, NC. 1983).

C. Uhl, R. B. Buschbacher and E. A. S. Serrao, *J. Ecol.* **76,** 663 (1988).

F. White,. *The Vegetation of Africa,* (UNESCO, Paris, 1983).

D. S. Wilcove, C. H. McLellan, and A. P. Dobson, in *Conservation Biology. The Science of Scarcity and Diversity*, M. E. Soule, Ed. (Sinauer Associates, Sunderland, MA, 1986), pp. 273-256.

B. A. Wilcox, in *Conservation Biology an Evolutionary-Ecological Perspective,* M. E. Soule and B. A. Wilcox, Eds. (Sinauer, Sunderland, MA 1980), pp. 95-117.

M. Williams, *Prog. Human Geogr.* **13**, 176 (1989).

M. Williams, in *The Earth as Transformed by Human Action*, B. L. Turner, R. W. Clark, R. W. Kates, J. F. Richards, J. T. Mathews, and W. B. Meyer, Eds. (Cambridge University Press, Cambridge,1990), pp. 179-201.

E. O. Wilson, *Issues Sci. Tech.* **2** (Fall), 20 (1985).

E. O. Wilson, *Science* **234**, 14 (1986).

E. O. Wilson, in *Biodiversity,* E.O. Wilson and F.M. Peters, Eds. (National Academy Press, Washington, DC, 1988), pp. 3-20.

E. O. Wilson and F. M. Peters, Eds., *Biodiversity* (National Academy Press, Washington, DC, 1988).

G. M. Woodwell, R. A. Houghton, T. A. Stone, R. F. Nelson, W. Kovalick, *J. Geophys. Res.* **92**, 2157 (1987).

World Resources Institute, *World Resources 1990-91, A Report by the World Resources Institute in Collaboration with the United Nations Environment Programme and The United Nations Development Programme* (Oxford Univ. Press, New York, 1990).

TOWARD GLOBAL PLANNING OF SUSTAINABLE USE OF THE EARTH
Development of Global Eco-Engineering
S. Murai (Editor)

Remote Sensing for Biosphere Modeling and Monitoring

Dennis G. Dye[a] and Samuel N. Goward[b]

[a]Global Engineering Laboratory, Institute of Industrial Science, University of Tokyo, 7-22-1 Roppongi, Minato-ku, Tokyo 106, Japan

[b]Department of Geography, 1113 Lefrak Hall, University of Maryland, College Park, MD 20742, USA

Abstract

Realizing the maximum potential of remote sensing for biosphere modeling and monitoring requires an understanding of both the relation between spectral measurements and biophysical attributes of the earth surface, and models that are suitable for taking advantage of the data that can be reliably derived from remote sensing. Recent progress toward this goal is evident in research directed toward using remote sensing to parameterize a radiation use "efficiency" type model of terrestrial primary production. Considerable success has been attained in developing and applying satellite methods to estimate and map the key variables required to implement the model on a global basis (incident PAR, vegetation PAR absorption capacity, vapor pressure deficit, temperature, soil moisture, biomass).

1. INTRODUCTION

Recognition of the altering influence of human activities in the biosphere and the prospect of climate change has highlighted the importance of an improved understanding of the role of biospheric processes in the global Earth system [1]. Biospheric activity as expressed through primary production is of particular importance because of its fundamental role in determining fluxes of energy (the surface radiation balance, the food chain) and matter (carbon dioxide, water) in the biosphere. Developing reliable methods to assess and monitor global primary production constitutes a major focal point and challenge in contemporary biospheric science [2].

Much of the difficulty faced in effectively utilizing remotely sensed data in biospheric studies arises from the disparity between the spatially integrative characteristics of remotely sensed measurements acquired at the landscape scale (e.g. 30 m to 1 km resolution), and the comparatively reductionist characteristics of vegetation process models that account for phenomena at a much finer scale (typically for an individual leaf or plant). Realizing the maximum potential of remote sensing for biosphere studies at landscape-to-global scales faces two parallel challenges: improving understanding of spectral measurements and their relation to biophysical variables on one hand, and introducing modeling approaches that are suitable for taking advantage of the type and range of data that can be reliably derived from remote sensing on the other. Strides toward meeting these challenges are being made in research to evaluate global primary production using remote sensed data in combination with a model that is based on the production "efficiency" concept [3, 4].

2. REMOTE SENSING OF BIOPHYSICAL VARIABLES FOR PRIMARY PRODUCTION MODELING

The production efficiency model (PEM) is based on concepts which originated in studies with field crops, where it was observed that biomass production is strongly correlated with the amount of photosynthetically active radiation (PAR) intercepted by the crop canopy during the growth period [5]. Production may thus be predicted from knowledge of the amount of intercepted or absorbed PAR (APAR) and the "efficiency" (ε) with which is converted to biomass. The possibility of applying this concept to natural vegetation to evaluate regional-to-global patterns of productivity has attracted increasing attention because 1) its relatively simple form provides a practical means, based on sound principals, to account for the production process as it is expressed at the landscape scale, which may not easily achieved with more detailed process models, and 2) substantial potential exists for directly estimating the key variables on a global scale by satellite remote sensing (e.g. [3, 6]. Application of the PEM concept requires measurements of:

1) incident PAR (photosynthetically active radiation)
2) PAR absorption capacity of the vegetation canopy
3) environmental constraints on ε, including
    a) air temperature
    b) vapor pressure deficit
    c) soil moisture
4) living biomass (to account for respiratory demands).

Recent studies have demonstrated considerable success in developing methods to estimate and map these variables using spectral measurements from satellite remote sensors.

2.1. Incident PAR

PAR (400-700 nm) is the fundamental energy source for photosynthesis. At least two satellite methods have been proposed and applied to estimate the amount of PAR that is incident over the global earth surface. The methods account for the effect of clouds and other scattering agents on incoming PAR based on reflectance measured at either visible [7] or ultraviolet wavelengths [8, 9].

2.2. PAR Absorption Capacity

The amount of incident PAR that is diverted to primary production is determined by the amount of photosynthetic biomass displayed in the landscape, its spectral properties and geometric arrangement. The capacity of a vegetation canopy to intercept and absorb incident PAR can be expressed quantitatively as the ratio of the incident PAR to the PAR absorbed by the photosynthetic canopy elements (APAR). This fraction (f_{APAR}) is highly correlated with spectral vegetation indices such as the normalized difference vegetation index (NDVI) (e.g., [10, 11]. By employing NDVI data such as is available from the Advanced Very High Resolution Radiometer (AVHRR), f_{APAR} may be estimated for the terrestrial globe [9]. Attaining maximum accuracy, however, requires methods to account for the effects of variable soil and litter reflectance on the f_{APAR}-NDVI relation [11].

2.3. Environmental Constraints on Production Efficiency

The theoretical maximum value for ε may be estimated from knowledge of the predominant photosynthetic pathway of the vegetation ([12]). Under field conditions, however, the actual value of ε is influenced by departures of temperature, soil moisture, and vapor pressure deficit from optimal values, which constrain the conversion of APAR to biomass. Notable success has been achieved estimating vapor pressure deficit by applying the "split-window" technique for estimating water vapor in a spatial or "contextual" setting [13]. The approach extracts information from the variations observed in the relation between sets of spatially coincident measurements acquired at two different thermal infrared wavelengths. The individual measurements are sampled as local blocks (e.g., 9 x 9 pixel matrix) within the larger satellite image arrays and evaluated in bivariate space. A contextual approach to analyzing coincident thermal infrared and NDVI measurements has likewise proven effective for estimating both soil moisture conditions and air temperature [13].

2.4. Living Biomass

Information on the amount of living biomass in an observed landscape is required to account for biomass lost through respiration. Active microwave observations offer potential for estimating biomass [14], but the current limited spatial coverage of data sets has so far been an obstacle to global scale applications.

One simple alternative approach employs global-coverage, optical wavelength data sets. Because shadowing in a landscape tends to increase with increasing biomass, the potential for estimating biomass from remotely sensed measurements of surface brightness appears promising. Such an approach was investigated by determining the seasonal minimum visible-band reflectance from AVHRR channel 1 data [4]. This simple method provides a useful, initial approximation of global biomass distribution, which may possibly be further improved with the inclusion of radar observations and additional optical wavelengths.

3. SUMMARY AND FUTURE PROSPECTS

The observations of the Earth available today from satellite remote sensors provide a global perspective on biospheric activity that is unprecedented in the history of science. Effective utilization of these global observations requires new insight and understanding concerning the relation between remotely sensed measurements and biophysical variables, as well as new avenues of thinking and research in the ecological sciences to address ecosystem processes at the spatially integrated, landscape scale. Although substantial research challenges remain, existing global-coverage satellite data sets provide a capability to estimate with reasonable accuracy the variables needed to model and monitor biospheric activity as expressed though global terrestrial primary production. Recent methods that employ measurements from two sensors, AVHRR and TOMS, were briefly reviewed here. Further refinements and improved accuracies can be expected with the enhanced spectral configurations of new earth observation sensors such as MODIS and VEGETATION. These capabilities represent only initial but nevertheless remarkable strides toward realizing the potential of remote sensing for analyzing and monitoring of the state and dynamics of the Earth's biosphere.

1 NRC, Toward an Understanding of Global Change, National Academy Press, Washington, D.C., 1988.
2 NRC, Remote Sensing of the Biosphere, National Research Council, National Academy Press, Washington, D.C., 1986.
3 S.N. Goward and D.G. Dye, (to be submitted).
4 S.D. Prince and S.N. Goward, Global Ecology and Biogeography Letters, (to be submitted).
5 J.L. Monteith, J. Appl. Ecol., 9 (1972) 747.
6 S.D. Prince, Int. J. Remote Sensing, 12 (1991) 1313.
7 R.T. Pinker and I. Laszlo, Journal of Climate, 5 (1992) 56.
8 T. Eck and D. Dye, Remote Sensing Environ., 38 (1991) 135.
9 D.G. Dye and S.N. Goward, Int. J. Remote Sensing, 14 (1993) 3361.
10 J.L. Hatfield, G. Asrar and E.T. Kanemasu, Remote Sensing of Environ., 14 (1983) 65.
11 S.N. Goward and K.F. Huemmrich, Remote Sensing Environ., 39 (1992) 119.
12 J.R. Ehleringer, HortScience, 14 (1979) 217.
13 S.N. Goward, *et al.*, Ecological Applications, 4 (1994) 322.
14 M.C. Dobson, F.T. Ulaby and T. LeToan, IEEE Trans. Geosci. Rem. Sens., 30 (1992) 412.

Spatial databases for global environmental issues

M.F. Goodchild[a]

[a]National Center for Geographic Information and Analysis, and Department of Geography, University of California, Santa Barbara, CA 93106-4060, U.S.A

Abstract

The paper begins with a review of the roles played by spatial data in global environmental research. While remote sensing provides much of the data for studies of the physical environment, other more problematic sources must be used for data on the human dimensions of global change, and for formulation of environmental policy. In addition to remote sensing, contemporary handling, storage, distribution, and analysis of spatial data is facilitated by a range of technologies that includes geographic information systems (GIS) and the global positioning system (GPS). The paper reviews the current status of these technologies, and the roles they are playing in global environmental research.

Current spatial data technology has inherited many of the limitations of its predecessors, particularly mapping. These include the difficulty of linking views at different levels of spatial resolution, or map scale; the essentially static or snapshot nature of much geographic data; the difficulty of dealing with continuous transition and fuzzy or uncertain position; and the varying quality of much data obtained by ground observation. Current research in geographic information science is attempting to remove or reduce many of these impediments, and the paper includes a brief status report in each of these areas. The capabilities of novel hierarchical data structures, automated generalization algorithms, object-oriented techniques, and spatial statistics are reviewed, and the paper concludes with a general assessment of the prospects for significant progress in these areas in the near future.

1. INTRODUCTION

The purpose of this paper is to provide an introduction to some of the issues surrounding the use of spatial databases in global environmental research. In recent years, the concept of a spatial, geographic, or geospatial database has achieved some degree prominence, due at least in part to the explosive growth of interest in computer technologies for handling geographic data, collectively termed geographic information systems (GIS). This growth began in the 1960s with the development of hardware specifically designed for geographic applications, including scanners, plotters, and map digitizers, and with the first attempts to develop computer codes to process representations of geographic features and distributions (Coppock and Rhind, 1991). It continued with research into algorithms and data structures in the 1970s, and by 1993 had grown to a software industry with annual sales of $450 million. Geographic information systems provide capabilities for input, storage, manipulation, analysis, and output, on platforms ranging from personal computer to mainframe. Texts by Burrough (1986), Star and Estes (1991), and Thompson and Laurini (1992) provide recent introductions to the field, while Maguire, Goodchild, and Rhind (1991) give a comprehensive synthesis.

Remote sensing of the Earth's environment from satellites began in the 1960s, and has developed in parallel with GIS software, and other parallel developments in image processing and pattern recognition. It offers the potential for high spatial, temporal, and spectral resolution, and uniform coverage of the Earth without regard to political boundaries, and is now widely used as a source of information for mapping, input to environmental models, and detection of change. The distinction between GIS and remote sensing is often blurred, although integration presents a variety of fundamental problems (Star, 1991).

A third recently developed geographic technology is the global positioning system (GPS) (Leick, 1990), a satellite-based system that allows reliable measurement of absolute position on the Earth's surface to some 100m accuracy (much higher accuracies can be achieved for relative positioning over short distances). As with remote sensing, GPS is increasingly integrated with GIS and used as a tool to support environmental science, particularly as a means of accurate positioning in the field.

With powerful techologies available for collection and processing, interest in the spatial component of data has increased dramatically in recent years. At the most fundamental level, spatial data provides the means to apply knowledge of physical processes to the Earth, taking into account the complex patterns of variation that exist on the planet's surface. In atmospheric modeling, for example, spatial data provides the key to disaggregating global climate models to provide local predictions. In ecology, it is the key to turning knowledge about dynamic ecological processes into policies for ecological conservation. The next section of the paper reviews the uses of spatial data in environmental modeling, and establishes a simple taxonomy. The subsequent section examines the implications of this review for spatial data policy, focusing on recent debates within the United States and their significance for global data. The final sections of the paper review some of the remaining impediments to use of spatial data, and ongoing research to remove them.

2. THE NATURE OF SPATIAL DATA

To the outsider, a focus on spatial data might seem somewhat surprising or non-intuitive. Why not, for example, focus on meteorological data, or ecological data? Essentially, the answer lies in experience that shows that spatial data have certain common characteristics, and working with them raises a series of common or generic issues that are significant to environmental modelers, whether they be atmospheric scientists, ecologists, or integrated modelers. One of these generic issues is accuracy, and the related modeling of error, an issue addressed in detail later in this paper. Another is the set of problems associated with representation of spatial data, often termed geographic data modeling.

Spatial, or more precisely geospatial or geographic data, are defined as data that contain explicit geographic reference, and thus are tied to locations on the surface of the Earth. The accuracy of the locational tie varies by many orders of magnitude, from the cm needed for analysis of crustal movement, to the 100m that is adequate for many environmental modeling purposes. For global climate modeling, or for international comparisons of UV exposure, a positional accuracy of tens or even hundreds of km may be adequate.

Within the domain established by this broad definition of spatial data, two major types are of importance to environmental modeling. They are termed here positional *framework* data sets, and sampled or measured *field* data sets. The defining characteristics of each type, and major issues associated with them, are discussed in the next two sections.

2.1. Framework data

Framework data sets serve to establish positions of activities, samples, features, and observations with respect to the *geoid*, a mathematical function approximating the shape of the Earth and anchored by the axis of rotation and the Greenwich Meridian. To measure geographic position directly, one must rely on astronomical observation coupled with accurate timekeeping, or more recently on the global positioning system (GPS), which provides civilian users with a positional accuracy of 40m for 50% of observations and 100m for 98%. Differential GPS is capable of cm accuracy, but only with respect to a fixed point whose position with respect to the geoid must have been established by some other means.

For most practical purposes, position is established not by direct measurement but by indirect measurement with respect to some fixed point of known location. Clearly this fixed point must be well-defined, a property more likely to be found in cultural than in natural features. Table 1 shows the sources of well-defined features and associated positions commonly used in the U.S. to measure position indirectly for various levels of positional accuracy; similar sources are available in other countries.

Table 1
Sources of U.S. framework data for establishing geographic position

Positional accuracy	Commonly used source of well-defined features
1cm	Geodetic control network
1m	Fixed monuments
10m	1:24,000 topographic mapping
100m	1:250,000 topographic mapping, GPS
1km	1:2,000,000 topographic mapping, Landsat imagery
10km	AVHRR imagery

The framework function of spatial data is to provide Earth locations of well-defined features, allowing other spatial data of various types to be registered to the Earth frame. Today, framework data sets are increasingly available in digital form on the Internet, allowing a fast, cheap, and reliable means whereby environmental scientists can build spatial data sets on a common framework. For many purposes the most important framework data sets contain the locations of political or local government boundaries, or centerlines of roads and streets. For scientific modeling, however, it is important that the framework contain features of environmental significance, such as hydrographic unit boundaries, or the locations of weather stations.

In summary, framework data sets allow one to find oneself, or the locations of samples points or observations, with respect to well-defined points, and hence the geoid. Thus they allow data to be registered to the geoid. They provide suitable sources of control or registration points for Earth imagery, or for transformations of coordinates or the merger of data sets. Finally, framework data sets provide a reasonably well-defined positional accuracy that can be matched to the needs of any application.

2.2. Sampled or measured fields

The second major class of spatial data includes sample measurements of continuous variation of some parameter over a two dimensional surface, or in three dimensions. Fields are often time-dependent, in which case sampling in time is also important. Fields provide the boundary conditions for many environmental models, as in the use of a field of topographic elevation in mesoscale atmospheric modeling. They provide the means to scale up or down, and to aggregate or disaggregate. Many fields are sampled remotely, either from the air or from space, using such instruments as AVHRR, Landsat, SPOT, airborne cameras, or airborne video. In other cases fields are sampled at points on the ground, for such parameters as surface temperature, elevation, soil moisture content, wind direction, or ground slope. In many cases the raw fields are processed, interpreted, and classified to produce fields of soil, land use, or land cover class.

In all such cases the focus of sampling or measurement is the more or less well-defined value of some variable which is conceived as having a unique value at every point on the two dimensional surface or in the three dimensional space. The variable is often a scalar but sometimes a vector, such as a wind field. It is represented digitally using a variety of methods [raster, triangular mesh (TIN), digitized contours, polygons (Goodchild, 1993)] depending on the context and the nature of the data. Fields are registered to the geoid using a framework data set, which provides a source of control points, and allows the sampling interval or spatial resolution of the measured field to be matched to the positional accuracy of the framework.

Technical problems abound in this process. For example, an Earth image can be registered to the geoid using only control points that are clearly visible in the image. This provides an effective upper limit of about 10m on positional accuracy, and much worse in unpopulated areas with no recognizable well-defined features, unless special targets are placed on the landscape.

2.3. Framework data sets: a paradox

Framework data sets have very little inherent scientific value to an environmental modeler. It is difficult, for example, to use the content of topographic maps as a data source for geomorphological studies of landforms (see Mark, 1983, for an excellent example of the problems this can cause). Almost all spatial data of significance in environmental modeling falls into the category of sampled or measured fields. Yet framework data sets are essential to the effective use of fields. Without them, for example, it would be impossible to find field areas, to register data sets to each other, to compare Earth images from different time periods, or to map the world's biological resources. Each of these processes requires the existence of framework data sets at appropriate levels of positional accuracy.

3. SUPPLYING THE NEED FOR FRAMEWORK DATA

Framework data sets are produced for numerous purposes, including a vast array of civilian and military activities in which science, and particularly environmental science, is a comparatively minor player.

Traditionally, the functions of framework data sets have been provided by paper maps, which have acted as readily available sources of position at approximately known levels of positional accuracy. The production of paper topographic maps has been a central government function in most countries, along with the maintenance of the primary geodetic control network that provides positional accuracies in the submeter range. However, the transition to digital dissemination of framework data has coincided with increasing pressure on central governments to recover costs. Traditional production of framework data sets is enormously costly and labor intensive, and it is difficult to build a broad base of popular support for this kind of expenditure. Moreover, science is traditionally cash poor and unable to contribute substantially to the cost of framework data set production. Another factor affecting the availability of framework data sets for science is national security, since accurate positioning of well-defined features is regarded in many countries as a valuable military resource. Finally, political stability is an obvious precondition to the creation and maintenance of national framework data sets.

Estes and Mooneyhan (1994) explore what they see as a series of popular misconceptions about mapping that ultimately affect the supply of framework data for environmental science. Although the American public and many other comparatively well-informed people in the world community believe that the world is well mapped, in reality much of the framework does not exist, is not available because of concerns for national security, or is old. This is a particularly worrisome problem for those fields of basic importance to science such as soils, land surface characteristics, hydrography, or climate, where fields are either not observable from space, or require spatial resolutions that are not currently available from space, or are too expensive to create from Earth images, or change significantly through time, requiring expensive periodic resampling or maintenance.

For example, in 1993 1:24,000 topographic map data, which potentially provides 10m positional registration, was available in digital form for only a fraction of the conterminous U.S. [for example, the transportation and hydrography layers were each available for 12% of the 54,145 quadrangles (Estes and Mooneyhan, 1994, p.522)]. Mapping at 1:100,000 or better (50m positional accuracy) existed in paper or digital form for only 59% of the Earth's land surface, and was available to environmental scientists for some undetermined fraction of that percentage. Digital elevation data was available at a 30m sampling interval for only 50% of the conterminous U.S. In summary, there are serious gaps in the framework both for the U.S. and for the globe as a whole, and there seems little prospect that the situation will improve dramatically in the near future. New technologies may provide an abundant and economical supply of digital elevation data and digital orthophotography, but the identification of cultural features and features invisible from space, an essential step in the creation of framework data sets, will remain costly and labor-intensive.

3.1. The U.S. National Spatial Data Infrastructure

Recently, an umbrella concept has emerged to embrace discussions of the national supply of framework data in the U.S. The National Spatial Data Infrastructure (NSDI) is defined by the National Research Council as "the means to assemble geographic information that describes the arrangement and attributes of features and phenomena on the Earth" (NRC, 1993). The concept of NSDI clearly mirrors many of the objectives of the National Information Infrastructure (NII), or "information superhighway", in the specific context of spatial data. NSDI is conceived as including the human resource issues of education and training; the development of standards for data exchange; and improvement in Internet access mechanisms through the development of digital catalogs and metadata. At the core of NSDI, however, are the digital framework data sets that allow users to tie their own information and observations to the Earth frame, and thus to integrate them geographically with other information.

Within the federal government, responsibility for the development of NSDI rests with the Federal Geographic Data Committee, an interagency group staffed by the U.S. Geological Survey and with representation from all major spatial data agencies. The creation of NSDI was mandated by an Executive Order signed by the President on April 11, 1994.

Although the precise contents of the NSDI framework data sets are still subject to intense debate, it is likely that they will include some combination of the following:

- the national geodetic control network of precisely defined point monuments, maintained by the National Geodetic Survey;

- Digital Line Graphs (DLGs) created from the 1:24,000 topographic map series of the U.S. Geological Survey (in June 1993 approximately 11% of these maps of the conterminous U.S. were available in DLG form);

- Digital Orthophoto Quads (DOQs), which are corrected panchromatic images with 1m pixel size and approximately 6m positional accuracy, each covering one quarter of the area of a 1:24,000 quadrangle (although only a small fraction of the conterminous U.S. is currently covered by DOQs, there are plans to produce them rapidly through a series of interagency and federal-state partnerships);

- street centerline data, which are digital representations of the centerlines of roads, highways, and streets, with a positional accuracy of 10m or poorer (the currently available TIGER data, produced by the U.S. Geological Survey and the Bureau of the Census, is being upgraded through a combination of public and private sector efforts); and

- Digital Elevation Models (DEMs), which are digital representations of terrain based on a 30m horizontal sampling interval (roughly 50% of the conterminous U.S. available in June 1993).

In the terminology of this paper, DEMs are measured fields, but they are frequently used as high-accuracy sources of such framework features as watershed boundaries.

Although the federal government is now committed to establishing NSDI, its potential costs are enormous. With modern technology it is possible to produce a DOQ for as little as $2,000, with a DEM as a byproduct, compared with the $40,000 cost of a conventional paper map quadrangle and its digital derivatives. But with 200,000 DOQs needed to cover the conterminous U.S., the cost of a single complete coverage is at least $400 million. One advantage of the relatively low cost of DOQs is that they can be updated frequently, but the cost of update is not significantly less than the cost of initial coverage. Constant efforts are being made by the mapping agencies to find funding for programs like DOQ coverage, but they face the harsh political reality that mapping is not the most effective way to buy votes. To return to an earlier point, the U.S. taxpayer believes by and large that the U.S. is well-mapped, and that mapping is a simple and cheap process. After all, it has been decades since the last blank parts of Africa and Antarctica were filled in, and centuries since "terra incognita" appeared on world maps.

3.2. Towards a global infrastructure

While recent debate over NSDI suggests a widely recognized need for a coordinated approach to framework data in the U.S., it seems appropriate at this stage to ask about parallel efforts at a global or international level. The Digital Chart of the World (DCW) provides a framework data set with a positional accuracy of 500m, based on U.S. 1:1,000,000 mapping, and has been distributed widely at low cost. Global DEM data is available with similar levels of resolution, and there has been discussion of the need for a new global DEM with 30m or even 10m sampling, based on novel remote sensing technologies. Although there are no truly international format standards for spatial data to promote data exchange, many national and commercial standards are widely recognized, and NATO has developed standards in the interests of common defense. There are many excellent efforts to establish spatial data clearinghouses, including the U.N. Environment Program's GRID, the efforts of CIESIN (Consortium for International Earth Science Information Network, headquartered in Saginaw, Michigan, U.S.A), and the World Conservation Monitoring Centre's spatial data resources in Cambridge, U.K.

These efforts notwithstanding, remote sensing remains virtually the only reliable source of truly global coverage. Dimensions of spatial variation that are not visible from space, including political boundaries, place names, and other cultural features, continue to pose enormous problems to environmental scientists needing reliable sources of framework data, and little progress has been made in utilizing remote sensing to detect social, demographic, industrial, or economic variation. The problems of building international spatial databases to address the social, economic, or political dimensions of global change are illustrated by CIESIN's CITAS (China in Time and Space) program, which aims to build a spatio-temporal database of Chinese demography. Although the Chinese population is arguably the world's best-counted, with accurate records extending back hundreds of years, on average one change per month is made in the internal boundaries of Chinese counties, making any longitudinal or spatial perspective virtually unattainable without complex spatial data handling technology; and variation and duplication of placenames merely adds to the difficulties.

Clearly, the problems being addressed nationally by NSDI need to be tackled also at the global level. Earth systems science requires a reliable source of framework data so that observations can be positioned with respect to the Earth frame and integrated with other data for modeling and policy formulation. Emerging environmental programs like GLOBE cannot succeed without a sound framework, and associated programs to address the needs of education and training, standards, and protection of copyright, national security, and intellectual property.

The remaining sections of the paper review some of the outstanding generic and fundamental research issues associated with widespread use of spatial data. Many of these issues are not new to the computing age, but rather have existed in various forms within conventional cartography, and other traditional methods for handling geographic information. But the emergence of GIS has focused attention on them, in part because of the high precision and mechanistic nature of computer representation and processing. The next section looks specifically at one particularly pervasive issue, that of uncertainty or data quality, in some detail, and is followed by a more general review of a wider cross-section of issues.

4. THE ACCURACY PROBLEM

As with any assemblage of scientific measurements, spatial data is subject to errors of measurement that depend on the accuracy of the measuring instrument, and propagate into the results of modeling. Two further characteristics of spatial data also contribute to the accuracy problem: first, the process of spatial averaging or generalization that is present in much spatial data of limited spatial resolution; and second, the use of data that embeds the subjective judgment of the observer. For example, measurements obtained from Earth imaging satellites, such as AVHRR, are averaged over areas as large as 1km square. When the land surface is characterized by assigning a class to each AVHRR pixel instead of a measured spectral response, the resulting class captures a comparatively sophisticated analysis of spatio-temporal variation, but it is probable that two analysts would produce somewhat different classifications, and consequently the pattern of classes is not entirely reproducible from one analyst to another. This lack of reproducibility of land surface characterization may be a substantial source of uncertainty in environmental modeling, particularly when land surface class is used as a surrogate for variability in soil moisture, evapotranspiration, albedo, or some other physical parameter.

In recent years significant advances have been made in modeling uncertainty in spatial data. Error models have been developed by Fisher (1991, 1992), Goodchild, Sun, and Yang (1992), Haining and Arbia (1993), Mark and Csillag (1989), and others, and the field of geostatistics also provides a comprehensive theoretical framework for discussion of spatial data uncertainty (Isaaks and Srivastava, 1989). But while suitable models exist, along with techniques for error analysis by propagating uncertainty through GIS processes into confidence limits on results, the implications of uncertainty are often difficult to understand, especially for users not familiar with the concepts of spatial statistics (Cressie, 1993). Beard, Buttenfield, and Clapham (1991) have proposed that visual techniques be developed to communicate information on uncertainty, and that data quality be an essential component of any spatial data display. Recent research on the visualization of spatial data quality is described by Buttenfield (1993).

A simple example is provided by a simple contour map. In a standard U.S. Geological Survey 30m digital elevation model (DEM), now widely used as a source of data in many types of environmental models, sample points form a square grid with a spacing of 30m on the ground. Using appropriate GIS functions, it is possible to convert a DEM into a contour map using a standard contouring algorithm. These 30m DEMs have a stated 7m root mean square error (RMSE), indicating that the root mean square difference between true and observed elevation at each sample point is 7m. For the purposes of this discussion, we assume that the distribution of elevation errors is Gaussian about the true value, with a standard deviation equal to the RMSE (see Hunter and Goodchild, in press, for a more detailed discussion of this example). Again using standard GIS functions, it is straightforward to create a visualization of uncertainty in the position of any given contour by displaying a map that shows at each point the probability that the elevation at that point is in reality greater than the elevation of the contour. For example, one can show the uncertainty in the position of the 350m contour as a map of probability that any point lies above the true 350m contour. By ignoring probabilities less than 2.5% or greater than 97.5%, one has in effect a display of the 95% confidence limits of the contour's position.

Although all of the ingredients are well-known, the results of this process can be somewhat surprising to people used to conventional contour maps, which show each contour with a constant width and make no attempt to communicate known uncertainty to the viewer. Positional uncertainty is inversely dependent on slope, and can be extreme in areas close to the horizontal. This lack of communication of knowledge about uncertainty is not particularly problematic as long as the contour map is viewed and used as a impression of spatial variation. But in the era of GIS, when it is increasingly likely that people will regard spatial databases as collections of measurements, lack of knowledge of uncertainty can be highly misleading and dangerous for decision-making.

Although the example and the illustration of uncertainty appear straightforward, closer examination shows that this is in fact far from the truth. Consider what happens when this example landscape is flooded to 350m, perhaps by construction of a reservoir. Had the elevations been lower, the consequences of sea level rise might have provided an alternate motivation.

Suppose first that the 7m RMSE refers not to the uncertain elevation of the ground surface, but to the uncertain elevation of the floodwater relative to an error-free ground surface. For example, the elevation of the reservoir surface on any given day might be sampled from a Gaussian distribution with a mean of 350m and a standard deviation of 7m. Water surface elevations of 347m or 355m are comparatively likely, but 371m lies 3 standard deviations from the mean, and is therefore highly unlikely. From the statistics of the Gaussian distribution we know that the risk of flood in areas mapped at 359m (1.29 standard deviations above the mean) is approximately 10%, and the appropriate areas with this risk of flood are clearly identified in the visualization discussed earlier.

Now suppose that the elevation of the water surface is exactly 350m, and that the uncertainty exists in the ground elevation. Although the same Gaussian distribution models the sampling of elevation, it is likely that some areas are characterized by positive errors (observed elevation greater than true elevation) and some by negative errors (observed elevation less than true elevation). In the extreme, the error at each sample point might be statistically independent. It seems more likely that errors are statistically correlated, such that a positive error at one sample point implies similar positive errors at neighboring sample points, but the correlation is likely less than perfect.

In these circumstances, the interpretation of the visualization changes. Although the visualization still shows correctly the probability that any given point is above 350m, the probability that an area is above 350m depends on the number of points lying in the area, and on the spatial autocorrelation of errors. If errors are statistically independent, an area of n sample points now has a probability of p^n of being above 350m, if p is the (constant) probability associated with each of the points in the area. In a visualization such as this, where sample point locations are not shown, it is virtually impossible to make this interpretation successfully. The risk that a patch containing 10 sample points will be *entirely* flooded lies somewhere between .10 and $.10^{10}$, if each contained sample point has a risk of flood of 10%, depending on the structure of spatial dependence of errors.

Unfortunately virtually nothing is known about the spatial dependence of errors in digital elevation models, or in any other type of spatial data. While this may be unimportant in many applications, the complex processing of spatial data that increasingly occurs in GIS and environmental modeling makes this issue hard to ignore. For example, the ability to estimate slope (Burrough, 1986, p.50) depends directly on spatial dependence in errors between neighboring sample points. If slope is calculated from adjacent pairs of elevation estimates, by taking the difference in elevation and dividing by 30m, a simple error analysis shows that the standard deviation of the uncertainty in slope estimates is given by:

$$\sigma^2_{slope} = 2\sigma^2_e \frac{(1 - r)}{h^2}$$

where σ_e is the standard deviation of elevation estimates (7m), h is the spacing, and r is the correlation between elevation errors at adjacent points. If the correlation is zero (adjacent estimates are independent), then the uncertainty in slope estimates is no less than 33%, far too high for slope calculations to have practical value. On the other hand, if errors at adjacent points are perfectly correlated (r=1) then slope estimates have no uncertainty. In practice, r is very close to 1 and slope estimates are reliable enough to be useful. But since r varies over the map, there may be some areas where slope errors are notably higher.

In summary, while our ability to model and propagate uncertainty has improved markedly in the past few years, lack of knowledge of fundamental parameters of spatial structure of errors makes it impossible to answer even the most simple question about propagation, and makes it very easy to misinterpret simple visualizations of uncertainty. Englund (1993) provides another illustration of this problem in the geostatistical context of GIS analysis using Kriging.

5. RESEARCH ISSUES IN SPATIAL DATA

Accuracy, as reviewed in the previous section, is only one of the generic issues that act as impediments to the widespread scientific use of spatial data. In this section several other issues are reviewed, with the objective of providing a brief overview of the breadth and importance of the current research effort. Goodchild (1992) has suggested that the generic problems of spatial data are sufficiently distinct that they merit specific identification, and has suggested the term *geographic information science.* In part, it was this set of issues that led to the establishment in 1988 of the U.S. National Center for Geographic Information and Analysis, and to the research agenda that it has followed and revised over the past six years

(NCGIA, 1989; 1992).

5.1. Multiple representations

The traditional technology of mapping is limited for most purposes to single views of regions or the world at uniform scale. Of course, the need to project the curved surface of the Earth onto flat paper results in variation in effective scale, and it is possible to use insets to present greater detail. But in the digital environment of a spatial database, where scale is undefined, it should be possible to escape many of these constraints. Unfortunately, the map has remained the dominant metaphor in GIS, and it has proven remarkably difficult to build multiscale databases that provide logical linkages between information at different scales.

Several research areas have potential promise. Automated generalization methods seek to overcome the multiple representations problem using algorithms that create coarse views from detailed ones. This is standard practice in many areas of image processing, where the passage of a low pass filter is sufficient to remove detail. But it is infinitely more complex in the handling of spatial data sets composed of representations of geographic features, where complex decisions must be made about the deletion or retention of features, merging of features, reduction in detail in their geometric representation, and the like (Buttenfield and McMaster, 1991).

Hierarchical data structures attempt to provide data set frameworks to link coarse and detailed representations. Among the first of these was the quadtree (Samet, 1990a,b), which represents spatial variation as a tree of nested squares, branching until each square is homogeneous in the represented variable. The quadtree data structure has been exploited in a number of systems, and extended to the spherical surface of the globe (Goodchild and Yang, 1992).

Recently, there has been much interest in wavelets (Chui, 1992) as a hierarchical method of storage of images. Rather than requiring a constant value within each leaf, as in a quadtree, wavelets extract spectral components at each level and within each leaf. A single data structure can be used to reconstruct a generalization of the image at any level of spatial resolution. To date, there has been no widespread application of wavelets to geographic data.

5.2. Time dependence and 3D

Spatial databases, and the associated GIS technology used to handle them, have developed as essentially static representations of geography. In part this is because of dependence on the map metaphor; in part it is because of the cost of collecting time-dependent data; and in part, it reflects a lack of interest in time dependence, and the absence of appropriate models and methods of analysis. Full three-dimensional, time-dependent data remains the greatest challenge facing environmental science.

Recently, there has been much interest in capitalizing on the momentum of GIS to introduce greater concern for time and the third dimension. Langran (1992) has reviewed the role of time in GIS, and Raper (1989) and Turner (1992) have discussed progress in the third dimension. Lack of data continues to present a problem, along with a lack of concepts for addressing either extension, but it is likely that progress will accelerate in the future.

5.3. Object orientation

The object oriented approach that emerged in computer science in the 1970s and 1980s includes a very rich collection of concepts, many of which are beginning to influence spatial data and GIS. The following sections review some of the areas in which the object oriented philosophy is having significant impact.

Spatial databases as collections of objects: Many influential early designs for spatial databases focused heavily on the representation of fields. For example, a map of land cover class provides a single value of class at every point in the plane. Within the database, such nominal fields are often represented as collections of *planar-enforced* polygons, which may not overlap and must exhaust the plane. In these early architectures, all area features must be represented in layers of such planar-enforced polygons. This limits the usefulness of the GIS in applications where area features may overlap, or where they do not exhaust the plane.

From an object-oriented perspective, planar enforcement can be seen as a series of constraints on the behavior of area features, or their digital representations. During edit, collections of polygons that are defined as planar enforced cannot be dragged apart, or dragged over each other. But a database can consist of many types of area features, each with its own set of behaviors.

Measurement-based GIS: In many spatial data sets, the connection between database contents and original field measurements can be particularly obscure. If field measurements are updated or improved, it can be virtually impossible to update the derived database. For example, suppose improved geodetic control becomes available for part of Antarctica, and it is desirable to merge this with existing data. Without knowledge of the connections between database contents and original measurements, it may be necessary to undertake a complete resurvey. In a measurement-based GIS, each object's position is derived explicitly, through a system of inheritance. Thus when one or more of the original field measurements is updated, the derivative parts of the database are automatically revised. Commercial systems with measurement-based architecture are beginning to appear, and promise great benefits in reducing the costs of updating spatial databases.

Encapsulation: In traditional GIS there has been a clear separation between the contents of the database, and the processes which operate on them. This is a commonly used approach in science, and is reflected in fields such as statistical analysis as well as in GIS. Often the database contains nothing but records, although increasingly the concept of metadata, or standardized data about the contents of a data set, is seen as the key to improving sharing of data among scientists over the Internet. Metadata is digital, travels with the data, and describes such aspects as its quality to potential users. Metadata is one example of the object oriented concept of encapsulation, or the tight coupling of an object with information on its behavior, in contrast to the traditional separation of data objects from the processes that act on them.

Consider a simple example of a spatial data set of weather records. Many uses of point observations of barometric pressure include the routine task of contouring, to create a set of isobars, or lines of constant barometric pressure. Within the context of GIS, contouring is achieved automatically by use of one of a number of alternative methods of *spatial interpolation*, such as Kriging (Isaaks and Srivastava, 1989). Traditionally, it has been the user of the data who selects the method of interpolation, and the values of any of its essential parameters. But increasingly, such users are widely separated, have little or no knowledge of the processes that control atmospheric pressure, and thus little objective basis on which to choose an appropriate method of interpolation. By contrast, selection of the method by the data producer would allow sophisticated and well-informed judgment.

From an object-oriented perspective, this represents a case of *encapsulation*, or the tight coupling of data objects with methods. To work, it requires the standardization of spatial interpolation methods, so that metadata fields can be used to control the processing of the data by its users.

5.4. Spatial analysis

Spatial analysis is defined as that set of techniques whose results are not invariant under changes in the geographic locations of the objects of analysis. As such, it is often considered a subset of statistics, although it includes many non-statistical methods (1981). GIS provides the computational basis for spatial analysis in a similar fashion to the role of statistical packages in statistics.

Recent developments in spatial analysis have been motivated both by the emergence and widespread adoption of GIS, and by increasing interest in exploratory methods in statistics (Tukey, 1977). The term *exploratory spatial data analysis* (ESDA) has been used to describe this approach, with its emphasis on providing tools that allow the user to visualize spatial data, and explore its characteristics to gain insight into its underlying processes. The recent books by Fotheringham and Rogerson (1994) and Hearnshaw and Unwin (1994) give an excellent cross-section of work in this area. Unfortunately, to date little of this work explicitly addresses data on the curved surface of the Earth.

6. CONCLUSIONS

The first two of the following six concluding points echo Estes and Mooneyhan (1994):

1. We need to dispel the myths that exist among the public, politicians, policy makers, key agencies, and our administrators and colleagues in the scientific community that the maps and spatial data needed to support environmental modeling and related science exist; that mapping is easy; and that there is no science left to do in mapping.

2. The availability of spatial data is often adversely affected by concerns over national security and sovereignty. We need to find a better balance between these concerns and the scientific need to understand the Earth as a system.

3. We must develop and maintain the framework data sets that are needed to register scientific measurements, models, and predictions to the surface of the Earth with accuracy appropriate to each application.

4. Besides the framework data sets themselves, the objectives of NSDI and its suggested global equivalents include the development of the means to assemble, process, store, analyze, and disseminate spatial data; and the human resource base that will make this feasible.

5. We need to implement the objectives of NSDI at the international and global levels,

and to find appropriate mechanisms for doing so.

6. Many of the objectives of NSDI and its proposed global equivalent raise questions about research issues such as data accuracy that will have to be solved before these programs can be fully implemented. We need a concerted effort to address the outstanding research issues of spatial data.

7. ACKNOWLEDGMENT

The National Center for Geographic Information and Analysis is supported by the U.S. National Science Foundation, grant SBR88-10917.

8. REFERENCES

Beard, M.K., Buttenfield, B.P., & Clapham, S. 1991. *Visualizing the quality of spatial information: scientific report of the specialist meeting.* Santa Barbara, CA: National Center for Geographic Information and Analysis, Technical Report 91-26.

Burrough, P.A. 1986. *Principles of Geographical Information Systems for Land Resources Assessment.* Oxford: Clarendon.

Buttenfield, B.P. (editor) 1993. Mapping data quality: a collection of essays. *Cartographica* 30(2&3): v-viii, 1-46.

Buttenfield, B.P. & McMaster, R.B. (editors) 1991. *Map Generalization: Making Rules for Knowledge Representation.* London: Longman.

Chui, C.K. 1992. *An Introduction to Wavelets.* Boston: Academic Press.

Coppock, J.T. & Rhind, D.W. 1991. The history of GIS. In D.J. Maguire, M.F. Goodchild, & D.W. Rhind (editors) *Geographical Information Systems: Principles and Applications.* London: Longman, pp. 21-43.

Cressie, N.A.C. 1993. *Statistics for spatial data.* New York: Wiley.

Englund, E. 1993. Spatial simulation: environmental applications. In M.F. Goodchild, B.O. Parks, & L.T. Steyaert (editors) *Environmental Modeling with GIS.* New York: Oxford University Press.

Estes, J.E. & Mooneyhan, D.W. 1994. Of maps and myths. *Photogrammetric Engineering and Remote Sensing* 60(5): 517-524.

Fisher, P.F. 1991. Modeling soil map-unit inclusions by Monte Carlo simulation. *International Journal of Geographical Information Systems* 5: 193-208.

Fisher, P.F. 1992. First experiments in viewshed uncertainty - simulating fuzzy viewsheds. *Photogrammetric Engineering and Remote Sensing* 58: 345-352.

Fotheringham, A.S. & Rogerson P.A. (editors) 1994. *Spatial Analysis and GIS.* London: Taylor and Francis.

Goodchild, M.F. 1992. Geographical information science. *International Journal of Geographical Information Systems* 6(1): 31-45.

Goodchild, M.F. 1993. Data models and data quality: problems and prospects. In M.F. Goodchild, B.O. Parks, & L.T. Steyaert (editors) *Environmental Modeling with GIS.* New York: Oxford University Press, pp. 94-104.

Goodchild, M.F., Sun, G.Q. & Yang, S. 1992. Development and test of an error model for categorical data. *International Journal of Geographical Information Systems* 6: 87-104.

Goodchild, M.F. & Yang, S.R. 1992. A hierarchical spatial data structure for global geographic information systems. *CVGIP-Graphical Models and Image Processing* 54(1): 31-44.

Haining, R.P. & Arbia, G. 1993. Error propagation through map operations. *Technometrics* 35: 293-305.

Hearnshaw, H.M. & Unwin, D.J. (editors) 1994. *Visualization in geographical information systems.* New York: Wiley.

Hunter, G.J., & Goodchild, M.F. in press. Dealing with uncertainty in spatial databases: a simple case study. *Photogrammetric Engineering and Remote Sensing.*

Isaaks, E.H. & Srivastava, R.M. 1989. *Applied geostatistics.* New York: Oxford University Press.

Langran, G. 1992. *Time in Geographic Information Systems.* New York: Taylor and Francis.

Leick, A. 1990. *GPS Satellite Surveying.* New York: Wiley.

Maguire, D.J., Goodchild, M.F., & Rhind, D.W. (editors) 1991. *Geographical Information Systems: Principles and Applications.* London: Longman.

Mark, D.M. 1983. Relations between field-surveyed channel networks and map-based geomorphometric measures, Inez, Kentucky. *Annals, Association of American Geographers* 73(3): 358-372.

Mark, D.M. & Csillag, F. 1989. The nature of boundaries on 'area-class' maps. *Cartographica* 26(1): 65-78.

National Center for Geographic Information and Analysis, 1989. The research plan of the National Center for Geographic Information and Analysis. *International Journal of Geographical Information Systems* 3(2): 117-136.

National Center for Geographic Information and Analysis, 1992. *Research Agenda for the NCGIA Renewal 1993-96*. Santa Barbara, CA: NCGIA, Technical Paper 92-7.

National Research Council 1993. *Toward a Coordinated Spatial Data Infrastructure for the Nation.* Washington, DC: National Academy Press.

Raper, J. 1989. *Three Dimensional Applications in Geographical Information Systems.* New York: Taylor and Francis.

Samet, H. 1990a. *Applications of Spatial Data Structures: Computer Graphics, Image Processing. and GIS.* Reading, MA: Addison-Wesley.

Samet, H., 1990b. *The Design and Analysis of Spatial Data Structures.* Reading, MA: Addison-Wesley.

Star, J.L. (editor) 1991. *The Integration of GIS and Remote Sensing.* Bethesda, MD: American Society for Photogrammetry and Remote Sensing.

Star, J.L. & Estes, J.E. 1991. *Geographic Information Systems: An Introduction.* Englewood Cliffs, NJ: Prentice Hall.

Thompson, D.R. & Laurini, R. 1992. *Fundamentals of Spatial Information Systems.* San Diego: Academic Press.

Tukey, J.W. 1977. *Exploratory Data Analysis.* Reading, MA: Addison-Wesley.

Turner, A.K. (editor) 1992. *Three-Dimensional Modeling with Geoscientific Information Systems.* Dordrecht: Kluwer.

Unwin, D.J. 1981. *Introductory Spatial Analysis.* London: Methuen.

TOWARD GLOBAL PLANNING OF SUSTAINABLE USE OF THE EARTH
Development of Global Eco-Engineering
S. Murai (Editor)

Determining Emissions of Carbon from Land: A Global Strategy

R. A. Houghton

The Woods Hole Research Center, P.O. Box 296, Woods Hole, Massachusetts 02543, U.S.A.

Abstract

Largest uncertainties in the global carbon balance are associated with the net flux of carbon to or from terrestrial ecosystems. The use of satellite data with high spatial resolution to determine rates of land-use change could significantly improve the accuracy of at least one component of this terrestrial flux. A further increase in accuracy would result from the ability to measure above-ground carbon stocks from satellite. The types of data required to determine emissions of carbon from land-use change and a global strategy for obtaining these data are described. The data on changes in land cover obtained in such an exercise would be valuable for investigation of other global environmental problems, such as emissions of methane and nitrous oxide, biodiversity, sustainability of land use, and the effects of climatic change on the earth's terrestrial ecosystems.

1. INTRODUCTION

The ecosystems covering the land surface of the earth vary. The variation extends far beyond the obvious differences between deserts, grasslands, and forests. Ecosystems vary with respect to their productivity, the numbers of species they hold, the amount of carbon stored in their vegetation and soils, the degree to which they stabilize soils and water flows, their exchanges of trace gases with the atmosphere, their climatological characteristics (surface roughness, albedo, and evaporation), and a number of other attributes. Their roles in affecting local and global climate, in regulating the chemical composition of water and air, and in providing a number of other services, without human intervention, are poorly understood. Human uses of terrestrial ecosystems are altering their distribution and their characteristics. Some uses are sustainable, others are not, but neither the services nor the changes underway at present are well described or recognized. The task of terrestrial ecologists is two fold: to define the functions and services of ecosystems, or how the world works, and to document the current status of ecosystems and how they are changing.

The tasks are difficult but, within the context of greenhouse gases and climatic change, both political and scientific responsibilities require that these tasks be addressed. The political and legal requirements are defined by the United Nations Framework Convention on Climate Change: every country must determine its emissions of greenhouse gases. In many countries, the major source of these gases is human induced changes in terrestrial ecosystems (deforestation, for example). Scientific justification for determining the role of terrestrial ecosystems in the global carbon cycle is based on an incomplete understanding of the cycle at present and the inability, therefore, to predict future atmospheric concentrations of CO_2 from alternative scenarios of future emissions.

2. THE ROLE OF TERRESTRIAL ECOSYSTEMS IN THE GLOBAL CARBON CYCLE

The purpose of this paper is to describe the data needed to determine the emissions of carbon to, and the accumulations of carbon from, the atmosphere as a result of changes in land use. The conversion of forests to agricultural and degraded lands in the tropics is releasing carbon to the atmosphere as trees are burned, as vegetation killed or damaged during logging and clearing decays, and as soil organic carbon is oxidized as a result of cultivation. The area of forests is increasing in some areas outside the tropics (in Europe, for example), but the net change in area of forests globally has been and continues to be a reduction in area. This net reduction causes a release of carbon to the atmosphere because the carbon lost through the burning and oxidation of trees and soils is not balanced by the small and temporary accumulation of carbon in crops and cropland soils. The net annual release of carbon from changes in land use has increased over the last century, although the regions responsible for the release have varied over time.

The net annual flux of carbon between the atmosphere and land as a result of land-use change is estimated to have been about 1.2±0.7 PgC/yr during the decade of the 1980s (Table 1).

Table 1
Net flux of carbon to (+) and from (-) the atmosphere as a result of changes in land use during the 1980s

Region	Net flux PgC/yr
Tropical Asia	0.7±0.4
Tropical America	0.6±0.3
Tropical Africa	0.3±0.2
Subtotal for tropics	1.6±0.5
Temperate and Boreal Zones	-0.4±0.5
Globe	1.2±0.7

The estimates of flux for the tropics are from Houghton [1]. Uncertainties are judged to be about 50%. Outside the tropics estimates vary from a net source of 0.1 PgC/yr [2,3] to a net sink of 0.7 PgC/yr or more [4]. The discrepancies between these estimates have not been fully resolved. Some of the estimates summarized by Dixon et al. [4] used approaches and data that included processes not necessarily related to land-use change. Uncertainty in the estimates of carbon flux for these regions is relatively high.

Despite the overwhelming evidence that rates of deforestation are higher than rates of reforestation, globally, and that changes in land use are responsible for a net flux of carbon to the atmosphere, the role of terrestrial ecosystems in the global carbon cycle is unclear. The reason for the uncertainty is that the net flux from land-use change may not represent the entire net flux from terrestrial ecosystems. Indirect evidence from geochemical models of the atmosphere and oceans suggests that terrestrial ecosystems are approximately in balance with respect to carbon [5-8]. If this is so, the net release from changes in land use must be offset by an accumulation of carbon either in undisturbed ecosystems or in disturbed ecosystems that are not adequately understood.

The geochemical evidence for carbon accumulation in terrestrial ecosystems can be summarized by considering the rest of the terms in the global carbon equation. According to the equation, more carbon is released from the combined effect of fossil fuel combustion (FF) and changes in land use (LUC) than is accumulating in the atmosphere (ATM) and oceans (SEA).

$$\begin{array}{ccccccc} \multicolumn{2}{c}{\text{RELEASES}} & = & \multicolumn{3}{c}{\text{ACCUMULATIONS}} & \\ \text{FF} \quad + & \text{LUC} & = & \text{ATM} \quad + & \text{SEA} \quad + & \text{TE} & \\ 5.5\pm0.5 & 1.6\pm0.7 & = & 3.4\pm0.2 & 2.0\pm0.8 & 1.7\pm1.2 & (1) \end{array}$$

The difference, or imbalance, is thought to be an accumulation of carbon in terrestrial ecosystems (TE) because all of the other terms are independently determined. Several independent studies support this conclusion, including an analysis of temporal and geographic variations in the $^{13}C/^{12}C$ ratio of atmospheric CO_2 [5,9]. The imbalance is approximately the magnitude of the net release from changes in land use.

An accumulation of carbon on land is not necessarily inconsistent with the finding of a loss of carbon from changes in land use. Analyses of land-use change have assumed that ecosystems not deliberated disturbed by humans are in steady state with respect to carbon. The frequencies of wildfires, disease, and pests are assumed not to have changed, and changes in climate and atmospheric CO_2 are assumed not to have changed the amount of carbon held in the world's vegetation and soils. These assumptions may not be valid. On the other hand, rates of tree growth over large areas are consistent with the rates expected as a result of land-use change (for example, agricultural abandonment and regrowth after logging) [10]. These rates of accumulation of carbon in growing forests have already been included in estimates of flux calculated from changes in land use, so the observed growth does not explain the global imbalance. Carbon may be accumulating in soils, but the accumulations expected as a result of deliberate use of land have again been calculated, and the natural rates of accumulation are too slow to account for the global imbalance [11]. The imbalance remains unexplained. Either analyses of land-use change overestimate the net release of carbon, measurements of tree growth underestimate actual rates, the terrestrial sink is distributed evenly over the surface of the earth in both trees and soils and hence is undetectable, or only a portion of the missing sink is on land. While it would be difficult to detect a 1 PgC/yr increase in the world's vegetation (500-600 PgC), the cumulative imbalance since 1920 represents between 50% and 70% of the current forest biomass in temperate zone forests [12]. Such a localized accumulation should be observed.

A resolution of the imbalance is important because projections of future CO_2 concentrations based on different scenarios of fossil fuel emissions are sensitive to the imbalance in the carbon cycle. The mechanisms believed likely to accumulate carbon on land (CO_2 fertilization, nitrogen mobilization, climatic variability) may or may not be expected to continue in the future. If CO_2 fertilization has been responsible, for example, it may be expected to continue and even to increase the storage of carbon in the future in proportion to atmospheric CO_2 concentrations. On the other hand, if today's carbon cycle is balanced by some other mechanism, or if the CO_2 fertilization factor is reduced in the future, concentrations of atmospheric CO_2 will be higher than projected by a model that assumes current and future terrestrial storage to be linked to atmospheric CO_2.

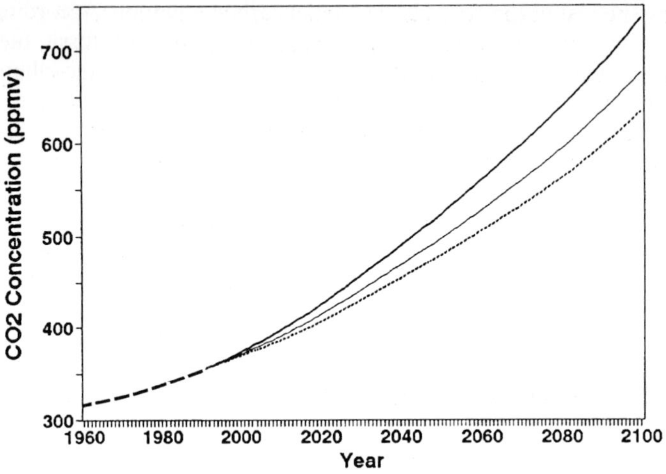

Figure 1. Differences in projected CO_2 concentrations resulting from CO_2 fertilization factors of 10, 26, and 40%. Based on the model of Wigley [13]; data supplied by T.M.L. Wigley (pers. comm.)

Figure 1 shows the effect of assumptions about CO_2 fertilization. The dashed line between 1960 and 1990 represents observed concentrations of CO_2. The three curves between 1990 and 2100 are projections of future concentrations. All three curves are based on the IS92A scenario from the 1992 IPCC Scientific Assessment [14], a projection intermediate in terms of future emissions of CO_2 and other greenhouse gases. The projections shown here (T.M.L. Wigley, pers. comm.) are based on the model of Wigley [13]. The three curves differ in the strength of the CO_2 fertilization factor, the ratio of net primary production (NPP) at 680 ppmv to NPP at 340 ppmv. The central curve assumes a CO_2 fertilization factor of 1.26, the value required to balance the mean emissions from land-use change during the 1980s (1.6 PgC/yr [12]) with mean values of oceanic uptake (2.0 PgC/yr). Under this scenario, CO_2 concentrations increase from 354 ppmv in 1990 to 676 ppmv in 2100. If some of the global carbon balance during the 1980s resulted from factors other than CO_2 fertilization, and the fertilization factor is assumed to be 10% rather than 26%, the concentration in 2100 becomes 736 ppmv. On the other hand, if the fertilization factor is assumed to be 40%, the concentration at the end of the 110-year projection is 634 ppvm. The uncertainties in the fertilization factor yield a 100 ppmv difference in projected concentrations. The difference is equivalent to 30% of the projected increase above 1990 concentrations of CO_2. Furthermore, these assumptions about CO_2 may be optimistic. If current trends in tropical deforestation continue, the areas where CO_2 fertilization might accumulate carbon will be significantly reduced. Loss of forests not only releases carbon to the atmosphere; it diminishes the capacity of terrestrial ecosystems to withdraw carbon from the atmosphere.

The need to understand the global carbon balance is crucial. The monitoring and modeling efforts that have made possible our current understanding of the other terms in the global carbon balance must be continued and the estimates improved. One of these efforts is determination of the net flux of carbon from changes in land use. Although analyses of land-use change may not resolve the current carbon imbalance, they help constrain its magnitude, and they may suggest the mechanisms responsible [1]. Current uncertainties in estimates of the flux of carbon from land-use change can be reduced considerably by the use of remote sensing to measure the rates of change.

Justification for determining the role of land-use change in the global carbon cycle is both scientific and political. The questions, scientifically, are *What and where are the terrestrial sources and sinks of carbon, and how are they changing on time scales of years to decades?* The question politically is *What are the annual emissions of CO_2 (and other greenhouse gases) from each country?*

3. A STRATEGY FOR DETERMINING EMISSIONS OF CARBON FROM CHANGES IN LAND COVER/LAND USE

3.1. What Should Be Measured?

There are at least two approaches for determining changes in the stocks of terrestrial carbon. One approach measures changes in the total amount of carbon in forests (including live and dead vegetation, litter, and soils) and products (wood in houses, paper, etc.). Analyses that use growing stocks reported in forestry statistics are generally of this approach [15,16]. A second approach uses changes in land use to calculate releases and accumulations of carbon in the affected ecosystems and accounts for all of the carbon initially held in those ecosystems. This is the approach taken in this paper. It has been described previously [2,17-19].

Two broad types of information are required for this second approach: rates of land-use change and the effects of such changes on the carbon held in vegetation and soils. A model is used to calculate the losses and accumulations of carbon per ha for different types of ecosystems undergoing different types of land-use change. The sum of these changes, along with changes in wood products removed from forests, yields the net flux of carbon to the atmosphere. Figure 2a shows an example of the changes in terrestrial carbon that result from clearing 1 ha of forest for agriculture and from subsequent abandonment of that agriculture 50 years later. Figure 2b shows the net annual flux of carbon to the atmosphere that results from these changes. Calculation of these changes in carbon requires that the curves shown in Fig. 2a be defined for each type of land use that is important in each type of ecosystem for each region analyzed.

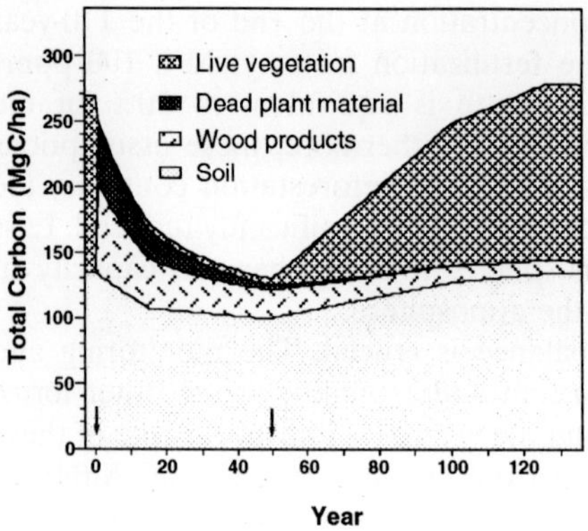

Figure 2a. Changes in the amount of carbon on 1 ha of land initially forested, cleared for agriculture, cultivated for 50 years, and then abandoned.

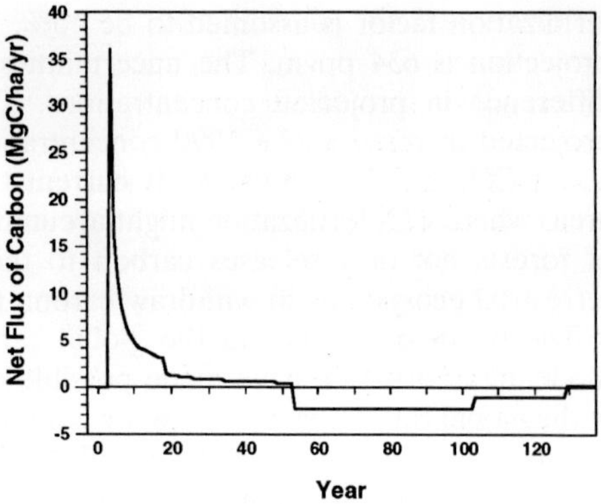

Figure 2b. The net flux of carbon to (+) or from (-) the atmosphere resulting from the changes in carbon stocks shown in Fig. 2a.

The two broad categories of information, rates of land-use change and changes in C/ha, can be further divided (Table 2). Methods for obtaining each of these types of information are discussed below.

3.1.1. Rates of land-use change

a. Rates of deforestation and reforestation, or rates of clearing, abandonment, and logging

The emphasis is on *changes* in land cover. The types of land-use change most important in affecting the net flux of carbon between terrestrial ecosystems and the atmosphere are rates of deforestation and reforestation. Per unit area, forests contain 20 to 50 times more carbon in their vegetation than cleared or agricultural lands. The difference in carbon stocks is released to the atmosphere when forests are cleared, and it is removed from the atmosphere when cleared lands return to forests. Measurement of changes in the area of forests is, thus, the critical requirement. This measurement requires, first, that forests and non-forests be distinguished, a requirement that is currently capable of being satisfied with data from satellites [20-22].

The most dramatic example of uncertainty in deforestation rates was, until recently, the rate for Brazilian Amazonia, where published estimates varied between 2.0 and 8.0 million ha/yr. Recent measurements with data from Landsat show the rate to have averaged about 1.5 million ha/yr in the 1980s [21]. The error associated with this estimate of deforestation is about 20%, considerably less than the uncertainty of 300% that existed before Landsat data were used.

The accuracy required for determining change in the area of forests is 15-20%, based on the following considerations. Estimates of the annual uptake of carbon by the oceans and annual release from fossil fuels are thought to be known within $\pm$0.8 and $\pm$0.5 PgC/yr, respectively (eq. 1). If the net flux of carbon from changes in land use is to be known as well, its precision must be about ±40% of the global estimate of 1.2 PgC/yr shown in Table 1. For the tropics, the relative accuracy must by ±30%. If as much as half of the uncertainty in this estimate of flux is from factors that land cover data will not address (e.g., carbon stocks) [23], the minimum accuracy for measurement of change in forest cover is half of 30-40%, or 15-20%. The current rate of tropical deforestation (15 X 10^6 ha/yr) [24] would have to be determined within ±2 to 3 X 10^6 ha/yr.

Table 2
Types of information required to calculate changes in terrestrial carbon from changes in land use/cover

1. Rates of Land-Use Change
   a. Rates of deforestation and reforestation, or rates of clearing, abandonment, and logging
   b. Fate of cleared land (permanent or temporary)
2. Changes in Carbon Stocks as a Result of Land-Use Change
   a. Initial stocks of carbon in vegetation and soil
   b. Fate of vegetation and soil on site
   c. Fate of vegetation removed from site (products)

It is important to note that net changes in forest and non-forest areas are not sufficient for determination of the net flux. Areas *and locations* are required so that initial carbon stocks and the dynamics of carbon loss and accumulation may be determined. A further reason for determining location is that forest biomass is dependant on age or the length of time since disturbance. The 22-year record of Landsat offers the opportunity to age a number of forests, and to distinguish between the clearing of young (low biomass) and old (high biomass) forests. The requirements to match estimates of carbon stocks to the specific areas cleared and to age forests argue for the use of a geographic information system (GIS) for recording the history of land use. Geographically specific data will also be useful for determining the fate of cleared land and for inferring changes in biomass and soil carbon (see below).

Of increasing importance, after the requirement for measurement of gross changes in forest area, is the need to recognize major disturbances within forests. In portions of the eastern Amazon Basin the area selectively logged each year is now equivalent to the area cleared [25]. Logging may be unsustainable, leading to a gradual reduction in the amount of carbon held in the forest, or it may be sustainable, such that the forests accumulate carbon following logging. Estimation of sources and sinks of carbon requires that logging and other types of major disturbances within forests be monitored. Fires fall into this category of disturbance, but are probably more appropriately addressed with data of coarse resolution, although data of high resolution may be required for calibration.

b. Fate of cleared land (permanent or temporary)

Deforested land may be cultivated or grazed indefinitely in some regions and may be used for only a few years in other regions before it is abandoned. About of third of the area deforested in Brazilian Amazonia between 1978 and 1988, for example, was in an early stage of regrowth [26]. The difference between cleared land and land with growing forests is obviously important in terms of carbon storage.

3.1.2. Changes in carbon stocks as a result of land-use change

a. Initial stocks of carbon in vegetation and soil

The amount of carbon released with deforestation depends in part on the amount initially held in the vegetation and soil of the forest. Estimates of the average amount of carbon per ha in forest vegetation (biomass) vary, and this uncertainty has contributed 30 to 50% of the uncertainty in estimates of the net flux of carbon [23]. The error may be even greater if the biomass of the forests actually cleared is systematically higher or lower than the average biomass. To reduce this uncertainty, the goal is to determine the biomass of forests actually cleared. The goal is all the more important in areas where biomass is changing, either increasing from regrowth or decreasing from selective logging, grazing, fuelwood gathering, or other activity. In South and Southeast Asia the reduction of biomass within forests over the last 100 years or more has been considerable [27,28]. This loss of biomass contributed about half as much carbon to the atmosphere as outright deforestation in the region [29].

Determining the biomass of forests from satellite is problematic. At most wavelengths, radar seems to saturate at values of standing biomass as low as 100-150 Mg/ha, or 50-75 MgC/ha [30-32]. Because most forests have above-ground densities in the range of 100-200 MgC/ha or more, an operational use of radar remains to be developed. Radar is also confounded by the water content of vegetation. Optical sensors have the potential to provide indirect estimates of standing biomass in forests where age and standing biomass are correlated with shadows [33]. With the use of satellites to determine rates of deforestation more accurately, biomass will become the factor contributing the greatest error to estimates of carbon flux. This conclusion identifies a major technological challenge: how can biomass be determined from space?

But, even if methods are developed for remote measurement of above-ground biomass, there will still be a need to determine below-ground carbon stocks, including roots and soil organic matter. Numerous measurements are reported in the literature; some of these have been collected, reviewed, and standardized. Further efforts are underway at present to put together a geographic data base for soils, and similar attempts are being developed for biomass and productivity. The means by which point measurements of carbon stocks are distributed over the globe need to be developed. Vegetation and soil maps exist, but data from satellites are likely to increase the accuracy with which different land covers are mapped.

b. Fate of vegetation and soil at the site

The major differences between recent estimates of carbon flux in temperate and boreal forests result from the ways in which changes in the wood and soils of disturbed ecosystems are calculated [10,15,16]. Estimates could be improved with geographically specific information on the fate of carbon on disturbed sites. This information includes the rates of mortality and decay of plant material following disturbance, oxidation of soil organic carbon, rates of recovery, and the proportion of the initial biomass burned or distributed among end products with different lifetimes.

c. Fate of vegetation removed from site (products)

Some of the wood harvested from forests is stored in long-lasting products. Net changes in this pool of carbon are determined by the balance between annual additions from harvest and annual losses from fire and decay. Changes may be difficult to measure directly, but they may be approximated from forestry statistics by finding the proportion of original biomass removed from the site during harvest and the distribution of this wood among products of different average lifetimes. Historical change in the proportion of original forest biomass going to products with different storage times is important to determine.

3.2. How Should Changes in Land Cover/Land Use Be Measured?

Because changes in the earth's atmosphere result from processes taking place over every part of the earth's surface, changes in land cover must be characterized globally if the total net flux is to be determined. The requirement for satellite data is crucial for this global characterization. The components of a long-term monitoring program include development and deployment of the appropriate satellite sensors; the means for ensuring the acquisition, correction and calibration, analysis, archiving, and distribution of data; the development of algorithms; and the use of data and algorithms in scientific investigation of the processes responsible for the climate and biogeochemistry of the earth.

Because the individual areas deforested or reforested in any specific location are commonly only a few ha/yr, satellite data must be of high spatial resolution. The U.S. Landsat series of satellites provides nearly global coverage with 80m (multispectral scanner, MSS) or 30m (thematic mapper, TM) resolution. With the loss of Landsat-6, the future of observations is less definite. Landsat 5 (and perhaps 4) will hopefully continue to provide coverage until Landsat-7 is functional in late 1998. In addition to Landsat, satellites from Japan (MOS and JERS-1 OPS), France (SPOT), and India (IRS) also provide optical data at a spatial resolution of 50m or less.

An issue here is the distribution of receiving stations. With enough stations, all satellite scenes can be acquired. If there are no receiving stations in the line-of-site, or if there is no communication with other satellites, acquisition may be limited by the storage capacity of the recorder on the satellite. Coordination of receiving stations and distribution of data are important aspects of a global monitoring strategy.

3.3. When Should the Measurements Be Made?

A goal of the strategy proposed here is to determine annual rates of land-use change in those regions where the changes are largest. Because of large year-to-year variations in the rate of CO_2 increase in the atmosphere, attempts to balance the global carbon equation should be carried out annually. The global net fluxes of carbon and hence measurement of changes in forest area should, therefore, be determined annually. Recent evidence from Brazilian Amazonia shows that rates of deforestation there have varied by more than a factor of two over short periods of 2 to 3 years [34]. Unfortunately, even though 15 X 10^6 ha of tropical forest may be lost in a single year [24], the distribution of this deforestation is such that the resolution of Landsat thematic mapper (30m) may be unable to distinguish much of this annual change from noise.

One of the best ways of dealing with this problem is with sampling. An optimum strategy might include complete coverage of the entire vegetated area of the earth (ignoring ice, sand, and rock) at intervals of once every 5 years, and an annual stratified sampling with an intensity corresponding to the intensity of change in land cover. The sampling should be based on a knowledge of where the activity is. The best short-term predictor of where change will occur next year is probably where it occurred last year. In large part these areas will be determined from complete coverage every 5 years. An annual sampling of 20-30% of the vegetated surface is expected to cover the regions of major change. Nine countries in the tropics are thought to have accounted for more than 70% of the net flux of carbon in 1989 [23]. Although these countries are large and include more than 70% of the forest area in the tropics, deforestation is often concentrated in specific administrative districts or along the borders of existing forests.

An alternative approach to complete coverage with high resolution data every 5 years might be to use satellite data of coarse resolution. Global coverage would be less expensive and might identify regions of change. Advanced Very High Resolution Radiometer (AVHRR) data with 1km resolution may now serve this purpose. In the future, MODIS data with 250m resolution may identify regions to be sampled with higher resolution. MODIS will not replace the higher resolution satellites, however.

3.4. Where Should the Measurements Be Made?

Emphasis should be on forests and other lands with high carbon stocks, such as temperate grasslands. Currently the largest net flux of carbon is from tropical forests where rates of deforestation are estimated to be about 15 million ha/yr [24]. In Amazonia, much of this deforestation seems to be concentrated along fronts, such as major roads or the forest/cerrado border of the east, south, and west. Similar patterns in tropical west and central Africa, and in south and southeast Asia may help define the regions to be sampled.

Outside the tropics, the emphasis should be on forests, grasslands, and croplands, particularly in the former Soviet Union, where existing estimates of change are often conflicting, and estimates of net carbon flux vary enormously. Current estimates of carbon flux for the former Soviet Union vary by almost 1 PgC/yr, from a net source of 0.1 PgC/yr [12] to a net sink of 0.7 PgC/yr [4].

Changes in forest cover do not necessarily include all of the significant changes in carbon storage. Cultivation of grassland soils in the temperate zones, for example, was responsible for a large release of carbon early in this century, and abandonment of agricultural lands would be expected to withdraw carbon from the atmosphere. Thus, measurement of changes in land cover for determining carbon flux may require that agricultural and non-agricultural lands be distinguished. Draining of wetlands and other changes may also be important. A preliminary list of the important ecosystems and regions is offered in Table 3.

One of the major obstacles to global monitoring of changes in land cover with Landsat is the inability to acquire new data in central and eastern Russia, one of the largest areas of forests. Landsat MSS data of the region were acquired in the early 1970s and in scattered areas throughout the 1980s, but the TDRSS (Tracking and Data Relay Satellite System) link that allowed for the transmission of data from this region to ground receiving stations was lost late in 1992. For the years between 1992 and the launch of Landsat 7 in 1998, high resolution data for the boreal forests of Russia must be acquired from other satellites, unless a new receiving station is installed in the region. The large areas recently leased to U.S., Korean, and Japanese timber companies and the construction of road and rail networks suggest that logging activities will be important here.

Table 3
Regions and ecosystems where changes in carbon storage are important

Highest Priority
Humid tropical forests
a. The Amazon Basin
b. Southeast Asia
c. Central and western Africa
Boreal forests and woodlands
a. Russia
High Priority
Humid tropical forests
a. South Asia
b. Central and South America outside the Amazon Basin
c. Eastern Africa
d. China
Boreal and temperate forests and woodlands
a. Canada and Alaska
b. The contiguous United States
c. Europe
d. Far east (China, Korea, Japan)
e. Australia and New Zealand
Seasonal and dry tropical forests and savannas
a. Africa
b. Latin America
Temperate agriculture and grasslands
Arid lands (desertification)
Tundra (changes in area, surface water, snow cover)

3.5. The Importance of Historical Changes in Land-Cover in Affecting the Current Flux of Carbon

While satellite data with high spatial resolution are ideal for determining changes in land cover, forests that were young during the early or mid portions of this century may not appear from space to have changed over the period of Landsat but are still accumulating carbon. Thus, changes initiated before 1972 (launch of Landsat) must be determined by other means. These other means include historical data on land use, such as agricultural and forestry statistics and maps [see 18]. These data are especially important in temperate and boreal regions. In these regions, rates of deforestation are low relative to the rates in the tropics, and the net flux of carbon is dominated by the accumulation of carbon in forests that were logged or abandoned well before 1972. Satellite data are capable of identifying regrowing forests while they are young, but as they age and the leaf area index increases above 6 or so, satellite optical sensors saturate. Nevertheless, these forests continue to grow and accumulate carbon. A 30-year-old temperate zone forest, for example, may be indistinguishable by optical data from a 200-year-old forest but may contain only half the carbon. The accumulation of carbon between 30 and 200 years will be missed if analyses use only satellite data.

Figure 3 shows an example of the importance of past logging to current estimates of carbon flux for North America. The actual flux from logging and regrowth is shown in the solid curve. It rarely exceeded 0.1 PgC/yr and between 1910 and 1970 shows an annual

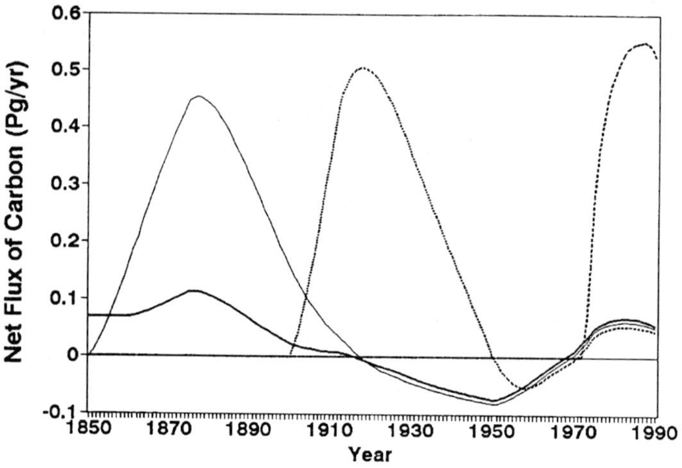

Figure 3. Estimates of net carbon flux from North America obtained by starting analysis of logging in different years.

accumulation of carbon on land. The other curves show the flux calculated as if data on land-use change became available in 1850, 1900, and 1972. Starting up the calculations causes a large release of carbon, as much as 0.5 PgC/yr, because the budgeting overlooks large areas of growing forests. The actual flux is not obtained until the historical data approximate the length of time it takes a forest to recover (50 years in this example). Initial estimates of flux are high because the emissions associated with logging occur over a brief period while the accumulations associated with regrowth persist for decades (Fig. 2b). The same artifact will occur if only satellite data are used. Satellites will identify logging but not the large areas of regrowing forests. Use of satellite data, alone, will bias estimates of flux toward greater releases or smaller accumulations of carbon. Because many northern temperate and boreal zone forests are recovering from changes in land use initiated before the existence of data from satellites, historical data are particularly important in these regions.

Another difference between the temperate and boreal regions and the tropics is the relative importance of logging and fires. Carbon lost to the atmosphere from logging and wildfire is generally accumulated again in new growth. In the long term the net flux of carbon from these activities approaches zero. For this reason, these activities are relatively unimportant in the tropics where clearing of forests for agriculture dominates the net flux. The exception is in Southeast Asia where rates of logging are apparently greater than rates of recovery, and biomass is decreasing [28]. Outside the tropics, however, changes in the area of forests are small, and logging and fire contribute relatively more to the net flux. Fires, that were assumed to be in steady state in earlier analyses, should now be considered.

Unlike the data used elsewhere to determine changes in land use, documentation of fires may be more feasible with data of coarse spatial resolution. The fires themselves need not be monitored, but fire scars (perhaps calibrated with high resolution data) will indicate the areas burned and the year of burning. This information will be useful in several ways. Age since burning will help constrain estimates of biomass as well as the net flux of carbon from regrowth and decay. Global information on fires will also be important to determine if the frequency or areal extent of burning are changing, perhaps in response to climatic change.

3.6. U.S. Projects Already Acquiring High Resolution Data for Use in Determining Emissions of Carbon

Several efforts are already underway in the U.S. to determine changes in land cover (Table 4). These high-spatial resolution data are being acquired (or planned) for the U.S., Mexico, and Central America, and a large part of the world's tropical humid forests. In addition to the projects listed here, the Canadian government is acquiring and processing Landsat data to produce a map of Canadian land cover. The largest gaps in a global acquisition of forest cover are the forested regions of Russia and China. Forests in other countries of the former Soviet Union, other northern Asian countries, and Australia are also not included in these studies. Areas of tropical humid forests not included in any of these efforts are western and southern tropical Asia, western and eastern Africa, and most of South America outside of the Amazon Basin.

The high resolution data described above provide a part of the global data required to determine terrestrial sources and sinks of greenhouse gases. But the data will also be useful for a number of other scientific questions critical to global change, including biotic feedbacks to climatic change, biodiversity, and sustainability of land use. For example, the IGBP/HDP (Human Dimensions Program) Project on Land Use and Land Cover Change is developing an international effort to understand the driving forces of land cover change.

4. OTHER REQUIREMENTS

4.1. Development of Algorithms

Several types of algorithms are needed. As discussed above, algorithms must be developed to distinguish between forests of different biomass. Because there is little chance of direct measurement of biomass with optical data from satellite, algorithms should be developed to identify logged forests or other disturbances. Pattern analysis might be useful for this purpose. If algorithms are eventually developed that permit evaluation of aboveground biomass with radar data, a strategy for a global inventory of land cover might also include the acquisition, calibration, correction, archiving, and use of radar or laser altimetry data . The goal identifies a need for research and development within the remote sensing community.

Although optical data from satellite will not provide direct estimates of biomass, they may be used with models (a second type of algorithm) to determine biomass indirectly. For example, the biomass of a forest is related to its age, which, in turn, can be obtained from multi-temporal satellite data if the temporal resolution is high enough to capture the disturbance and the fate of the cleared land. The development and use of models that simulate forest growth (and decay), together with geographic information systems, can be used to estimate carbon sinks (and sources).

Table 4
Projects using Landsat data to determine changes in land cover

* North American Landscape Characterization (NALC)
  The U.S. Environmental Protection Agency (EPA) is acquiring and processing complete coverage of the continental U.S., Mexico, and Central America with Landsat multi-spectral scanner data (MSS) (resolution 80m) for three dates: mid-1970s, mid-1980s, and early 1990s.

* Multi Resolution Land Characterization (MRLC)
  The U.S. EPA is acquiring complete coverage of the U.S. with Landsat thematic mapper data (TM) (30m resolution).

* Landsat Pathfinder Humid Tropical Forest Project (HTFP)
  This project is a part of the Landsat Pathfinder program, funded by the National Aeronautics and Space Administration (NASA) and EPA. It is acquiring and processing complete coverage of major areas of closed tropical forests (Amazonia, Central Africa, Southeast Asia) with Landsat MSS and TM at three points in time from the 1970s to the mid 1990s.

* Landsat Pathfinder Global Land Cover Test Sites.
  The NASA Landsat Pathfinder Project, with support from the EPA, is developing an acquisition plan for a set of about 40 globally distributed test sites using Landsat TM data. These data will be used to calibrate coarse resolution data such as AVHRR, and to develop site-specific models of ecological processes in places where detailed in-situ measurements have been and are being made.

* High Resolution Data Exchange Project.
  This is a joint project of the Committee on Earth Observation Satellites (CEOS), an international affiliation of space agencies, and the International Geosphere-Biosphere Programme (IGBP). The project is building a set of data containing several hundred SPOT, MOS, ERS-1, and IRS images from the same locations as the Test Sites, described above. Data from the different sensors are designed to enable the testing and development of improved multi-sensor methods of analysis and to enable an intercomparison of sensors.

A third need is for algorithms of classification and change detection. Whether change detection is calculated before or after classification depends on the rate of land-cover change relative to other factors, such as differences in phenology, sun angle, atmospheric interference, and sensor drift. The lack of ground control points in much of the world's tropics makes registration difficult, but registration is required for a GIS whether the change detection is pre- or post classification. The important point is that the spatial information be recorded in a GIS of some sort.

Algorithms need to be developed for the compositing of cloud-free images.

Finally, the volume of data required to measure year-to-year changes in land cover suggests that automated techniques will be helpful. Can data be stored so that the user can define the scene size? Should the data be acquired and stored as orbital strips, for example, rather than as scenes?

4.2. Requirements for Infrastructure

Ultimately, the end-to-end flow of satellite data must include acquisition, processing, analysis, archiving, distribution, and feeding of results back into the system. Probably at least two products are advisable, one a low level product, close to "raw data", that would be archived so as to be available for improved algorithms developed in the future. A second, higher level product, such as a map indicating rates of deforestation and reforestation, would probably serve much of the scientific community, although institutional differences in capability and approaches might argue for a series of variously processed data. Central archives where global data are to be stored should have product generation systems that allow the user to define the level and format of the product.

Other requirements for infrastructure include cooperation, for example, between field sites and remote sensing laboratories. There is a need for international cooperation so that receiving stations will have standard formats, easy-to-use browse facilities, and readily available data.

4.3. Costs

The cost of a global program to monitor changes in land cover in order to determine the annual net flux of carbon between terrestrial ecosystems and the atmosphere is difficult to estimate precisely. Too many variables remain to be resolved. The costs would include those associated with acquisition of data, with processing, archiving, and distribution of data, and with scientific analysis and modeling. It seems likely that the cost would be tens of millions of $U.S. per year. Such a cost is small relative to the total budget of the U.S. Global Change Research Program and smaller still relative to the money spent by the nations of the world on the production of energy, food, and fiber, the major activities responsible for emissions of greenhouse gases.

The cost is small relative to the usefulness of data on land cover. Such data would not only help determine emissions of carbon from changes in land use, and emissions of other biogenic greenhouse gases, methane and nitrous oxide, as well, but would show the distribution of natural and managed ecosystems and changes occurring in those resources. The data would help governments, local to national, evaluate sustainability of agriculture and help identify degraded as well as productive lands. Fragmentation of the landscape could be determined to help estimate loss of biodiversity. Over time, the data might also reveal the effects of climatic change on ecosystems.

There are probably numerous other uses as well, many of them unforeseen. In fact, many of the uses of a monitoring program are discovered after the first data have been obtained and analyzed. The difficult part of the program is the start, obtaining a baseline. If spatially specific global data on land-cover already existed, justifying a second global inventory would be easy because changes would then be apparent. The time for obtaining the baseline never comes, yet the need for one requires that it already have been acquired. Obtaining a baseline is a little like Mark Twain's observation about reading the classics: everyone wants to have read them, but no one wants to do so. In this case, however, the justification exists: uncertainties in the global carbon balance leave projections of future concentrations of atmospheric CO_2 very uncertain.

5. CONCLUSIONS

In June 1992 at the United Nations Conference on Environment and Development in Rio de Janeiro, the nations of the world signed the United Nations Framework Convention on Climate Change and the Biological Diversity Convention. They also set up an intergovernmental negotiating committee to draft a convention to combat desertification, and they agreed to a statement of principles on forests. About 160 nations signed the U.N. Framework Convention on Climate Change, and more than 100 are expected to have ratified the Convention by the first Meeting of the Conference of the Parties in 1995. By ratifying the Convention these nations have agreed to work toward the ultimate objective of achieving "stabilization of greenhouse gas concentrations in the atmosphere at a level that would prevent dangerous anthropogenic interference with the climate system". The Convention seeks to meet this objective "within a time-frame sufficient to allow ecosystems to adapt naturally to climate change, to ensure that food production is not threatened, and to enable economic development to proceed in a sustainable manner".

Much remains to be done in agreeing to the next steps, defining joint implementation, and finding financial mechanisms; but a program for monitoring global changes in land cover would contribute significantly to both the political and scientific aspects of discussions concerning these agreements. From a political perspective, monitoring land cover will identify changes in the areas of forests, agriculture, and deserts, areas with the greatest changes in biodiversity, carbon storage, and productivity. For the Framework Convention on Climate Change, the data will help determine the sources and sinks of carbon for individual countries. From a scientific perspective, this knowledge will help constrain unknowns in the global carbon balance, and help determine the reductions in emissions required for a smooth transition to a sustainably managed planet.

Acknowledgements

Portions of this paper were prepared while the author was a visiting scientist at the Office of Mission to Planet Earth, National Aeronautics and Space Administration, Washington, D.C. Much of what is reported here resulted from discussions with A. Janetos, C.O. Justice, M. Maiden, D.L. Skole, T.A. Stone, J. Townshend, D.E. Wickland, and G.M. Woodwell. Contributions from these individuals are gratefully acknowledged. The author is also grateful to T.M.L. Wigley for providing the results of his model projections.

6. REFERENCES

1 R.A. Houghton, in: G.M. Woodwell and F.T. Mackenzie (eds.), Biotic Feedbacks in the Global Climate System: Will the Warming Speed the Warming?, Oxford University Press, New York, in press.

2 R.A. Houghton, R.D. Boone, J.R. Fruci, J.E. Hobbie, J.M. Melillo, C.A. Palm, B.J. Peterson, G.R. Shaver, G.M. Woodwell, B. Moore, D.L. Skole, and N. Myers, Tellus, 39B (1987) 122.

3 J.M. Melillo, J.M., J.R. Fruci, R.A. Houghton, B. Moore and D.L. Skole, Tellus, 40B (1988) 116.

4 R.K. Dixon, S. Brown, R.A. Houghton, A.M. Solomon, M.C. Trexler, and J. Wisniewski, Science, 263 (1994) 185.

5 C.D. Keeling, R.B. Bacastow, A.F. Carter, S.C. Piper, T.P. Whorf, M. Heimann, W.G. Mook, and H. Roeloffzen, in: D.H. Peterson (ed.), Aspects of Climate Variability in the Pacific and the Western Americas, Geophysical Monograph 55, American Geophysical Union, Washington, D.C. (1989) 165.

6 P.P. Tans, I.Y. Fung, and T. Takahashi, Science, 247 (1990) 1431.

7 I.G. Enting, and J.V. Mansbridge, Tellus, 43B (1991) 156.

8 J.L. Sarmiento and E.T. Sundquist, Nature, 356 (1992) 589.

9 U. Siegenthaler and H.Oeschger, Tellus, 39B (1987) 140.

10 R.A. Houghton, Global Biogeochemical Cycles, 7 (1993) 611.

11 W.H. Schlesinger, Nature, 348 (1990) 232.

12 R.A. Houghton, in preparation.

13 T.M.L. Wigley, Tellus, 45B (1993) 409.

14 J. Leggett, W.J. Pepper, and R.J. Swart, in: J.T. Houghton, B.A. Callander, and S.K. Varney (eds.), Climate Change 1992. The Supplementary Report to the IPCC Scientific Assessment, Cambridge University Press, Cambridge (1992) 69.

15 P.E. Kauppi, K. Mielikainen, and K. Kuusela, Science, 256 (1992) 70.

16 R.A. Sedjo, Ambio, 21 (1992) 274.

17 R.A. Houghton, J.E. Hobbie, J.M. Melillo, B. Moore, B.J. Peterson, G.R. Shaver, and G.M. Woodwell, Ecol. Monogr., 53 (1983) 235.

18 R.A. Houghton, D.S. Lefkowitz, and D.L. Skole, For. Ecol. Manage., 38 (1991) 143.

19 R.A. Houghton, D.L. Skole, and D.S. Lefkowitz, For. Ecol. Manage., 38 (1991) 173.

20 G.M. Woodwell, R.A. Houghton, T.A. Stone, R.F. Nelson, and W. Kovalick, Jour. Geophys. Res., 92 (1987) 2157.

21 D. Skole and C. Tucker, Science, 260 (1993) 1905.

22 S.W. Running, T.R. Loveland, and L.L. Pierce. Ambio, 23 (1994) 77.

23 R.A. Houghton, Climatic Change, 19 (1991) 99.

24 FAO, Forest Resources Assessment 1990. Tropical Countries, FAO Forestry Paper No. 112, FAO, Rome (1993).

25 P.A. Lefebvre and T. A. Stone, Int. Archives Photogramm. Remote Sens. Proc., in press.

26 D.L. Skole, pers. comm.

27 S. Brown, L.R. Iverson, and A.E. Lugo, in: V.H. Dale (ed.), Effects of Land Use Change on Atmospheric CO_2 Concentrations: South and Southeast Asia as a Case Study. Springer-Verlag, New York (1994) 117.

28 E.P. Flint and J.F. Richards, in: V.H. Dale (ed.), Effects of Land Use Change on Atmospheric CO_2 Concentrations: South and Southeast Asia as a Case Study. Springer-Verlag, New York (1994) 201.

29 R.A. Houghton and J.L. Hackler, in: V.H. Dale (ed.), Effects of Land Use Change on Atmospheric CO_2 Concentrations: South and Southeast Asia as a Case Study. Springer-Verlag, New York (1994) 301.

30 S.A. Sader, Photogramm. Eng. Remote Sens., 53 (1987) 193.

31 M. Moghaddam, S. Durden, and H. Zebker, Remote Sens. Environ., 47 (1994) 154.

32 K.J. Ranson and G. Sun, IEEE Trans. Geosci. Remote Sens., 32 (1994) 388.

33 W.B. Cohen and T. Spies, Remote Sens. Environ., 41 (1992) 1.

34 P.M. Fearnside, Ambio, 22 (1993) 537.

TOWARD GLOBAL PLANNING OF SUSTAINABLE USE OF THE EARTH
Development of Global Eco-Engineering
S. Murai (Editor)

Global Potential Natural Vegetation: Dynamic Benchmark in the Era of Disruption

Elgene O. Box

List of Tables

List of Figures

PREAMBLE

Sustainable use of resources implies that they be used in a way which causes no continuing net degradation, such as to landscape productivity, maintenance functions or diversity patterns. The Earth's surface has been changed by humans more in this century, however, largely in the last 30-40 years, than in all of previous human history. It will be changed and degraded much more in the next 50-100 years, in the coming **Era of Disruption**. Consider, for example, the well known ecological parable of the water weed that doubles in area every day until it covers the pond after 30 days: on the 28th day only 25% of the pond is covered and open water still seems plentiful; on the 29th day only 50% is covered. The world is already overpopulated, for current economic conditions, by a factor of perhaps three, and the doubling time for most of the world is less than 50 years. Even without global warming, the biosphere is being so disrupted by mindless population growth, economic and political competition, and unrestrained greed that not only "natural" but also other productive, functional landscapes will soon be completely gone from most of the world, replaced by landscapes which are degraded and deteriorating further. Unrestrained market forces and other "accepted" economic mechanisms accelerate the destruction by driving up the value of the ever smaller, more valuable resource remnants. As already noted (Meadows, herein), it is already too late for sustainable development -- now we can only try for survivable development. The message of "The Limits to Growth" was irrelevant to the vested interests and was not heeded by the governments which must control them (tragedy of the commons, cf. Hardin & Baden 1977).

1. INTRODUCTION

Current and future ecological disruption requires that global planning and comprehensive global modeling be employed in order to estimate the changing ecological potentials of the Earth and to minimize its degradation. Global ecological modeling efforts are coordinated largely by the International Geosphere-Biosphere Program (IGBP), which has identified the production of Dynamic Global Vegetation Models as a high priority, since these could show land potentials under changing conditions (IGBP 1992, Steffen et al. 1992). Other scientific efforts include global modeling of primary productivity and other aspects of the global carbon budget, including the emissions of greenhouse gases from man's activities. Many of these models, however, operate differently for different biome types and thus require accurate global maps of the Earth's terrestrial biomes, the potential locations of which will also change as climatic conditions change. For the above reasons, an initial requirement seemed to be the development of an ecologically and geographically accurate map of the world's potential natural vegetation.

2. POTENTIAL NATURAL VEGETATION

The type of vegetation found in an area depends on environmental factors and the development of the vegetation over time. Over large areas, natural vegetation types are determined mainly by climatic conditions and are sometimes called "climatic climax" or zonal vegetation types, or biomes, e.g. tropical rainforest, summergreen deciduous forest or temperate grassland. At more local scales, non-climatic factors become more important, such as topography, soil or other substrate characteristics, fire or other natural disturbance regimes, and disturbance by human activities. For a given bioclimatic region, there may thus be various topogenic, edaphic (soil-related), pyrogenic or other vegetation types, which are "natural" in the sense that they are also stable or quasi-stable end-points of vegetation development. All are floristically related to and can be represented, in a geographic framework, however, by the climatic potential vegetation.

The development of vegetation over time, i.e. vegetation dynamics, is based on competitive relationships for major plant requirements such as light, water and nutrients. In particular, the light requirements of different plant types seem to drive vegetation dynamics in many cases. In humid-temperate East Asia, for example, the light-demanding pines of early secondary succession are eventually replaced by less light-demanding deciduous trees which can grow under the pine canopy and then, once they reach the canopy, shade out the pines and replace them. These deciduous trees are in turn replaced by the even more shade-tolerant evergreen broad-leaved ("laurel forest") trees, which cannot be replaced by any other (i.e. even more shade-tolerant) form and thus represent the stable final stage of succession.

The ecological status of vegetation can be summarized as follows. Original vegetation is the vegetation which covered an area before major disturbance by mankind, e.g. primary tropical rainforest. After cutting and subsequent land abandonment, the subsequent secondary vegetation develops (in the absence of major disturbance), through various successional stages, to reach a final stage called the potential natural vegetation. **Potential natural vegetation** is that stable (or quasi-stable), terminal vegetation type which could occur on an area, given its environmental conditions (climate, soil, etc.), if all outside (i.e. human) disturbances were removed. It is not necessarily the same as the original vegetation but may be similar if the physical conditions have not been greatly changed. Potential natural vegetation is determined largely by climate but may also be influenced by topography, soil, natural disturbance regimes, or other more local factors. Actual vegetation is the vegetation currently occupying an area, often resulting from disturbance to the "natural" vegetation. It may include stable natural vegetation but more often involves recovering secondary vegetation and/or managed substitute vegetation, such as tree plantations, cropland, or ruderal weeds.

Potential natural vegetation is a good basis for land-use planning and biological inventories since it indicates the ecological potential of an area, including flora, fauna, functional patterns, and biodiversity. Altered actual vegetation may, however, also be an important indicator of ecological conditions, and both may be the object of vegetation mapping. Before satellite technology was well developed, climatic data from meteorological stations represented the only global data-base with which to model vegetation patterns over large areas. Climatic data, however, can only suggest the climatic potential vegetation, not the altered vegetation which now covers much of the globe. Satellite sensors cannot estimate potentials but can see what is actually on the ground and thus represent a means of monitoring actual vegetation and land uses. Methodologies using both types of data are needed for mapping vegetation and its functional attributes. Such methodologies must be coordinated, however, so that maps of potential and actual vegetation use the same vegetation units and can be directly compared.

3. A MODEL OF POTENTIAL NATURAL VEGETATION

Until recently, climatic data represented the only available large-area environmental data-base with sufficient geographic coverage for geographic modeling purposes. Fortunately, climate also poses the overall environmental constraints within which metabolism and other aspects of plant/vegetation ecology must operate, including other physical as well as various biotic factors. A good review of climatic approaches to vegetation modeling, within the context of global modeling, is provided by Cramer and Leemans (1993).

3.1. Limiting Factors and Envelope Models

In the absence of more complete information and more mechanistic models, simple geo-correlational approaches such as envelope models, which predict site occurrences of biotic entities governed by environmental variables, may provide the most useful methods for predicting plant/vegetation occurrences. A climatic envelope represents the climatic space corresponding to the geographic range within which a plant or vegetation taxon grows and reproduces under natural conditions. Such climatic spaces, however, must be defined in terms of available climatic data, mainly temperature and moisture variables for which relatively long-term records are available at many sites. The envelopes are defined by particular climatic variables and their limiting values, inferred from the climatic values at range boundaries. Plants and vegetation types are assumed not to grow at a place if the local value of any envelope variable is outside the envelope limits (Holdridge 1947, Box 1981, Box et al. 1994).

The main climatic constraints on plants and vegetation generally involve temperature and water availability, and their interrelationships. Some limits, such as 'cardinal' extreme temperature limits and tissue tolerance to desiccation, are based on apparently identifiable if not completely understood physiological mechanisms and seem to be relatively universal (Larcher 1976, Levitt 1969, 1972; Woodward 1987). Some apparent values for cardinal temperatures are shown in Table 1 (cf. Larcher 1976), some of which have been corroborated by results from studies with envelope models (e.g. Box 1981).

Table 1.
Main Temperature Limits (Cardinal Temperatures) for Most Plants.

1. Upper limits:

40-45°C	most species

2. Lower limits:

5°C	many tropical species; also for death of soft leaf tissue in general
-2°C	many subtropical species
-15°C	temperate evergreen broad-leaved species
-40°C	buds, etc. of non-conifers

Cardinal temperatures represent limitations on plant physiology, often involving single events such as short-term exposure to extreme cold (absolute minimum temperature). The values shown are from Larcher (1976) and Woodward (1987).

Other limiting values must be inferred from geographic correlations of range boundaries and corresponding climatic values, especially water-balance variables. Moisture indices, based on ratios of water availability (e.g. precipitation) to potential water loss (e.g. potential evapotranspiration, PET), have proven to be especially useful in biogeographic studies (cf. Holdridge 1947, Mather and Yoshioka 1966, Box 1981). It was shown, for example, that the forest/non-forest boundaries in Argentina, in the subtropical north, in the problematic pampa region, and in subpolar Tierra del Fuego, all seem to follow a P/PET value of about 0.9 to 1.0, once a PET estimator was employed which does not geographically bias tropical/subtropical versus temperate/polar climates (Box 1986). Such a correlational approach to estimation of limiting values is necessarily the main calibration method for most climatic-envelope models.

3.2. A Minimal Global Set of Basic Vegetation Types

The world model of potential natural vegetation presented here was developed partly in order to identify the minimum set of vegetation types (biomes) necessary to cover the main functional, structural and geographic patterns of the world's land vegetation. The entities are potential natural vegetation (PNV) types defined by pheno-physiognomic features and delimited by climatic envelopes in a manner similar to those in an earlier world model of plant growth forms (Box 1981, 1987). These PNV types are predicted from relatively conservative climatic envelopes, however, which represent the "ecological limits" of the vegetation types (i.e. including competition, cf. Walter 1985) rather than the physiological limits. The procedure for identifying the PNV types and calibrating their climatic envelopes consisted of the following steps:

1. An initial minimal set of PNV types was developed from combinations of the main pheno-physiognomic characters of vegetation: dominant structural types (trees, grasses, etc.), seasonality types (evergreen, deciduous, etc.), and other structural features (e.g. broad versus needle leaves). Well recognized geographic sub-types were added as necessary, e.g. tropical rainforest versus extra-tropical evergreen broad-leaved forest.

2. Climatic envelopes for predicting the geographic locations of these PNV types were then developed, based on the envelope variables and parameterization of the growth-form model.

3. The PNV types were then run with a global climatic data-base (1600 sites) to predict their geographic occurrence, and envelope limits were refined as necessary to improve the resulting predictions, a form of "geographic regression". New envelope variables were identified at this time, and new PNV types were added when no type in the list appeared to represent the vegetation at particular sites.

4. The model was tested by checking predicted and actually occurring "natural" vegetation at the climatic data-sites, using especially the Schmithüsen (1976) biogeographic atlas and extensive field experience for deciding the correct vegetation type.

The main criteria for acceptance of the final list of PNV types and the respective envelope values involved completeness of global geographic coverage and accuracy of site predictions.

The climatic variables used for envelope models of plant and vegetation types generally involve different annual and seasonal aspects of long-term average temperature, precipitation and climatic moisture balance. From these variables, the following eight were chosen for the model of PNV types:

BT = biotemperature, i.e. the sum of monthly mean temperatures above 0°C, divided by 12 (from Holdridge 1959)

TMAX = mean temperature of the warmest month

TMIN = mean temperature of the coldest month

TABMIN = absolute minimum temperature

DTY = annual range of monthly mean temperature (= TMAX - TMIN)

MI = annual moisture index (= average precipitation / potential evapotranspiration)

PMIN = lowest average monthly precipitation amount

PMTMAX = average precipitation of warmest month

Two variables used in the growth-form model, namely annual and peak-month average precipitation, were considered unnecessary and were deleted; two temperature variables, representing cumulative growing-season warmth (biotemperature, BT) and absolute minimum temperature (TABMIN), were added. The annual moisture index (MI), which expresses the ratio of precipitation to potential evapotranspiration, gives an immediate indication of potential forest versus drier, non-forest vegetation types and may be the most useful single variable for predicting general vegetation structure and landscape classes. Initially it was thought that a heat-sum variable (i.e. BT) could be used instead of warmest-month mean temperature (TMAX) to unify the representation of polar treeline in the Northern and Southern Hemispheres. This did not seem to be the case however. The lower limit of TMIN (coldest-month mean temperature) represents tolerance to low temperatures, but this relatively available datum does not actually represent the mechanism of limitation, so absolute minimum temperature was added. The PNV model thus hypothesizes that summer temperature (BT, TMAX), winter temperature (TMIN, TABMIN), overall moisture balance (MI), dry-season and summer precipitation (PMIN, PMTMAX), and annual range of mean temperature (DTY) control the natural distribution of the potential vegetation types identified.

Development of the envelope model began with a minimal set of about 20 PNV types and proceeded by refining the envelope limits and by adding new PNV types when sites were found which were not well described by any existing PNV type. This geographic trial-and-error procedure is similar to that used to develop other envelope models but focused specifically on identification of the vegetation types needed to cover world vegetation patterns and used more conservative envelope limits, representing the results of natural vegetation dominance relationships. Model development continued until all areas of the world seemed to be covered with appropriate PNV types and until predictive errors were eliminated as far as possible, given the variables available.

3.3. Potential Dominant Vegetation

Since biological ranges overlap, it often happens that multiple PNV types are predicted for a site. The true PNV is the vegetation type which is stable, i.e. cannot be replaced naturally by any other type. Usually, that vegetation type which is the most shade-tolerant but can still produce a foliar layer above that of all other types in the area will control the all-important light resource and become the stable "dominant" or "climax" vegetation. After envelope processing, a competition model is employed which first attempts to identify the dominant vegetation by means of a simple physiognomic hierarchy (cf. Box 1981), in which potentially taller PNV types are ranked higher than shorter types (unless the taller type has little light-intercepting area, such as succulents). If this does not produce a unique prediction, certain special situations are considered more directly (e.g. grassland versus open woodland). If this still fails to resolve the issue, the type finally chosen is that which is farthest from its closest climatic limit. The entire two-step PNV modeling procedure is summarized in Figure 1.

3.4 World PNV (Biome) Types

The combined envelope+competition modeling procedure resulted in a list of 41 potential natural vegetation types needed to represent world vegetation patterns, as shown in Table 2. This number is larger than the roughly 20 types often seen in world biome classifications (e.g. Holdridge 1959, 1967; Lieth 1975, Whittaker & Likens 1975, Walter 1977, 1985; Prentice et al. 1992; cf. Bailey 1990) but includes some types which are climatically, structurally and perhaps functionally different. These types generally cover smaller areas and are thus often not recognized as distinct types at global scale, such as tropical alpine puna and páramo, tropical montane and cloud forests, tropical dry evergreen forest, subpolar and subalpine krummholz, Southern Hemisphere evergreen woodlands, cool-maritime grassland, maritime tundra, and polar/subnival cold-deserts. The 41 PNV types also include separately some vegetation types which have very similar pheno-physiognomies but occur under different climatic conditions (different climatic envelopes), such as dry-continental versus mediterranean conifer forests and cool-temperate versus subpolar evergreen broad-leaved forests.

Figure 1. Structure of the Global Model for Potential Natural Vegetation Types.

1. **Envelope Model**: check inclusion/exclusion and distance to closest limit.

	BT	TMAX	TMIN	DTY	MI	PMIN	PMTMAX
Tropical Rainforest							
Maximum	30	35	30	8	****	***	***
Minimum	18	21	18	0	1.00	20	20
Tropical Deciduous Forest							
Maximum	31	35	30	18	****	25	***
Minimum	16	17	13	0	0.58	0	30

.
. (41 types)
.

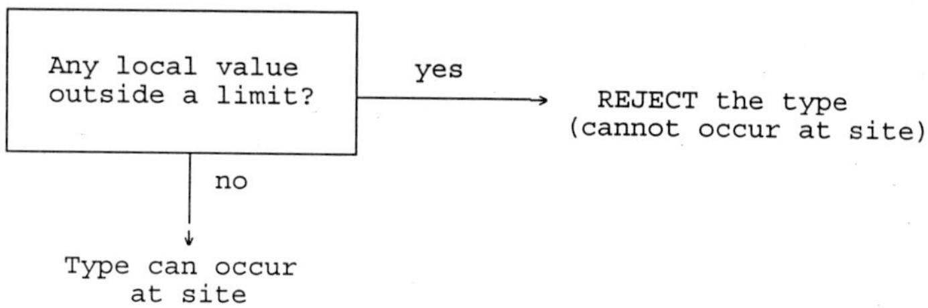

Sample result: Haerbin (China)	Limiting factor	Distance to closest limit
1. Summergreen broad-leaved woodland	MI	0.15
2. Temperate grassland	MI	0.08
3. Summergreen broad-leaved forest	MI	0.02

2. **Competition Model** (if envelopes yield > 1 PNV possibility):

Forest > Woodland > Grassland ---> Forest

BUT: at Haerbin, forest is very close (0.02) to dryness (MI) limit

Woodland > Grassland -- and Woodland farthest from any limit, so

Final prediction is: **Summergreen broad-leaved woodland**

<u>Model Development/Checking Results</u>

Check 3 kinds of results:
a. results at each site
b. sites predicted for each vegetation type
c. near-misses for each vegetation type

If results not acceptable (e.g. if a near-miss should be included or a prediction excluded), modify the climatic envelopes accordingly and return to step 1; if no changes needed, accept the results as final.

For definition of envelope variables, see main text.

Table 2.
Minimal World Set of Biome-Level Terrestrial Potential Vegetation Types.

1. Tropical Rainforests
    - Tropical Rainforest (lowland)
    - Tropical Montane Rainforest
    - Tropical Subalpine (Cloud) Forest
    - Subtropical Rainforest

2. Raingreen/Semi-Evergreen Forests/Woodlands/Scrub
    - Tropical Semi-Evergreen Forest
    - Raingreen Forest
    - Raingreen Scrub (incl. montane)

3. Evergreen Broad-Leaved Forests
    - Tropical Dry Evergreen Forest
    - Evergreen Broad-Leaved Forest
    - Mediterranean Evergreen Forest
    - Temperate Rainforest (BL/mixed)
    - Cool-Temperate Evergreen BL Forest
    - Subpolar Evergreen BL Forest

4. Summergreen BL Forests and Woods
    - Summergreen Broad-Leaved Forest
    - Summergreen Broad-Leaved Woodland
    - Subpolar Summergreen BL Forest

5. Needle-Leaved Evergreen Forests/Woods
    - Dry Conifer Forest
    - Mediterranean Conifer Forest
    - Temperate Conifer Rainforest
    - Boreal Conifer Forest
    - Subpolar/Subalpine Conifer Woodland

6. Summergreen Needle-Leaved (Larch) Forest/Woods

7. Subhumid Woodlands/Scrub
    - Semi-Evergreen Dry Woodland/Scrub
    - Mediterranean Woodland/Scrub

8. Shrublands (seasonal/evergreen)

9. Grasslands
    - Tropical Savanna
    - Temperate Grassland
    - Cool-Maritime Grassland

10. Tropical Alpine Vegetation
    - Páramo
    - Puna

11. Tundra and related Krummholz/Dwarf-Scrub
    - Cool-Evergreen BL Scrub/Krummholz
    - Cool-Summergreen BL Scrub/Krummholz
    - Subalpine Conifer Krummholz
    - Polar/Alpine Tundra
    - Maritime Tundra
    - Polar/Subnival Cold-Desert

12. Semi-Desert Scrub

13. Deserts (extreme)
    - Arid Desert
    - Cold Arid Desert/Semi-Desert

14. Ice Desert

3.5. Checking Model Results

Checking model performance, i.e. model validation, involves comparing model predictions against reality, both at sites used in model development and at new sites. Initial checking of model results was done in two ways:

1. by visually comparing the predicted list of occurrences (sites) for each PNV type against vegetation maps; and
2. by computerized checking of predicted dominant PNV type (unique result) against codes for true vegetation type contained in the site data-base (based mainly on Schmithüsen 1976 and personal field experience).

The first procedure checks the envelope limits directly; the second checks the limits (indirectly) as well as the dominance model.

Application of the site-derived PNV model to the much denser global pixel raster also represents a useful method of validation. Some envelope limits were indeed found to be inadequate as the model was subjected to new combinations of variable values not seen at the 1600 climatic sites. Almost all of the consequent envelope changes involved necessary widening of limits.

4. GLOBAL MAPPING OF POTENTIAL NATURAL VEGETATION

Potential natural vegetation, despite having been destroyed over large areas, represents the ecological potential of the landscape and thus not only a baseline for comparison and but also a geographic framework. The satellite-based remote-sensing capabilities which provide new methods for studying actual vegetation and other land covers have also fostered improvements in computer cartographic methodologies, including those which permit global mapping of potential natural vegetation using climatic data. PNV results must be reduced to a small enough number that the types can all be readily distinguished on the map. The main obstacle to global pixel-raster mapping of most model results, however, involves the construction of geographically accurate global fields for the input variables, necessarily interpolated from available site data, in this case climatic data from meteorological stations.

4.1. Global Data Fields and Mapping Methodology

All of the climatic variables needed for the PNV model can be derived from long-term average monthly temperature and precipitation data, except for absolute minimum temperature (TABMIN). Global data fields for monthly mean temperature and average monthly precipitation (albeit without Antarctica!) have been provided on CD-ROM by Leemans and Cramer (1990), saving the international global-modeling community much time and effort. For the PNV map shown here, however, these Leemans-Cramer fields could not be used since they do not include TABMIN, which could only be produced for deadline using a different cartographic projection.

As a result, global fields were produced for four of the input variables (TMAX, TMIN, PMIN, and PMTMAX), from long-term average climatic data (3000 sites), using an in-house interpolation routine which does not extrapolate as well in mountain areas. The DTY field was produced directly from the interpolated TMAX and TMIN fields. Biotemperature (BT) was calculated directly from interpolated monthly temperature fields. The field for the annual moisture index (MI) was calculated by dividing an interpolated field for annual precipitation by that for annual potential evapotranspiration. Finally, the global field for TABMIN was developed in a three-step procedure which involved: 1) collection of TABMIN data for as many meteorological stations as possible in the site data-base (about half); 2) extrapolation of these values to the sites which did not have TABMIN; and 3) interpolation to create the global pixel field, using the in-house program. The resulting eight data fields were used to run the PNV the model, producing a PNV map (integer raster) with 41 types.

4.2. Simple Biome Types

Since the individual PNV types are too numerous to be distinguished on a single map, they were grouped by ecological similarity into broad classes which essentially represent natural biomes. Some types covering quite small areas are necessarily subsumed into more different types, e.g. orobiomes into adjacent lowland biomes (cf. Walter & Box 1976) and krummholz into the adjacent polar/alpine types. This reduces the total number to 14 simple biome types, shown by the groupings in Table 2. The global field of simple biome types was derived from the PNV field, based on these groupings. This was then colored to produce a preliminary map of biome-level potential natural vegetation, which is shown in Figure 2.

Figure 2: Preliminary Mapping of Climatically Estimated, PNV-Based World Terrestrial Biomes.

Medium green	Tropical Rainforests
Red	Raingreen/Semi-Evergreen Forests/Woodlands/Scrub
Dark green	Evergreen Broad-Leaved Forests
Light green	Summergreen BL Forests and Woods
Blue	Needle-Leaved Evergreen Forests/Woods
Lightest green	Summergreen Needle-Leaved (Larch) Forest/Woods
Olive-brown	Subhumid Woodlands/Scrub
Orange	Shrublands (seasonal/evergreen)
Yellow	Grasslands
Lighter yellow	Tropical Alpine Vegetation
Gray	Tundra and related Krummholz/Dwarf-Scrub
Tan	Semi-Desert Scrub
Dark brown	Deserts (extreme)
White	Ice Desert

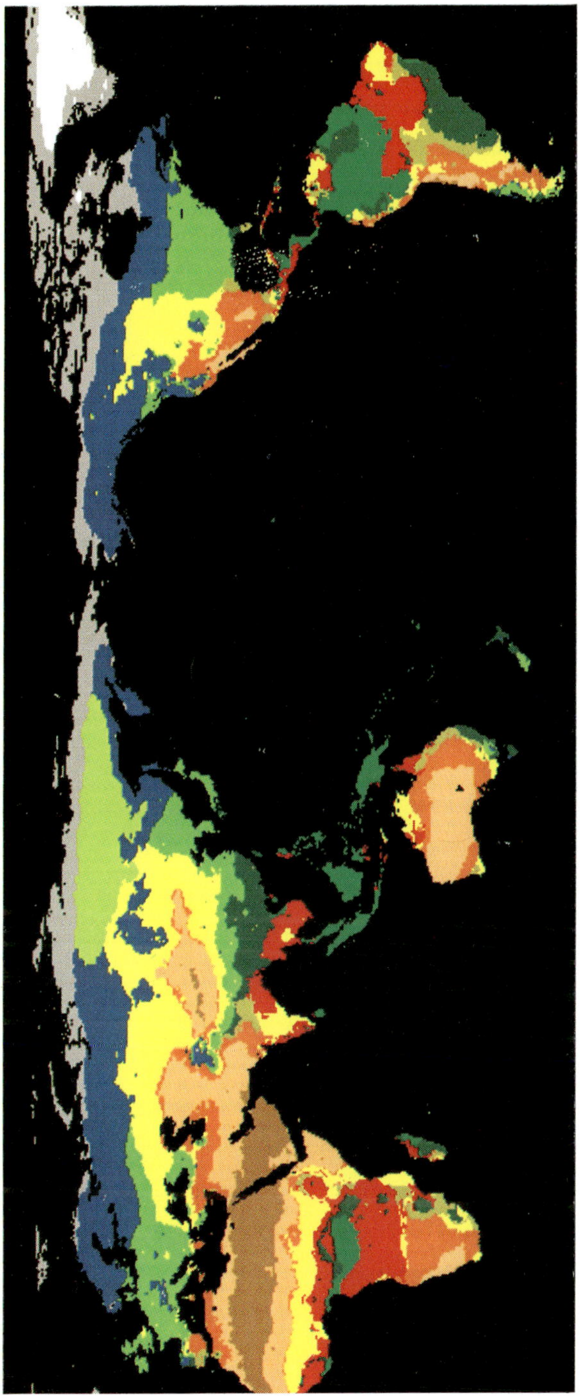

Since topography (especially mountain chains) is not well represented in some areas, Figure 2 appears patchy and boundaries must be considered only approximate. This map shows general patterns reasonably but should not be considered an accurate basis for pixel-scale modeling or comparisons in its present form. It is, however, already better than previous computerized world vegetation maps in several respects. These include not only the more complete definition of the world's main vegetation units but also better representation of vegetation types in:

1. some smaller areas, such as Japan and New Zealand; and
2. some paired regions with apparently conflicting values for presumed, limiting factors, such as southern Japan versus the southeastern USA (cf. Box 1988);

plus better representation of even a few boundaries, such as

3. the clearly artificial four-point intersection of vegetation zones (central USA) on the map of Prentice et al. (1992).

An improved final version of this PNV map, with better representations of mountain ranges (based on the Leemans-Cramer data fields if possible) as well as some vegetation boundaries, is being developed.

4.3. Actual Vegetation

The 41 PNV types can also be grouped, for better recognition by satellite data, into classes based entirely on structural and phenological characteristics, for example all extra-tropical evergreen broad-leaved forest types. Such pheno-physiognomic classes correspond closely to biome types and facilitate mapping without really sacrificing ecological integrity. A world map of actual vegetation types, based on cluster analysis of seasonal NDVI greenness patterns, was presented by Tateishi et al. (1991). Their units of actual vegetation and 15 purely pheno-physiognomic PNV groupings similar to those in Table 2 were compared and seen to correspond largely but not completely (Box 1993). Development of a common global set of biome types is needed in order to compare different models as well as maps of potential and actual vegetation. The largest problems with the satellite approach seem to involve discrimination of various scrub/shrubland versus grassland types, as well as extra-tropical evergreen forests.

5. CONCLUSION

The type of vegetation found in an area depends on its limiting environmental factors and its dynamic status. Biome-level potential natural vegetation is a useful index of land potential, not only for itself but also as a representation of the full suite of edaphic, topogenic, secondary and other associated vegetation types in the biome. A climate-based world model of biome-level potential natural vegetation has been produced using pheno-physiognomically defined PNV units and their climatic limits, following the envelope approach used by other vegetation models (e.g. Holdridge 1947, Box 1981). Since biome-level PNV is largely determined by climate, it is possible to produce an adequate PNV model based entirely on available climatic data. The model has been geographically validated for sites and shown to provide a reasonable representation of world geographic patterns of natural vegetation. The pheno-physiognomic PNV units may be recognizable by satellite data and thus also represent a basis for models and maps of actual vegetation.

The PNV map presented here is preliminary but better than previous computerized world vegetation maps in several respects, including the definition of the PNV units. Since climate change may also change basic plant-climate relationships, however, models of future landscape potentials may require new biocalibrations and eventually even new vegetation units. For this reason, efforts are continuing to develop dynamic global vegetation models based on global classifications of plant functional types (e.g. Smith 1993), which can react to environmental conditions, compete, and combine in various ways to form new PNV types and represent the wider range of actual land-cover types.

Potential natural vegetation, like most ecological concepts, normally assumes stable climatic conditions. The concept of PNV is still useful in an era of changing climatic conditions, however, perhaps even more so, since it represents the potential of the landscape, even if that potential is changing and may require many centuries to attain. It thus serves as a dynamic benchmark, providing continuity with the past and one of few possible bases for scientific comparison, for understanding the changing actual vegetation, and for geoecological planning.

REFERENCES

Bailey, R. G. 1990. "Ecoregions Map of the Continents" (world map with explanatory supplement). *Envl. Conserv.*, 16(4):307-309, plus map.

Box, E. O. 1981. *Macroclimate and Plant Forms: An Introduction to Predictive Modeling in Phytogeography.* Tasks for Vegetation Science, Vol. 1. The Hague: Dr. W. Junk BV, Publishers. 258 pp.

Box, E. O. 1986. Some Climatic Relations of the Vegetation of Argentina, in global perspective. In: *Contributions to the Knowledge of the Flora and Vegetation of Northern Argentina* (U. Eskuche and E. Landolt, eds.), pp. 181-216. Zürich: Veröff. Geobot. Inst. ETH, Stiftung Rübel.

Box, E. O. 1987. Plant life forms and mediterranean environments. *Annali di Botanica*, 45(2):7-42.

Box, E. O. 1988. Some similarities in the climates and vegetation of central Honshu and central eastern North America. *Veröff. Geobot. Inst. Rübel*, 98:141-168.

Box, E. O. 1993. Coordinating theoretical and satellite-based global vegetation maps. In: *Proceedings, Autumn 1993 Meeting, Japanese Society for Photogrammetry and Remote Sensing, Kanazawa*, pp. 55-56.

Box, E. O., D. W. Crumpacker, and E. D. Hardin 1994. A climatic model for plant species locations in Florida. *J. Biogeography*, 20:629-644.

Cramer, W. P., and R. Leemans 1993. Assessing Impacts of Climate Change on Vegetation using Climate Classification Systems. In: *Vegetation Dynamics and Global Change* (A. M. Solomon and H. H. Shugart, eds.), pp. 190-217. London: Chapman and Hall. 338 pp.

Hardin, G., and J. Baden (eds.) 1977. *Managing the Commons.* San Francisco: W. H. Freeman.

Holdridge, L. R. 1947. Determination of world plant formations from simple climatic data. *Science*, 105:367-368.

Holdridge, L. R. 1959. A simple method for determining potential evapotranspiration from temperature data. *Science*, 130:572.

Holdridge, L. R. 1967. *Life Zone Ecology.* San Jose (Costa Rica): Tropical Science Center.

IGBP 1992. Global Change: Reducing Uncertainties. Stockholm: IGBP Secretariat. 40 pp.

Larcher, W. 1976. *Ökologie der Pflanzen.* 2nd edition. Stuttgart: Verlag Eugen Ulmer. 320 pp.

Leemans, R., and W. P. Cramer 1990. The IIASA Data-Base for Mean Monthly Values of Temperature, Precipitation and Cloudiness on a Global Terrestrial Grid. Working Paper no. WP-90-41, Biosphere Dynamics Project. Laxenburg (Austria): International Institute for Applied Systems Analysis. 28 pp + CD-ROM.

Levitt, J. 1969. Growth and survival of plants at extremes of temperature -- a unified concept. *Symp. Soc. Experim. Biol.*, 23:395-448.

Levitt, J. 1972. *Responses of Plants to Environmental Stresses.* New York: Academic Press. 697 pp.

Lieth, H. 1975. Primary Production of the Major Vegetation Units of the World. In: *Primary Productivity of the Biosphere* (H. Lieth and R. H. Whittaker, eds.), pp. 203-215. New York: Springer-Verlag.

Mather, J. R., and G. A. Yoshioka 1966. The Role of Climate in the Distribution of Vegetation. In: *Climatic Classification for Environmental Biology* (D. B. Carter and J. R. Mather, eds.), Publicat. in Climatol. (Univ. Delaware), 19(4):372-384.

Prentice, I. C., W. Cramer, S. P. Harrison, R. Leemans, R. A. Monserud, and A. M. Solomon 1992. Global biome model: predicting global vegetation patterns from plant physiology and dominance, soil properties and climate. *J. Biogeogr.*, 19:117-134.

Schmithüsen, J. 1976. *Atlas zur Biogeographie*. Meyers Grosser Physischer Weltatlas, vol. 3. Mannheim/Wien/Zürich: Bibliographisches Institut.

Smith, T. M., H. H. Shugart, F. I. Woodward, and P. J. Burton 1993. Plant Functional Types. In: *Vegetation Dynamics and Global Change* (A. M. Solomon and H. H. Shugart, eds.), pp. 272-292. London: Chapman and Hall. 338 pp.

Steffen, W. L., B. H. Walker, J. S. Ingram and G. W. Koch (eds.) 1992. *Global Change and Terrestrial Ecosystems: The Operational Plan*. Global Change Report no. 21. Stockholm: Internat. Geosphere-Biosphere Programme. 95 pp.

Tateishi, R., K. Kajiwara and T. Odajima 1991. Global land cover classification by phenological methods using NOAA GVI Data. *Asian-Pacific Remote Sensing Journal*, 4(1):41-50.

Walter, H. 1977. *Vegetationszonen und Klima*. 3rd ed. Stuttgart: Eugen-Ulmer-Verlag. 309 pp.

Walter, H. 1985. *Vegetation of the Earth*. 3rd ed. Berlin, New York, Tokyo: Springer-Verlag. 318 pp.

Walter, H., and E. O. Box 1976. Global classification of natural terrestrial ecosystems. *Vegetatio*, 32:75-81.

Whittaker, R. H., and G. E. Likens 1975. The Biosphere and Man. In: *Primary Productivity of the Biosphere* (H. Lieth and R. H. Whittaker, eds.), pp. 305-328. New York: Springer-Verlag.

Woodward, I. 1987. *Climate and Plant Distribution*. Cambridge University Press. 174 pp.

Acknowledgement

This research was supported by the Toyota Corporation through its Global Engineering Program at the Institute of Industrial Science, Tokyo University. This support by Toyota and Tokyo University is quite important to global ecology and is greatly appreciated.

TOWARD GLOBAL PLANNING OF SUSTAINABLE USE OF THE EARTH
Development of Global Eco-Engineering
S. Murai (Editor)

A Simulation Study of Grassland Carbon Dynamics as Influenced by Atmospheric CO_2 Concentration

Takehisa Oikawa

Institute of Biological Sciences, University of Tsukuba, Ibaraki, Japan

Abstract

The effects of atmospheric CO_2 concentration on a savanna grassland ecosystem were analysed by modifying a previously developed microcomputer model (Oikawa 1985, 1986), which was based on production and reproduction processes following Monsi's (1960) guiding principles. In this study, the two photosynthetic pathways utilized by grasses, i.e. C_3 and C_4 photosynthesis, were evaluated in terms of surplus production and resulting carbon dynamics.

Surplus production (*SP*), which was calculated using a modified version of Kuroiwa's (1966) production formula, was enhanced by a higher atmospheric CO_2 concentration in both C_3 and C_4 grasslands. The fertilization effect of CO_2 on *SP* was more considerable in a C_3 than in a C_4 grassland. This was due to impoved carboxylation efficiency at higher CO_2 concentration in the C_3 grassland, although C_3 grasses show a negative photorespiration process.

On the basis of predictions from microcomputer simulations of grassland carbon dynamics conducted at three atmospheric CO_2 concentrations ranging from 200 to 600 ppm, species changes in grasslands utilizing the C_3 and C_4 photosynthetic pathways from the last glacial period through the present to the near future are discussed.

Using light energy from the sun, carbon dioxide from the atmosphere, and water and nutrients from the soil, green plants have produced a wide variety of complex organic compounds that have evolved and sustained the tremendous richness of terrestrial biomes throughout geological history.

The types of biome existing on Earth are closely dependent on the two major prevailing climatic elements, temperature and humidity, which are, respectively, determined mainly by global distribution patterns of solar energy reaching the surface at each latitude and the uplift frequency and intensity of air resulting in cloud formation due to adiabatic cooling. Thus, as a rule, mean temperature is higher in lower latitudes and precipitation is usually high in regions where there is orographic uplift along high mountain ranges lying across the path of prevailing winds, frontal uplift along weather fronts, nonfrontal uplift brought about by local convection and nonfrontal uplift associated with large-scale convergence in low pressure systems such as typhoons.

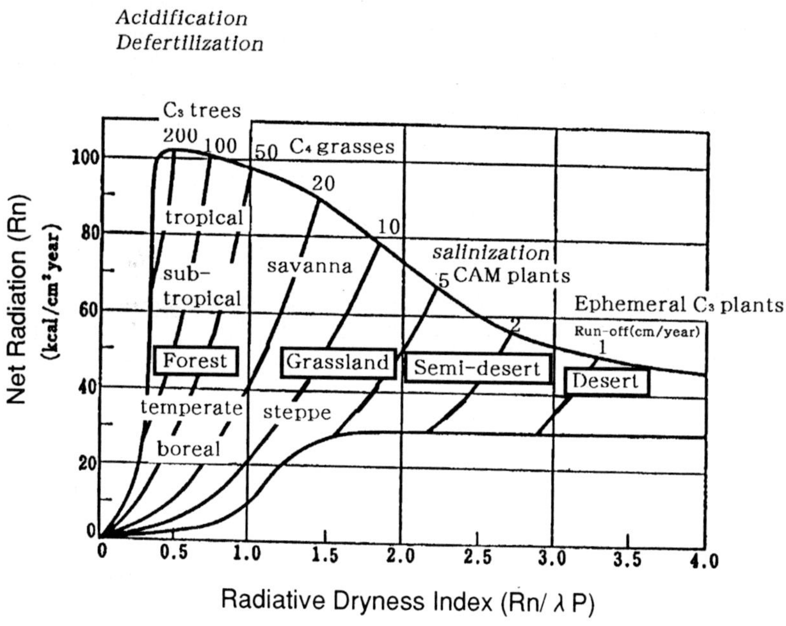

Figure 1 The dominant biome types plotted against radiative dryness index (x-axis) and net radiation (y-axis). Countor lines showing runoff rate in cm/yr are also illustrated. Representative carbon fixation pathways and edaphic conditions are indicated for each biome (modified from Budyko 1971).

Figure 1 shows a diagram of the dominant natural biome types in the world plotted against the radiative dryness index on the abscissa and the annual net radiation on the ordinate. The radiative dryness index, which was proposed by Budyko (1971), seems to be the most suitable parameter for representing the moisture condition available for plants. This index is defined as a dimensionless unit, being the ratio of the annual net radiation ***Rn*** to the amount of annual precipitation ***P*** multiplied by λ, where λ is the latent heat of water vaporization. Grassland ecosystems, which are the main focus of the present study, occupy semi-arid to arid habitats with moisture availability ranging approximately from unity to 2 in the radiative dryness index. Among them, savanna ecosystems distributed in tropical or sub-tropical regions are located in a higher ordinate and temperate grasslands such as prairie and steppe ecosystems in a lower ordinate. It is worth noting that the drought tolerance of grasslands is superior to that of forest ecosystems ranging from tropical rain forests to boreal coniferous forests, because they cannot establish themselves in arid climates where the radiative dryness index is higher than unity, as shown in Figure 1.

Under the climatic constraints mentioned above, the productivity of each biome is undoubtedly a major underlying determinant of the biome type that is finally established and maintained stably, as Lieth (1975) has emphasized. It should be noted that three different types of photosynthetic carbon fixation mechanism operate among vascular plants, i.e. the C_3, C_4 and CAM pathways, where the C_3 pathway is essential for CO_2 fixation in all green plants, and the C_4 and CAM pathways have been evolutionally derived from the C_3 pathway. Many grass species distributed in warm to hot climates, and practically all grass species from non-flooding tropical savanna biomes are C_4 plants, although the vast majority of species making up the world's vegetation are C_3 plants.

Plant communities in temperate regions sequester much CO_2 in the atmosphere during the growing season from spring to summer by actively operating a photosynthetic mechanism, and, conversely, evolve CO_2 to the atmosphere during the resting season from autumn to winter when the community respiration including the decomposition in the soil exceeds the photosynthetic CO_2 fixation. As shown in Figure 2, such metabolic activities of plant communities are apparently seen from the monthly changes in atmospheric CO_2 concen-

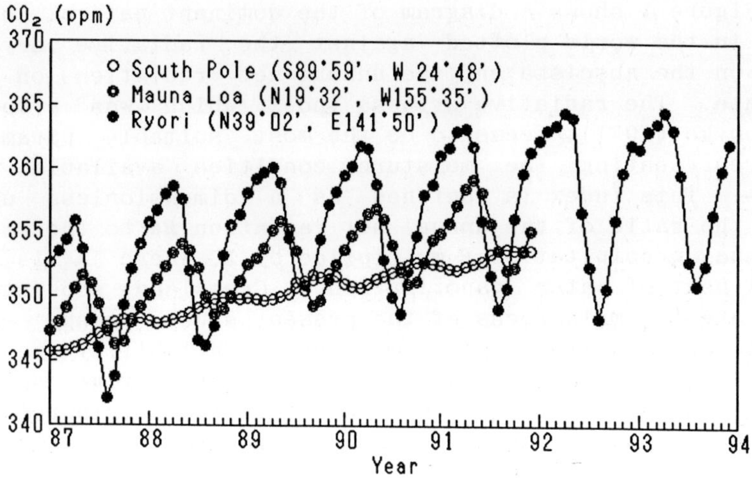

Figure 2 Monthly changes in atmospheric CO_2 concentration since 1987 monitored at three meteorological stations, Ryori in Iwate prefecture, Japan, Mauna Loa in Hawaii, USA and the South Pole. After from World Data Centre for Greenhouse Gases (WDCGG).

tration measured at two representative monitoring stations in the northern hemisphere, Ryori in Iwate prefecture, Japan (39°02´N) and Mauna Loa in Hawaii, USA (19°32´N). In addition to a steady yearly increase in background concentration of around 1.5 ppm caused by anthropogenic activities, a regular seasonal oscillation of atmospheric CO_2 concentrations, which attains a maximum in spring and a minimum in late summer to autumn, is found at both stations, being especially marked in Ryori, located at a higher latitude than Mauna Loa; the yearly differences between the maximum and the minimum since 1987 have ranged from 10 to 16 ppm at Ryori and around 6 ppm at Mauna Loa. At the South Pole, in contrast to the stations in the northern hemisphere, the difference is 2 ppm at most, being close to the background increase, because metabolic activity in and around Antarctica is tiny throughout the year.

CO_2 concentrations in the atmosphere covering a full glacial-interglacial cycle during the past 160,000 years have been clarified by Barnola et al. (1987). By determinations on ice core samples from Vostok, Antarctica, they have provided new reliable data on natural variations in CO_2; their measurements show that CO_2 concentrations fluctuated in close accordance with air temperature, ranging from approximately 180-200 ppm at the glacial maximum (18,000 and 150,000 yr before present) to 280 ppm or more at the peak of the interglacial period (5,000 and 130,000 yr before present).

Independent of any anthropogenic or natural increase, atmospheric CO_2 concentration should have a profound effect on competition among plant species and ecosystem composition through photosynthetic production. It is now possible to integrate a wide range of physiological and ecological observations which relate plant productivity to increasing CO_2 levels, distinguishing the C_3 and C_4 photosynthetic pathways.

Outline of the model

In order to understand the carbon dynamics of a plant community, it is useful to construct a plant community model based on dry-matter production and reproduction processes following Monsi's (1960) guiding principles. Because the detailed structure of the model employed here has been described previously (Oikawa 1985, 1986), only a brief description is presented below.

For the mathematical representation of community carbon dynamics, a compartment modelling approach has been adopted. Three compartments describe the living biomass: **F**, **C**, and **U** represent foliage, woody, and root compartments, respectively. Only one compartment, **M**, describes the dead biomass in the soil. Carbon is distributed among these functional compartments connected by fluxes.

The carbon flux from atmosphere to foliage, i.e. gross primary production (*GPP*), is the sole carbon input into a plant community and is thus the driving force for all the other carbon fluxes. As formulated in Eq.1, this flux has been closely computed by Kuroiwa's (1966) production formula as a function of leaf area index *LAI*.

$$GPP(A) = \frac{2 \cdot D \cdot p_{max} \cdot G \cdot \kappa}{K} \left[ln\{1+\sqrt{1+K \cdot \beta \cdot I_{o,max}/p_{max}}\} - ln\{1+\sqrt{1+K \cdot \beta \cdot I_{o,max} \cdot exp(-K \cdot A)/p_{max}}\} \right] \quad (1),$$

where A is the leaf area index (LAI), G is the growth period (days), D is the day-length (h), $I_{o.max}$ is the maximum light intensity at noon (klux), p_{max} is the light-saturated photosynthesis rate of a single leaf (mg CO_2 dm^{-2} h^{-1}), β is the light-use efficiency (quantum yield of photosynthesis, mg CO_2 dm^{-2} h^{-1} $klux^{-1}$), K is the light attenuation coefficient of a canopy, and κ is the conversion constant from CO_2 assimilated by photosynthesis into dry matter synthesized (κ=0.60). This formula has been extended from the original Monsi and Saeki equation (1953) so as to give daily productivity directly by assuming a sinusoidal variation of daily illuminance over a canopy.

In order to distinguish the effect of different photosynthetic pathways, a C_3 and a C_4 fixation mechanism, on primary productivity of a grassland, the photosynthetic relation for single leaves should be extended. A C_3 photosynthetic pathway inevitably accompanies photorespiratory CO_2 evolution. Here, the light-saturated photosynthesis rate p_{max} and light-use efficiency β were modified into Eq.2 and Eq.3, respectively, as a function of ambient CO_2 concentration Ca;

$$p_{max}(Ca)=(Ca-Cc)/300 \cdot p_o \qquad (2)$$

where Cc is the CO_2 compensation point for photosynthesis in ppm. As is well known, C_4 plants having an efficient internal CO_2 concentrating mechanism in their mesophyll cells show virtually no photorespiration, and thus their Cc value is close to 0 ppm. On the other hand, C_3 plants without a concentrating mechanism show a fairly high Cc value, being usually around 50 ppm under normal atmospheric conditions.

Light-use efficiency of C_3 plants increases with increasing ambient CO_2 concentration, since photorespiration is directly related to the CO_2/O_2 ratio in their leaf tissues. On the basis of many measurements, Osmond et al. (1980) formulated a generalized light-use efficiency relation of C_3 plants as a hyperbolic function:

$$\beta(Ca)=0.0081 \times Ca/(1+0.0048 \times Ca) \times \beta_o \qquad (3\text{-}1) \quad \text{for } C_3 \text{ plants}$$

where β_o is the light-use efficiency of C_4 plants, independent of ambient CO_2 concentration, that is;

$$\beta=\beta_o \qquad (3\text{-}2) \quad \text{for } C_4 \text{ plants}$$

Eq.3-1 shows that, when compared with that of C_4 plants, the light-use efficiency of C_3 plants changes from 0 at $C\alpha$=0 ppm to approximately 25% higher at $C\alpha$=600 ppm. The boundary CO_2 concentration, where the efficiency becomes equal in C_3 and C_4 plants, is close to 300 ppm. When an ambient CO_2 concentration exceeds this boundary, C_3 plants have an advantage over C_4 plants, because C_4 plants need the additional energy to operate the CO_2 concentrating mechanism in their mesophyll cells.

The three significant production variables, surplus production *SP*, net primary production *NPP* and net ecosystem production *NEP*, can be defined as:

$$SP(A)=GPP(A)-R_F(A) \quad (4),$$
$$NPP(A)=SP(A)-(R_C+R_U) \quad (5),$$
$$NEP(A)=NPP(A)-R_M \quad (6).$$

where each symbol *R* in Eqs.4-6 represents the respiratory or decomposition carbon flux from the compartment designated by its subscript, and the rate of each flux (ton d.w. ha^{-1} y^{-1}) can obtained from

$$R_F(A)=0.001x24\cdot G\cdot\kappa\cdot r_A\cdot A \quad (7),$$
$$R_C=0.001x24\cdot G\cdot\kappa\cdot r_C\cdot W_C \quad (8),$$
$$R_M=0.001x24\cdot G\cdot\kappa\cdot r_M\cdot W_M \quad (9),$$

where each symbol *r* in Eqs.7-9 represents the respiration or decomposition rate per unit amount of the compartments designated by each subscript: r_A is the unit respiration rate of a single leaf in mg CO_2 dm^{-2} h^{-1} and r_C, r_U, r_M are the unit respiration rates of respective compartments in mg CO_2 g $d.w.^{-1}$ h^{-1}. Multiplying by the numbers 0.001 and 24, and the constants κ and G, these hourly values are converted into the model units of ton d.w. ha^{-1} y^{-1}.

Results and Discussion

Surplus productivity of a C_3 and a C_4 plant canopy

Surplus production *SP* values of a C_3 and a C_4 plant canopy, which were calculated from Eq.1, are displayed on a contour map and three-dimensional diagram in Figs.3 and 4, respectively, in terms of ton dw ha^{-1} y^{-1} by a contour plot with an interval of 10 ton dw ha^{-1} y^{-1} for each graph of the

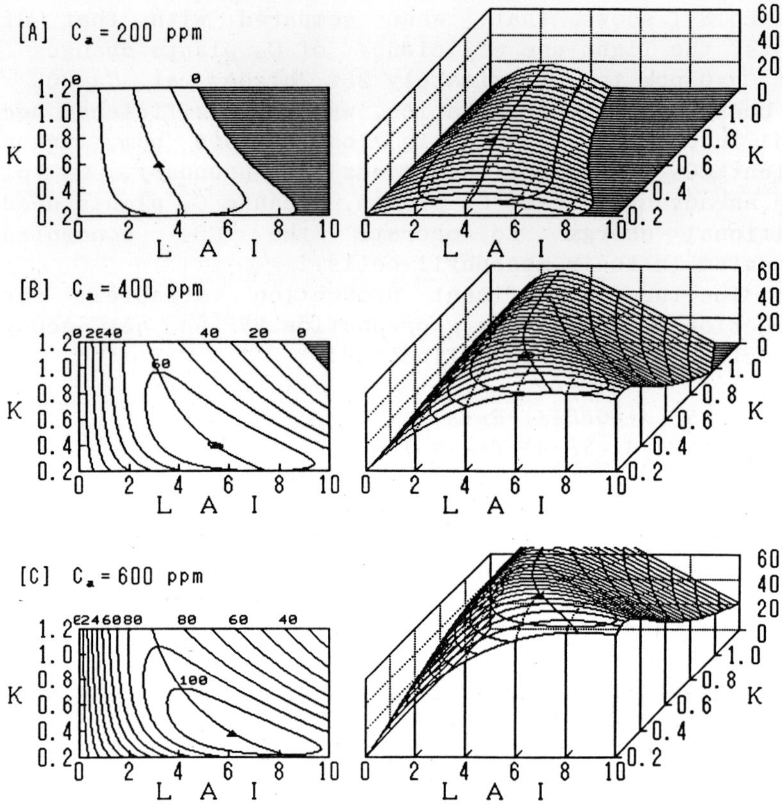

Figure 3 Rates of surplus production (*SP*) of a C_3 plant canopy calculated at three levels of atmospheric CO_2 concentration. [A], $C\alpha$=200 ppm; [B], $C\alpha$=400 ppm; [C], $C\alpha$=600 ppm. Rates are illustrated as a function of leaf area index (*LAI*) and attenuation coefficient of light (*K*), by a contour plot with an interval of 10 ton dw ha^{-1} y^{-1} for each graph of the series in the left-hand column and are further plotted three-dimensionally against the *z*-axis in the right-hand column.

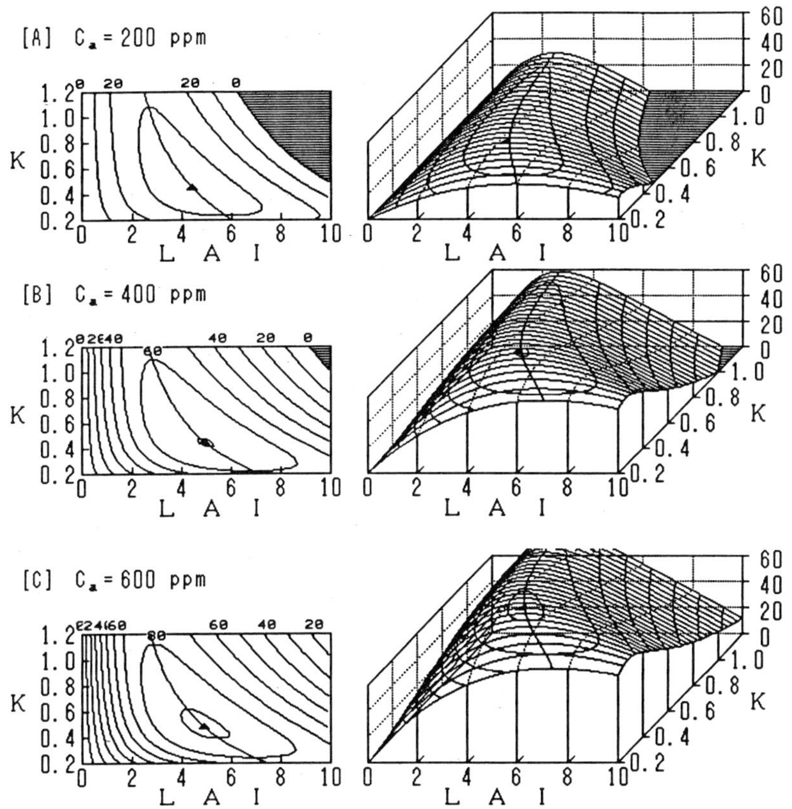

Figure 4 Rates of surplus production (*SP*) of a C_4 plant canopy calculated at three levels of atmospheric CO_2 concentration. [A], Ca=200 ppm; [B], Ca=400 ppm; [C], Ca=600 ppm. Other explanations are the same as those for Fig.3.

series in the left-hand column and *SP* is further plotted against the z-axis in the right-hand column. In both figures, the x-axis represents LAI (A) values from 0 to 10, and the y-axis values of the light attenuation coefficient (K) from 0.2 to 1.2.

In both figures, ambient CO_2 concentrations are changed at three levels: 200 in [A], 400 in [B] and 600 ppm in [C]. Carbon dioxide compensation points for photosynthesis (Cc)

are assigned at 50 ppm in a C_3 plant canopy (Fig.3) and 0 ppm in a C_4 plant canopy (Fig.4), respectively. The values for the other six parameters included in Eqs.1-3, namely the maximum light intensity at noon ($I_{o,max}$), day-length (D), the single-leaf light-saturated photosynthetic rate (p_o), the light-use efficiency of a C_4 plant (β), the unit respiratory rate of a leaf (r_A) and the conversition factor from CO_2 to carbohydrate (κ) were fixed, respectively, at 100 klux, 12 h, 25 mg CO_2 dm^{-2} h^{-1}, 2.0 mg CO_2 dm^{-2} h^{-1} $klux^{-1}$, 2.0 mg CO_2 dm^{-2} h^{-1} and 0.60 g dw/g CO_2.

Because the light-saturated photosynthesis rate of a single leaf (p_{max}) is assumed to increase linearly to the ambient CO_2 concentration (Ca) as formulated in Eq.2, all SP values corresponding to K and A should increase approximately linearly as shown in Figs.3 and 4, where a triangle shows each SP_{top}, which is the highest of the SP_{max} values.

It is worth noting that, taking into account the positive CO_2 compensation point (Cc=50 ppm in Eq.2) and the dependence of light-use efficiency β on Ca (Eq.3-1), SP values of a C_3 plant canopy change considerably with Ca, compared with those of a C_4 plant canopy. For instance, SP_{top} values of a C_3 canopy increase 5.7-fold from 18.8 ton dw ha^{-1} y^{-1} at Ca=200 ppm to 108 ton dw ha^{-1} y^{-1} at Ca=600 ppm. On the other hand, the effects of CO_2 enrichment on a C_4 canopy are less than 1/2, compared with those for a C_3 canopy; that is, SP_{top} values for a C_4 canopy increase 2.6-fold from 35.3 ton dw ha^{-1} y^{-1} at Ca=200 ppm to 91.1 ton dw ha^{-1} y^{-1} at Ca=600 ppm. Figure 4 illustrates clearly the dual effect of ambient CO_2 concentration on surplus productivity of a C_3 plant canopy. In addition to the considerable CO_2 enrichment effect on a C_3 canopy, the overall pattern of SP behavior in the A-K coordinate changes with Ca; when it is 200 ppm, SP_{top} is located on K_{opt}=0.60 and A_{opt}=3.2, and when it is 600 ppm, it is located on K_{opt}=0.38 and A_{opt}=6.1.

In C_3 plants which evolve CO_2 due to photorespiration under light, it is well known that CO_2 compensation points of a single leaf are different among species or due to environmental conditions such as temperature. The reported compensation points range from 40 to 65 ppm CO_2 in 21% O_2 and at 20 °C. At 30 °C the compensation point considerably increases, being over 110 ppm CO_2. Such variations of CO_2 compensation point will affect surplus productivity of a C_3 plant canopy. Figure 5 demonstrates surplus productivity

(SP) values as influenced by CO_2 compensation points (Cc), where the Cc values were changed at three levels, 30 in [A], 50 in [B] and 80 ppm in [C] under three levels of ambient CO_2 (Ca) conditions, 200, 400 and 600 ppm from the left-hand to the right-hand column.

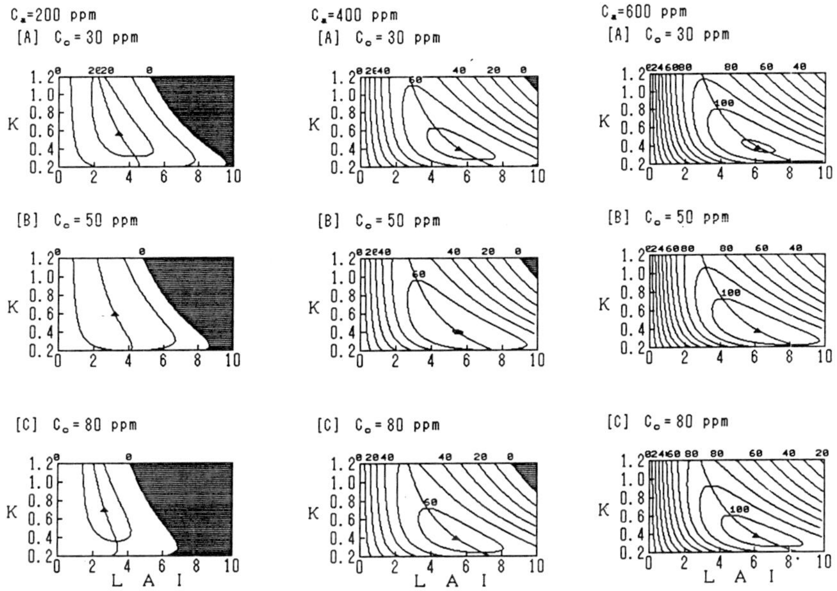

Figure 5 Effects of CO_2 compensation point on surplus production (SP) of a C_3 plant canopy calculated at three levels of atmospheric CO_2 concentration. [A], Ca=200 ppm; [B], Ca=400 ppm; [C], Ca=600 ppm. CO_2 compensation points are 30 (top), 50 (middle) and 80 ppm (bottom).

As summarized in Fig.6, the *SPtop* values were approximately linearly reduced by increasing *Cc* values (Fig. 6A); the lower the ambient CO_2 concentration, the steeper the gradient and *vice versa*. For instance, when the ambient CO_2 condition was 200 ppm, *SPtop* values were 23.2 and 11.9 ton dw ha^{-1} y^{-1}, respectively, at Cc=30 and Cc=80 ppm; the former being as high as 95% compared with the latter. On the other hand, when the ambient CO_2 condition was 600 ppm, they were 110 and 104 ton dw ha^{-1} y^{-1}, respectively; the former being only 5% higher than the latter. From this calculation, we can understand that higher atmospheric CO_2 condition will ameliorate production environment of a C_3 canopy considerably, even if global warming induced by higher ambient CO_2 would increase *Cc* values of a C_3 plant to a considerable extent.

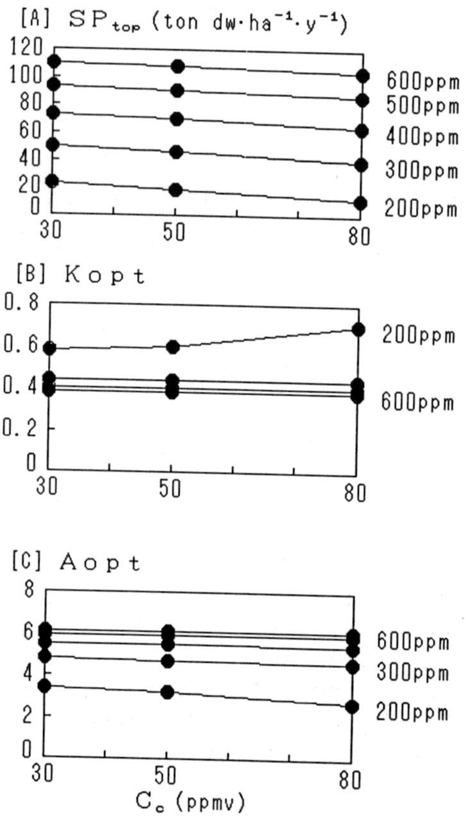

Figure 6 Effects of CO_2 compensation point (*Cc*) on [A] *SPtop*, [B] *Kopt* and [C] *Aopt* under different ambient CO_2 conditions.

As shown in Figs.3 and 4, the magnitude of atmospheric CO_2 enrichment on surplus productivity differs greatly between a C_3 and a C_4 plant canopy, and this difference will affect carbon dynamics of these plant communities to different degrees, and will likely result in changes in ecosystem type with different photosynthetic pathways through iterative growth processes of production-reproduction in a plant community.

Carbon dynamics simulation of a C_3 and a C_4 grassland

Figure 7 shows a series of the carbon dynamics of a grassland simulated at three levels of ambient CO_2 concentration Ca, 200 in [A], 400 in [B] and 600 ppm in [C]. The graphs on the left-hand side and the right-hand side illustrate the biomass of each compartment on a normal scale and the relative illuminance beneath the canopy (RI) on a log scale, of a C_3 and a C_4 grassland ecosystem, respectively. The overall time trends of each biomass attained a steady state independent of photosynthetic pathways, and each biomass at the steady state was influenced greatly by the ambient CO_2 level.

Figure 8, which summarizes the carbon dynamics simulation experiments, shows that the CO_2 fertilization effect on total biomass Wp and net primary productivity NPP at the steady state was much markedly in a C_3 grassland, and that the effect on a C_3 grassland was particularly conspicuous at lower ambient CO_2 levels; for instance, the present simulation experiments suggest that a treble increase in CO_2 concentration from 200 to 600 ppm leads to more than a tenfold increase in Wp and NPP of a C_3 grassland. At Ca=200 ppm, which is close to the minimum concentration in the recent two glacial maxima, Wp and NPP of a C_3 grassland were as low as 13% relative to those of a C_4 grassland, because of the significantly higher CO_2 compensation point of a C_3 canopy (Cc=50 ppm) and lower light-use efficiency under reduced CO_2 conditions (Eq.3-1). Around Ca=400 ppm, which is close to the present-day level, Wp and NPP of a C_3 grassland are predicted to become approximately equal to those of a C_4 grassland, and exceed them at a Ca over 400 ppm owing to the great CO_2 enrichment effect on surplus productivity of a C_3 canopy (cf. Fig.3).

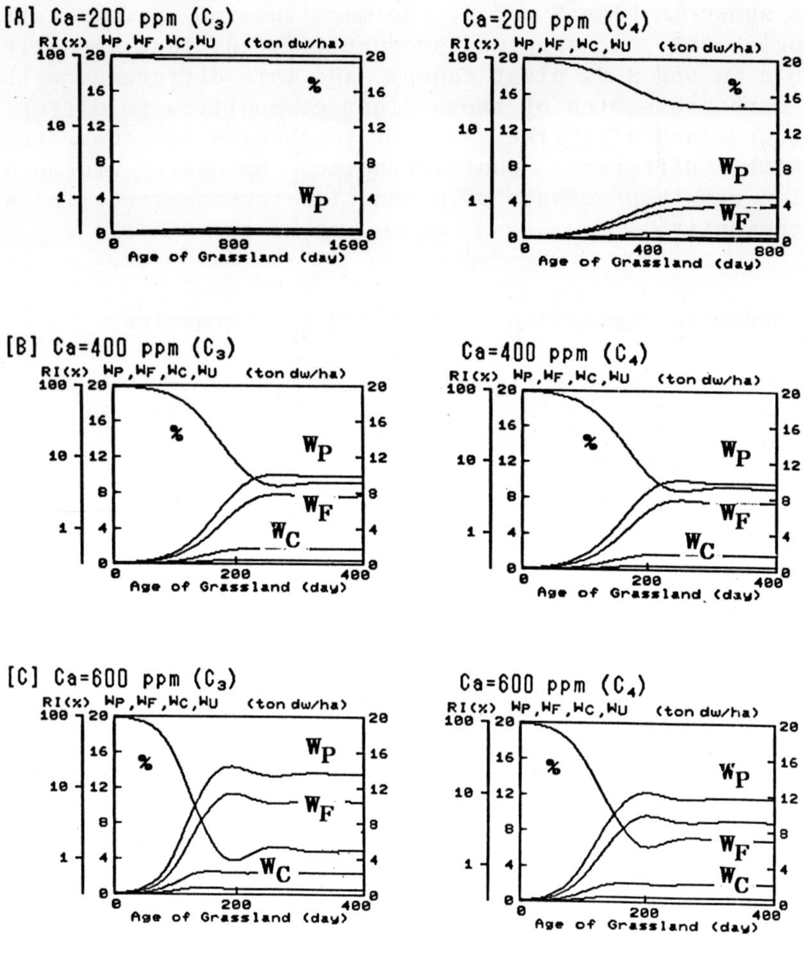

Figure 7 Simulated dry-matter accumulation patterns of a C_3 and a C_4 grassland in the left-hand and right-hand columns, respectively, at three levels of atmospheric CO_2 concentration ($C\alpha$). [A], $C\alpha$=200 ppm; [B], $C\alpha$=400 ppm; [C], $C\alpha$=600 ppm. Each graph illustrates variations of illuminance (RI, %) under the canopy in log scale and the total plant biomass (W_P), leaf biomass (W_F), and stem biomass (W_C) in normal scale.

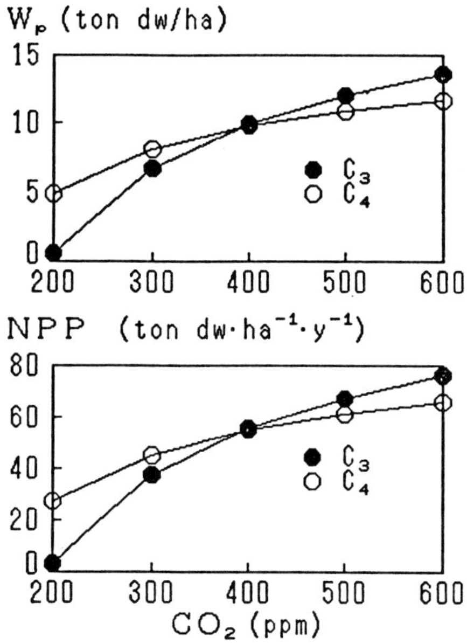

Figure 8 Changes in plant biomass (*W_P*, top) and net primary Production (*NPP*, bottom) of C_3 and C_4 grasslands at the steady state, with increasing atmospheric CO_2 concentration (*Ca*) ranging from 200 to 600 ppm.

The present predictions will be confirmed partly by paleo-ecological evidence when the ambient CO_2 concentration is as low as 200 ppm. By compiling the glacial section of the sediment core from a lake, Street-Perrott (1994) showed that there was a substantial increase in the global abundance of C_4 plants at the last glacial maximum (LGM). The compilation of $\delta^{13}C$, carbon/nitrogen ratio and pollen analysis data suggests that the present montane rain forest on Mt. Kenya, East Africa, was largely replaced by a grassy moorland vegetation rich in C_4 plants at the LGM. This is consistent with the predictions demonstrating the overwhelming superiority of a C_4 grassland in a lower CO_2 level, and may suggest that, although the present-day boundary between a forest and a grassland lies at unity in the radiative dryness index as shown in Fig.1, the past boundary at the LGM would have shifted towards a lower radiative dryness index value.

On the other hand, it may be rational to predict from the present simulation that ongoing global increases in atmospheric CO_2 concentration demonstrated in Fig.2 may favor a shift towards C_3 plant communities, at the expense of C_4 grasslands. This suggests that the forested area may extend to more arid climates having higher radiative dryness index. While the paleo-ecological evidence provides important insights on transient responses of plant communities to a changing climate, present to future rates of climate change will be very much greater than have occurred previously, because they are driven by human activities which have become a significant force affecting the functioning of the Earth system. It is urgent to improve confidence in the predictive abilities of the present simulation model. Such an effort will deepen our understanding of the distribution and abundance of different community types in the world, and stimulate the development of a new and promising scientific field, global ecology.

Original atmospheric CO_2 concentration data shown in Fig.2 were provided by WMO World Data Centre for Greenhouse Gases (WDCGG). This research is part of the outcome of the Special Research Project on Global Environmental Change at the University of Tsukuba.

References

Barnola, J.M., Raynaud, D., Korotkevitch, Y.S. and Lorius, C. (1987) Nature, **329**, 408-414.

Budyko, M.I. (1971) Climate and Life. [Japanese translation: Z. Uchijima & S. Iwakiri (1973), Tokyo Univ. Press, Tokyo, pp488].

Kira, T. (1987) *In*: M.R. Sethuraj and A.S. Raghavendra (ed.) Tree Crop Physiology, pp99-119. Elsevier Science Publishers, Amsterdam.

Kuroiwa, S. (1966) *In*: A Series of Modern Biology Vol.9, pp71-100. Iwanami Shoten, Tokyo (in Japanese).

Lieth, H. (1975) *In*: H. Lieth & R.H. Whittaker (ed.) Primary Productivity of the Biosphere. Speringer-Verlag, New York.

Monsi, M. (1960) Bot. Mag. Tokyo, **73**, 81-90.

Monsi, M. & Saeki, T. (1953) Jap. J. Bot., **14**, 22-52.

Oikawa, T. (1985) Bot. Mag. Tokyo, **98**, 225-238.

Oikawa, T. (1986) Bot. Mag. Tokyo, **99**, 419-430.

Osmond, C.B., Bjorkman, O. & Anderson, D.J. (1980) Physiological processes in plant ecology. Toward a synthesis with *Atriplex*. Springer-Verlag, Berlin, pp468.

Street-Perrott, F.A. (1994) Ambio, **23**, 37-43.

TOWARD GLOBAL PLANNING OF SUSTAINABLE USE OF THE EARTH
Development of Global Eco-Engineering
S. Murai (Editor)
1995 Elsevier Science B.V.

Predicting natural and anthropogenic climate changes

By Taroh Matsuno
Graduate School of Environmental Earth Science
Hokkaido University
Kita 10-jo Nishi 5-chome, Sapporo 060 Japan

Introduction

In recent years human society becomes more and more vulnerable to climate variations because expanded human living and industries are possible only under usual climate conditions hence slight swings of climate from its normal state (the so-called anomalous weather) give rise to significant damages to them. On the contrary expanded human activities have reached such a level that causes anthropogenic climate changes such as global warming due to greenhouse gases, regional climate change due to deforestation and so on. Under these circumstances predicting climate variations and changes both due to natural and anthropogenic origins is crucial for well being of mankind and efficient conducts of industries and especially for finding what is the permissible amount of fossil fuel combustion retaining reasonable living standards and simultaneously keeping global climate change within a manageable limit, i.e., for "sustainable use of the earth", the theme of this conference. In what follows I will explain the present status of climate prediction putting emphasis on the prediction of global warming.

1. Climate Model

Climate is usually defined as average state of weather : Climate of any location is described by mean temperature, precipitation and so on. So, climate prediction is considered to be an extension of weather forecast. Even though the final form is just the same as extended weather forecast the methodology and underlying concept of climate prediction is fairly different from those of weather forecast. Since weather is essentially atmospheric phenomena weather forecast is made by predicting future state of the atmosphere based on physical laws. In the case of climate prediction, however, we need to know the future states of oceans and land surfaces because long-term average of the physical state of the atmosphere is strongly affected by the sea surface temperature and the land surface conditions rather than the atmosphere itself. The reason is that the atmosphere has a short memory time of about a month or so while the ocean thermal conditions persist for much longer time because of its huge thermal inertia, and the land surface conditions such as snow cover or soil moisture state last longer than the atmospheric memory time. The physical states of the ocean and land surface are, in turn, affected by the atmosphere. Thus the atmosphere, ocean and land surface together with snow and ice therein form a mutually interacting system ; they interact with each other by exchanging energy, water and momentum. This system is called "climate system". Therefore for predicting climate we have to predict the evolution of the whole climate system based on the physical laws which describe interactions among each members of the system.

A (global) climate model is a computer model to calculate quantities representing physical state of each member of the climate system at a big number (more than 10,000) of grid points covering the whole globe, by applying the equation of motion, the thermodynamic equation and so on. The calculation is carried out from arbitrary initial state following the time evolution of the climate system for a long time (more than ten years) with a time interval of 15 minutes or so. Therefore the amount of computation is enormous and it is not easy to perform a simulation of global warming. So variants or simplified versions of the full climate model are used depending on the nature or purpose of a study. As a matter of fact, in the scientific assessment of climate change by the Intergovernmental Panel on Climate Change (IPCC, 1990) different kinds of models were used for different problems as listed in the Table below.

	Uptake of CO2 by the oceans	Equilibrium climate at 2 x CO2	Climate change and / or global temperature rise
IPCC (1990)	1-D diffusion model of the oceans.	Coupled atmosphere-ocean mixed layer models with coarse resolution.	Only one coupled atmosphere ocean model. 1-D upwelling diffusion model of the ocean.
Present status	Same as above but small number of 3-D ocean models are now available.	Same as above but for medium resolution.	Several coupled models. 1-D upwelling diffusion models are still useful.

2. Projection of climate change due to greenhouse gas emissions

One of the most important objectives of developing climate models is to use them for prediction of climate change which may be caused by the increase of greenhouse gases. This is, however, not "prediction" because emissions of anthropogenic greenhouse gases may be changed for avoiding the predicted climate change if it is a dangerous one. So, climate models are used to "project" the future climate according to a specific scenario of greenhouse gas emissions. However, the word "prediction" will be used in the following when it naturally sounds in the context.

The projection of the future climate will be done by three steps as following :

(1) Prediction of concentrations of greenhouse gases given a scenario of emissions. Especially the CO_2 concentration is predicted by estimating the absorption of excess CO_2 by the ocean and the terrestrial ecosystem.

(2) Estimation of the global climate as an equilibrium state under the condition of doubled CO_2 concentration. This provides detailed information about climate changes both on

global and regional scales under higher CO_2 conditions.

(3) Prediction of the time evolution of climate when the CO_2 concentration increases following the result obtained by the step (1). The effect of ocean's thermal inertia to delay and/or to moderate climate changes is incorporated by using fully coupled atmosphere-ocean models or by simplified models.

Major findings and problems to be solved in the above procedures are listed in the following including the present author's view.

(i) A feature particular to the current global warming issue is that in addition to the well known warming agent, CO_2 other minor greenhouse gases such as methane and clorefluorocarbons (CFCs) contribute significantly to trapping of thermal radiation. As a consequence projected "radiative forcing", the magnitude of trapping outgoing thermal radiation will reach the level equivalent to doubling of CO_2 by about 2025 and it will exceed the quadraple-CO_2 equivalent by the year 2100.

(ii) There are significant uncertainties in the future projection of the atmospheric CO_2-concentration because carbon cycle among the atmosphere, ocean and the terrestrial biosphere is not well understood and the current carbon budget has a "missing sink" of about 30% of the emission due to fossil fuel combustion. Now many researchers consider that most of it could be attributed to the carbon sequentasation by the terrestrial biosphere. However, the present author considers that some part, say 10% out of 30% may be additionally absorbed by the ocean more than the widely accepted value, based on a careful examination of circulations and mixing processes in the world ocean.

(iii) In the past decade many computer experiments were carried out at many research centers all over the world to estimate global climate under the doubled CO_2 conditions. Since our understanding of many individual physical processes to determine the terrestrial climate is yet premature and computer resources are insufficient for computer experiments with reasonable resolution, results scattered over a wide range. The estimated differences of global mean temperature between the present and doubled CO_2 scatter over a climates range from 2°C to 5°C. At the present stage it is difficult to make such predictions of regional climate change those can be used for impact assessments and subsequent planning of response strategies.

One of major deficiencies lies in treatment of cloud formation and clouds' effects on radiation, whose solution may need a long-time research efforts. The difficulty of predicting regional climate change is due to insufficiency of computer power to some extent and this problem can be solved in the near future, at least partly.

(iv) Predicting gradual climate change corresponding to a specified scenario of time-dependent CO_2 concentration is done by coupled atmosphere-ocean general circulation model. However, those models now developed have many problems especially in ocean processes. There are many unknown or less-known physical processes in ocean dynamics. The present author considers that at present simplified ocean models such as one-dimensional upwelling-diffusion model which is designed to fit observational facts give more reliable results. It is difficult to go beyond the level of the IPCC prediction of the future temperature rise following the Business as Usual Scenario of the greenhouse gas emissions. According to it most probable temperature rise will be about 1°C at the time of CO_2 doubling (2025) and reach about 3°C by the end of the 21st century.

In the present author's view key factors to improve 3-D ocean general circulation models are to clarify the bottom friction effects on the Antarctic bottom water spreading and to understand the nature of the so-called isopicnal diffusion (diffusion along equal-density surfaces) and to quantify its effect.

(v) Verification of the total results of the best estimate model calculation for the past global-mean temperature rise shows a significant over estimate, i.e., about 0.9°C after industrialization while observation in the past 120 years indicates only 0.5°C rise. Now the difference is attributed to the effect of reflection of solar radiation by anthropogenically produced sulfate aerosols which inevitably accompany fossil fuel combustion. The present author would like to point out another overlooked problem, the side effect of forest clearing. In the past a considerable amount of CO_2 was released by deforestation in the Northern Hemisphere. But at the same time deforestation (change from forest to grassland or bare soil) causes an increase of the surface albedo. From a simple calculation it is shown that an increase of albedo of 0.1 would well cancel the effect of increase of the greenhouse effect due to the emission of CO_2 by forest clearing. Thus IPCC's best estimate of the global mean temperature rise corresponding to a doubling of CO_2, 2.5°C is not inconsistent with the observed temperature rise in the past and hence its prediction for the future still remain as a reasonable one.

3. Toward the realization of "controlled" global warming

Since a cut of CO_2 emission to such a level that stabilizes the CO_2-concentration seems almost impossible for a while in view of the continuation of present industrial activity and accommodating the rapidly increasing world population, climate change seems unavoidable to some extent. On the other hand it may not be so unrealistic to expect that new innovative technologies to reduce CO_2 emission drastically may emerge by the middle of the next century or so. Thus it may be more realistic not to hurry to implement severe regulation of CO_2-emission at the expense of economic development instead to make long-term planning for a large reduction of emission by developing new energy technologies in the future. In this case we have to adapt our society to the unavoidable but still not so dangerous up-coming climate change. In order to realize such a controlled (or planned) global warming, so to say, we have to predict climate change both natural and anthropogenic origins by which the human society can adapt to the change.

Development of the climate science is a prerequisite for this and now we have a scope that it can be achieved after 10~20 years through an international cooperative research "World Climate Research Programme".

TOWARD GLOBAL PLANNING OF SUSTAINABLE USE OF THE EARTH
Development of Global Eco-Engineering
S. Murai (Editor)
1995 Elsevier Science B.V.

Human influences on the global nitrogen budget and their implications for the global carbon budget

Jerry M. Melillo

The Ecosystems Center, Marine Biological Laboratory, Woods Hole, Massachusetts, USA, 02543

Abstract

Humans have dramatically altered the global nitrogen cycle. Through fossil fuel combustion, fertilizer production, and legume and rice cultivation, they are converting about 145 Tg N (10^{12} g N) each year from biologically unreactive forms to biologically reactive forms and introducing most of the reactive nitrogen into terrestrial ecosystems. While a large fraction of the reactive nitrogen escapes to the atmosphere and oceans, as much as 55 Tg N appear to remain on land each year. Some of this nitrogen makes its way into the ground water and some of it is stored in vegetation and soils in organic form. Nitrogen storage in vegetation and soils implies carbon storage. I estimate that this nitrogen-stimulated carbon storage is between 0.6 and 0.9 Pg C/y in the temperate and boreal forests of North America, Europe and Russia. The nitrogen-stimulated carbon storage, when added to carbon storage associated with forest regrowth and CO_2 fertilization, may help us to better understand the contemporary global carbon budget.

Climate change has the potential for altering the global nitrogen budget with a positive feedback to carbon storage on land. The mechanism underlying this feedback involves moving nitrogen from soil which has a narrow C:N ratio to vegetation with a wide C:N ratio. This movement results from enhanced soil decomposition in response to warming. This mechanism is particularly important in the mid- and high-latitude regions where carbon storage is thought to be nitrogen limited.

Key words: Global change, global carbon budget, global nitrogen budget, nitrogen-stimulated carbon storage, soil warming

1. INTRODUCTION

The terrestrial biosphere is being eutrophied. This was one of the points highlighted in the classic "Biosphere" issue of Scientific American that appeared in September, 1970. In his article on mineral cycles, Deevey (1970) noted that one result of this eutrophication is that "the biosphere is becoming larger." He attributed its growth to several factors including increased inputs of nitrogen to terrestrial ecosystems. A quarter of a century ago, Deevey made the connection between the human alteration of the global nitrogen budget and follow-on consequences for the global carbon budget. Since that time, the human impacts on the global nitrogen budget have increased dramatically and we are just beginning to understand the consequences for a sustainable biosphere.

Humans are altering the global nitrogen budget directly through their industrial and agricultural activities. They are converting unreactive nitrogen to reactive nitrogen and adding it to both terrestrial and marine ecosystems. They are also causing the redistribution of reactive nitrogen among ecosystems.

Humans also have the potential to indirectly affect the global nitrogen budget. Through fossil fuel use and changes in land cover and land use, they are increasing the atmosphere's content of carbon dioxide, methane and nitrous oxide. The loading of the atmosphere with these greenhouse gases is expected to result in a warming of the Earth. The warming is predicted to accelerate soil organic matter decay and make more reactive nitrogen available to plants.

In this paper I will explore how an increase in the availability of reactive nitrogen affects carbon storage in terrestrial ecosystems. The paper has four parts. In the first part I review how humans are altering the global nitrogen budget. Next I discuss the direct effects of alterations of the global nitrogen budget on carbon storage in terrestrial ecosystems. In the third part of the paper I consider the indirect effects of climate change on the global nitrogen budget and consequences for the global carbon budget. And finally, I set out a blueprint for future research designed to explore more completely how alterations in the global nitrogen budget affect the global carbon budget. The proposed research includes a range of activities from whole-ecosystem ^{15}N labeling experiments to regional and global terrestrial ecosystem model simulations.

2. HUMAN INFLUENCES ON THE GLOBAL NITROGEN BUDGET: DIRECT EFFECTS

Three processes, fossil fuel combustion, fertilizer production, and legume and rice cultivation, are responsible for converting unreactive nitrogen to various forms of reactive nitrogen. In recent years, this conversion rate annually has been about 145 Tg of N; 20 Tg N associated with fossil fuel combustion (Levy and Moxim 1989), 80 Tg N in fertilizer production (FAO 1989), and 45 Tg N resulting from legume and rice cultivation (Burns and Hardy 1975). Throughout this section of the paper all nitrogen fluxes and pools are rounded to the nearest 5 Tg.

The nitrogen from fertilizer production and legume and rice cultivation enters terrestrial ecosystems each year. The nitrogen from fossil fuel combustion first enters the atmosphere, mostly in oxidized forms - NO_x. Based on my interpretation of recent model calculations of atmospheric residence times of NO_x (e.g., Levy and Moxim 1989; Penner et al. 1991), about 75% of the oxidized nitrogen or about 15 Tg N is deposited on the land and the remaining 25% or about 5 Tg N is deposited in the ocean. In summary, 140 of the 145 Tg N of the newly reactive nitrogen enters terrestrial ecosystems each year (Figure 1).

Annually, terrestrial ecosystems lose a large fraction of this 140 Tg N to the atmosphere and the ocean. Galloway and colleagues (1994) estimate that about 50 Tg N of the 140 Tg N are transferred from the land to the atmosphere as N_2 in the process of denitrification. The estimate of N_2 flux to the atmosphere is probably the most speculative value in the global nitrogen budget I have presented in Figure 1. Another 5 Tg N is transferred from the land to

Figure 1. The "anthropogenic" component of the global nitrogen budget: human influences on major nitrogen fluxes and the increment in the terrestral nitrogen pool. Units are Tg N/y. All values rounded to nearest 5 Tg.

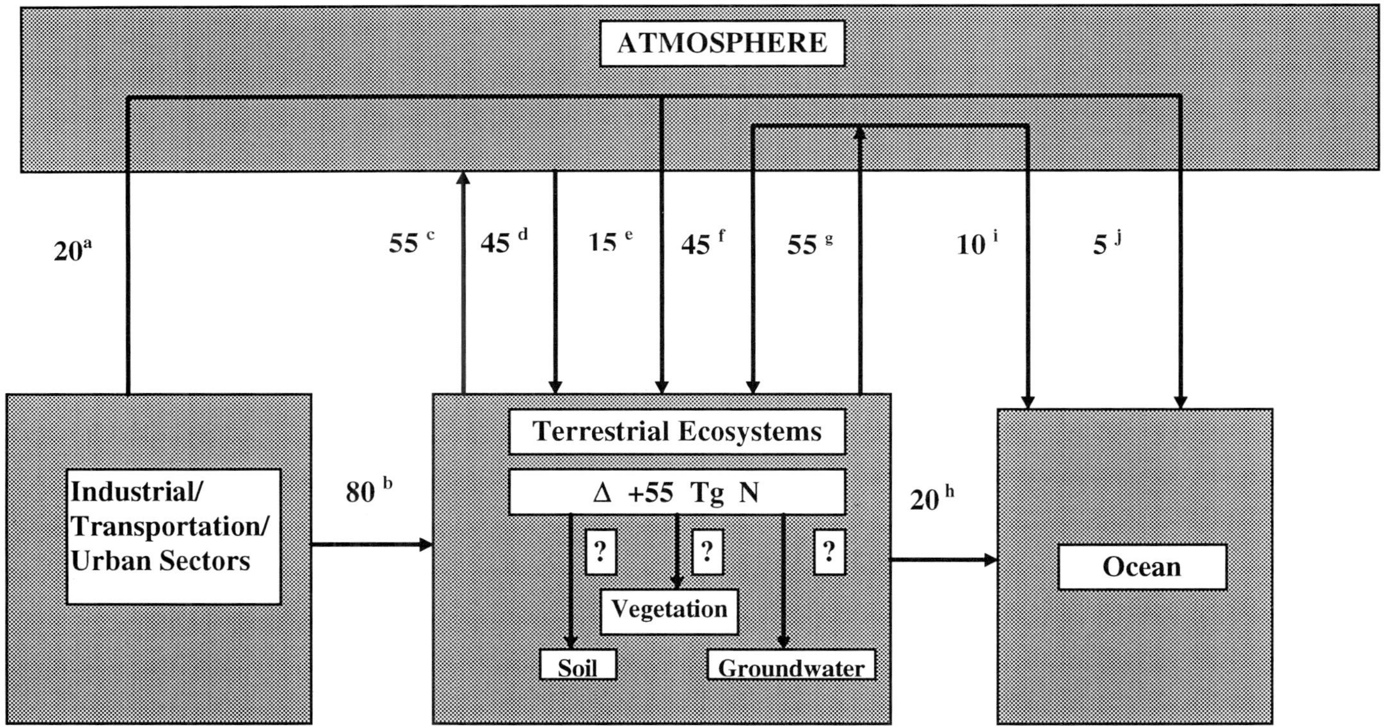

Notes: The fluxes are as follows: a) NO_x released in fossil fuel burning; b) fertilizer nitrogen applied to land; c) denitrification and other gaseous N losses from land; d) nitrogen fixation associated with agriculture; e) deposition on land of NO_x derived from fossil fuel burning; f) deposition on land of NO_x and NH_x derived from a variety of human activities including crop fertilization, animal husbandry and biomass burning; g) loss from land of NO_x and NH_x associated with human activities listed in "f" above; h) flux of nitrogen from the land to the ocean via rivers; i) deposition on the ocean of NO_x and NH_x derived from human activities listed in "f" above; j) deposition on the ocean of NO_x derived from fossil fuel burning. See the text for sources of the estimates.

the atmosphere as N_2O in a variety of processes associated with human activities including biomass burning. Both N_2 and N_2O are generally unreactive and so remain in the atmosphere for long periods.

Human activities also cause the redistribution of nitrogen among the Earth's ecosystems. Each year about 55 Tg N enter the atmosphere from the land in forms that are reactive such as ammonia (NH_3) and nitric oxide (NO). Animal husbandry and biomass burning are the major sources of NH_3, while biomass burning is the major source of NO (Table 1). Ammonia has shorter residence time in the atmosphere than NO. Because of this, I estimate that about 85% of the NH_3 injected into the atmosphere as a consequence of human activities returns to the land, as compared to 75% for NO. Based on these assumptions, it appears that of the 55 Tg of the reactive N that enters the atmosphere from human activities including animal husbandry, biomass burning and soil fertilization, 45 Tg N returns to the land. The remaining 10 Tg N are transferred to the ocean.

Table 1. Redistribution of reactive nitrogen as a result of human activities. The NH_3 data is from Schlesinger and Hartley (1992) and the NO data is from Levy et al. (1991).

Nitrogen form (s)	Activity	Annual flux to the Atmosphere (Tg N)
NH_3	Fertilization of fields Animal husbandry Biomass burning	10 32 5
NO	Biomass burning	8
NH_3 + NO		55

Besides receiving a nitrogen input through the atmosphere, the ocean receives an input through rivers. About 20 Tg N move from the land to the ocean each year through rivers (Duce et al. 1991).

Using a simple mass balance calculation, I estimate that there is an annual net accumulation of about 55 Tg N in the globe's terrestrial ecosystems. Galloway and colleagues (1994), who have done a similar analysis after which I patterned mine, have concluded that there is a net annual accumulation of 50 Tg N.

A critical question to ask is - Where is this "new" nitrogen stored in terrestrial ecosystems? There are three places in terrestrial ecosystems where the "new" nitrogen may be stored - groundwater, vegetation and soils. In the past decades, there has been a dramatic accumulation of nitrogen as nitrate in ground water in many parts of the world. Recently Spalding and Exner (1993) reviewed a number of site-specific studies of ground-water pollution and concluded that many of the most severe problems are found in agricultural areas where nitrogen is added to croplands as industrially-produced fertilizer or as animal wastes.

Vegetation and soils can also be sites of nitrogen accumulation. Recently, my colleagues and I have conducted several studies of the fate of nitrogen inputs to forest ecosystems of the northeastern United States. We found that while both vegetation and soils do accumulate added nitrogen, the soil is the major nitrogen sink (Aber et al. 1993, Nadelhoffer et al. 1994).

How do these site-specific results translate to regional and global estimates of the fate of "new" nitrogen added to terrestrial ecosystems? The answer is, we do not know for sure. We do not yet have a carefully validated georeferenced data set on nitrogen inputs to terrestrial ecosystems and we do not know, for various ecosystems, how much of the nitrogen will be retained in the ecosystems and where it will be stored. This kind of information is important to our understanding not only of the global N budget, but also of the global C budget. This is so because in many ecosystems of the world, the storage of carbon is thought to be limited by a lack of nitrogen.

3. NITROGEN-STIMULATED CARBON STORAGE IN TERRESTRIAL ECOSYSTEMS: OUR CURRENT UNDERSTANDING

Carbon storage in many temperate and boreal forests of the northern hemisphere is nitrogen limited, (e.g., Aber et al. 1982, Miller 1981, Safford and Filip 1974, Tamm 1990). Many of these forests are receiving substantial inputs of nitrogen each year in precipitation. Some areas of western and central Europe have annual nitrogen inputs of more than 50 kg/ha each year. In 1989, I used the summaries of nitrogen deposition prepared by Barrie and Hales (1984), Buijsman and Erisman (1988), and Bonis et al. (1980), in conjunction with a global vegetation map, to produce estimates of the spatial patterns of wet plus dry nitrogen deposition in the temperate and boreal zones of North America and Europe (Melillo et al. 1989). I assumed dry deposition to be equivalent in magnitude to wet deposition. Europe was broadly defined by me to include all of the temperate and boreal areas in western and eastern Europe, and all of the temperate and boreal areas in the former Soviet Union; that is, the vast forested landscape east of the Ural Mountains as well as "European Russia" west of the Urals.

My estimate of nitrogen deposition in precipitation for these temperate and boreal zones of the Northern Hemisphere was about 18 Tg N per year for the period between 1975 and 1980, with North America receiving about 6 Tg N annually and Europe receiving about 12 Tg N annually (Melillo et al. 1989). I further divided the deposition between temperate and boreal zones on the two “continents.” In North America, the temperate forests received about 5 Tg N annually and the boreal forests about 1 Tg N. In Europe, temperate forests received about 10 Tg N annually and the boreal forests about 2 Tg N.

For this study, I have used my 1989 estimates of nitrogen deposition to the temperate and boreal forest regions of the Northern Hemisphere with the methodology described by Melillo and Gosz (1983) and Peterson and Melillo (1985), to estimate the magnitude of nitrogen-stimulated storage of carbon in these regions. I have also used a model of the global carbon and nitrogen cycles, the Terrestrial Ecosystem Model (TEM, Melillo et al. 1993), to estimate whole-ecosystem C:N ratios for the temperate and boreal forests. The whole-ecosystem C:N ratios were weighted by area to account for the differences in C:N ratios among the deciduous, coniferous and mixed forests of the temperate zone, and between the boreal forests and boreal woodlands. In this study, I have assumed that nitrogen entering an ecosystem is stored in the vegetation and soil in the same proportions as it is found in mature forests. This assumption is consistent with our recent research on the fate of nitrogen added to forests (Aber et al., 1993, Nadelhoffer et al., 1994).

The maximum annual nitrogen-stimulated carbon storage I estimate for the temperate and boreal zones of the Northern Hemisphere is about 0.9 Pg C (Table 2). In making this estimate, I have assumed that all of the nitrogen that enters these ecosystems is retained within them; that is, none of the nitrogen entering the temperate and boreal forests in precipitation is lost through processes such as denitrification or leaching - a nitrogen retention efficiency of 100%. I have previously argued that this is a questionable assumption, and that it is more realistic to use a 60% retention efficiency for these calculations (Melillo and Gosz 1983, Peterson and Melillo 1985). Doing this, I estimate a nitrogen-stimulated carbon storage of 0.6 Pg annually in the forests of the temperate and boreal zones of the Northern Hemisphere (Table 2).

Table 2. Estimates of nitrogen-stimulated carbon storage in the forests and woodlands of North American and Europe - this study.

Biome Type	C:N Ratio	Nitrogen Input (Tg N)	Maximum # estimate of C Storage (Pg C)	Likely * C Storage (Pg C)
Boreal Forests and Woodlands	45	3	0.13	0.08
Temperate Forests	54	15	0.81	0.49
Total		18	0.94	0.57

Assumes 100% retention efficiency
* Assumes 60% retention efficiency

Recent estimates of nitrogen-stimulated carbon storage range from 0.2 to 6 Pg C annually (Table 3). The analyses producing this range differ in many ways including: 1) portion of the terrestrial biosphere considered; 2) magnitude and pattern of the nitrogen input; 3) nitrogen retention efficiency; 4) allocation of the nitrogen among ecosystem storage pools (e.g., vegetation, surface litter, soil); and 5) the C:N ratio of the storage pools. The more complete of the analyses (Kohlmaier et al. 1988, Hudson et al. 1994 and Townsend 1994) give estimates of between 0.2 to 1.7 Pg C for annual nitrogen-stimulated carbon storage in terrestrial ecosystems in recent years. My "best estimate" from this study is towards the lower end of this range.

How important is nitrogen-stimulated carbon storage to the global carbon budget? Recent analyses which balance CO_2 accumulation in the atmosphere against fossil fuel release, the effects of changing land use, and ocean uptake, calculate a substantial residual term that is often called the "missing sink." The Intergovernmental Panel on Climate Change (IPCC) has recently argued that the missing sink is mostly on land and estimated that its magnitude, expressed as an average annual value for the 1980's, is 1.4±1.5 Pg C (Table 4; Schimel et al. 1994). In this context, my estimate of an annual storage of between 0.6 to 0.9 Pg C in terrestrial ecosystems that is stimulated by nitrogen input is important in the global carbon budget.

Table 3. Recent estimates of nitrogen-stimulated carbon storage in terrestrial ecosystems - a literature survey.

	Region(s) considered	N input Tg N/yr	C Store Pg C/y
Kohlmaier et al. (1988)	30 - 60° N	21 ± 7	up to 0.7
Schindler & Bayley (1993)	globe	13	~ 0.7 - 2.0
Hudson et al. (1994)	northern hemisphere	30	~ 1.7
Galloway et al. (1994)	globe	50	2 - 6[a]
Townsend et al. (1994)	global (excluding cultivated areas)	7[b]	0.2 - 1.2[c]
Melillo (this study)	temperate & boreal forests of North America and Europe	18	0.6 - 0.9[d]

[a] maximum range
[b] NO_y only
[c] best estimate 0.3 - 0.6
[d] best estimate 0.6

Direct confirmation of nitrogen-stimulated carbon storage is difficult for two reasons. First, we are looking for a small change in a large pool; an annual change in the range of one part per thousand. Second, and equally important, several mechanisms are probably operating simultaneously and interactively to account for terrestrial carbon storage in Northern Hemisphere ecosystem including; forest regrowth following harvest (e.g., Dixon et al. 1994), fertilization by carbon dioxide (Gifford 1994), climate variability (Dai and Fung 1993), and, of course, nitrogen stimulation. Kauppi and his colleagues (1992) have suggested that nitrogen-stimulated carbon storage may be the cause of the increases in European forest resources observed between 1950 and 1980. Houghton (1993) challenged this interpretation and argued that it is difficult to separate nitrogen stimulation from forest regrowth. Hudson and his colleagues (1994) have reviewed both positions and have concluded that the nitrogen stimulation argument cannot be dismissed for the reasons set forth by Houghton (1993). Clearly, additional research is needed to document the magnitude of the nitrogen-stimulated carbon storage.

If, as I have argued, nitrogen-stimulated carbon storage does account for about half of the "missing sink", then it is important that projections of future atmospheric CO_2 levels take this into account. Environmental protection policies and the evolution of the world economy in the 21st century may reduce nitrogen-stimulated carbon storage and so lead to an increased rate of CO_2 accumulation in the atmosphere if fossil fuel use remains at current levels or increases.

Table 4. Average annual budget of CO_2 perturbations for 1980-1989. Fluxes and reservoir changes of carbon expressed in Pt C/y, error limits correspond to an estimated 90% confidence interval (adapted from Schimel et al. 1994)

CO_2 sources	Pg C/y
Emissions from fossil fuel combustion and cement production	5.5 ± 0.5
Net emissions from changes in tropical land use	1.6 ± 1.0
Total anthropogenic emissions	7.1 ± 1.1
CO_2 sinks	
Storage in the atmosphere	3.2 ± 0.2
Oceanic uptake	2.0 ± 0.8
Uptake by Northern Hemisphere forest regrowth	0.5 ± 0.5
Additional terrestrial sinks (CO_2 fertilization, nitrogen fertilization, climatic effects)	1.4 ± 1.5

Scientists in both Europe and North America have warned that chronic nitrogen inputs can adversely affect terrestrial, fresh water and coastal marine ecosystems (e.g., Aber et al. 1989, Fisher et al. 1988, Galloway et al. 1987, Johnson et al. 1991, Melillo et al. 1989, Schindler and Bayley 1993, Schultze 1989, Tamm 1990). The effects include forest dieback, lake and stream acidification and eutrophication of coastal estuaries and bays. In Europe and North America, there may be government-mandated regulation of fossil fuel emissions to reduce nitrogen inputs to ecosystems. If the regulatory efforts are successful, nitrogen-stimulated carbon storage in these regions will be reduced.

Fossil fuel emissions in the developing countries, including many in the tropics, are likely to increase as their economies develop in the 21st Century (IPCC 1992). In many of the ecosystems in the tropics, including forests, carbon storage appears to be phosphorus limited rather than nitrogen limited (Sanchez et al. 1982). Therefore nitrogen-stimulated carbon storage per unit of fossil fuel combustion in the developing countries of the tropics will be less than it is today in the mid-latitude region of the Northern Hemisphere.

4. CLIMATE CHANGE, THE GLOBAL NITROGEN BUDGET AND CARBON STORAGE IN TERRESTRIAL ECOSYSTEMS

The atmospheric concentrations of the major long-lived greenhouse gases continue to increase because of human activity (Watson et al. 1992). Current models of climate change predict that as part of the greenhouse effect, the global mean temperature will increase between 1.5°C and 4.5°C during the next century (IPCC 1992). One result of global warming may be changes in the rates of temperature-dependent soil processes such as decomposition (Swift et al. 1979) and net nitrogen mineralization and nitrification (Focht and Verstraete 1977). In recent years, my research group has conducted a series of field, laboratory and simulation modeling studies to explore, in a quantitative way, how predicted temperature increases will affect the nitrogen cycle, interactions between the nitrogen and carbon cycles, and carbon storage in terrestrial ecosystems. In this section I highlight some of the important results from these studies.

Field study - For the past three years we have been conducting a soil warming experiment in a deciduous forest stand at the Harvard Forest in central Massachusetts. At the research site we established 18, 6x6 m plots in April, 1991. The plots are grouped into 6 blocks and the 3 plots within each block were randomly assigned to 1 of 3 treatments. The treatments are: 1) heated plots in which the average soil temperature is elevated 5°C above ambient using buried heating cables; 2) disturbance-control plots that are identical to heated plots except they receive no electrical power; and 3) undisturbed-control plots that have been left in their natural state. We are measuring a variety of parameters at the plots including CO_2 emissions from the soil and indexes of N availability. Details of the experimental design and the field and laboratory methods have recently been reported by Peterjohn et al. (1994).

During the first year of study (July 1991 through June 1992) we observed that heating increased the emission of CO_2 and the rate of nitrogen mineralization. We estimated CO_2 fluxes to be approximately 7,100 kg C/ha y, 7,900 kg C/ha y and 11,100 kg C/ha y in the control, disturbance control and heated plots, respectively; an increase due to warming of approximately 3,200 kg C/ha y. Nitrogen mineralization was doubled in the surface soil which we defined as the forest floor plus the top 10 cm of the mineral soil. During the first growing season, we observed that in the forest floor, heating increased the average net mineralization rates from 1.02 to 2.47 mg N/kg d. In the mineral soil, heating increased the average net mineralization rates from 0.09 to 0.19 mg N/kg d. No nitrification was observed in either the forest floor or the mineral soil in any of the treatments.

In the second year of study (July 1992 -June 1993) we found that warming had a much less dramatic effect on CO_2 flux, but a sustained large effect on nitrogen mineralization. We estimated CO_2 fluxes in the second year to be approximately 6,800 kg C/ha, 7,600 kg C/ha and 8,700 kg C/ha for control and disturbed control and heated treatments. Nitrogen mineralization rates remained doubled. Again, no nitrification was observed in either the forest floor or the mineral soil.

These results can be interpreted in several ways. One interpretation of the differences in the CO_2 response between the first and second years versus the constant response in nitrogen mineralization is as follows. There are two major soil organic matter (SOM) pools; a fast turnover pool and a slow turnover pool. The fast pool contains high C:N ratio material (e.g., recent litter) and the decay of fast-pool material results in large CO_2 losses per unit of material processed and small net nitrogen release in the mineralization process. In contrast to the fast pool, the slow pool contains low C:N ratio material (e.g., meta-stable humus) and the decay of material in the slow pool results in a smaller loss of CO_2 per unit of material processed and a large net nitrogen release in the mineralization process. If this is an appropriate interpretation, it suggests that the slow pool is key to ecosystem response to climate change over the long term. As elevated soil temperature increases the rate of decay in the slow pool, both carbon (as CO_2) and nitrogen (as inorganic N) are released. The nitrogen then becomes available to be taken up by and stored in plants. Since the C:N ratio of plant material is substantially larger than the C:N ratio of SOM in the slow pool, warming may lead to increased carbon storage in the ecosystem. The magnitude of such an increase would, of course, depend on how plant carbon balance is affected by other components of climate change including the availability of water and the effect of increased temperature on both photosynthesis and respiration.

A coupled plant-soil model - To evaluate how simultaneous changes in plant and soil processes affect ecosystem carbon balance and other terrestrial ecosystem responses to global change, my research group has developed a coupled plant-soil model that considers both carbon and nitrogen cycling (Raich et al. 1991, McGuire et al. 1992, Melillo et al. 1993). We call the model TEM, an acronym for the Terrestrial Ecosystem Model. It uses information on soils, vegetation, and climate variables such as temperature, precipitation, cloudiness and solar radiation input to make monthly estimates of important transfers of carbon and nitrogen in terrestrial ecosystems. The data sets used in TEM are currently organized into 0.5° latitude by 0.5° longitude units or grid cells that cover the vegetated surface of the Earth.

We applied the model to more than 56,000 grid cells in order to develop estimates of terrestrial plant and soil carbon stocks for current climate conditions and atmospheric CO_2 levels. Our model-based estimate of the total terrestrial carbon stocks is 2279 Pg, of which 977 Pg C is in the vegetation and 1301 Pg C is in soil (Table 5).

We did not include land-use change in this version of the model, and so the estimates are for potential plant and soil carbon stocks, not for present-day stocks. The conversion of natural ecosystems that have large woody plant components into agricultural systems has certainly reduced the Earth's plant carbon pool (e.g. Houghton et al. 1983, Melillo et al. 1988). This pool has also been reduced by firewood gathering, especially in poor tropical countries. In addition, the present-day soil carbon stock may be less than the potential stock because of the initial effects of converting natural ecosystems to agricultural systems that are tilled.

When both land area and carbon density are considered, tropical ecosystems account for about 47% of the total terrestrial carbon pool: 56% of the vegetation carbon and 41% of the soil carbon. With most of the world's population growth expected to occur in the tropics over the next century, these large carbon stocks are likely to be reduced through land-use change activities, and perhaps other aspects of global change.

I used the output from three climate models to run TEM (Melillo et al. 1993) to simulate changes in terrestrial carbon stocks under various climatic conditions. The three climate models, more correctly referred to as general circulation models (GCMs), were: the Goddard Institute of Space Studies (GISS) GCM; the Oregon State University (OSU) GCM; and the Geophysical Fluid Dynamics Laboratory (GFDL) GCM. These GCMs predict that with a CO_2 doubling: 1) mean global temperature will increase between 2.8°C and 4.2°C; 2) global precipitation increases between 7.8% and 11.0%; and 3) cloudiness will decrease between 0.4% and 3.4%.

Changes in climate with no change in CO_2 concentration are predicted by TEM to result in substantial losses of carbon from terrestrial ecosystems (Table 5). For all three GCM climates the predicted carbon storage reductions are about 200 Pg and the losses come mostly from soils. This represents a 15% loss from the soil carbon pool.

Table 5. Response of carbon storage (Pg C) by region for experiment involving two levels of atmospheric CO_2 and four levels of climate

CO_2 concentration		312.5 ppmv			625.0 ppm			
Climates	Contemporary*	GFDL	GISS	OSU	Contemporary	GFDL	GISS	OSU
Vegetation carbon								
Northern ecosystems[+]	140	+43	+32	+31	+4	+71	+42	+40
Temperate ecosystems[&]	290	-1	+24	+13	+33	+67	+80	+63
Tropical ecosystems[%]	547	-37	-50	-74	+108	+116	+110	+68
Total vegetation carbon	977	+5	+6	-30	+145	+254	+232	+171
Soil carbon								
Northern ecosystems	327	-42	-17	-8	+11	-7	+2	+8
Temperate ecosystems	441	-66	-47	-34	+73	+28	+36	+48
Tropical ecosystems	533	-103	-140	-138	+89	+8	-31	-37
Total soil carbon	1301	-211	-204	-180	+173	+29	+7	+19
Total carbon								
Northern ecosystems	466	+1	+15	+22	+15	+64	+44	+48
Temperate ecosystems	733	-67	-23	-20	+105	+95	+118	+112
Tropical ecosystems	1080	-140	-190	-212	+198	+124	+78	+30
Total terrestrial carbon	2279	-206	-198	-210	+318	+283	+240	+190

* Pool size that is used as the baseline for responses.

[+] Northern ecosystems include polar desert, alpine tundra, wet/moist tundra, boreal woodland, and boreal forest. These ecosystems occupy a combined area of 28.2 million km^2.

[&] Temperate ecosystems include forests, savannas, grasslands, shrublands and deserts. These ecosystems occupy a combined area of 56.7 million km^2.

[%] Tropical ecosystems include forests, savannas and shrublands. These ecosystems occupy a combined area of 42.5 million km^2.

Tropical ecosystems account for much of the global response. With the GFDL climates they account for about 75% of the carbon loss, and with the GISS and OSU climates they account for virtually all of the loss. In contrast to tropical systems, TEM predicted that northern ecosystems respond to climate change alone by increasing their carbon storage, albeit by relatively small amounts (Table 5). The increase in ecosystem-wide carbon storage occurred because the increase in vegetation carbon was larger than the decrease in soil carbon. The sequence of process-level changes leading to this increased carbon storage in the vegetation at the expense of the soil is thought to be as follows. As temperature increases, the soil loses carbon, and nitrogen mineralization is accelerated (see the earlier discussion of soil warming). The increased availability of inorganic nitrogen results in a plant fertilization response and an increase in plant productivity and plant carbon storage. This sequence represents an important carbon-nitrogen interaction at the ecosystem level that has global-scale consequences..

For doubled CO_2 with no climate change, TEM predicts an increase in global terrestrial carbon storage of 318 Pg, which is a 14% increase in terrestrial carbon storage relative to the estimated contemporary condition (Table 5). In absolute terms, tropical ecosystems are the most responsive to a CO_2 doubling alone. The model predicts that they will account for almost two-thirds of the increased carbon storage with doubled CO_2, but no climate change. The carbon storage in tropical ecosystems under this scenario is predicted to be almost equally divided between vegetation and soils. Northern ecosystems, including tundra, boreal woodlands and boreal forests are predicted to be the least responsive to doubled CO_2 without climate change. The small response occurs because TEM predicts that productivity in nitrogen-limited ecosystems does not substantially respond to elevated CO_2 alone.

In response to changes in both CO_2 and climate, TEM predicts substantial increases in terrestrial carbon stocks, although the magnitude of the increases vary according to GCM climate (Table 5). The OSU climate combined with CO_2 doubling resulted in the smallest increase in terrestrial carbon storage globally, 190 Pg (an 8% increase), and the GFDL climate combined with a CO_2 doubling yielded the largest increase, about 280 Pg C (a 12% increase). Changes in vegetation carbon stocks accounted for about 90% of the increased storage. The vegetation of northern and temperate ecosystems is very responsive to the combination of CO_2 doubling and climate change because the nitrogen limitation is lessened by warming, through an increase in soil nitrogen mineralization (McGuire et al. 1992).

5. BLUEPRINT FOR FUTURE RESEARCH

New research is needed to improve our understanding of how human activities affect terrestrial carbon storage through direct and indirect alterations of the global nitrogen budget. This new research should include both field research and modeling studies.

Research on direct effects - To better estimate the magnitude of nitrogen-stimulated carbon storage through the addition of "new" nitrogen to terrestrial ecosystems, we need to improve our knowledge in five areas: 1) nitrogen deposition; 2) ecosystem nitrogen retention efficiency; 3) distribution of the retained nitrogen among ecosystem components (e.g., vegetation, litter, soils); 4) C:N ratios in ecosystems at various stages of maturity; and 5) land-use history.

To begin with, we need georeferenced estimates of nitrogen deposition for the world's terrestrial ecosystems. Models that estimate NO_y and NH_x deposition exist (NO_y - Levy and Moxim 1989, Penner et al. 1991; NH_y - Crutzen, personal communication). They use information on georeferenced sources, atmospheric residence times and, atmospheric transport models to estimate deposition fields. With NH_x it will be important to separate natural from anthropogenic sources and keep track of them for the deposition calculations. Estimates of past and future deposition rates, as well as current deposition rates, will eventually be needed. Deposition estimates from past and current measurements should be used to validate the deposition models. Future deposition rates will have to be estimated from projected energy and fertilizer use.

Previously, I used the input-output budgets from small watershed studies to estimate nitrogen retention efficiencies for forest ecosystems (Melillo and Gosz 1983). This kind of analysis will have to be done in a more rigorous way so as to include: 1) finer stratification by ecosystem type; 2) differences resulting from nitrate versus ammonium input because the former is more mobile in soils and therefore more likely to be lost from the system; and 3) possible differences associated with land-use history since "ecological memory" may influence retention capacity.

Once we have determined the nitrogen retention efficiencies of various ecosystems, we will need to know how the "retained" nitrogen is distributed among the ecosystems' components (i.e., vegetation, litter, soils). The distribution will be affected by a variety of factors including nitrogen form, (i.e., nitrate versus ammonium) maturity of the ecosystem, and land-use history. The chronic nitrogen input study that we are conducting at the Harvard Forest (Aber et al., 1993) and the ^{15}N addition studies that we are carrying out at the Harvard Forest and in Maine (Nadelhoffer et al., 1994) are providing us with this information. In the ^{15}N studies we introduce a small amount of nitrogen, that is highly enriched in ^{15}N, to a forest stand and then follow its movement through time among the ecosystem's nitrogen pools (e.g. soil, litter and plant).

A number of literature reviews of C:N ratios in plants, litter and soils are available (e.g., Schindler and Bayley 1993). These can be compared to model-derived estimates like the ones we used from TEM for this study. A comprehensive data base on C:N ratios should be developed. It must consider how the C:N ratio changes in plants, litter, and soil with stages of ecosystem maturity and land-use history.

As noted above, a good knowledge of land-use history is essential for our understanding of nitrogen-stimulated carbon storage because ecosystems have "memory;" that is, past events affect current and future behavior. Land-use history reconstructions at regional and global scales have been developed, albeit at relatively coarse scales (Melillo et al., 1988; Houghton et al., 1991). These early efforts used tabular records of land cover and land use. More recently, time series of satellite-acquired data on land-cover change have been used to calculate major shifts such as forests to pastures in tropical ecosystems (Skole and Tucker 1993). This work is essential for many scientific purposes and should be encouraged. In addition, work to refine "pre-satellite" land-use change estimates from tabular records should continue. This will involve an interdisciplinary effort that includes geographers and historians as well as ecologists.

All of this data, georeferenced when appropriate, should be drawn together, organized, and made available to the research community for use in model studies to estimate nitrogen-stimulated carbon storage. At least one study has been made using "high-resolution" geographically referenced data on NO_y deposition with a dynamic ecosystem model for this purpose (Townsend et al., 1994). While Hudson and his colleagues (1994) did not use "high resolution" georeferenced data on nitrogen input, they did include considerations of retention efficiency, variable distribution of "new nitrogen" among ecosystem compartments depending upon ecosystem maturity, and C:N ratios that varied among ecosystems. With this much progress in synthesis efforts already, I expect rapid advances in our understanding of nitrogen-stimulated carbon storage once the data sets I have called for become available to the research community.

Research on indirect effects - The research on this issue of indirect effects requires both biogeochemistry models such as TEM (Melillo et al., 1993) and vegetation redistribution models such as BIOME (Prentice et al., 1992) that operate at the global scale. Ideally, these models would be transient models, capable of running with time series data on climate change and changes in nitrogen deposition rate. Several models of each type already exist with transient versions, although they have not been coupled and run specifically to address this "indirect effects" issue. This work should be encouraged.

Laboratory and process-level field studies on the "indirect effects" issue should also be encouraged. These studies will be useful for developing modeling concepts and for providing parameters for models. Whole-ecosystem manipulations will also be useful for model parameterization and model validation. Here I am thinking about manipulations like the soil warming experiment I described earlier or like the FACE of free air circulation experiment currently being conducted at the Duke Forest in North Carolina (Schlesinger, personal communication).

Prospects for action - The blueprint for future research proposed above is totally consistent with the International Geosphere-Biosphere Program (IGBP). Several of the core projects and task groups of the IGBP are undertaking the data collection, laboratory and field experiments, and modeling activities required to further our understanding of how human activities affect terrestrial carbon storage through alterations of the global nitrogen budget. This work is needed by us to develop a scheme for managing the land systems of the Earth in a sustainable way.

ACKNOWLEDGMENTS

This research is being supported by funds from the United States Environmental Protection Agency (Cooperative agreement CR817734-010, Athens Environmental Research Laboratory), the Department of Energy's National Institute for Global Change (Northeast Regional Center, subagreement 901214-HAR), and the National Aeronautics and Space Administration (contract NAGW-1825, subcontract 91-14).

REFERENCES

Aber, J.D., J.M. Melillo, C.A. Federer. 1982. Predicting the effects of rotation length, harvest intensity, and fertilization on fiber yield from northern hardwood forests in New England. For. Sci. 28:31-48.

Aber, J.D., K.J. Nadelhoffer, P.A. Steudler and J.M. Melillo. 1989. Nitrogen saturation in northern forest ecosystems - hypotheses and implications. BioScience 39:378-386.

Barrie, L.A. and J.M. Hales. 1984. The spatial distributions of precipitation acidity and major ion wet deposition in North America during 1980. Tellus 36B:333-335.

Bónis, K., E. Mészáros and M. Putsay. 1980. On the atmospheric budget of nitrogen compounds over Europe. Idójárás (periodical of the Hungarian Meteorological Service) 84:57-68.

Buijsman, E. and J.W. Erisman. 1988. Wet deposition of ammonium in Europe. J. Atmos. Chem. 6:265-280.

Burns, R.C. and R.W.F. Hardy. 1975. Nitrogen fixation in bacteria and higher plants, 54-58. Springer-Verlag, New York.

Dai, A. and I.Y. Fund. 1993. Can climate variability explain the "missing" CO_2 sink? Global Biogeochemical Cycles 7:599-610.

Deevey, E.S., Jr. 1970. Mineral cycles. Sci. Am. 223:148-158.

Duce, R.A., P.S. Liss, J.T. Merrill, E.L. Atlas, P. Buat-Menard, B.B. Hicks, J.M. Miller, J.M. Prospero, R. Arimoto, T.M. Church, W. Ellis, J.N. Galloway. L. Hansen, T.D. Jickells, A.H. Knap, K.H. Reinhardt, B. Schneider, A. Soudine, J.J. Tokos, S. Tsunogai, R. Wollast, and M. Zhou. 1991. The atmospheric input of trace species to the world ocean. Global Biogeochemical Cycles 5:193-259.

Fisher, D., J. Ceraso, T. Mathew and M. Oppenheimer. 1988. Polluted coastal waters: the role of acid rain. Environmental Defense Fund, New York.

Focht, D.D. and W. Verstraete. 1977. Biochemical ecology of nitrification and denitrification, 135-214. In: M. Alexander (ed.) Advances in Microbial Ecology. Plenum Press, New York.

Food and Agriculture Organization (FAO). 1989. FAO Fertilizer Yearbook. Vol. 43, Rome.

Galloway, J.N., G.R. Hendrey, D.L. Schofield, N.E. Peters and A.H. Johannes. 1987. Processes and causes of lake acidification during spring snowmelt in the west-central Adirondack Mountains, New York. Can. J. Fish. Aquat. Sci. 44:1595-1602.

Galloway, J.N., W.H. Schlesinger, H. Levy II, A. Michaels and J.L. Schnoor. 1994. Nitrogen fixation: Anthropogenic enhancement-environmental response. Global Biogeochemical Cycles (submitted).

Gifford, R.M. 1994. The global carbon cycle: a viewpoint on the missing sink. Aust. J. Plant Physiol. 21:1-15.

Houghton, R.A., J.E. Hobbie, J.M. Melillo, B. Moore, B.J. Peterson, G.R. Shaver and G.M. Woodwell. 1983. Changes in the carbon content of terrestrial biota and soils between 1860 and 1980: a net release of CO_2 to the atmosphere. Ecol. Monogr. 53:235-262.

Houghton, R.A. 1991. Tropical deforestation and atmospheric carbon dioxide. Clim. Change 19:99-118.

Houghton, R.A. 1993. Is carbon accumulating in the northern temperate zone? Global Biogeochemical Cycles 7:611-618,

Hudson, R.J.M., S.A. Gherini, R.A. Goldstein. 1994. Modeling the global carbon cycle: nitrogen fertilization of the terrestrial biosphere and the "missing" CO_2 sink. Global Biogeochemical Cycles 8:307-333.

IPCC. 1992. Climate Change 1992: The Supplementary Report to the IPCC Scientific Assessment, 200 pp. Cambridge University Press, Cambridge, UK.

Johnson, D.W., M.S. Cresser, S.I. Nilsson, J. Turner, B. Ulrich, D. Binkley and D.W. Cole. 1991. Soil changes in forest ecosystems: evidence for and probable causes. Proc. R. Soc. Edinburgh, Ser. B. 97:81-116.

Kauppi, P.E., K. Mieliainen and K. Kuusela. 1992. Biomass and carbon budget of European forests. Science 256:70-74.

Kohlmaier, G.H., A. Janecek and M. Plochl. 1988. Modelling response of vegetation to both excess CO_2 and airborne nitrogen compounds within a global carbon cycle model, 207-234. In: Marani, A. (ed.) Advances in Environmental Modelling. Elsevier, New York.

Levy II, H. and W.J. Moxim. 1989. Simulated global distribution and deposition of reactive nitrogen emitted by fossil fuel combustion. Tellus 41:256-271.

Levy II, H., P.S. Kasibhatla, W.J. Moxim and J.A. Logan. 1991. The global impact of biomass burning on tropospheric reactive nitrogen. In: J. Levine (ed.) Proceedings of the Chapman Conference on Global Biomass Burning. MIT Press, Cambridge, MA.

McGuire, A.D., J.M. Melillo, L.A. Joyce, D.W. Kicklighter, A.L. Grace, B. Moore III and C.J. Vorosmarty. 1992. Interactions between carbon and nitrogen dynamics in estimating net primary productivity for potential vegetation in North America. Global Biogeochemical Cycles 6:101-124.

Melillo, J.M. and J.R. Gosz. 1983. Interactions of biogeochemical cycles in forest ecosystems, 177-222. In: B. Bolin and R.B. Cook (eds.) The major biogeochemical cycles and their interactions. John Wiley & Sons, New York.

Melillo, J.M., J.R. Fruci, R.A. Houghton, B. Moore III and D.L. Skole. 1988. Land-use change in the Soviet Union between 1850 and 1980: Causes of a net release of CO_2 to the atmosphere. Tellus, Ser. B. 50B:116-128.

Melillo, J.M., P.A. Steudler, J.D. Aber and R.D. Bowden. 1989. Atmospheric deposition and nutrient cycling, 263-280. In: M.O. Andreae and D.S. Schimel (eds.) Exchange of trace gases between terrestrial ecosystems and the atmosphere. John Wiley & Sons, Ltd., Dahlem, Konferenzen.

Melillo, J.M., A.D. McGuire, D.W. Kicklighter, B. Moore III, C.J. Vorosmarty and A.L. Schloss. 1993. Global climate change and terrestrial net primary production. Nature 363:234-240.

Miller, H.G. 1981. Forest fertilization: some guiding concepts. Forestry 54:157-167.

Nadelhoffer, K.J., M.R. Downs, B. Fry, J.D. Aber, A.H. Magill and J.M. Melillo. 1994. The fate of ^{15}N labeled nitrate additions to a northern hardwood forest in eastern Maine, USA. Oecologia (submitted).

Penner, J.E., C.S. Atherton, J. Dignon, S.J. Ghan, J.J. Walton and S. Hameed. 1991. Tropospheric Nitrogen: A three-dimensional study of sources, distributions, and deposition (Paper 90JD02228). J. Geophys. Res. 96:959-990.

Peterjohn, W.T., J.M. Melillo, P.A. Steudler, K.M. Newkirk, F.P. Bowles and J.D. Aber. 1994. Responses of trace gas fluxes and N availability to experimentally elevated soil temperatures. Ecolog. Appl. 4:617-625.

Peterson, B.J. and J.M. Melillo. 1985. The potential storage of carbon caused by eutrophication of the biosphere. Tellus 37B: 117-127.

Prentice, I.C. W. Cramer, S.P. Harrison, R. Leemans, R.A. Monserud and A.M. Solomon. 1992. Predicting global vegetation patterns from plant physiology and dominance, soil properties and climate. J. Biogeography 19:117-134.

Raich, J.W., E.B. Rastetter, J.M. Melillo, D.W. Kicklighter, P.A. Steudler, B.J. Peterson, A.L. Grace, B. Moore III and C.J. Vorosmarty. 1991. Potential net primary productivity in South America: Application of a global model. Ecol. Appl. 1:399-429.

Sanchez, P.A., D.E. Bandy, H. Villachica and J.J. Nicholaides. 1982. Amazon basin soils: management for continuous crop production. Science 216:821-827.

Schimel, D., I. Enting, M. Heimann, T. Wigley, D. Raynaud, D. Alves and U. Siegenthaler. 1994. The carbon cycle, IPCC Interim Assessment.

Schindler, D.W. and S.E. Bayley. 1993. The biosphere as an increasing sink for atmospheric carbon: estimates from increased nitrogen deposition. Global Biogoeochemical Cycles 7:717-734.

Schulze, E.D. 1989. Air pollution and forest decline in a spruce (Picea abies) forest. Science 244:776-783.

Skole, D.L. and C. Tucker. 1993. Tropical deforestation and habitat fragmentation in the Amazon: Satellite data from 1978-1988. Science 260:1905-1910.

Spalding, R.F. and M.E. Exner. 1993. Occurrence of nitrate in groundwater - A review. J. Environ. Qual. 22:392-402.

Stafford, L.O. and S.M. Filip. 1974. Biomass and nutrient content of 4-year-old fertilized and unfertilized hardwood stands. Can. J. For. Res. 4:549-554.

Swift, M.J., O.W. Heal and J.M. Anderson. 1979. Decomposition in Terrestrial Ecosystems. University of California Press, Berkeley, CA.

Tamm, C.O. 1990. Nitrogen in Terrestrial Ecosystems, 116 pp. Springer-Verlag, New York.

Townsend, A.R., B.H. Braswell, E.A. Holland and J.E. Penner. 1994. Nitrogen deposition and terrestrial carbon storage: Linking atmospheric chemistry and the global carbon budget. Geophys. Res. Letters (submitted).

Watson, R.T., L.G. Meira Filho, E. Sanhueza and A. Janetos. 1992. Sources and sinks, 25-46. In: J.T. Houghton, B.A. Callander and S.K. Varney (eds.) Climate Change 1992: The Supplementary Report to the IPCC Scientific Assessment. Cambridge University Press, Cambridge, UK.

TOWARD GLOBAL PLANNING OF SUSTAINABLE USE OF THE EARTH
Development of Global Eco-Engineering
S. Murai (Editor)

Carbon Dynamics Model for Terrestrial Ecosystems

Motoyuki Suzuki and Naohiro Goto*

Institute of Industrial Science, University of Tokyo, 7-22-1 Roppongi, Minato-ku, Tokyo, 106, Japan

* Present address: Research Institute of Innovative Technology for the Earth, 9-2, Kizugawadai, Kizu-cho, Soraku-gun, Kyoto, 619-02, Japan

Abstract

The recent increase of atmospheric carbon dioxide is mainly attributed to the disturbances of global carbon dynamics caused by anthropogenic activities including excessive usage of fossil fuel and deforestation. The increase of carbon dioxide will bring about climate change in the global system. Also the climate change itself may affect the carbon dynamics cycles in terrestrial ecosystems through change of respiration rates and photosynthetic rates of vegetation as well as fertilization effects due to increase of carbon dioxide.

To make clear the overall scheme of the dynamics of carbon storage and carbon exchange between atmosphere and terrestrial ecosystems, a carbon dynamics model is proposed which is, in principle, a substantial model based on mass balance and kinetics of carbon in terrestrial vegetation systems. Vegetation was divided into five typical types and the growth rate of each vegetation was described in terms of local soil types, solar densities and precipitation as well as temperature. The terrestrial surface was divided into meshes of latitude 1° x longitude 1°. Calculations were carried out for all the meshes and for each mesh, a dominant type of vegetation judged from the comparison of the accumulated amount of biomass for each vegetation type. This procedure provides the vegetation map based on the present geographical information. The model could simulate the present vegetation map with reasonable accuracy and it can be used to predict the dynamic changes of vegetation when climate conditions alter.

1. INTRODUCTION

Increase of atmospheric carbon dioxide concentration is caused by anthropogenic activities such as burning fossil fuel and deforestation. These activities disturb natural carbon dynamics and then the subsequent increase of atmospheric carbon dioxide concentration is considered to be a main cause of anticipated global warming in the future. Numbers of meteorological works[1] have pointed out that serious climate changes may occur when the atmospheric carbon dioxide concentration doubles. For the prediction of transient behavior of carbon dioxide in the atmosphere, carbon exchanges between the atmosphere and the terrestrial vegetation and between the atmosphere and the ocean are to be clearly defined. The carbon dynamics in terrestrial ecosystems, however, are expected to be affected by the climate change. Hence, in order to estimate carbon exchanges between the atmosphere and the terrestrial ecosystems and to assess the impact of climate changes on land vegetation, it is necessary to describe the holistic picture of carbon dynamics in the terrestrial ecosystems by means of a mathematical model that accounts for the dominant factors in ecosystems.

There have been mathematical models so far, which describe global carbon dynamics including carbon dynamics in ocean[2-9]. These models provided quantitative estimates of carbon dynamics in the terrestrial ecosystems, carbon storage and net primary production (NPP) They also predicted their changes when climate conditions alter. Also, Gate proposed a model that shows that global vegetation distribution may change by the global warming[10]. One of the serious impacts of climate changes is the change of vegetation in the terrestrial ecosystems, which eventually affects global carbon storage and NPP. Therefore it is necessary to construct the model that can express global carbon dynamics and vegetation distribution at the same time. However most of the previous models do not include the change of global vegetation distribution when climate conditions alter.

Global vegetation distribution is determined by a number of environmental factors. Especially climatic factors, such as temperature, precipitation and so on, play an important role. Numbers of works have been carried out on the relationship between vegetation distribution and climatic conditions. These works were mainly based on statistic or regressive methods[11,12,13,14,15]. These methods can predict the change of the global vegetation distribution from the climate change by using regressive relations between types of vegetation and climate conditions. Global carbon storage or NPP, however, cannot be assessed only by this method. Smith and Shugart[16] used the climate-vegetation classification scheme proposed by Holdridge[17] to estimate the effect of climate change. Holdridge scheme is suitable for assessing a global vegetation distribution from the climate condition, but not for the estimation of carbon storage. Then they used other relationships between carbon storage and vegetation[18] and between soil carbon storage and Holdridge scheme[19] to predict the changes of global carbon storage. The method that employs the relationship between vegetation and climate condition can assess changes of global carbon dynamics. It cannot be used, however, to predict the process of changes because it does not include the concept of time.

A general carbon dynamics model is suitable for the assessment of the process. In this paper, a carbon dynamics model is proposed which makes it possible to estimate not only global carbon dynamics but also global vegetation distribution.

2. MODEL CONCEPT

2.1 Carbon Dynamics Model

A carbon dynamics model in terrestrial ecosystems is based on the simplified material balance concept as shown in Fig.1. Terrestrial ecosystem is composed of several types of vegetation. Each vegetation is divided into two compartments and subsequent four sub compartments as follows:

(I) Biomass compartment composed of living cells

(I-a) Leaf where photosynthesis occurs

(I-b) Trunk

(II) Soil Organics compartment composed of dead cells

(II-a) Dead Biomass which has high decomposition rate

(II-b) Humus which has rather low decomposition rate

Carbon exchanges kinetics among these four sub compartments and the atmosphere compartment[20,21] are described. Detailed equations of the model are shown in Appendix 1.

2.2 Vegetation Distribution Model

In the present model, only three vegetation types are considered: Tropical Forest, Temperate Forest, and Boreal Forest. Each vegetation has different kinetic parameters defined in equations in Appendix 1 as shown in Appendix 2. A dominant vegetation at a certain point is determined from the climate parameters, such as temperature, precipitation, solar intensity, soil characteristics, etc. As a first approximate, the growth rates of vegetation are considered here to be the determining factor.

For calculation of distribution on the earth, the global land area is divided into meshes, each of which is composed of latitude 1° x longitude 1°. In all meshes, calculations were carried out for the Biomass carbon storage ($=C_1+C_2$; see Appendix 1) of the three vegetation types as shown in Fig. 2 based on the climate condition and soil distribution data for each mesh until reaching the steady state.

By comparing the calculated carbon storage of the three vegetation types in one mesh, we assumed that the vegetation with the maximum carbon storage is the vegetation that dominates the mesh. If the carbon storage of the dominant vegetation is less than 0.5 (kgC/m^2), the mesh is classified into No Vegetation and if less than 5.0 (kgC/m^2) and above 0.5 (kgC/m^2) the mesh is classified into Grassland. This judgment is executed in all meshes and the global vegetation map can be provided.

3. ESTIMATION OF PRESENT TERRESTRIAL ECOSYSTEMS

3.1 Present Terrestrial Ecosystems

For model validation we calculated global vegetation distribution, carbon storage and NPP ($=R_1{}^{P}-R_2{}^{R}$; see Appendix 1). Calculation was done by using the current climate condition data and soil distribution data (from EPA[22]) until reaching the steady state. As the initial conditions, carbon storage of

Leaf and that of Trunk are set to be 0.01 (kgC/m^2). Fig. 3 shows the current global vegetation map from the model calculation. Apparently, this model can represent well the global vegetation distribution.

Tables 1-(a) and (b) show the comparison of global vegetation area, carbon storage and NPP from several sources. Vegetation classifications shown in the previous sources differ from each other but they were modified into the classification of this work since it gives most simplified definition.

The vegetation areas of Holdridge[17], Olson[23] and Honda et al.[24] are estimated from the global vegetation map. Holdridge's vegetation map is derived from Holdridge Life Zone scheme that describes the relationship between climate condition and global vegetation. Olson's vegetation data set is derived from investigation by airplane and biomass measurement in objective places. Honda et al. made a vegetation map from satellite data. Estimation of Whittaker and Likens[25] is not derived from vegetation map but from collecting local measurement data.

Table 1-(a) Comparison of global vegetation area (10^6 km^2)

	Tropical forest	Temperate forest	Boreal forest	Grass-land	No Vege-tation	Total
This Work	33.6	10.5	10.6	31.6	57.7	136.1
Holdridge[17]	45.2	15.2	8.1	21.3	36.3	136.0
Olson[23]	28.2	18.5	16.3	39.1	33.7	135.8
Honda et al.[24]	8.0	37.7		65.5	22.8	134.5
Whittaker and Likens[25]	22.5	12	12	47	50	143.5

Grassland area of each estimation is different remarkably from each other as shown in Table 1-(a). Two major reasons for this difference can be pointed out. One is the difference in definition in vegetation classification. This work classified global vegetation into 5 types, but the others were based on more detailed classifications. Holdridge, Olson and Whittaker and Likens originally classified vegetation into 37 types, 44 types and 14 types respectively. Grassland collected from their definitions contains Tropical Grassland or Boreal one, wet Grassland or dry Grassland, and Grassland that has high carbon storage (savanna, etc.) or low carbon storage (desert shrub, etc.). It might be difficult to represent all these various vegetation types by a single category as Grassland.

The other reason is the difference in the meaning of a map. The map derived in this work and Holdridge's one are drawn only on the basis of environmental and climatic data. The results, therefore, do not take into account the effects of human activities, such as agriculture and deforestation.

Hence, the map derived in this study is considered to give potential vegetation distributions on the ground. On the other hand, the maps drawn by Olson and Honda et al. are based on actual vegetation on the ground.

Table 1-(b) Comparison of global carbon storage (GtC) and NPP (GtC/yr)

	Tropical forest	Temperate forest	Boreal forest	Grass-land	No Vege-tation	Total
Carbon Storage						
This Work	219.4	77.7	85.4	88.5	1.3	472.5
Whittaker and Likens	460	175	108	62.3	8	813.3
NPP						
This Work	30.2	4.6	3.9	11.0	0.2	49.8
Whittaker and Likens	20.4	6.7	4.3	13	1.14	45.54

Estimation of Whittaker and Likens, though they do not give in the form of map, falls in the same category as Olson's since it is based on the direct measurement. The difference between the potential map and the actual one is emphasized when the area of Grassland is estimated. According to the definition of vegetation in this study, actual agricultural land is included in Grassland. The climate conditions in the agriculture area, however, is generally moderate and suitable for plant growth and the area may be originally existent as Forest (Tropical, Temperate or Boreal forest), but by human efforts it has been changed to the area of agriculture.

It is noteworthy from Table 1-(b) that the carbon storage calculated from this model is less than that from Whittaker and Likens while this model gives about the same order of magnitude of NPP with that from Whittaker and Likens. This means that this model may overestimate the amount of litterfall rate. NPP means difference between the fixation rate of carbon by photosynthesis and the release rate by respiration while the difference between NPP and the litterfall rate is accumulated in the form of Biomass. Then this model may somewhat underestimate the carbon storage.

3.2 Model Structure Sensitivity

It is desirable to clarify a sensitivity of each parameter included in the model equation on global carbon dynamics and vegetation distribution, since this model employs number of parameters for calculations. Also functional forms to describe effects of climatic conditions might affect the final results as well. For prediction of the change of the global vegetation, the most significant points are those to describe the dependence of growth rates on temperature and precipitation (soil water).

A simplified examination is made here to check the effects of temperature and precipitation. In the growth equation shown in Appendix 1, $\eta_{T\text{-}P}$ and $\eta_{W\text{-}P}$ respectively represent dependence of photosynthesis rate on temperature and soil water content in the form of efficiencies; namely at ideal conditions, η's become unity.

Three extreme cases are compared with the original model calculation to see model responses of steady state vegetation distribution for the present climate conditions: Case (1) is the original calculation where both temperature and precipitation dependence are taken into account by η_{T-P} and η_{W-P}, Case (2) shows the case where η_{W-P} = 1 where only effect of temperature on growth rate is considered, Case (3) corresponds to the case of η_{T-P} = 1, where only effect of precipitation is considered, and Case (4) where $\eta_{T-P} = \eta_{W-P} = 1$ where effects of temperature and precipitation are set to the maximum. Figs. 4-(a)-(c) show the global vegetation area, carbon storage and NPP for these cases at present climate conditions. From Fig. 4-(b) the total carbon storage is more than 1500 [GtC] for Case (4), while it is about 500 [GtC] for Case (1). Obviously, the big difference is caused by neglecting the effects of temperature and precipitation in Case (4). From Fig. 4-(a) global vegetation is divided into Tropical Forest, Grassland and No Vegetation in Case (4), because the limitation factor in this case is only solar radiation or daytime length. From Fig. 4-(a) global vegetation is divided into Tropical forest, Grassland and No vegetation in Cases (3) and (4)while it is divided into five types in Cases (1) and (2). From these results, clearly the precipitation condition influences the formation of Forest or non-Forest (Grassland and No Vegetation) and the temperature condition influences the area ratio of Tropical, Temperate and Boreal Forest.

Budyko[26] used two indexes for representing global vegetation zone: radiative dryness index and net radiation. As the radiative dryness index increase, vegetation changes from forest, steppe, semi desert to desert. On the other hand, as the net radiation increase, forest becomes Tropical Forest from Temperate Forest. In the present definition, temperature and precipitation, respectively, correspond loosely to net radiation and to radiative dryness index defined by Budyko. As stated earlier, Fig. 4-(a) shows that Forest types depend on temperature in this work and precipitation condition determines the relative area of Forest and Grassland. This is in good agreement with the tendency given by Budyko, suggesting the properness of the model structure in the present work.

4. PREDICTION OF CHANGES IN TERRESTRIAL ECOSYSTEMS

4.1 Equilibrium Prediction

The global carbon dynamics and vegetation distribution in future climate condition were predicted by this model. Three climate change scenarios of GCM were used (Houghton et al.[1]): Geophysical Fluids Dynamics Laboratory (GFHI), Canadian Climate Center (CCC) and United Kingdom Meteorological Office (UKHI). The initial conditions employed for the calculation are that the carbon storage of Leaf and Trunk are the steady state values in current climate condition as shown in Fig. 3. Fig. 5 shows the change of global vegetation distribution derived from GFHI GCM scenario. As shown in Fig. 5 the area of Forest declines and the area of Grassland and No Vegetation increases. Figs. 6-(a)-(c) show the change of global carbon dynamics and vegetation distribution by using the three GCM scenarios. By the climate changes it is anticipated that the area of forest will decline and the area of

Grassland and No Vegetation will increase, especially Temperate Forest and Boreal Forest will lose their areas. These trends are the same in the three GCM scenarios.

4.2 Non-Equilibrium Prediction

In the previous section the results of GCM were used as the climate conditions and the steady state calculations were made. Actual climate change is expected to occur gradually from the current climate condition since carbon dioxide in atmosphere is expected to increase year by year. It is reasonable to assume that it takes 100 years to reach doubling of carbon dioxide concentration. Then the climate can be assumed to change linearly and reach the GCM output after a hundred years. Change of vegetation during the gradual climate change can be made by using the current kinetic model.

Figs. 7-(a)-(c) show the change of global vegetation area, carbon storage and NPP in this transient scenario that eventually reaches GFHI scenario after a hundred years. From Fig. 7-(a), it is noteworthy that the area of Grassland decreases and changes to No Vegetation in the first 10 years. After 10 years, however, Grassland increases again. Forest areas gradually decrease and are changed to be Grassland area after the 10 years, which later is changed to No Vegetation. As a result, Grassland area is in dynamic equilibrium and seems to stay almost constant after fifty years. It can be pointed out that climate change first affects Grassland at the first stage of this scenario and later Forest decline to Grassland, which eventually is changed to No Vegetation.

Figs. 7-(b) and (c) also show total carbon storage and NPP derived from GFHI scenario in equilibrium condition as shown in Figs. 6-(b) and (c). Both calculations show carbon storage and NPP decline to the same degree.

5. CONCLUSION

Carbon dynamic model was proposed to describe carbon storage, net primary production and vegetation distribution. Carbon balance equation and growth kinetic equations for three types of forests were provided accounting for photosynthesis rate, litterfall rates and respiration rate of each forest type. Type of vegetation is then determined from the by comparing the calculations for the three forest type sub models with different parameters on the basis of local environmental conditions such as solar radiation, temperature, and soil water condition as a results of precipitation. The global vegetation distribution and carbon dynamics in the current climate condition were described well by this model.

By using this model, prediction was made on the change of the global vegetation distribution, carbon storage and NPP for doubling carbon dioxide climate condition by using the steady state assumption and by using transient.

On the basis of the results of the model calculation, it is anticipated that Forest area declines when the carbon dioxide concentration increases. The most significant reason for such results might be the temperature change caused by the greenhouse effect.

In the present work, special attention was paid upon the vegetation distribution and carbon storage in the terrestrial ecosystems. By using the present model, effects of environmental conditions on the carbon exchange rate between atmosphere and the terrestrial vegetation can be quantitatively estimated. For the future prediction of atmospheric carbon dioxide concentrations, the model can be employed for the precise estimation by accounting for the fertilizing effect and the temperature effect on the exchange rate.

6. REFERENCES

1 J.T. Houghton, G.J. Jenkins and J.J. Ephraums (Eds.): *Climate Change*; The IPCC Scientific Assessment, WMO/UNEP, Cambridge Univ. Press, Cambridge, 1990.
2 C.M. Gowdy, R.J. Mulholland and W.R. Emanuel, *Int. J. Systems Sci.*, 6 (1975) 965.
3 W.R. Emanuel, G.G. Killough, M.M. Post and H.H. Shugart, *Ecology*, 65 (1984) 970.
4 J. Goudriaan and P. Kenter, *Climatic Change*, 6 (1984) 167.
5 G. Esser, *Tellus*, 39B (1987) 245.
6 L.D.D. Harvey, *Climatic Change*, 15 (1989) 343.
7 V.F. Krapivin and L.P. Vilkova, *Ecological Modelling*, 50 (1990) 57.
8 V.F. Krapivin, *Ecological Modelling*, 67 (1993) 103.
9 J.D. Guo and G.Y. Shi, *Proceedings of the International Symposium on Global Cycles of Atmospheric greenhouse Gases*, Sendai, Japan, (1994) 265.
10 D.M. Gate, *Climate Changes and its Biological Consequences*, Sinauer Associates, Inc., Massachusetts, 1993.
11 R.A. Monserud and R. Leemans, *Ecological Modelling*, 62 (1992) 275.
12 R.A. Monserud, N.M.Tchebakova and R. Leemans, *Climatic Change*, 25 (1993) 59.
13 I.C. Prentice, W. Cramer, S.P. Harrison, R. Leemans, R.A. Monserud and A.M. Solomon, *Journal of Biogeography*, 19 (1992) 117.
14 K.C. Prentice, *J. Geophysical Res.*, 95, D8 (1990) 11811.
15 P.J. Guetter and J.E. Kutzbach, *Climatic Change*, 16 (1990) 193.
16 T.M. Smith and H.H. Shugart, *Water, Air and Soil Pollution*, 70 (1993) 629.
17 L.R. Holdridge, *Science*, 105 (1947) 367.
18 J.S. Olson, J.A. Watts, and L.J. Allison, *Carbon in Live Vegetation of Major World Ecosystems*, ESD Pub. No. 1997, Oak Ridge National Laboratory, TN. ,1983.
19 W.M.Post, W.R. Emanuel, P.J. Zinke and A.G. Stangenberger, *Nature*, 298(1982) 156.
20 M.Suzuki, N. Goto and A. Sakoda, *Ecological Modelling*, 70 (1994) 161.
21 N.Goto, A. Sakoda and M. Suzuki, *Ecological Modelling*, (1994) in press
22 EPA, *Global Ecosystems Database*, EPA Contract No. DW13934786-01-0, 1992.
23 J.S.Olson, Carbon Dioxide Information Center, Oak Ridge National Laboratory, Oak Ridge, NDP-017, 1983.
24 Y. Honda, S. Murai, E.O. Box and S. Goto, *Report of the Institute of Industrial Science, The University of Tokyo*, 38, 1993.
25 R.H.Whittaker and G.E.Likens, *Carbon and Biosphere*, U.S.Atomic Energy Commission, 281, 1973.
26 M.I. Budyko, *Climate and Life*, Academic Press, New York, 1971.

Appendix 1

Carbon balance equations

LEAF	$\frac{dC_1}{dt} = \alpha R_1^{\mathrm{P}} - R_1^{\mathrm{L}}$
TRUNK	$\frac{dC_2}{dt} = (1-\alpha) R_1^{\mathrm{P}} - R_2^{\mathrm{R}} - R_2^{\mathrm{L}}$

C_1	carbon storage in LEuF	[kgC/m^2]
C_2	carbon storage in TRUNK	[kgC/m^2]
t	time	[day]
R_1^{P}	photosynthesis rate	[kgC/m^2/day]
R_1^{L}	litterfall rate of LEuF	[kgC/m^2/day]
R_2^{R}	respiration rate of TRUNK	[kgC/m^2/day]
R_2^{L}	litterfall rate of TRUNK	[kgC/m^2/day]
α	allocation ratio for LEuF	[-]

Carbon flux equations

photosynthesis rate	$R_1^{\mathrm{P}} = \eta_{\mathrm{T-P}} \eta_{\mathrm{W-P}} \eta_{\mathrm{CO-P}} r_1^{\mathrm{P}}$
litterfall rate of LEAF	$R_1^{\mathrm{L}} = k_1^{\mathrm{L}} C_1$
respiration rate of TRUNK	$R_2^{\mathrm{P}} = \eta_{\mathrm{T-R}} k_2^{\mathrm{R}} C_2$
litterfall rate of TRUNK	$R_2^{\mathrm{L}} = k_2^{\mathrm{L}} C_2$

r_1^{P}	basic photosynthesis rate	[kgC/m^2/day]

$$r_1^{\mathrm{P}} = D_L \frac{p_0}{K} \ln\left[\frac{1+(b/p_0)KI_m}{1+(b/p_0)KI_m \exp(-KA_L C_1)}\right]$$

A_{L}	specific leaf area	[m^2/kgC]
b	initial tangent of photosynthesis–light intensity relation	[kgC/m^2/hr/klux]
D_{L}	day time length in a day	[hr/day]
K	attenuation coefficient of light	[-]
p_0	photosynthesis rate at saturated light intensit	[kgC/m^2/hr/klux]
I_{m}	illumination intensity	[klux]

$$I_m = \frac{1}{10.5}\frac{\Sigma S}{D_L} \qquad S = S_0(1-C_d)\sin\beta$$

S	solar radiation	[W/m^2]	S_0	solar constant	(=1.38x10^3) [W/m^2]
Cd	cloud density	[%]	β	elevation angle	[°]

k_2^R	reapiration rate constant of TRUNK	[1/day]
k_1^L	litterfall rate constant of LEuF	[1/day]
k_2^L	litterfall rate constant of TRUNK	[1/day]

η_{T-P}	photosynthesis efficiency depending on temperature	[-]
η_{W-P}	photosynthesis efficiency depending on water condition	[-]
η_{CO-P}	photosynthesis efficiency depending on CO_2 concentration	[-]
η_{T-R}	respiration efficiency depending on temperature	[-]

$$\eta_{T-P} = \begin{cases} 0 & T \le T_{opt} - \frac{\Delta T}{2}, T \ge T_{opt} + \frac{\Delta T}{2} \\ \cos\left(\frac{T-T_{opt}}{\Delta T}\pi\right) & T_{opt} - \frac{\Delta T}{2} \le T \le T_{opt} + \frac{\Delta T}{2} \end{cases}$$

$$\eta_{W-P} = \begin{cases} 0 & W \le W_{wp} \\ \left(\frac{1}{W_{fc}-W_{wp}}\right)(W-W_{wp}) & W \ge W_{wp} \end{cases}$$

$$\eta_{CO-P} = \frac{R_2^R + (R_{1,0}^R - R_2^R)(1+0.5\ \ln(CO_2\ /CO_2^0))}{R_{1,0}^R}$$

$$\eta_{T-R} = 2^{((T-20)/10)}$$

T	temperature	[°C]
T_{opt}	optimum temperature for photosynthesis	[°C]
ΔT	temperature range for active photosynthesis	[°C]
W	soil water content	[cm]
W_{wp}	wilting point	[cm]
W_{fc}	field capacity	[cm]
$R_{1,0}^P$	photosynthesis rate without effect of CO_2 concentration	[kgC/m^2/day]
CO_2	carbon dioxide concentration	[ppm]
CO_2^0	present carbon dioxide concentration	[ppm]

Appendix 2

Vegetation parameters

	Tropical Forest	Temperate Forest	Boreal Forest
p_0	1.2×10^{-3}	6.0×10^{-4}	3.0×10^{-4}
b	5.8×10^{-5}	5.8×10^{-5}	5.8×10^{-5}
K	0.40	0.45	0.30
A_L	110	110	180
α	0.90	0.95	0.95
k_2^R	8.0×10^{-3}	8.4×10^{-3}	9.0×10^{-3}
k_1^L	7.0×10^{-4}	7.0×10^{-4}	4.6×10^{-4}
k_2^L	7.0×10^{-5}	7.0×10^{-5}	6.0×10^{-5}
T_{opt}	25	15	9
ΔT	50	30	18

Soil parameters

	W_{fc}	W_{wp}		W_{fc}	W_{wp}
heavy clay	32	17	silty clay	32	16
clay loam	31	15	silty clay loam	31	15
light clay	31	14	silty loam	29	12
sandy clay	27	12	loam	27	10
sandy clay loam	22	7.5	sandy loam	16	5
loamy sand	12	3.5	sand	7.5	2.5

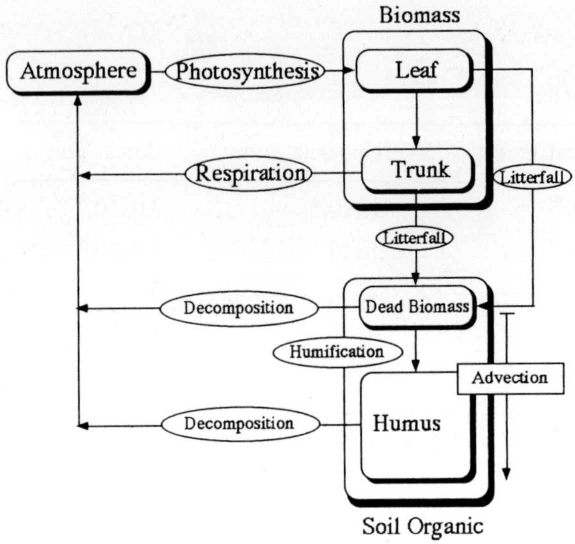

Carbon balance equations

LEAF	$\frac{dC_1}{dt}=\alpha R_1^{\mathrm{P}}-R_1^{\mathrm{L}}$
TRUNK	$\frac{dC_2}{dt}=(1-\alpha)R_1^{\mathrm{P}}-R_2^{\mathrm{R}}-R_2^{\mathrm{L}}$
DEAD BIOMASS	$\frac{\partial C_3}{\partial t}=F\frac{\partial C_3}{\partial z}+R_1^{\mathrm{L}}+R_2^{\mathrm{L}}\ (z=0)$
	$\frac{\partial C_3}{\partial t}=F\frac{\partial C_3}{\partial z}-R_3^{\mathrm{H}}-R_3^{\mathrm{D}}\ (z>0)$
HUMUS	$\frac{\partial C_4}{\partial t}=F\frac{\partial C_4}{\partial z}+R_3^{\mathrm{H}}-R_4^{\mathrm{D}}$

C_1	carbon storage in LEAF	[kgC/m^2]	C_2	carbon storage in TRUNK	[kgC/m^2]
C_3	carbon storage in DEAD BIOMASS	[kgC/m^2]	C_4	carbon storage in HUMUS	[kgC/m^2]
R_1^{P}	photosynthesis rate	[kgC/m^2/day]	R_1^{L}	litterfall rate of LEAF	[kgC/m^2/day]
R_2^{R}	respiration rate	[kgC/m^2/day]	R_2^{L}	litterfall rate of TRUNK	[kgC/m^2/day]
R_3^{D}	decomposition rate of DEAD BIOMASS	[kgC/m^2/day]	R_3^{H}	humification rate	[kgC/m^2/day]
R_4^{D}	decomposition rate of HUMUS	[kgC/m^2/day]	α	allocation ratio for LEAF	[−]
t	time[day] z	soil depth[m]	F	advection rate[m/day]	

Fig.1 Model framework and balance equations

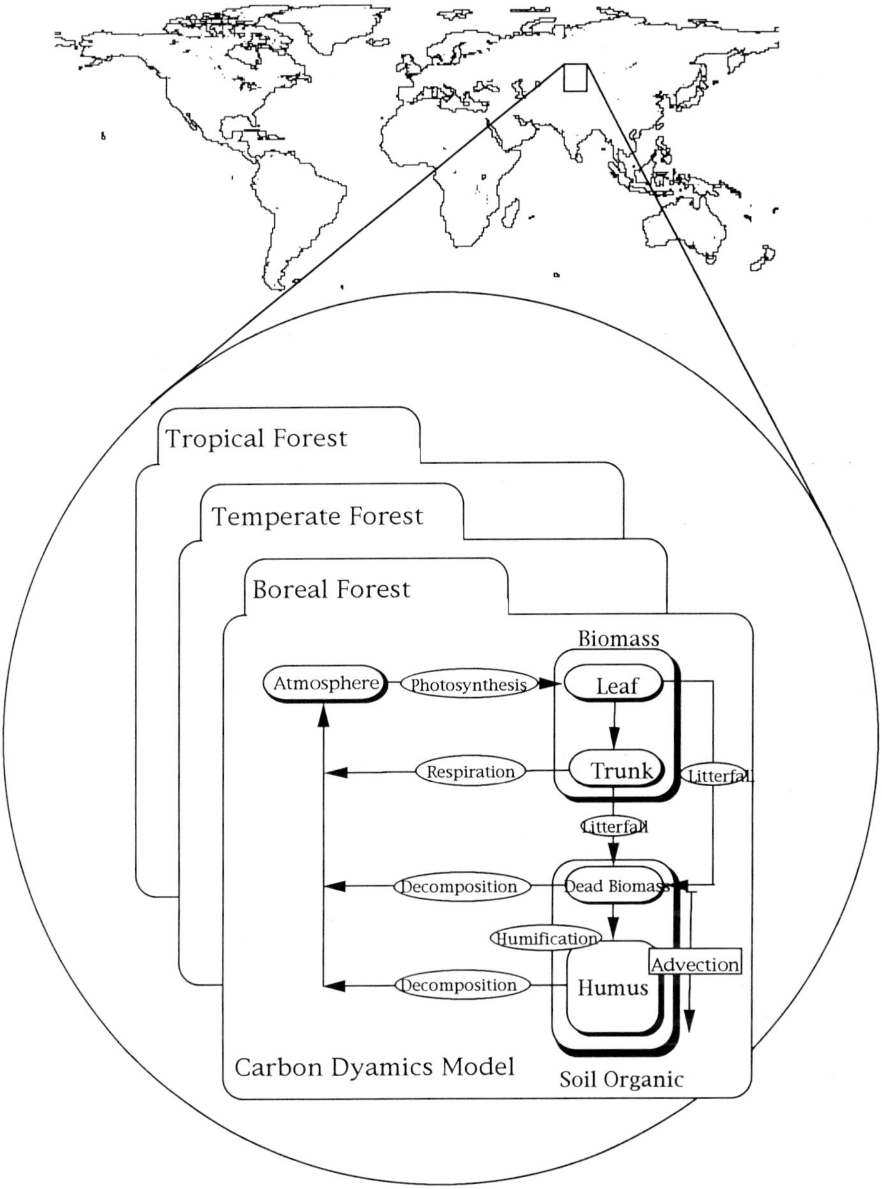

Fig. 2 Model concept: calculations were done for three types of forest at each $1^\circ \times 1^\circ$ mesh and vegetations type was selected

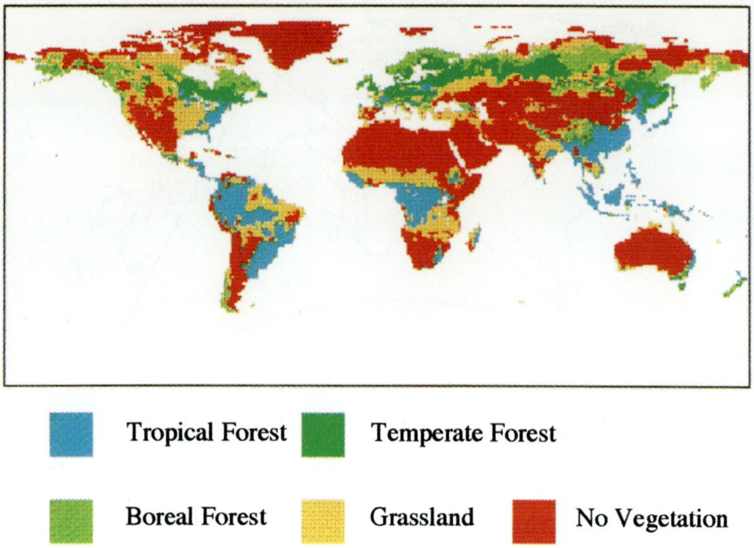

Fig.3 Global vegetation distribution derived from current climate conditio

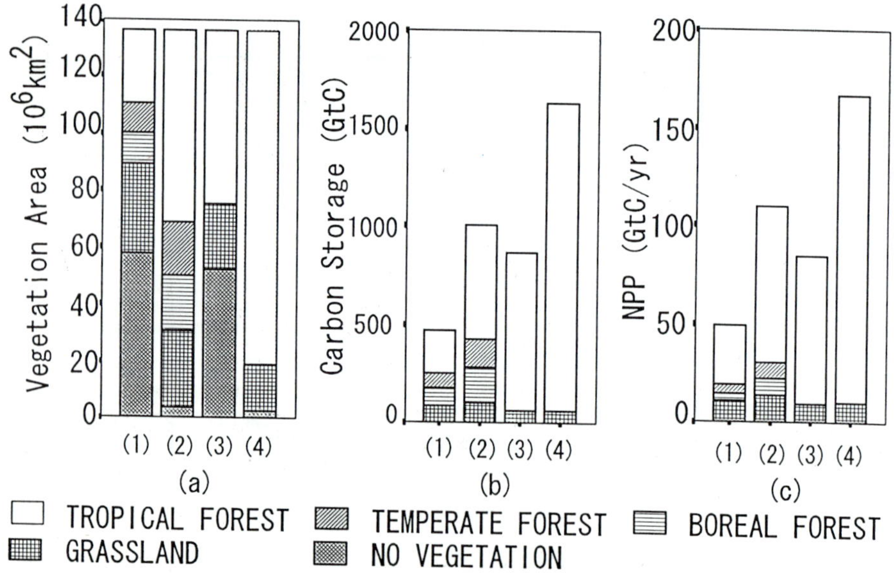

Fig. 4 Effect of environmental conditions on (a)Vagetation area, (b) Carbon storage, and (c) NPP for Case (1): original calculation, Case (2):considering only effect of temperature on growth rate (η_{W-P}=1), Case (3): consideringonly effect of precipitation on growth ate (η_{T-P}=1), and Case (4):effects of temperature and precipitation are set to the maximum (η_{W-P}=1 ,η_{T-P}=1)

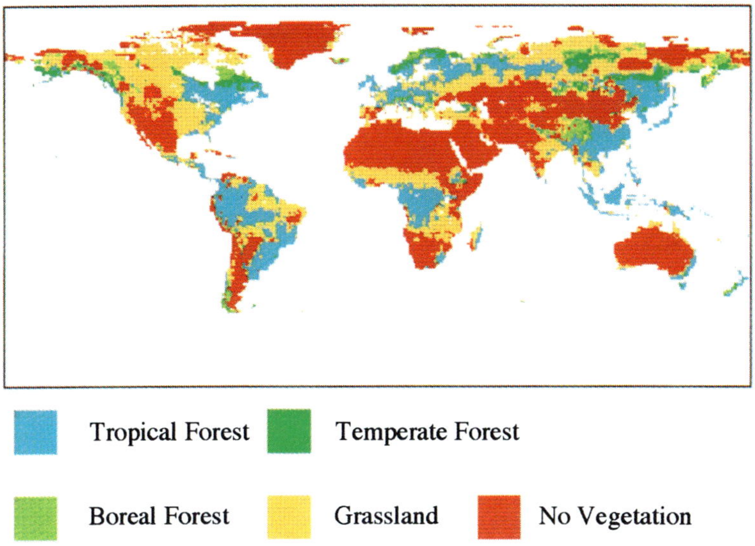

Fig.5 Global vegetation distribution derived from GFHI scenario

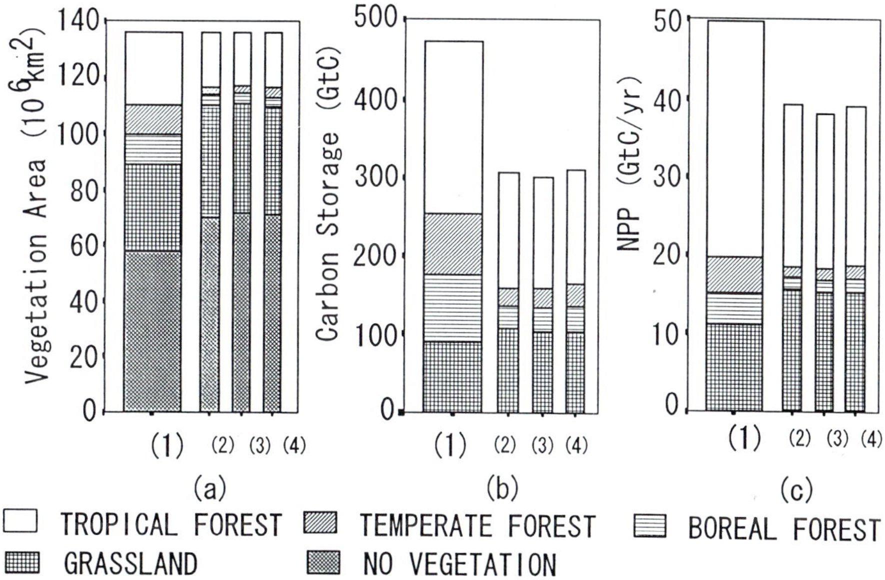

Fig. 6 Changes of (a) Vegetation area, (b) Carbon storage, and (c) NPP derived for (1) present climate condition, (2) GFHI scenario, (3) UKHI scenario, and (4) CCC scenario

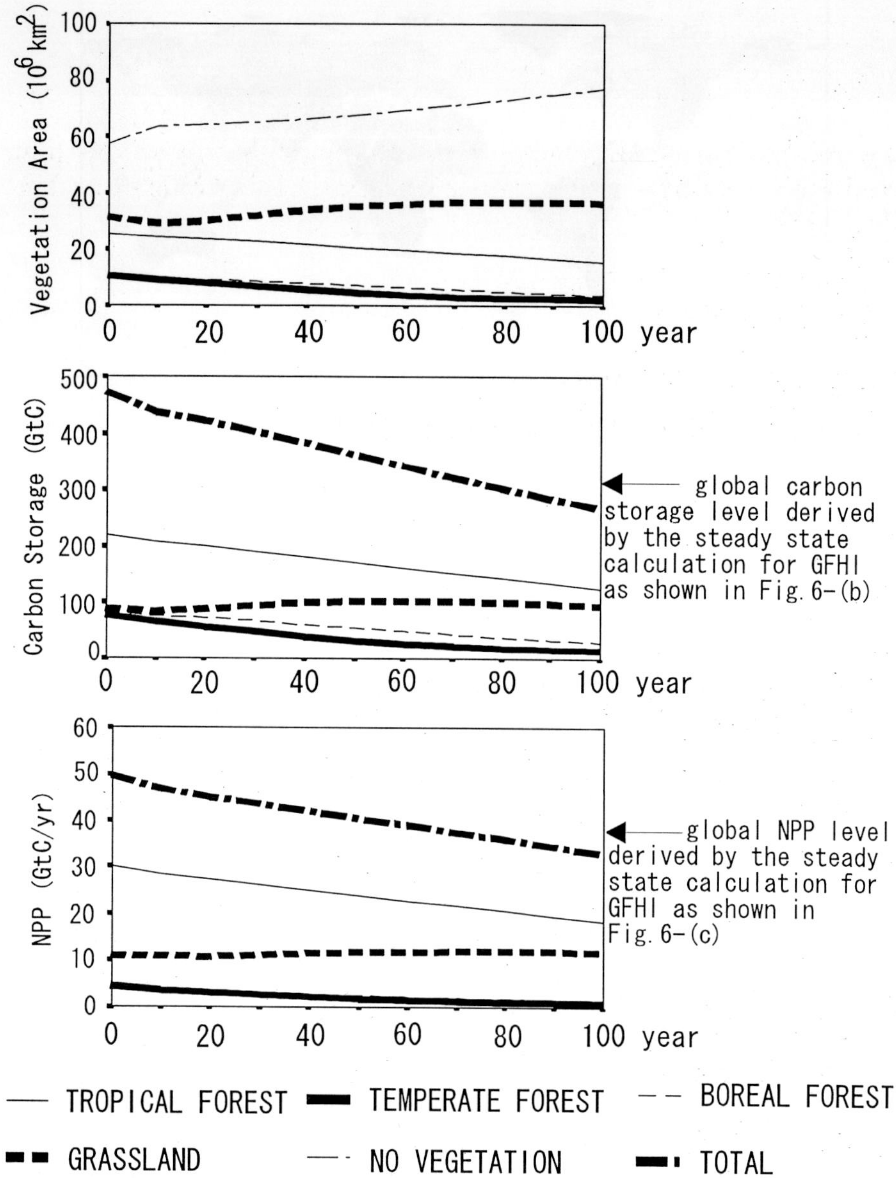

Fig. 7 Changes of (a) Vegetation area, (b) carbon storage, and (c) NPP when climate conditions linearly move from the current climate conditions to the GFHI results in a hundred years

TOWARD GLOBAL PLANNING OF SUSTAINABLE USE OF THE EARTH
Development of Global Eco-Engineering
S. Murai (Editor)

Asian-Pacific Integrated Model for Evaluating Policy Options to Reduce Greenhouse Gas Emissions and Global Warming Impacts

M. Kainuma[a], Y. Matsuoka[b], T. Morita[a],
K. Kai[a], H. Harasawa[a] and D. Lee[a]

[a] National Institute for Environmental Studies, Tsukuba, 305 Japan

[b] Faculty of Engineering, Kyoto University, Kyoto, 606-01 Japan

Abstract

The objectives of this study are to develop the Asian-Pacific Integrated Model (AIM), so that policy options for stabilizing global climate, particularly in the Asian-Pacific region, can be assessed from the two perspectives of reducing greenhouse gas emissions and avoiding the impacts of climate change. AIM is an integrated 'top-down, bottom-up' model with regional models and a major global model.

The three linked models are an emission model, a climate model, and an impact model. The emission model combines an end-use energy model and a technological selection model. More than 100 technologies are evaluated for their potential to improve energy efficiency, and energy demand estimates are linked to a top-down economic model. One component of the emission model is the forest resources alternation model to estimate the global greenhouse gas emissions from land-use changes. The climate model was created by developing original linkages to join other established models. The impact model has a spatial water balance model, an ecological matching model and a malaria distribution model. The impact models were used to estimate the increased risks of droughts, floods, vegetation changes and malaria. Other related models are being developed.

1. INTRODUCTION

It is predicted that global warming will have a significant impact on the society and economy of the Asian-Pacific region, and that adoption of measures to tackle global warming will force the region to carry a very large economic burden. Also, if the Asian-Pacific region fails to adopt such countermeasures, it has been estimated that its greenhouse gas emissions will increase to become half of all global emissions by the end of the next century.

The global warming issue has been recognized as one of the most important policy issues to be solved for the region's development. Japan's role in the region has increased through greater contributions in the fields of ODA, technology transfer, researches and joint implementation of countermeasures.

In order to promote adoption of countermeasures, it is necessary to predict precisely the greenhouse gas emissions in the region and also the impact of global warming. The effects of countermeasures on emission reduction and impact abatement also need to be determined, taking into account international cooperative efforts. Such predictions and analyses require an integrated simulation model for the region. The role of this study is to establish that integrated model.

There are only three such models in the world; the GCAM model developed by the Battelle-Pacific Northwest Laboratory in USA (Edmonds *et al.*, 1993), the IMAGE2 model by the National Institute for Public Health and Environment (RIVM) (Alcamo *et al.*, 1993) and our AIM model. Other models such as the MIT model by Massachusetts Institute of Technology (MIT, 1994) are also in the process of being developed.

The AIM model integrates an emission, a climate and an impact models and their component processes, as well as preparing country modules and a global integrated module. It focuses on policy assessment in the Asian-Pacific region and has the ability to prepare regional overviews. A high-quality spatial data-base support these functions. It is a collaborative project, and is being conducted with institutes of various countries in the region. Another characteristics of the AIM model is that it is directly linked to, and takes account of, technological changes. It has a bottom up type energy model and can estimate future energy demand based on demand side energy strategies taking into accounts of various energy technologies. This model can evaluate effects of policy options such as a carbon tax and subsidies. Energy demand estimates are linked to a top-down economic model.

The climate model estimates atmospheric concentrations of greenhouse gases and temperature increase. It was created by developing original linkages to join other established models, such as the box-diffusion oceanic model, the IPCC radiative forcing model, the General Circulation Models (GCMs) for regional climate change and the AMAC model (Prather, 1988) for atmospheric composition. The outputs of the emission model is linked to the impact model through the climate model.

Simulation models to estimate water resources changes, vegetation changes and malaria spread caused by global warming have been developed. Using those scenarios developed by the climate model, we have simulated changes in catchment water run-off, the survival status of vegetation and the reproduction rates of malaria. These will be linked to higher-order impact models to estimate impacts on the regional economy.

2. OUTLINE OF AIM

The AIM comprises three main models - the AIM/emission model for predicting greenhouse gas emissions, the AIM/climate model for estimating global and regional climate change, and the AIM/impact model for estimating the impacts of global warming. Figure 1 shows the relationships between these models (Matsuoka *et al*, 1994, Morita *et al.*, 1993).

The AIM/emission model is comprised of Asian-Pacific country models and a World model which ensures interactions between these regional models are consistent. The major

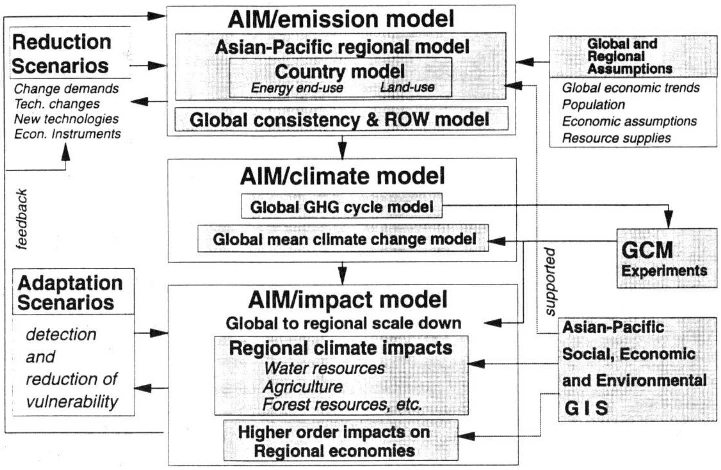

Figure 1. A summary of AIM: Asian-Pacific Integrated Model.

component of the country model is an end-use energy demand model. The emission model combines an end-use energy model and a technological selection model. More than 100 technologies are evaluated for their potential to improve energy efficiency, and energy demand estimates are linked to a top-down economic model. So far, a prototype model has been completed and applied to Indonesia, China, and Japan.

The AIM/climate model is designed to link to other established models for calculating global and regional climate change. It comprises the revised AMAC model for atmospheric composition, the IPCC radiative forcing model, a box-diffusion ocean uptake model, the Rashof feedback model, and a model for regional scenarios of climate based on GCM outputs.

The AIM/impact model is designed to calculate the primary impacts on water supply, agricultural production, wood supplies, human health, etc., and then make predictions of higher-order impacts on the regional economy. The AIM/impact model is linked to the AIM/emission model through GCM experiments which are based on the greenhouse gas emission scenarios calculated by the AIM/emission model.

Various economic and environmental data for the region have been gathered and filed in a spatial data-base. Simulation models to estimate water resources changes, vegetation changes and malaria spread caused by global warming have been developed.

3. AIM/EMISSION MODEL

The AIM/emission model is comprised of Asian-Pacific country models which are integrated into a regional model. This in turn is linked to a Rest of the World model that ensures interactions between these regional models are consistent. A variety of global and regional assumptions about such things as population, economic trends, as well as

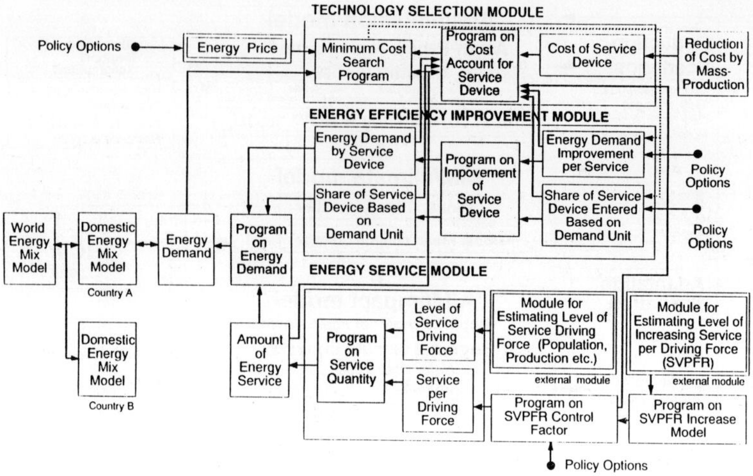

Figure 2. Outline of AIM End-use Energy Demand Model.

government policies, are then entered into the model and interact with the regional and country models to provide estimates of energy consumption, land-use changes, etc. which ultimately give predictions of greenhouse gas emissions.

3.1. Country energy demand model

Figure 2 shows the outline of the country energy demand model. In each country model, energy demand is calculated by multiplying the energy service (calculated by the energy service sub-module) by an energy efficiency factor. This factor is determined by the energy efficiency sub-module, and is the product of assumptions made about the introduction of new technologies for energy conservation as influenced by energy prices. The technology selection sub-module decides which technologies will be introduced.

3.1.1. Indonesian model: simple technological assumptions

The Indonesian model uses an energy service model comprising four sectors; a transport sector, an industrial sector, a residential sector, and a commercial sector. In a "Business-as-Usual (BaU)" scenario, the improvement in energy efficiency was simply assumed to be linear, at the rate of about 1% per annum. Other major assumptions were: a population of 340 million at 2100; a GNP increase of 4.1% per annum up to 2025, followed by 3.3% per annum; an urbanization rate of 73%; and a deforestation rate of 2.2 million ha per year. Figure 3 and 4 show the predicted emissions of CO_2 and CH_4 over the next 100 years. CO_2 emissions from deforestation increase until the early part of the next century and then decline as the remaining forests disappear. Emissions from other

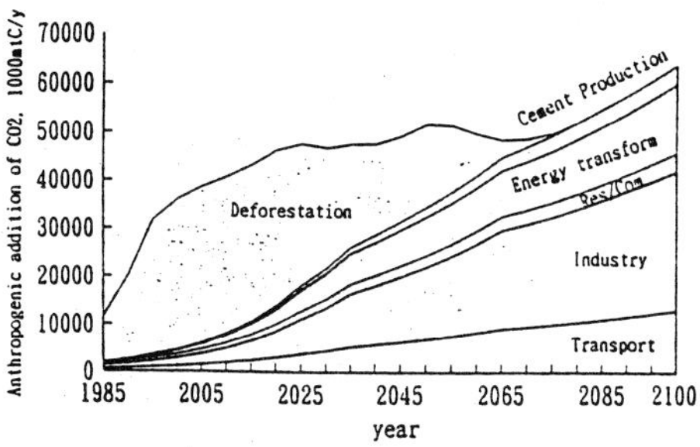

Figure 3. A prediction of CO_2 emissions from Indonesia (BaU case).

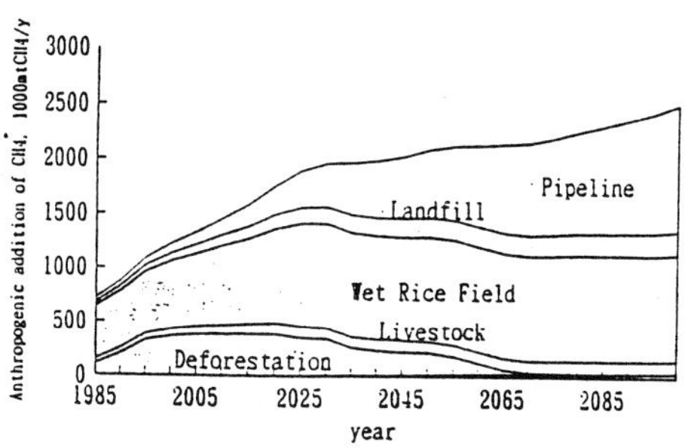

Figure 4. A prediction of CH_4 emissions from Indonesia (BaU case).

sources increase steadily, but at different rates. The main contributors of CH_4 emissions will be rice cultivation and leaks from pipelines.

3.1.2. Chinese model: complex technological assumptions

As with the Indonesian model, an energy service model of four sectors was used. However, estimates of energy efficiency improvements were the result of individual scenarios for each of 30 different technologies. In the "Business as Usual" case, these improvements ranged from 0.5-1.0% per annum. When allowances were made for energy conservation policy implementation, the improvements increased to a little more than 1% per annum. The population was assumed to grow to 1.66 billion at 2100, while the GNP growth rate was 5.3% per annum until 2025, then 3.9% per annum. Figure 5 illustrates the "Business as Usual" forecast, with energy consumption increasing from 25 EJ in 1985, to 170 EJ in 2050 and 250 EJ in 2100. If energy conservation policies are implemented as assumed, total energy consumption growth is half of the "Business as Usual" case. See Figure 6.

3.1.3. Japanese model: technological selection model

Development of the Japanese model is well-advanced. One main feature of this model is the integration of the technological selection module into the energy demand model. As described in the AIM outline, energy efficiency improvements depend on the introduction of energy conservation technologies, which in turn is influenced by energy price. More than 100 technologies for energy conservation used in the industrial, residential, commercial and transport sectors were examined, and their initial cost, maintenance cost, market share, energy conservation potential, pay back period, etc. were input into the model.

Scenarios up to the year 2010 were prepared for population growth, industrial production, as well as the expansion of the average household area and office floor space, and used as assumptions for predicting energy service demand. Computer simulations were

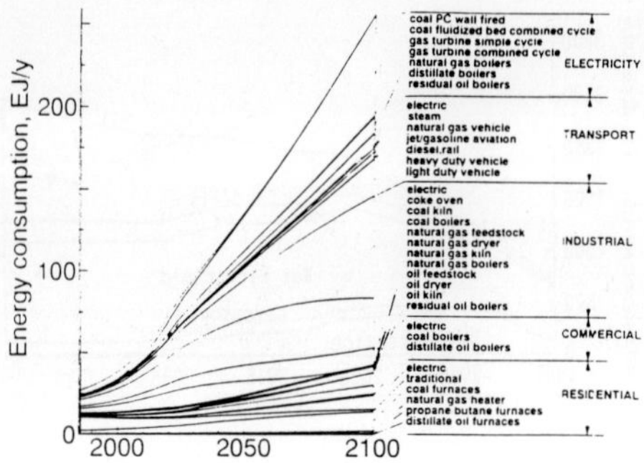

Figure 5. Projection of energy consumption of China (BaU case).

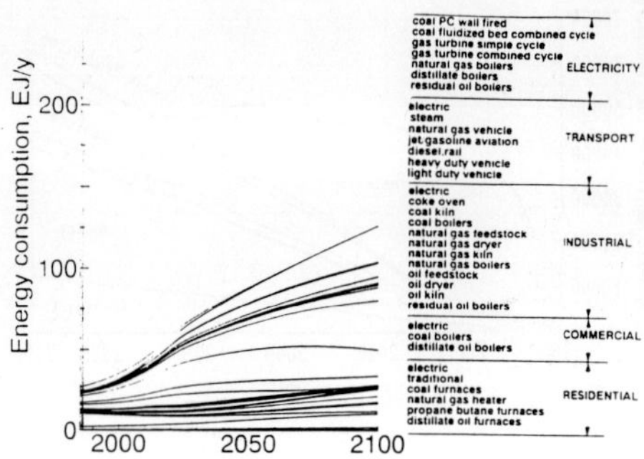

Figure 6. Projection of energy consumption of China (Improved efficiency case).

made of five different situations:

1) No changes in technology; no carbon tax or subsidies.
2) Technology selection with no carbon tax or subsidies - Base Case.
3) Base case with a carbon tax of 30,000 yen /tC.
4) Base case with a carbon tax of 30,000 yen/tC, and a subjective extended period of pay back over a maximum of 20 years.
5) Base case with a carbon tax of 3,000 yen/tC and recycling of the carbon tax revenue as a subsidy for the introduction of energy conservation technologies.

These simulations gave several Japan's alternative scenarios of CO_2 emissions shown in Figure 7. The simulation results indicate that if the Japanese people come to understand the economic benefits of energy conservation, then energy conservation technologies will be able to be promoted and introduced without any tax or subsidy. As a result, it would be possible to stabilize CO_2 emissions in the industrial sectors, but impossible to stabilize them in other sectors without a carbon tax and subsidies, because of the rapid increase of energy service demand.

If a carbon tax is introduced, energy consumption would be reduced, but this alone cannot stabilize total CO_2 emissions. The combination of a carbon tax and a subsidy from recycling tax revenue would have a significant effect over the short term.

In summary, to stabilize CO_2 emissions in Japan, we must introduce mutually-reinforcing policies, such as a combination of a carbon tax, extensions of pay back period, and subsidies.

3.2. Regional and global model

All of the country models for the region have not yet been completed, but we have developed a rough emission model, not only for the regions of the Asian-Pacific, but

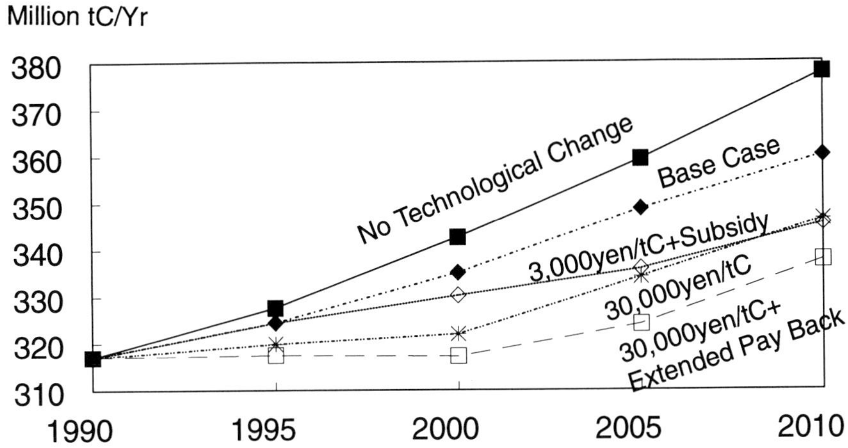

Figure 7. A prediction of CO_2 emissions in Japan

also for six regions of the World. This regional emission model was developed using the Edmonds-Reilly model (1983), which is a partial equilibrium world model.

Using the rough model for these countries in conjunction with the detailed completed models, and making them consistent with the Rest of the World model, we prepared preliminary estimates of regional greenhouse gas emissions for the year 2100.

CO_2 emission intensities are shown in Figure 8, When compared with the recent situation, the areas where major growth in CO_2 emissions will occur can easily be identified. They include, Korea, China, Thailand, Malaysia, Indonesia, India and Bangladesh.

4. AIM/CLIMATE MODEL

The AIM/climate model estimates the atmospheric concentration of greenhouse gases and the temperature increase. We have developed a world climate model using original linkages to join other established models. It comprises the revised AMAC model for atmospheric composition, the IPCC radiative forcing model, a box-diffusion ocean uptake model, the Rashof feedback model, and a model for regional scenarios of climate change based on GCM outputs. The AIM/climate model was linked to the AIM/emission model to evaluate the effects of greenhouse gas emission increases on the climate. It is also planned to link to the AIM/impact model.

The regional climate change scenarios to be used in AIM/impact are mainly prepared by the GCMs. In order to apply these GCM experiments, their outputs were normalized by the difference between the global means of the equilibrium $2\times CO_2$ experiment and the $1\times CO_2$ experiment. This procedure eliminates equilibrium sensitivities of different models, and also allows the normalized result to be applied to time-dependent scenarios of global mean warming produced by the AIM/emission model. These normalized climate change patterns are basically prepared in the GCM spatial resolutions. However, they

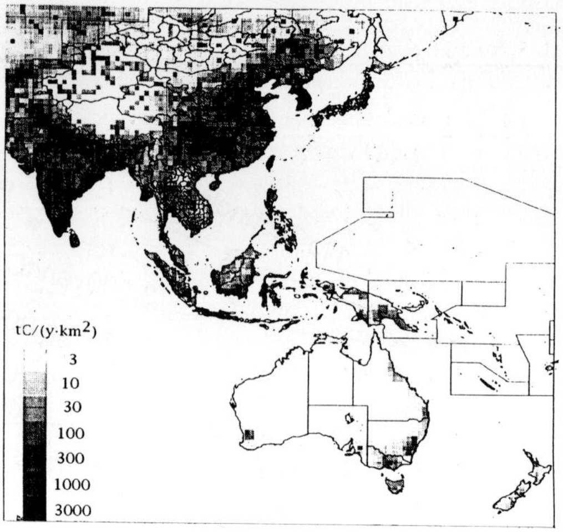

Figure 8. CO_2 emission intensity from the Asian-Pacific region in 2100.

are subsequently interpolated to a finer grid in order to correspond with the regional impact models of each sector. As the sub-grid interpolation of GCM outputs has a lot of problems, we are also intensively examining methods based on statistical relationships between broad-scale climatic data and small scale observations from regional climate records.

The GCM experiments we prepared now are shown in Table 1 and greenhouse gases transient scenarios used in GCM experiments are shown in Table 2. Some of them are equilibrium experiments and others are transients. We normalize them and couple them with our simplified global GHG cycle model in order to calculate the spatially detailed climatic parameters under an arbitrary emission scenario.

5. AIM/IMPACT MODEL

The AIM/impact model estimates impacts of global warming in the Asian-Pacific region. AIM/impact is comprised of a sea level rise model, a water balance model, a vegetation change model, a health impact model, a natural disaster model, an agriculture production model, and an economic evaluation model of global warming damage.

The outputs of the AIM/emission model will be entered into the AIM/climate model which will provide a variety of scenarios used in GCM experiments. Data from the GCM experiments will be used as the basic assumptions for the AIM impact model, and will reflect estimates of regional climate impact. The AIM/impact model will calculate the primary impacts on water supply, agricultural production, wood supplies, human health, etc., and then make predictions of higher order impacts on the regional economy.

Table 1
GCM experiments used in our model

climate model	Date Calculated	Resolution (lat ×long)	Base 1×CO_2 ppmv	Δ T (C)	ΔP (%)	No
		Equilibrium Experiment				
GISS	1982	7.83 ×10.0	315	4.2	11.0	1
GFDL	1984-85	4.44 ×7.50	300	4.0	8.7	2
OSU	1984-85	4.00 ×5.00	326	2.84	7.8	3
UK Met	June 1986	5.00 ×7.50	320	5.2	15.0	5
GFDL Q-flux	Feb.1988	4.44 ×7.50	300	4.0	8.3	6
GFDL R30	May 1989	2.22 ×3.75	300	4.0	8.3	7
CCC	Nov.1989	3.75 ×3.75	330	3.5	3.8	8
UKHC	Nov.1989	2.50 ×3.75	323	2.7	9.0	10
		Transient				
GISS-A	1984-1985	7.83 ×10.0	315	4.2	11.0	101
GISS-B	1984-1985	7.83 ×10.0	315	4.2	11.0	101
GFDL	1990	4.44 ×7.50	300	4.0	8.3	106
UKHC	Nov.1989	2.50 ×3.75	320	3.5	9.0	110
ECHAM2/OPYC	1990	5.63 ×5.63	390	2.8		109
ECHAM2/LSG	1990	5.63 ×5.63	390	2.8		129
NCAR		4.50 ×7.50	330	4.5		104

Table 2
GHGs transient scenarios used in GCM expriments

climate model	No	Note
GISS-A	101	Own scenario. In 2060, CO2 becomes 527ppm
GISS-B	101	Own scenario. In 2060, CO2 becomes 485ppm
GFDL	106	1%/y increase, GHG compounded
UKHC	110	1%/y increase, GHG compounded
ECHAM2/OPYC	109	SA90A
ECHAM2/LSG	129	SA90A
NCAR	104	1%/y increase, linear

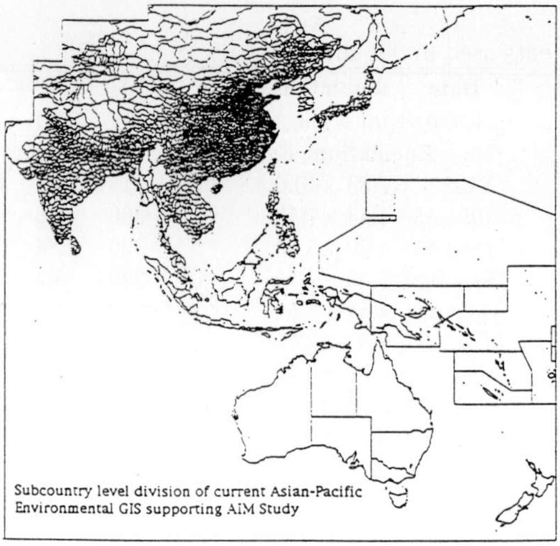

Figure 9. Subcountry level division of an Asian-Pacific GIS supporting the AIM study.

5.1. Geographical Information System

To function properly, our models require a high-quality spatial data-base. As such, we are developing the Asian-Pacific environmental, social and economic Geographical Information System (GIS). This GIS has a spatial segmentation at the sub-country level as shown in Figure 9. Information which comes from many international organizations, including the UNEP/GRID system as well as individual governments, is integrated with our originally constructed data, and the information is organized in a format useful to support the AIM study.

5.2. Impacts on water resources

Hydrological impacts are one of the most important aspects of the coming climate change. Changes in the magnitude, frequency and duration of hydrological factors influence the availability of water resources, flooding intensity as well as agricultural and natural terrestrial ecosystems. A rainfall-runoff process submodel was developed as one of the basic submodules of the AIM/impact model. This submodule consists of water balance and water transport components, and it is intended to provide basic hydrologic information to the impacts models of other sectors. Specifically, it creates gridded high resolution datasets of surface runoff, soil moisture, evapotranspiration and river discharge.

The major parameters of the hydrological model are elevation, soils, vegetation as well as precipitation, temperature, and potential evapotranspiration. Except for the first three, these parameters are endogenous variables of the total AIM system. As the coupling

Figure 10. Estimates of current river flow intensity.

of the total system is not complete, we set soil and vegetation characteristics as well as elevation conditions at their current situations.

The water balance component of the model is based primary on the models of Thornthwaite and Mather (1955) and their successors. Bookkeeping of the water balance among precipitation, snowmelt, evapotranspiration and streamflow is calculated for each grid cell in the simulation region. A number of climatological and geographical data sets were prepared from various sources. Precipitation and temperature values were taken from interpolated results of GCM experiments by GFDL. Soil moisture capacities were estimated using current vegetation classes and soil textures (Vorosmarty *et al.*, 1989, Webb *et al.*, 1993).

In the water transport component, the network topology of streams was determined from digital elevation data and modified with various hydrological maps of analyzed regions. Modelling of surface water retention time in each cell followed Vorosmarty's model (1989). Figure 10 is a result of estimating annually-averaged accumulated runoff under current climate conditions. The calculation was conducted with 1/4 degree grid cells. Climate data was taken from the monthly average data sets of Legates and Willmott (1989) with 1/2 degree resolution. The degree of shading in the figure expresses the intensity of discharge from a cell to the next one downstream. Black indicates the highest flow areas, and white indicates the ares of lowest flow.

Using this model as the base condition, we then applied outcomes of GCM experiments (GFDL Q-flux experiment as the perturbed climate scenario), which provide precipitation, temperature and soil humidity data. These experiments were based on a future CO_2 level which is twice that of the Pre-industrial Revolution level. After interpolating the daily and monthly results of the GCM experiment, so they could be overlaid with Legates climatic pattern which has a higher resolution than GCM outputs, these input conditions

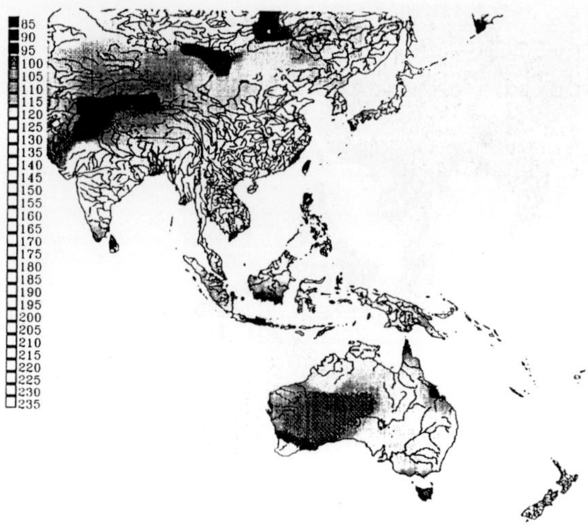

Figure 11. Prediction of monthly high-flow discharge for a 10 year return period.

were used to prepare simulations of variability in the water discharge of each river basin. These simulations were conducted for a 10 year period after a 2 year initial simulation under each condition, and then the probability distributions of the monthly simulated discharges were identified in each grid cell. Based on these distributions, we estimated flood and drought severities over a 10 year period in the $2\times CO_2$ condition.

Figure 11 shows an output of such a simulation for flood discharge. The light gray shading indicates the area where the highest flow discharge level for a 10 year return period may be expected to exceed twice the current level. Parts of India, China, and Japan could experience much higher flood levels. Figure 12 shows the changes in low flow conditions over a 10 year period. The dark shading indicates areas where the lowest flow levels decrease by 40 or 50 %. As shown in the figure, large areas of the region are forecasted to experience much drier periods. The spatial pattern of influence was not sensitive to the selection of a return period. Intensification of flood discharge does not mean the relief of drought, and in fact, an increase in the incidence of both events is anticipated for some regions.

In order to assess the more direct impact of hydrological changes on the human dimension, we have to consider the effect of water control and mangement devices, as well as the intensity and style of water consumption. So far, we have estimated only population weighted average values of drought and flood intensities. It is still too incomplete to comprehend the overall impacts of hydrological changes on human society. To clarify these impacts, we are now accumulating information about vulnerabilities to water related factors at the regional and subcountry levels.

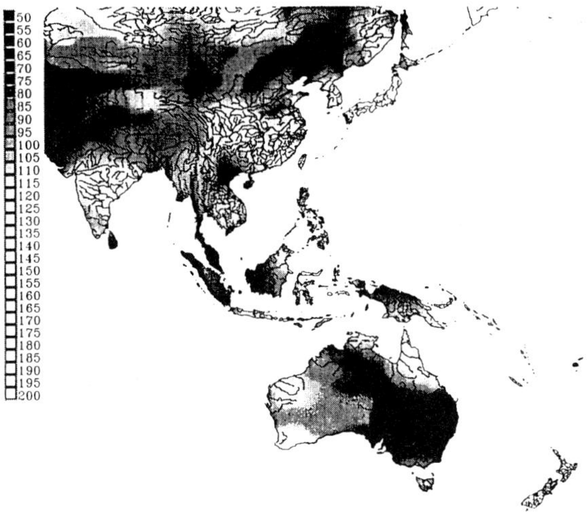

Figure 12. Prediction of monthly low-flow discharge for a 10 year return period.

5.3. Impacts on natural ecosystems

The spatial distribution of natural vegetation is strongly controlled by climate. Future climate change should have profound impacts on the distribution of vegetation patterns in the Asian-Pacific area. As a preliminary assessment of this impact, we established a simple model of global vegetation change caused by the change of related climate elements. In this model, a vegetation type is assumed to change if any climate element in its habitat exceeds the current global minimum or maximum level found in that vegetation habitat. As for climate elements, we used mean annual precipitation, mean annual temperature, coldest monthly temperature, hottest monthly temperature, and degree-days above 5°C.

Figure 13 shows the potential changes in vegetation under the several scenarios coupled with the results of GFDL-R30 GCM experiment. Significant changes in northern China are observed. Boreal conifer forests and larch taiga in this area are predicted to be significantly influenced. Tibetan and Himalayan alpine tundra would also be influenced. Evergreen-deciduous areas in Southeast China, drought-deciduous forests in India, Indo-China Peninsula and Northern Australia would also adversely affected by climate change. These changes should be considered as potential shifts under equilibrium assumptions. The actual mechanism of succession is far more complex and dynamic. For a preliminary analysis, the above method is useful. However, we consider such a static model to be unsatisfactory and of limited use, so we are preparing dynamic population-based succession models.

5.4. Impacts on malaria incidence

Global warming will result in increases in the temperature and changes to the vegeta-

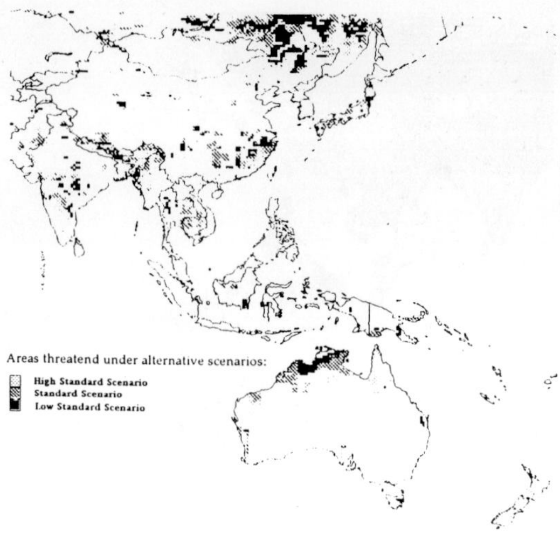

Figure 13. Potential changes in vegetation under different scenarios.

tion close to the ground. This will allow the habitat of the Anopheles mosquito, which is the malaria vector, to expand. As well, the development period of the malaria protozoan will shorten and its reproductive potential will increase. As a result, it is predicted that the global malaria risk will increase. In order to estimate the risk quantitatively, a global model has been developed.

The malaria risk increase is estimated in the model using the following steps:

1) The adaptability of Anopheles to basic eco-climate parameters, such as annual and daily temperature variability plus soil moisture is identified.
2) The future temperature variability based on GCM outputs, as well as soil moisture based on the water resource impact model developed in our study is estimated.
3) Using the values predicted in 2, the changes in the habitat area of Anopheles is calculated.
4) The rate of transfer of the disease between patients (the rate of spread of the disease in each climatic grid) is calculated taking into account the shortening of the sporozyte's development phase within the mosquito. The risk based on the reproduction rate in each region, as well as at the national and global levels is estimated. The sensitivity of GCM outputs in relation to risk outcomes is then determined.

Figure 14 shows the predicted expansion of the area in which malaria will be endemic if the atmospheric CO_2 concentration doubles. Parts of northern, central and southern America, Australia, China and south-east Asia, India and Africa will have an increased risk of malaria infection. It was concluded that the area in which malaria will become endemic will increase by 6 - 20 % (i.e. there will be a 6 - 20 % increase in the human population at risk of malaria infection).

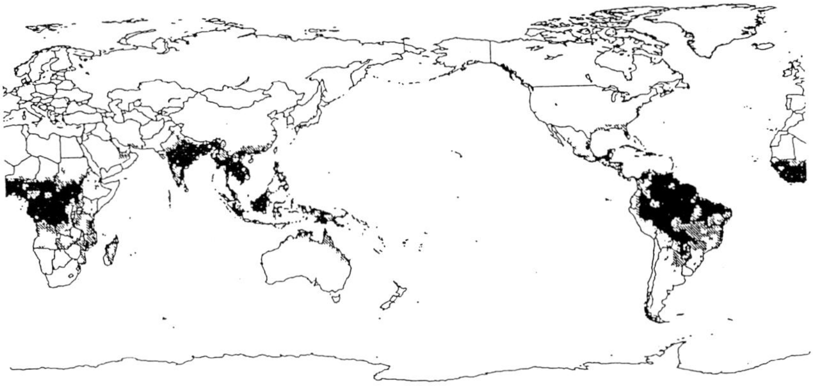

Figure 14. Extend of malaria endemic area caused by global warming.

6. CONCLUDING REMARKS

We have developed the integrated model for the policy analysis of global warming. This model comprises the emission, climate and impact models. We completed the World models for emissions and climate and also developed some prototype countrywide models, and applied them to emission scenario analysis and policy evaluation. We plan to prepare country models for the Asian-Pacific region in cooperation with institutes in each country and develop the project on a regional basis.

We also have developed a water resource change model, a vegetation change model, and a health impact model. These models and the database used for simulation need to be improved to explain the complex and dynamic process of real mechanisms.

It is also necessary to develop some models to estimate other primary impacts caused by global warming. Based on these primary impact estimates, we plan to estimate higher order impacts on regional economies while taking into account the effects of international relations.

7. REFERENCES

Alcamo, J., Kreileman, G.J.J., Krol, M. and Zuiema, G. (1993): Modelling the global society-biosphere-climate system: Part1: model description and testing, *Papers of International Workshop on Integrative Assessment of Mitigation, Impacts and Adaptation to Climatic Change*, 13-15, October, 1993, Laxenburg, Austria.

MIT (1994): *Joint Program on the Science and Policy of Global Change*, Center for Global

Change Science and Center for Energy and Environmental Policy Research, MIT, MA, USA.

Edmonds, J.A., Pitcher, H.M., Rosenberg, N.J. and Wigley, T.M.L. (1993): Design for the global change assessment model -GCAM, *papers of International Workshop on Integrative Assessment of mitigation, Impacts and Adaptation to Climate Change*, 13-15 October, 1993, Laxenburg, Austria.

Edmonds, J.A. and Reilly, J.M. (1983): Global energy and CO_2 to 2050. *The Energy Journal*, Vol. 4, No. 3, pp.21-47.

Legates, D.R. and Willmott, C.J. (1989): *Monthly Average Surface Air Temperature and Precipitation, Digital Data on a 0.5 degree Geographic 361 × 721 Grid*, NCAR, Boulder, CO, USA.

Matsuoka, Y., Kainuma, M. and Morita, T. (1994): Scenario Analysis of Global Warming Using the Asian-Pacific Integrated Model (AIM), *Integrative Assessment of Mitigation, Impacts and Adaptation to Climate Change (N. Nakicenovic et al. eds.)*, CP-94-9, IIASA, Austria, pp.309-338.

Morita,T., Matsuoka, Y., Kainuma, M., Harasawa, H. and Kai, K. (1993) AIM- Asian-Pacific Integrated Model for Evaluation policy options to reduce greenhouse gas emissions and global warming impacts. *Proceedings of the Workshop on Global Warming Issues in Asia*, Bangkok, 8-10 September, 26 pages.

Prather, M. (1988): *Assessment Model for Atmospheric Composition*, NASA Conference Publication 3023, NASA Goddard Institute for Space Studies, NY.

Thornthwaite, C.W. and Mather, J.R. (1955): The Water Balance, *Publ. Climatol. Lab. Climatol. Drexel Inst. Technol.*, Vol. 8, No. 1, pp.1-104.

Vorosmarty, C.J., Morre III, B., Grace, A.L. and Gildea, M.P. (1989): Continental Scale Model of Water and Fluvial Transport: an Application to South America, *Global Biogeochemical Cycle*, Vol. 3, No. 3, pp.241-256.

Webb, R.S. and Rosenzweig, C.E. (1993): Specifying Land Surface Characteristics in General Circulation Models: Soil Profile Data Set and Derived Water-holding Capacities, *Global Biogeochemical Cycle*, Vol. 7, No. 1, pp.97-108.

TOWARD GLOBAL PLANNING OF SUSTAINABLE USE OF THE EARTH
Development of Global Eco-Engineering
S. Murai (Editor)

Valuing climate change: An economic assessment of global warming impacts

Samuel Fankhauser*

*Centre for Social and Economic Research on the Global Environment (CSERGE), University College London and University of East Anglia, Gower St, London WC1E 6BT, UK.

Abstract

The paper provides an overview on work dealing with an economic assessment of global warming impacts. While there is a relatively large body of literature on the physical impacts of climate change, the question of an *economic valuation of* greenhouse damages has gained less attention so far. The paper surveys the results of existing studies in this area and discusses their main shortcomings. It concludes by sketching out the main implications of global warming damage estimates for the design of an optimal climate policy.

Most studies on climate change impacts have concentrated on the case of atmospheric CO_2 concentration doubling ($2xCO_2$). An emerging benchmark figures appears to be that exposing a world with today's structure to $2xCO_2$ would reduce GNP by about 1-2%. There are several problems with this type of assessment, however, as will be discussed in the paper. The *total cost* assessment for $2xCO_2$ can serve as a basis for calculations of the *marginal* social costs per tonne of emission. Marginal cost estimates tend to be in the order of 10-20 $/tC, but are highly sensitive and could be much higher. A caveat is required with respect to the policy implications of such figures. While they give a good indication about the most desirable level of a carbon tax from a pure cost-benefit point of view, several other factors, such as equity considerations or the precautionary principle, may also play a role in the design of a suitable climate change policy.

1. INTRODUCTION[1]

There is a considerable interest in the potential impacts of global warming. Working Group 2 of the Intergovernmental Panel on Climate Change (IPCC) is exclusively devoted to this topic (see IPCC, 1990). The comprehensive analysis of IPCC has been complemented by a multitude of additional studies on sectoral or regional impacts (for a survey see e.g. Fankhauser, 1994a). In the United States, for example, the Environmental Protection Agency has commissioned an extensive analysis on climate change impacts for the country (see Smith and Tirpak, 1989). Similar endeavours were also undertaken for other countries, although usually on a much smaller scale (e.g. Parry and Duncan, 1994, for the UK; Nishioka *et al.*, 1993, for Japan; Asaduzzaman, 1994, for Bangladesh).

Most of these assessments concentrate on a description of impacts in physical terms, though. Attempts at a monetary quantification of physical impacts are still comparatively rare, even though damage estimation has recently been boosted by the development of a series of *integrated assessment* models (e.g. Manne *et al.*, 1993; Hope *et al.*, 1993; Dowlatabadi and Granger, 1993). Partly this slow start can be attributed to the fact that proper economic valuation requires a reasonably accurate knowledge of impacts in physical terms, something which even now is available only to a limited extent. It should also be clear that the valuation of such damage aspects as human hardship will be beyond the reach of economic valuation techniques. Yet, many aspects *are* quantifiable in monetary terms, and damage valuation is thus probably the one area in greenhouse research where most catching up is required.

In classifying greenhouse damage studies we can distinguish between two approaches (Fankhauser, 1994a). The first is a partial equilibrium approach. Each damage aspect is analysed separately and total greenhouse damage is simply the sum of the individual categories. This method is often called the *enumerative approach* to greenhouse damage estimation (Cline, 1994). The second method analyses greenhouse damage in a more *integrated*, general equilibrium set up. That is, climate change is imposed on a system of interacting markets. Initial impacts on one sector may then also have higher order effects and spill over to other sectors of the economy.

The paper is structured as follows. Section 2 briefly surveys the main impacts usually associated with climate change. Section 3 then deals with monetary damage estimates from both enumerative and integrated studies. It also outlines the main shortcomings of each of these methods. Section 4 extends the previous analysis of *total* damage costs by asking about the *marginal* impact caused by each tonne of greenhouse gas emitted. Section 5 contains policy implications and conclusions.

2. GREENHOUSE IMPACTS

Global warming will have a variety of effects. They can be classified as either market related - i.e. effects which will become manifest in the national accounts - or non-market related - i.e. impacts affecting "intangibles", such as ecosystems or human amenity.

The IPCC foresees a shift in the current agricultural production pattern away from current production areas to more northern latitudes. Together with changes in soil water availability, the increased occurrence of climatic extremes and crop diseases this may lead to an overall reduction in agricultural yields, and could result in serious regional or year-to-year food shortages. The IPCC further predicts that the increased stress on ecosystems may lead to the extinction of species unable or too slow to adapt.

The rise in sea levels connected with a warmer climate will threaten low lying coastal areas. Sea level rise will particularly affect densely populated coastlines and small island states. Extreme events like storms, floods and droughts may occur more frequently. Health experts expect a rise in climate related diseases such as heat strokes and a spread of vector borne diseases like malaria into so far unaffected areas. Others have warned about the consequences of increased water shortages in summer. Climate dependent economic activities like construction, transport and tourism will be affected in an unknown way. Finally, global warming may trigger a stream of climate refugees away from the worst affected regions and coasts (IPCC, 1990).

The scientific research on global warming impacts has almost entirely focused on the case of CO_2 concentration doubling, that is on the impacts of an atmospheric CO_2 concentration of twice the preindustrial level ($2xCO_2$). $2xCO_2$ is a completely arbitrary benchmark, chosen solely for analytical convenience. It is neither an optimal point nor a steady state, and warming will continue, and in fact aggravate beyond $2xCO_2$. This point has repeatedly been made in the literature, most prominently by Cline (1992a). Nevertheless, as a consequence of the scientific focus studies on the economic costs of global warming have tended to concentrate on $2xCO_2$ as well. By far the best studied aspects are the impacts on agriculture (e.g. Reilly *et al.*, 1994; Rosenzweig and Parry, 1994) and the costs of sea level rise (e.g. Fankhauser, 1994c; Titus *et al.*, 1991). Several authors have tried to provide a first order assessment of total global warming damage, including non-market aspects (Fankhauser, 1994a; Cline, 1992a; Nordhaus, 1991; Titus, 1992; Tol, 1994a).

Monetary damages in both market and non-market sectors are ideally expressed in the form of willingness to pay (WTP) or willingness to accept compensation (WTA) estimates. Measuring the amount of income a person is willing to sacrifice in exchange for an improved state of the world, or the compensation needed in order to accept a deterioration, WTP and WTA measures are used in welfare economics as a money metric for individuals' preferences. A high WTP hints at a strongly preferred state of the world, a low WTP indicates a less significant improvement in individual welfare, and so on (see e.g. Pearce and Turner, 1991).

Unfortunately, WTP/WTA measures are not always available for the estimation of global warming impacts. Reductions in revenues, changes in factor payments and other indicators are frequently used to approximate the welfare impacts of climate change. In addition, several damages could so far not be estimated and were ignored in existing studies. The damage estimates surveyed below are therefore neither accurate nor complete, a fact which will have to be borne in mind when using them as a basis for policy recommendations.

Table 1 summarises the expected impacts from global warming and assesses how accurately they have been evaluated in literature so far.

3. GLOBAL WARMING DAMAGE ESTIMATES

3.1 Enumerative Studies

The pioneering paper trying to provide a first order assessment of greenhouse damage in economic terms is Nordhaus (1991). Concentrating mainly on the costs of agriculture and sea level rise, Nordhaus estimated an overall damage of global warming in the order of a quarter percent of GNP for the United States. To allow for the many non-market impacts neglected in the study this value is raised to 1%, with a range of error of 0.25-2%. Improvements on Nordhaus' back-of-the-envelope estimate have been provided by Cline (1992a) and Titus (1992), two studies again focusing on the US, and also Fankhauser (1994a) and Tol (1994a), who distinguish between several world regions. These studies all perform the same hypothetical exercise, and ask what the damage would be if today's economy were exposed to $2xCO_2$. An overview of results is provided in Tables 2 (for the United States), 3 and 4 (for several world regions).

Despite differences in individual damage categories - and not all papers deal with all aspects - the five US studies display a surprisingly high degree of consistency in the overall result. The emerging picture from enumerative studies seems to be that, to the extent that it can be monetised, $2xCO_2$ damage in the United States could lie *in the order of 1-2% of GNP, for a 2.5-3°C warming*. A similar, if slightly higher range is predicted for other OECD countries and the world as a whole. Cline (1992a) and Nordhaus (1991) have both asserted

Table 1 Overview on Climate Change Impacts

Damages	*Market Impacts*				*Non-Market Impacts*		
	Primary sector damage	Other sector damage	Property loss	Damage from extreme events	Ecosystem damage	Human amenity impacts	Damage from extreme events
Fully and suitably estimated	agriculture[1]		dryland loss coastal protect.		wetland loss		
Fully estimated, using approximations	forestry	water supply		hurricanes	forest loss		hurricanes
Partially estimated	fisheries[2]	energy demand leisure activity	urban infrastructure		species loss	human life air pollution water pollution migration	
Not estimated		insurance construction transport energy supply		non trop. storms droughts river floods hot/cold spells other catastrophes	other ecosystems loss	morbidity physical comfort polit. stability human hardship	non trop. storms droughts river floods hot/cold spells other catastrophes

[1] widespread disagreement
[2] often included in wetland loss

Source: Pearce *et al.* (1995).

Table 2 Economic Damages from $2xCO_2$: US Economy at Present Scale (1990bn $)

	Cline (2.5°C)	Fankhauser (2.5°C)[1]	Nordhaus (3°C)[1]	Titus (4°C)	Tol (2.5°C)[2]
Agriculture	17.5	8.4	1.1	1.2	10.0
Forest Loss	3.3	0.7	small	43.6	-
Sea Level Rise	7.0	9.0	12.0	5.7	8.5
Electricity	11.2	7.9	1.1	5.6	-
Non-Electric Energy	- 1.3	-		-	-
Mobile Air Cond.	-	-		2.5	-
Construction	± ?	-		-	-
Urban Infrastructure	0.1	-		-	-
Leisure/Tourism	1.7	-		-	-
Water Supply	7.0	15.6		11.4	-
Hurricanes	0.8	0.2	} [3]	-	0.3
Human Amenity	+ ?	-		-	12.0
Life/Morbidity	5.8	11.4		9.4	37.4
Migration	0.5	0.6		-	1.0
Species Loss	4.0 + ?	8.4		-	5.0
Water Pollution	-	-		32.6	-
Air Pollution	3.5 + ?	7.3		27.2	-
Total	61.1 + ?	69.5	54.5	139.2	74.2
(% of GDP)	(1.1)	(1.3)	(1.0)	(2.5)	(1.5)[2]

[1] transformed to 1990 base
[2] USA and Canada, base year 1988
[3] not assessed categories, estimated at 0.75% of GNP

Sources: Cline (1992a), Fankhauser (1994a), Nordhaus (1991), Titus (1992), Tol (1994a).

that their US estimates would extend to the entire world, a claim which is only partly contradicted by the Tol and Fankhauser estimates (Table 3).

In as far as figures are comparable, the above benchmark result is roughly confirmed by the outcome of a poll of impact experts conducted by Nordhaus (1994), although the Nordhaus study exhibits a much wider range of opinions. Asked about the likely impacts of a 3°C warming by the year 2090, experts predicted damages in the order of 0.7 - 8.2% of GNP (90% confidence interval). The median value was 1.9% of GNP.

A caveat is however necessary. The high degree of convergence between studies should not necessarily be seen as a sign of a high confidence in the figures. Since all studies build

Table 3 Monetary $2xCO_2$ Damage in Different World Regions

	Fankhauser (1994a) bn$	%GDP[1]	Tol (1994a) bn$	%GDP[1]
• European Union	63.6	1.4		
• United States	61.0	1.3		
• Other OECD	55.9	1.4		
• OECD America			74.2	1.5
• OECD Europe			56.5	1.3
• OECD Pacific			59.0	2.8
Total OECD	*180.5*	*1.3*	*189.5*	*1.6*
• E. Europe / Former USSR	18.2[2]	0.7[2]	- 7.9	-0.3
• Centrally Planned Asia	16.7[3]	4.7[3]	18.0	5.2
• South and South East Asia			53.5	8.6
• Africa			30.3	8.7
• Latin America			31.0	4.3
• Middle East			1.3	4.1
Total Non-OECD	*89.1*	*1.6*	*126.2*	*2.7*
World	*269.6*	*1.4*	*315.7*	*1.9*

[1] Note that the GDP base may differ between the studies
[2] Former Soviet Union only
[3] China only

Source: Pearce *et al.* (1995).

on similar primary sources, in particular the US results of Smith and Tirpak (1989), the figures may be biased towards convergence, as was pointed out by Pearce *et al.* (1995). The actual robustness of the figures may thus be much lower than implied by the rather narrow confidence interval.

There are considerable differences with respect to individual damage categories. Cline's agricultural damage, for example, is more than an order of magnitude higher than the respective figures of Nordhaus and Titus. The Titus study, on the other hand, reports pollution and forestry damages significantly higher than the other assessments.

There are broadly two reasons for such differences. Firstly, there are discrepancies because of different assumptions on quantitative impacts, due to the prevailing scientific uncertainty. Agricultural damage is probably the best example for this. Secondly there are cases where the authors agree on the quantitative impacts but value them differently. The mortality estimates are a prime example here. Cline's assessment, for example, is based on a value of a statistical life of 0.6 m$, while Tol (1994a) uses figures of over 3 m$ for developed countries. This discrepancy is still rather modest. A recent survey by Viscusi (1993) has found value-of-life estimates ranging from 0.2 m$ to 16 m$.

Table 4 2xCO_2 Damages in Different World Regions (2.5°C warming)

(a) Quantitative Estimates

Type of damage	Damage indicator	EU	USA	Ex USSR	China	Non-OECD	OECD	World
agriculture	welfare loss (% GNP)	0.21	0.16	0.24	2.10	0.28	0.17	0.23
forestry	forest area lost (km^2)	52	282	908	121	334	901	1,235
fishery	reduced catch (1000t)	558	452	814	464	4,326	2,503	6,829
energy	rise in electricity demand (TWh)	54.2	92.0	54.6	17.1	142.7	211.2	353.9
water	reduced water availability (km^3)	15.3	32.7	24.7	32.2	168.5	62.2	230.7
coastal protect.	annual capital costs (m$/yr)	133	176	51	24	514	493	1,007
dryland loss	area lost (1000 km^2)	1.6	10.7	23.9	0	99.5	40.4	139.9
wetland loss	area lost (1000 km^2)	9.9	11.1	9.8	11.9	219.1	33.9	253.0
ecosystems loss	number of protected habitats lost, assuming 2% loss	16	8	?	4	53	53	106
health/mortality	number of deaths (1000)	8.8	6.6	7.7	29.4	114.8	22.9	137.7
air pollution	equivalent increase in emissions							
troposph. O_3	(1000 t NO_x)	566	1,073	1,584	227	2,602	1,943	4,545
SO_2	(1000 t sulphur)	285	422	1,100	258	1,864	873	2,737
migration	additional immigrants (1000)	229	100	153	583	2,279	455	2,734
hurricanes								
casualties	number of deaths	0	72	44	779	7,687	313	8,000
damages	m$	0	115	1	13	124	506	630

Table 4 (cont.)

(b) Monetary Estimates (bn$)

Type of damage	EU	USA	Ex USSR	China	Non-OECD	OECD	World
agriculture	9.7	7.4	6.2	7.8	16.0	23.1	39.1
forestry	0.1	0.6	0.4	0.0	0.2	1.8	2.0
fishery	-[1]	-[1]	-[1]	-[1]	-[1]	-[1]	-[1]
energy	7.0	6.9	-0.7	0.7	2.6	20.5	23.1
water	14.0	13.7	3.0	1.6	11.9	34.8	46.7
coastal protect.	0.1	0.2	0.0	0.0	0.5	0.5	1.0
dryland loss	0.3	2.1	1.2	0.0	5.9	8.1	14.0
wetland loss	4.9	5.6	1.2	0.6	14.7	16.9	31.6
ecosystems loss	9.8	7.4	2.3	2.2	15.0	25.5	40.5
health/mortality	13.2	10.0	2.3	2.9	14.8	34.4	49.2
air pollution	3.5	6.4	2.1	0.2	3.5	11.9	15.4
migration	1.0	0.5	0.2	0.6	2.3	2.0	4.3
hurricanes	0.0	0.2	0.0	0.1	1.7	1.0	2.7
TOTAL (bn$)	63.6	61.0	18.2	16.7	89.1	180.5	269.6
(% GNP)	(1.4)	(1.3)	(0.7)	(4.7)	(1.6)	(1.3)	(1.4)

[1] fishery loss is included in wetland loss

Source: Pearce *et al.* (1995) and Fankhauser (1994a).

The point to note is that ignorance with respect to valuation may cause a similar range of error as scientific uncertainty, and that the quality of estimates can be ameliorated by improving either side. This conclusion is confirmed by Peck and Teisberg (1993), who have calculated the value of improved information, and found similar figures for a better knowledge about the costs of warming as for an improved scientific understanding.

Greenhouse impacts are likely to vary between regions, and could be particularly severe in developing countries. Tables 3 and 4 provide damage estimates for different world regions. Leaving the special case of the former Soviet Union aside, Fankhauser (1994a) predicts damages of about 70 bn$ in the non-OECD regions (China and Rest of the World). This corresponds to about 2.2% of GNP in these regions, some 60% higher than the OECD average. Tol (1994a) obtains regional figures which are up to five times the OECD average.

The main causes for these high estimates are health impacts, which dominate the Tol estimates, and the high portion of natural habitats and wetlands found in developing countries. Poor nutrition and health standards as well as a higher vulnerability to extreme events also play a role (Pearce *et al.*, 1995). Developing countries could further suffer from a limited capacity for adaptation. For example, most poor countries will probably lack the funds necessary to provide an adequate protection of their coastlines. Although the data are weaker in the case of non-OECD countries, it seems evident therefore that global warming will clearly have its worst impacts in the developing world.

3.2 Integrated Damage Assessments

In the enumerative approach total damage is the sum of individual damage categories, i.e. the estimates are based on a partial equilibrium approach. *Inter alia* this means that they neglect the higher order effects which, for example, a change in agricultural yields will induce on the food, tobacco or textile industry. Integrated, or general equilibrium assessments correct for this shortcoming. However, while more comprehensive with respect to the treatment of market based damages, they usually ignore non-market impacts.

General equilibrium effects have been incorporated in the so-called MINK study, probably the most comprehensive regional damage estimation project undertaken so far (see Rosenberg, 1993). Concentrating on agriculture, forestry, water and energy, the study analyses the impacts of climate change on the four US states Missouri, Iowa, Nebraska, and Kansas (hence the acronym MINK). First order damage effects, e.g. on agriculture, are fed back into the economy through an input-output module. The study has other interesting features. Rather than working with 2xCO_2 data, it uses the weather pattern of the 1930s as a climate analogue, thus allowing for more temporal and spatial variability. Estimates are provided for both today's economy and that expected for 2030. Particular care is further taken to the issue of adaptation. The study concludes that global warming would probably not reduce regional income by more than 1% (for roughly 1°C warming). The impacts on agriculture and water would be most strongly felt.

In a similar attempt, Scheraga *et al.* (1993) have used the Jorgenson and Wilcoxen (1990) general equilibrium model to estimate the macroeconomic effects of climate change for the US. The study is limited to three types of impacts: A change in agricultural yields, a change in energy demand, and an increase in the sea level. Estimates are based on a rather pessimistic 2xCO_2 scenario of 5.1°C warming, assumed to occur gradually between now and the year 2060. The study illustrates how the structure of an economy may adjust to climate change by moving away from agriculture and consumption related activities towards investment and capital related industries. The increase in public expenditures on sea level rise protection, that is an increase in government purchases from the construction sector, leads to a redirection of spending away from consumption towards investment. At the same time higher prices for agriculture-based products such as food and tobacco lead to a fall in the demand for these goods. Overall, GNP in the year 2060 is 0.8% lower than in the base case without warming. In comparison, agriculture, sea level rise and energy account for about 0.2 - 0.6 % of GNP in enumerative studies, even though they assume only about half as much warming.

3.3 Shortcomings and Extensions

Valuation: Comprehensive damage assessments have been fiercely criticised by many authors (e.g. Grubb, 1993; Ayres and Walter, 1991). Probably the main objection concerns valuation, in particular the validity of economic valuation techniques. Grubb (1993) for example criticised 2xCO_2 damage estimates as being based on a "largely subjective valuation of non-market impacts" (p.153). However, while many of the existing figures are indeed based on approximations (see Table 1), it would be wrong to take this as an indication of the ineptitude of economic valuation techniques in general.

Economic valuation of non-market goods is controversial. Nevertheless, the problem of greenhouse damage estimates is currently perhaps not so much the accuracy of valuation methods as such, but the fact that they have not yet been applied to the problem to a sufficient degree. This is not to say that a full and complete valuation of all greenhouse impacts will ever be possible. Given the size of the problem and the uncertainties involved, it will probably not be, at least not within reasonable time. However, existing estimates are clearly far from perfect and the policy debate would gain from their improvement. Further, and more detailed valuation studies are thus warranted. What is particularly needed is a broadening of the scope from the emphasis on agriculture and sea level rise to the inclusion of other damage aspects such as ecosystems loss, climate amenity, health and morbidity.

Better estimates are also needed with respect to the damage costs to developing countries, where only little is known e.g. about the willingness to pay for non-market goods like wetlands, or the value of a statistical life. The confidence in developing country estimates is distinctly lower than that in the respective OECD figures.

Adaptation: Most authors in principle agree that damage estimates should be based on the assumption of a cost efficient response. That is, damage should be kept at a minimum through appropriate adaptation measures, like the erection of sea defences, or adjustments in agricultural and forest management. In the presence of adaptation, *the costs of global warming consist of two parts, the cost of adaptation, and the costs of remaining, unmitigated damages*. Unfortunately, while undisputed as a general idea, data limitations do not always allow the incorporation of optimal adaptation strategies when it comes to actual damage calculations. For simplicity it is then often assumed that no adjustments are taken at all, or that current service levels (e.g air quality standards) are maintained at all costs.

Adaptation can be a very powerful option. Hope *et al.* (1993), for example, found benefit-cost ratios of more than 20:1 in favour of an "aggressive adaptive policy". The question of adaptation is particularly prominent with respect to sea level rise protection. In the absence of any response even a modest rise in sea level would be disastrous. The inundation of low lying coasts would threaten highly valuable assets and displace millions of people. Nicholls *et al.* (1990) and IPCC (1994), for example, predict that a 1 metre sea level rise could displace nearly 100 million people in Bangladesh, Brazil, China, Egypt, Malaysia, Nigeria and Senegal alone. Damage will, however, be much lower once appropriate protection measures are incorporated. In a series of country studies IPCC (1994) found that through appropriate adaptation measures the number of people at risk from flooding could on average be reduced by about 65%. For the countries of the OECD, Fankhauser (1994c) calculated optimal coastal protection levels between 50 and 100%.

Agricultural damage provides a similar picture. While earlier studies have often ignored the impact of managerial responses (the so called "dumb farmer" assumption), recent work has tried to incorporate adaptation. The results so far seem to imply that through appropriate adaptation agricultural damage can on average be reduced to rather small amounts, at least for the temperature levels and regions considered. In the MINK study mentioned above it was estimated that low-cost adaptation measures, like earlier planting or increased irrigation, could succeed in reducing agricultural damages to the region by 30 to 60% (Easterling *et al.*, 1993). A comparable range was found in the global studies by Rosenzweig and Parry (1994) and (using the same yield data) Reilly *et al.* (1994). In the Rosenzweig and Parry study a change of -1.2 to -7.6% in worldwide cereal production without adaptation is reduced to 0% to -5% with moderate farm-level adaptation and +1% to -2.5% with a more comprehensive managerial adjustment. As a consequence, the global welfare loss reduces from -0.1 to -61.2 $bn without adaptation to +7.0 to -37.6 $bn in the case of a moderate response (see Pearce *et al.*, 1995).

Worst Case vs. Best Guess: The estimates of Tables 2 to 4 concentrate on the most probable damage scenario, i.e. they merely provide a *best guess* assessment of what damages are most likely to be. Given the complexity of the climatic system and the unprecedented stress imposed on it, this focus may be too narrow, though. Other, more disastrous scenarios cannot be excluded with certainty. Rather than with only one point, we are confronted with an entire damage probability distribution. Unfortunately, only little is known about the shape of this distribution, and in particular about the probability of an extremely adverse outcome. Several catastrophe scenarios have been portrayed in the literature so far (Howarth and Monahan, 1992; Cline, 1992a):

* The melting of the polar ice caps. A possible *disintegration of the west-antarctic ice sheet* would rise sea levels by up to 6 meters. This process is however slow and would take place over a time span of 300-500 years.

* A *shut-down of the ocean conveyer belt* may lead to changes in ocean circulation patterns. A redirection of the gulf stream would - somewhat ironically - cause significant cooling in Western Europe, with temperatures comparable to those currently observed in Canada (abstracting from the amount of warming which will already have occurred at that date).

* The *runaway greenhouse effect*: Initial warming levels may be amplified through massive feedback effects, e.g. through the liberation of methane from previously frozen sediments into the atmosphere.

* *Abrupt, non-linear changes in climate patterns*. There is paleo-climatic evidence from ice-cores pointing at the prospect of a highly unstable climate with temperature

changes of several degrees Celsius within only a few years.

In addition to such worst case impacts, there may also be *surprises*, events which are impossible to predict before hand and for which no probability of occurrence therefore exists. Evidently, by their very nature nothing can be said about the direction or magnitude of such events. However, in neither considering the entire damage distribution, nor the possibility of surprises, existing estimates are clearly incomplete. This has to be remembered in the decision about the optimal greenhouse policy response. Arguably, decision makers may be inclined to take precautionary measures to reduce the risk of a catastrophe or an unhappy surprise (see Section 5).

Long Term Impacts: Existing estimates are predominantly concerned with the case of $2xCO_2$. For practical purposes, however, $2xCO_2$ is only of limited relevance. Atmospheric CO_2 concentrations could double as early as 2030-2050. It would therefore be important to know more about damages beyond $2xCO_2$. Regrettably, only very little work has been done in this respect, both with respect to impacts in physical terms and with respect to the economic costs. The only economic assessment of long run damage costs is found in Cline (1992a), whose estimates are, however, in parts rather speculative. An attempt has also been made by Nordhaus (1994), in his poll of impact experts. The interviewed experts predicted an average damage of 6.7% of GNP for 6°C warming by 2175, with a median value of 4.1%. Experts felt generally rather uncomfortable with respect to long run predictions, though. In the absence of better information, greenhouse damage costs are usually specified as a polynomial (typically linear to cubic) function of temperature rise, calibrated around the available $2xCO_2$ estimates. Damage is further assumed to rise in proportion to GNP. Clearly, these assumptions are rather *ad hoc*, and much further effort will be required for damage assessments to reach a similar level of sophistication as has already been achieved in carbon abatement modelling.

Predicting the Future: Forecasting the structure of the future societies confronted with climate change is extremely difficult. Predicting the preferences of future generations for commodities affected by warming is a near impossibility. For this reason, damage estimates will of necessity always be uncertain, even if scientific uncertainty is resolved. Enumerative estimates usually circumvent this problem by imposing $2xCO_2$ on an economy with today's structure. In most studies, base period damage is then assumed to simply grow in proportion to GNP. This is clearly an approximation, though, and we can expect some damage aspects (e.g. the value of endangered ecosystems) to grow faster than others.

Schelling has illustrated the difficulties of predicting future impacts by deliberating on how forecasters in the year 1900 would have predicted possible greenhouse impacts for the year 1992. "There would have been no way to assess the impact of changing climates on air travel, electronic communication, the construction of skyscrapers, or the value of Californian real estate" (Schelling, 1992, p.4).

However, while Schelling's observation is correct, the difficulty of the task is in itself no reason not to undergo the analysis. The current discussion about the best possible policy response suggests that a damage assessment of some sort is urgently needed, and the fact that it will be based on uncertain data should not prevent us from obtaining it. There is nothing unusual about expectations being formed subject to a limited information set. It would only be unusual not to make use of all the information available. Of course, the quality of the forecast will also play a role when deciding about the optimal greenhouse policy response.

4. MARGINAL DAMAGE COSTS AND ABATEMENT BENEFITS

4.1 Social Cost Estimates and the Benefits from Avoided Warming

Section 3 has dealt with the economic costs of greenhouse concentration doubling ($2xCO_2$). Despite the attention the $2xCO_2$ case enjoys in the literature, it is not directly relevant for practical purposes, though. For the appraisal of abatement projects it is more important to know the *marginal social costs per tonne of emission*, or, which is equivalent, the marginal benefit per tonne abated.

Social cost estimates are conceptually different from the $2xCO_2$ damage discussed above. A tonne of CO_2 emitted today has its impact not only today, but over several time periods - as long as the gas, or fractions of it, remains in the atmosphere. The social costs from increased emissions thus occur as a stream of higher damages over time. The damage costs of a tonne of emission are then defined as *the discounted sum of additional future damages*, which occur as a consequence of the emission increase. $2xCO_2$ damages, on the other hand, only concern one particular point in time: the period when concentration doubling has actually occurred (see Fankhauser, 1994a). $2xCO_2$ assessments can at best provide a rough indication of the actual marginal costs of emission - or the benefits of abatement. A careful analysis would require much more information than is available from $2xCO_2$ estimates, in particular about the size of damage outside the $2xCO_2$ benchmark, and how it is affected by economic and population growth. Little is known about these factors and assumptions are often rather *ad hoc*.

Two different methods are available to provide social cost estimates. Most studies estimating the social costs of greenhouse gas emissions do so in an intertemporal cost-benefit framework. That is, their primary concern is to calculate socially optimal emission levels at each point in time. In such a model the social costs are equal to the *shadow price* of emissions, that is the pollution tax necessary to keep emissions at the optimal level in each time period (see Fankhauser, 1994a). A second way is to calculate social costs directly as the difference in future damage levels caused by a marginal change in baseline emissions. Both methods have been used so far.

The pioneering paper on the social costs of CO_2 emissions is Nordhaus (1991). Using a simplified version of a dynamic optimisation model, he calculates social costs of 7.3 $ per tonne of carbon emitted. Imposing different assumptions on the rate of discount and $2xCO_2$ damage leads to a range of 0.3 $/tC to 65.9 $/tC. Implying that abatement should only be undertaken as long as costs do not exceed $ 7.3 per tonne of carbon abated, the estimates formed the backbone of Nordhaus' claim that global warming may not, after all, be such a big problem, and may justify only a modest policy response.

The Nordhaus estimates have been strongly criticised by several authors, including for example Ayres and Walter (1991), Cline (1992a) and Grubb (1993). While the main objection concerned Nordhaus' $2xCO_2$ damage estimate, the more substantial shortcomings are probably those concerning the model itself (see Cline, 1992a). Especially controversial is the assumption of a *resource steady state*, which *inter alia* implies a constant level of CO_2 emissions over time. Obviously this is unrealistic. The IPCC for example predicts an increase in annual CO_2 emissions from about 7 GtC in 1990 to about 9-14 GtC by 2025 (IPCC, 1992). The simple (linear) structure of the climate and damage sectors also implies that costs will remain constant at 7.3 $/tC throughout. Climate processes are clearly non-linear, and the costs of CO_2 emissions will thus depend on future concentration and warming levels, i.e. they

Table 5 The Social Costs of CO_2 Emissions in Different Decades (current value $/tC)

Study	Type	1991-2000	2001-2010	2011-2020	2021-2030
Nordhaus (1991)	MC		7.3 (0.3-65.9)		
Ayres/Walter (1991)	MC		30 - 35		
Nordhaus (1993)	CBA	5.3	6.8	8.6	10.0
Cline (1992b, 1993)	CBA	5.8 - 124	7.6 - 154	9.8 - 186	11.8 - 221
Peck and Teisberg (1992)	CBA	10 - 12	12 - 14	14 - 18	18 - 22
Fankhauser (1994a, b)	MC	20.3 (6.2 - 45.2)	22.8 (7.4 - 52.9)	25.3 (8.3 - 58.4)	27.8 (9.2 - 64.2)
Maddison (1993)	CBA/ MC	5.9 - 6.1	8.1 - 8.4	11.1 - 11.5	14.7 - 15.2

Note: MC = marginal social cost study,
CBA = shadow value in a cost-benefit study.
Figures in brackets denote 90% confidence intervals.

Source: Fankhauser (1994a).

will vary over time. Subsequent estimates suggest that they may in fact rise over time. That is, a tonne of CO_2 added to an already large stock of atmospheric CO_2 is likely to cause a higher damage than a tonne emitted under a low concentration level.

These objections are also relevant to the study by Ayres and Walter (1991), whose calculations are based on the Nordhaus model. The paper has additional shortcomings. Particularly questionable is the fact that the authors use uniform commodity values, e.g. for land, for the whole world. This is unrealistic. Land prices clearly differ between regions. By considering both the costs of sea level rise protection and the costs of climate refugees from coastal regions Ayres and Walter further appear to double count at least some of the sea level rise impacts. On the whole, their cost estimate of 30-35 $/tC is therefore doubtful.

The shortcomings of the earlier model were recognised and corrected in Nordhaus' subsequent approach, the DICE (Dynamic Integrated Climate Economy) model (Nordhaus, 1993). DICE is an optimal growth model in the Ramsey tradition, extended to include a climate module and a damage sector which feed climate changes back to the economy. The shadow values of carbon following from DICE are in the same order as Nordhaus' previous results, starting at 5.3 $/tC in 1995 and gradually rising to 6.8 $/tC in 2005 and 10 $/tC in 2025 (see Table 5). Note that figures for future periods are current value estimates, i.e. they denote the social costs valued at the time of emission.

The DICE model was also used by Cline (1992b), who concludes that Nordhaus' choice of parameter values may have lead to an underestimation of the true costs. Of particular importance is the discount rate. The shadow values of the Cline replication are reported in Cline (1993), and are also shown in Table 5.

Figures within the same order of magnitude as the original DICE results were suggested by Peck and Teisberg (1992, 1993) and Maddison (1993). Peck and Teisberg came up with a shadow value of carbon of about 10\$/tC in 1990, rising to about 22 \$/tC by 2030. The CETA (Carbon Emission Trajectory Assessment) model, on which their calculations are based possesses a similar climate and damage sector as DICE, but is more detailed on the economy side by incorporating a carefully modelled energy sector. Differences between the estimates appear to be mainly due to different assumptions about the size of $2xCO_2$ damage. Common to all three papers is the assumption of a 3% utility discount rate. The results are highly sensitive to this assumption, as Cline's results have underlined.

In comparison, Fankhauser (1994a, b) calculates a flatter trajectory with costs rising from 20 \$ per tonne of emissions in 1991-2000 to 28 \$/tC by 2021-2030. An important feature of this model is its stochastic character. All major parameters, including the discount rate, are treated as random. Conceptually, Fankhauser's figures are based on a direct calculation of social costs, rather than the shadow value method.

4.2 Non-Climate Related Benefits

The benefits of greenhouse gas abatement will not be limited to reduced climate change costs alone, but are likely to spill over to other sectors. For example, if abatement is brought about by an environmental tax (e.g. a carbon tax), the resulting revenues can be used to lower other taxes, such as those on income - the so called double dividend of carbon abatement. Efforts to halt deforestation will reduce the emission of CO_2 as well as contributing to the conservation of the world's biological diversity. Other ancillary benefits could occur in the form of local and regional air quality improvements, as a reduction in traffic related externalities like accidents or congestion, and as a reduced risk of tanker accidents and oil spills (Pearce *et al.* 1995).

Secondary Benefits: Because no economical CO_2-removal technologies currently exist, attempts to limit CO_2 emissions will by and large concentrate on cutting down the use of fossil fuels. A reduction in CO_2 emissions will therefore also reduce other environmental problems related to fuel combustion. In particular, CO_2 abatement will curb the emission of classic air pollutants like SO_2, NO_x and particulates. These effects are often called the secondary benefits of carbon abatement. In studies for Norway, the UK and the United States researchers tentatively found average secondary benefits ranging from about \$2 per tonne of carbon abated to over 500 \$/tC. Depending on the pollutant and baseline air quality standards, reductions in air pollution levels varied from 1% to over 30%, for carbon emission cuts between 8% and 15% (see Pearce *et al.*, 1995). In some countries, most prominently the UK, air quality improvements appear to be sufficient to justify carbon abatement measures independent of global warming considerations.

Double Dividend: If greenhouse gas abatement is achieved through tax measures (e.g. a carbon tax), governments will raise a substantial amount of revenues. This would enable governments to lower other, distortionary taxes such as those on labour, and lead to an additional, non-environmental benefit in the form of a reduced deadweight loss from government revenue raising. Taxing economic activities generally leads to distortions of the economic system and causes efficiency losses. Income taxes, for example, create a disincentive for people to work, biasing their preferences toward the untaxed activity leisure. Replacing an income tax with a carbon tax can help to ease this distortion. The size of this so called double dividend could potentially be considerable. Distortionary losses due to taxation were estimated at 20-50 cents per tax dollar raised (Ballard *et al.*, 1985).

The exact size of the double dividend depends on the chosen mode of revenue recycling. Different recycling modes, including personal income tax cuts, deficit reduction and investment tax credits, have been analysed by Shackleton *et al.* (1992). They found that the greatest double dividend will be achieved under schemes involving reductions in capital and labour costs which encourage capital formation, and thus long run growth, rather than immediate consumption. In some cases, the imposition of a carbon tax could actually lead to welfare *improvements*. Although not without conceptual problems (see Fankhauser, 1994a), the double dividend idea is thus clearly important, and deserve further attention.

5. POLICY IMPLICATIONS AND CONCLUSIONS

5.1 Frameworks for Decision Making

What are the policy implications of the above global warming damage results? There is hardly an aspect of greenhouse economics which is more fiercely disputed than the question of the optimal policy response. Several approaches to the problem can be distinguished on a methodological level, and not all of them require a knowledge of greenhouse impacts in monetary terms.

The most prominent approach, at least among economists, is probably the cost-benefit approach. In the cost-benefit approach the optimal policy is determined through a trade off between the costs of policy action and the benefits from greenhouse damage avoided. This need not necessarily imply a strict cost-benefit analysis in the traditional sense, with costs and benefits expressed in monetary units, though. The cost-benefit philosophy is more broadly centred around the general notion of weighing up "goods" against "bads".

Alternatively emission targets could emerge from ethical, political or precautionary considerations, more or less independently of the resulting costs and benefits. This is for example the stance taken by Howarth and Monahan (1992), who propose a rule for greenhouse action based on the *sustainability principle*. Today's generation has a moral obligation to defend the safety and wellbeing of future generations, "if doing so would not noticeably diminish ... [today's] quality of life" (p. 6-85). Even if climate impacts were not catastrophic overall, obligations could still occur towards a minority of severely affected people, e.g. the inhabitants of small island states. Carbon targets based on the *precautionary principle* have for example been proposed by Krause *et al.* (1989) and Swart and Vellinga (1994).

It is not the purpose of this paper to decide on the respective merits of each of these approaches. Our interest rests with the economic assessment of greenhouse impacts, and consistent with this scope, the following analysis will concentrate on those approaches which require some knowledge about the economic costs of global warming.

5.2 Cost-Benefit Results

The first application of the cost-benefit method to global warming is the influential paper by Nordhaus (1991). On the basis of a marginal CO_2 damage of 7.3 \$/tC in the best guess case (see Table 5), the paper concludes that only a limited amount of greenhouse abatement would be warranted, most of it achieved through reforestation and the phasing out of CFCs. This controversial conclusion has provoked much criticism, and the model has indeed several shortcomings, as noted above (see Section 4).

An alternative to the Nordhaus model has been provided by Cline (1992a). Owing to the inclusion of features like no regrets options and risk aversion, and by using a different discount rate Cline found favourable benefit-cost ratios for an aggressive abatement plan of freezing CO_2 emissions at 4 GtC per annum, about two thirds of their 1990 level. His analysis is still not based on an optimal control model, but assesses the merits of one particular project.

Optimal control models developed since provide a similarly differentiated picture, see for example the results from the two models CETA (Carbon Emission Trajectory Assessment - Peck and Teisberg, 1992, 1993) and DICE (Dynamic Integrated Climate Economy - Nordhaus, 1993). While Nordhaus' results are very similar to his earlier conclusions - he proposes optimal CO_2 reductions of only about 15% off baseline projections by 2100 -, the results are in general quite heterogeneous and appear to be extremely sensitive to the underlying assumptions.

This has been noted by Cline (1992b) who replicated the DICE model and observed that the Nordhaus result of only modest abatement "does not stem inherently form the optimization model and approach used, but hinges on the particular assumptions applied" (p.31), the most important one being the choice of the discount rate. In some of Cline's alternative calculations CO_2 emissions are virtually phased out by the end of the 21st century. Other critical features include the availability of a carbon free backstop technology and the treatment of non-market damages. By interpreting non-market damages as a direct element in the utility function, rather than a production cost, Tol (1994b) found significantly higher abatement levels than Nordhaus.

A similar, although less extreme picture also arises from the sensitivity analyses done by Peck and Teisberg (1992, 1993). Over the first forty or fifty years the various optimal emission trajectories calculated in the model barely differ from each other and all closely follow the baseline. The subsequent trajectory, however, strongly depends on the chosen damage parameters, particularly on the slope of the damage function. Significant emission reductions only occur for a steep damage-temperature relationship (Peck and Teisberg, 1993).

5.3 Hedging Against Catastrophes

It is one of the main weaknesses of most greenhouse cost-benefit models that they tackle the problem as if all parameters were known with certainty (exceptions are Hope *et al.*, 1993; and Dowlatabadi and Morgan, 1993). Clearly, this is not the case, and an optimal greenhouse policy has to take uncertainty into account. Many advocates of immediate action interpret greenhouse gas abatement as an insurance, and argue that for the same reasons as individuals are willing to secure themselves against hazards of all sorts, society should be willing to spend some money to protect itself against adverse climatic effects, particularly since these may be irreversible and potentially disastrous. This view is implicit in much of the literature on the precautionary principle, and it has also been issued by e.g. Schneider (1993) and Schelling (1992). An explicit analysis of the insurance argument is Manne and Richels (1992).

Much of the persuasiveness of the insurance argument hinges on the fear of a possible climate catastrophe. The question of how to handle such high impact/low probability events is however far from settled. Two archetypal methods of decision theory have been applied so far. The first is the *expected utility* approach, as used by e.g. Heal (1984) and Peck and Teisberg (1994). The second is a *maximin* approach, as proposed by e.g. Krause *et al.* (1989). To minimise the risk of a climate catastrophe the approach requires that an emission target be set at the maximum level of emissions under which a climate catastrophe can reasonably be excluded .

In Heal's model the atmosphere is interpreted as an exhaustible resource of unknown stock. Exhausting the resource (passing a climate threshold) would trigger a climate catastrophe, and utility would fall to zero. People take precautions against this event by reducing emissions, and the optimal rate of emissions is determined in a trade off between abatement costs and the risk of catastrophe. The amount of abatement undertaken crucially depends on the degree of people's risk aversion, as well as on their perception of the likelihood of a catastrophe.

Peck and Teisberg (1994) have implemented Heal's idea using the probability assessments of the Nordhaus (1994) poll of experts. Surprisingly, their results hardly differ from those achieved with a conventional damage function (e.g. Peck and Teisberg, 1992). In the main scenario - with catastrophe probabilities calculated using data from all respondents - CO_2 emissions continue to rise throughout the next century and reach almost 40 GtC/year by the year 2100, about six times 1990 levels. In a more pessimistic scenario based on the responses from natural scientists only, emissions still rise to about 20 GtC/year in 2100, and are only gradually reduced below 1990 levels thereafter.

With a maximin approach the picture is completely different. Under a maximin strategy society is only concerned with the worst possible outcome (i.e. an early catastrophe) and implements the policy which would maximise the payoff under this scenario. As a consequence the resulting abatement targets are rather high. In the analysis of Krause *et al.* (1989) it leads to the imposition of a maximum warming target of 0.1°C/decade, and 2.5°C in total. Working backwards they found that this would translate into a reduction in emissions of 20% below 1985 levels by 2015 and 75% by 2050.

5.4 Policy Implications

The variability of methodologies and the high sensitivity of results makes it difficult to derive policy conclusions. Existing optimal control models on the one hand have a tendency to favour relatively moderate abatement levels. A risk minimisation strategy on the other hand would require significant emission cuts within the next two decades or so. Outside the cost-benefit paradigm a case for more stringent abatement could also be made for reasons of intergenerational equity, or on the grounds that the fate of the worst affected nations, e.g. small island states, should merit particular attention.

Fortunately, the problem is less severe once it is recognised that decisions can be taken sequentially (Manne and Richels, 1992). While model predictions heavily differ over the medium and long term, there is far less divergence with respect to the immediate future - the time period with which current abatement decisions are concerned. Cline (1993, p.28) even observes "a surprising convergence of the various analyses" for the first decade, with a 10-15% emission cut emerging as a possible consensus policy for the next ten years or so.

For subsequent periods, there are signs that the inclusion of secondary benefits and revenue recycling into cost-benefit analyses could tilt the balance in favour of a more stringent abatement policy in the medium run. This would be consistent with the precautionary and equity views held by many authors, and would hopefully leave enough time to learn more about the emission policy required in the long term. Insuring the world against global warming may not be too expensive, after all.

Notes

1. In many sections the paper draws on the more detailed expositions in Fankhauser (1994a) and Pearce *et al.* (1995).

References

Asaduzzaman, M. (1994). *Economic and Social Impacts of Climate Change: A Case Study of Bangladesh Coastal Zone*, Mimeo, Bangladesh Institute of Development Studies, Dhaka.

Ayres, R. and J. Walter (1991). The Greenhouse Effect: Damages, Costs and Abatement, in: *Environmental and Resource Economics* 1: 237-270.

Ballard, C.L., J.B. Shoven, and J. Whalley (1985). General Equilibrium Computations of the Marginal Welfare Costs of Taxes in the United States, in: *American Economic Review* 75(1): 128-138.

Cline, W.R. (1992a). *The Economics of Global Warming*, Washington, DC: Institute for International Economics.

Cline, W.R. (1992b). *Optimal Carbon Emissions over Time: Experiments with the Nordhaus DICE Model*, Mimeo, Institute for International Economics, Washington, DC.

Cline, W.R. (1993). *Modelling Economically Efficient Abatement of Greenhouse Gases*, Paper presented at the United Nations University Conference on "Global Environment, Energy and Economic Development", September 1993, Tokyo.

Cline, W.R. (1994). The Costs and Benefits of Greenhouse Abatement: A Guide to Policy Analysis, in: OECD/IEA, *The Economics of Climate Change*, Paris: OECD/IEA.

Dowlatabadi, H. and M.G. Morgan (1993). A Model Framework for Integrated Studies of the Climate Problem, in: *Energy Policy* 21(3): 209-221.

Easterling III, W.E., P.R. Crosson, N.J.Rosenberg, M.S. McKenney, L.A. Katz and K.M. Lemon (1993). Agricultural Impacts of and Responses to Climate Change in the Missouri-Iowa-Nebraska-Kansas (MINK) Region, in: *Climatic Change* 24(1/2): 23-62.

Fankhauser, S. (1994a). *Valuing Climate Change. The Economics of the Greenhouse Effect*, London: Earthscan, forthcoming.

Fankhauser, S. (1994b). The Social Costs of Greenhouse Gas Emissions: An Expected Value Approach, in: *Energy Journal* 15(2): 157-184.

Fankhauser, S. (1994c). Protection vs. Retreat. The Economic Costs of Sea Level Rise, in: *Environment and Planning A*, forthcoming.

Grubb, M. (1993). The Costs of Climate Change: Critical Elements, in: Y. Kaya, N. Nakićenović, W.D. Nordhaus and F. Toth (eds.), *Costs, Impacts and Benefits of CO_2 Mitigation*, IIASA Collaborative Paper Series, CP93-2, Laxenburg, Austria.

Heal, G. (1984). Interactions between Economy and Climate: A Framework for Policy Design under Uncertainty, in: *Advances in Applied Microeconomics* 3: 151-168.

Hope, C. J. Anderson, and P. Wenman (1993). Policy Analysis of the Greenhouse Effect. An Application of the PAGE Model, in: *Energy Policy* 21(3): 327-338.

Howarth, R.B. and P.A. Monahan (1992). *Economics, Ethics, and Climate Policy*, Mimeo, Lawrence Berkeley Laboratory, Berkeley, CA.

Intergovernmental Panel on Climate Change (1990). *Climate Change, the IPCC Impacts Assessment*. Report from Working Group II, Canberra: Australian Government Publishing Service.

Intergovernmental Panel on Climate Change (1992). *Climate Change 1992. The Supplementary Report to the IPCC Scientific Assessment*, Cambridge: Cambridge University Press.

Intergovernmental Panel on Climate Change (1994). *Preparing to Meet the Coastal Challenges of the 21st Century*. Conference Report of the World Coast Conference, Nordwijk, November 1993, Geneva: WMO and UNEP.

Jorgenson, D.W. and P.J. Wilcoxen (1990). *Global Change, Energy Prices, and US Economic Growth*, HIER Discussion Paper 1511, Harvard Institute of Economic Research, Cambridge, MA.

Krause, F., W. Bach, and J. Kooney (1989). *Energy Policy in the Greenhouse*, London: Earthscan.

Maddison, D. (1993). *The Shadow Price of Greenhouse Gases and Aerosols*, Draft, Centre for Social and Economic Research on the Global Environment, University College London and University of East Anglia, Norwich.

Manne, A.S., R. Mendelsohn, and R.G. Richels (1993). *MERGE - A Model for Evaluating Regional and Global Effects of Greenhouse Gas Reduction Policies*, Mimeo, Electric Power Research Institute, Palo Alto, CA.

Manne, A.S. and R.G. Richels (1992). *Buying Greenhouse Insurance*, Cambridge MA: MIT Press.

Nicholls, R., K. Dennis, and S. Leatherman (1990). *Impacts of Sea level Rise on Selected Developing Countries*, Proceedings of 3rd European Workshop on Coastal Zones, Cyprus, September 1990.

Nishioka, S, H. Harasawa, H. Hashimoto, T. Ookita, K. Masuda, and T. Morita, eds. (1993). *The Potential Effects of Climate Change in Japan*, Centre for Global Environmental Research, and National Institute for Environmental Studies, Tsukuba (Japan).

Nordhaus, W.D. (1991). To Slow or not to Slow: The Economics of the Greenhouse Effect, in: *Economic Journal* 101(407): 920-937.

Nordhaus, W.D. (1993). Rolling the "DICE": An Optimal Transition Path for Controlling Greenhouse Gases, in: *Resources and Energy Economics* 15(1): 27-50.

Nordhaus, W.D. (1994). Expert Opinion on Climate Change, in: *American Scientist*, Jan/Feb: 45-51.

Parry, M.L. and R. Duncan, eds. (1994). *Economic Implications of Climate Change in Britain*, London: Earthscan.

Pearce,D.W, W.R. Cline, R.K. Pachauri, P. Vellinga, S. Fankhauser and R.S.J Tol (1995). The Social Costs of Climate Change: Greenhouse Damage and the Benefits of Control, in: Intergovernmental Panel on Climate Change, *The IPCC Second Assessment Report: Working Group III*, Geneva: WMO and UNEP, forthcoming.

Pearce, D.W. and R.K. Turner (1990). *Economics of Natural Resources and the Environment*, London: Harvester Wheatsheaf.

Peck, S.C. and T.J. Teisberg (1992). CETA: A Model for Carbon Emissions Trajectory Assessment, in: *Energy Journal* 13(1): 55-77.

Peck, S.C. and T.J. Teisberg (1993). Global Warming Uncertainties and the Value of Information: An Analysis Using CETA, in: *Resource and Energy Economics* 15(1): 71-97.

Peck, S.C. and T.J. Teisberg (1994). *Note on Optimal CO_2 Control Policy with Temperature Dependent Disaster Probabilities*, Mimeo, Electric Power Research Institute, Palo Alto, CA.

Reilly, J., N. Hohmann and S. Kane (1994). Climate Change and Agricultural Trade: Who Benefits, Who Loses?, in: *Global Environmental Change* 4(1): 24-36.

Rosenberg, N.J. (1993). A Methodology Called MINK for Study of Climate Change Impacts and Responses on the Regional Scale. An Introductory Editorial, in: *Climate Change* 24(1/2):1-6.

Rosenzweig, C. and M.L. Parry (1994). Potential Impact of Climate Change on World Food Supply, in: *Nature* 367(13 January): 133-138.

Schelling, T. (1992). Some Economics of Global Warming, in: *American Economic Review* 82(1): 1-14.

Scheraga, J.D., N.A. Leary, R.J. Goettle, D.W. Jorgenson, and P.J. Wilcoxen (1993). Macroeconomic Modelling and the Assessment of Climate Change Impacts, in: Y. Kaya, N. Nakićenović, W.D. Nordhaus and F. Toth (eds.), *Costs, Impacts and Benefits of CO_2 Mitigation*, IIASA Collaborative Paper Series, CP93-2, Laxenburg, Austria.

Schneider, S.H. (1993). Pondering Greenhouse Policy, in: *Science* 259(5 March): 1381.

Shackleton, R., M. Shelby, A. Cristofaro, R. Brinner, J. Yanchar, L. Goulder, D. Jorgenson, P. Wilcoxen, P. Pauly, and R. Kaufmann (1992). *The Efficiency Value of Carbon Tax Revenues*, Paper submitted to the Stanford Energy Modelling Forum Report 12 (EMF 12), US Environmental Protection Agency, Washington, DC.

Smith, J.B. and D.A. Tirpak, eds. (1989). *The Potential Effects of Global Climate Change on the United States*, Washington, DC: US Environmental Protection Agency.

Swart, R.J. and P. Vellinga (1994). The "Ultimate Objective" of the Framework Convention on Climate Change Requires a New Approach in Climate Change Research. An Editorial, in: *Climatic Change* 26(4): 343-350.

Titus, J. (1992). The Cost of Climate Change to the United States, in: S.K. Majumdar, L.S. Kalkstein, B. Yarnal, E.W. Miller, and L.M. Rosenfeld (eds.), *Global Climate Change: Implications Challenges and Mitigation Measures*, Pennsylvania: Pennsylvania Academy of Science.

Titus, J.G., R.A. Park, S.P. Leatherman, J.R. Weggel, M.S. Greene, P.W. Mausel, S. Brown, C. Gaunt, M. Trehan, and G. Yohe (1991). Greenhouse Effect and Sea Level Rise: The Cost of Holding Back the Sea, in: *Coastal Zone Management* 19: 172-204.

Tol, R.S.J. (1994a). *The Damage Costs of Climate Change: Towards More Comprehensive Calculations* (revised version), Working Document W94/?, Free University Amsterdam, forthcoming.

Tol, R.S.J. (1994b). The Damage Costs of Climate Change. A Note on Tangibles and Intangibles, Applied to DICE, in: *Energy Policy* 22(5): 436-438.

Viscusi, W.K. (1993). The Value of Risks to Life and Health, in: *Journal of Economic Literature*, XXXI(December): 1912-1946.

TOWARD GLOBAL PLANNING OF SUSTAINABLE USE OF THE EARTH
Development of Global Eco-Engineering
S. Murai (Editor)

Economic interdependence and eco-balance: accounting for the flow of environmental loads associated with trade

Hidefumi Imura[a] and Yuichi Moriguchi[b]

[a]Institute of Environmental Systems, Faculty of Engineering, Kyushu University, 6-10-1, Hokozaki, Higashi-ku, Fukuoka, Japan

[b]National Institute for Environmental Studies, 16-2 Onogawa, Tsukuba, Japan

Abstract

All human beings have common but differentiated responsibilities for protecting the global environment. Through trade in goods and services, the citizens' life and economic activities in one country are linked to the rest of the world, giving rise to environmental costs in other places. Responsibilities for causing impacts on the global environment should be assessed by taking into account such global interdependence. This paper discusses the economic and ecological links arising from trade in goods and services between nations, regions, and differing economic sectors. It focuses on the use of energy and the associated environmental impact using CO_2 emissions as an indicator. Using international trade statistics and energy data, it presents an estimate of the export and import balance of energy and CO_2 embodied in the transacted goods and services among nations and regions. By and large, there already exists a large gap in direct energy use per capita between industrialized and developing countries. It is shown that this gap is larger if the indirect input of energy associated with trade is taken into account. Thus the trade under the present world economic structure tends to increase the disparity between North and South in terms of resource consumption and impacts on the global environment. The paper also discusses the interdependence between economic sectors and regions within a country.

1. INTRODUCTION

The protection of global assets such as the climatic system, the oceans, the ozone layer, and biological diversity is the collective responsibility of all human beings. However, there are many views on the responsibilities of individual nations, communities, and economic sectors. Conflicts exist between industrialized and developing countries. The former enjoy material affluence and cause larger environmental impact, while the latter have a large population but each of their people uses a smaller amount of environmental resources. The Rio Declaration on Environment and Development therefore proclaims that States have "common but differentiated responsibilities" [1].

In today's globalized world economy, countries closely depend upon each other for transactions of goods and services. Through trade, economic activities and their private lives, even citizens of a small local community are linked to the rest of the world. Responsibilities for causing environmental impacts should be discussed taking into account such economic interdependence between nations, regions, and economic sectors.

A simple example of the problems of trade and environment is where hazardous materials are exported to other countries and cause damage to the environment of the importing countries. In general, however, the relationship between trade and environment is more indistinct and complicated. Traded goods and services give rise to environmental impact throughout the process of their production, transportation, consumption and disposal. The imported goods may not even be consumed in the country, but be changed into other forms and exported to other countries. Hidden within the trade there may exist disparities between countries in the cost and the required level of environmental countermeasures as in the price of products; these may reflect differing policy priorities, regulatory systems and standards.

Similar considerations also apply to the relation between regions within a country. Industrial cities manufacture products and generate electricity at their factories and power plants, burning large amounts of fossil fuels and emitting carbon dioxide (CO_2). Commercial cities, on the contrary, use smaller amounts of energy and reduce their emissions of CO_2 by importing goods and services from outside.

This argument can be extended to the roles and responsibilities of producers and consumers. Energy is used at various stages of the production and distribution of goods and services. Such consumption is nothing but the indirect use of the energy embodied in them. Consumers should be aware of their responsibilities for the total (direct and indirect) environmental impact caused by the goods and services they consume.

This paper discusses the economic and ecological interdependence through trade in goods and services between nations, regions, and various economic sectors. It focuses on the use of energy and the associated environmental load in terms of CO_2 emissions. Using international trade statistics and energy data, it presents an estimate of the export and import balance of energy and CO_2 embodied in the transacted goods and services between nations and regions in the world. It also discusses the interdependence between economic sectors and regions within a country.

2. FRAMEWORK OF ANALYSIS

2.1. Trade and environment implications of the gaps in energy intensities of products among countries

Energy is indispensable for every kind of economic activity and human life. Use of fossil fuels, however, is inevitably accompanied by the emission of carbon dioxide and other air pollutants. The amount of energy used in the production and consumption of goods and services can serve as a good indicator of the environmental loads caused by human activities.

In line with the Framework Convention, many industrialized nations presented national action plans for stabilizing the emission of CO_2 and other greenhouse gases in order to protect the climatic system of the earth [2]. Questions, however, may be raised as to the definition of national stabilization targets. Firstly, most of the targets are determined with reference to the baseline CO_2 emissions in the year 1990. The baseline CO_2 emission per capita exhibits remarkable disparity between North and South regions as well as among countries: for example, 6.15 tC in the United States, 2,69tC in Japan, and 0.60 tC in China [3]. From the viewpoint of "equity," the reduction of such disparities should be a matter for international discussion. The second question concerns the definition of CO_2 emissions: if a country moves its production facilities to other countries and imports finished products as much as possible, it can reduce its apparent CO_2 emissions. What should be considered here is the "embodied energy" or the total energy input for the production of goods and services traded between countries.

Primary energy demand per unit of monetary production (e.g. per US $ 1 million) greatly differs even among industrialized countries: 151 TOE (tons oil equivalent) in Japan, 249 TOE in Germany, 288 TOE in the United Kingdom, and 396 TOE in the United States for 1990. The disparity further increases if the former centrally-planned economies and developing countries are included. The energy intensity of production in China, for example, is 1763 TOE / million $ US, twelve times larger than that in Japan[3]. The gaps among countries in resource endowment, energy prices, technologies and climatic conditions give rise to such differences, and they govern the international balance of environmental loads associated with the trade of goods and services. Import of products from countries having larger energy intensity per unit of monetary production may have an environmentally significant meaning even if the amount of trade, expressed in monetary terms, is marginal.

2.2. Review of the methodology

Methodologies have been developed for quantifying the environmental loads associated with the production and consumption of goods and services, and for allocating the responsibilities for such loads to the different economic entities concerned. The most typical method is based on the input-output model originally developed by Leontief. The method was widely applied to the analysis of energy consumption induced by the final demand of goods and services in an economy [4]. Costanza[5] applied this model to the US economy and showed that there is a strong relation between embodied energy values and market-determined dollar values for most economic sectors with the appropriate perspective and boundaries. Moreover, quite a few studies recently have been undertaken to analyze the induced emission of CO_2 from differing economic sectors in a national economy [6, 7].

The input-output model can be adapted to analyze the environmental implications of international trade if relevant international I-O tables are made available. However, application of the model to the World economy is not practical mainly due to the lack of consistent data. There are several approaches to overcome the difficulty. Moriguchi *et al* [8] adopted a method of using national input-output tables and energy data available for a number of countries. They calculated the embodied energy intensities of major export products in major countries, and analyzed the export-import balance of embodied environmental loads between countries. Imura *et al* [9] presented another method which assumes that each country and/or region consists of one economic sector. They dealt with the world economy in a simplified manner and presented the trade balance of embodied environmental loads between countries and regions.

On the other hand, since the early 1990s studies have been undertaken on the life cycle assessment (LCA) of products, aiming to analyze their "cradle to grave" environmental loads and impacts. If LCA could be applied to every product manufactured around the world, it would enable us to assess the environmental implications of traded goods in a more consistent manner.

2.3. Embodied energy and environmental loads

According to the standard methodology of input-output analysis [4], the embodied energy intensity – the total energy required by the economy directly and indirectly in supporting one unit monetary value final demand – can be calculated by the following equation:

$$\varepsilon = E\,(X^{*} - X)^{-1} \tag{1}$$

Here, $\varepsilon = (\varepsilon_1, \varepsilon_2, ..., \varepsilon_n)$ is the embodied energy intensity vector, where ε_i denotes the total (direct and indirect) energy required per dollar's worth of final demand presented to sector i ; $E = (E_1, E_2, \cdots, E_n)$ is the energy input vector, where E_i being the external input of energy to sector i ; $X = \{X_{ij}\}$ is the input-output table or the transaction matrix between industry sectors; X^* is a diagonal matrix, the diagonal element X^*_{ii} being the total output of the i-th sector.

Eq.(1) can be expressed in a somewhat different form:

$$\varepsilon = \varepsilon_0\,(I - A)^{-1} \tag{2}$$

Here, $\varepsilon_0 = E\,X^{*-1}$ is the vector of direct energy consumption, I is a unit matrix, and $A = X\,X^{*-1}$. Eq.(2) can be expanded as

$$\varepsilon = \varepsilon_0\,(I + A + A^2 + A^3 + \cdots\cdots) \tag{3}$$

$$= \varepsilon_0 + \varepsilon A \tag{4}$$

The first and second terms of eq.(4) represent, respectively, the direct and indirect energy requirements.

The international trade table has a similar structure to that of input-output tables, if we take it that each country or region is represented by one economic sector. Then the above model can be directly adapted with minor modifications to analyze the total (direct and indirect) energy required for the production of goods and services by countries. Here, the above eq.(1) can be used with several adjustments. Firstly, the suffix "i" is replaced by "α" which denotes country α; $X = \{X_{\alpha\beta}\}$ is the international trade matrix expressed in dollar values; and X^* is the diagonal matrix expressing the total final demand of the countries and/or regions:

$$X^*_\beta = Z_\beta + \sum_{\gamma=1}^{N} X_{\beta\gamma} \tag{5.1}$$

$$= Y_\beta + \sum_{\alpha=1}^{N} X_{\alpha\beta} \tag{5.2}$$

where Z_β is the sum of the consumption and investment by private and government sectors of the country β, and Y_β is the gross domestic product (GDP) of the country β.

3. ECO-BALANCE OF WORLD TRADE

3.1. Import-export balance of embodied energy among regions

The above model is applied to the world divided into twelve regions: Western Europe, Canada, the USA, Japan, Australia & New Zealand, Developing Market Economies in Africa, Developing Market Economies in America, the Middle East, Developing Market Economies in Asia, Centrally Planned Economies in Asia, Eastern Europe and the former USSR. Calculations are made by using the world trade tables for the years 1975, 1980, 1986 and 1989 [10]. Table 1 demonstrates the estimated trade balance of the energy flow embodied in traded goods and services between the regions for the year 1989, while Table 2 shows the breakdown of the balance. Here,

$$E_\beta^{(Z)} = \varepsilon_\beta Z_\beta \tag{6.1}$$

is the net domestic input of energy or the total (direct and indirect) energy required to support the domestic consumption and investment of region β, while

$$E_\beta^{(e)} = \sum_{\gamma} \varepsilon_\beta X_{\beta\gamma} \tag{6.2}$$

and

$$E_\beta^{(i)} = \sum_{\alpha} \varepsilon_\alpha X_{\alpha\beta} \tag{6.3}$$

show the total energy embodied in the goods and services exported from and imported to region β. Then,

$$E_\beta = E_\beta^{(Z)} + E_\beta^{(e)} - E_\beta^{(i)} \tag{6.4}$$

is the direct input of energy into region β.

In addition, Tables 3 and 4 demonstrate the balance in terms of energy requirement per capita and per GDP respectively. As a whole, Western Europe, which includes all EU and EFTA nations, recorded the largest flow both in terms of export and import, reflecting the large size of its economy and the vigorous trade promoted within the region.

Table 1
Trade balance of the embodied energy flow between regions in 1989 (million TOE)

Export to / Export from	Western Europe	Canada	USA	Japan	Australia & NZ	DME Africa	DME America	Mid. East	DME Asia	CPE Asia	Eastern Europe	Former USSR	Total
Western Europe	243.3	3.7	26.1	7.3	3.1	10.1	4.5	10.0	12.9	2.8	4.9	5.0	333.6
Canada	4.5	0.0	33.7	2.9	0.4	0.3	0.6	0.4	1.8	0.4	0.1	0.2	45.4
USA	34.1	27.6	0.0	15.8	3.4	2.2	13.9	3.8	17.5	2.9	0.4	1.6	123.0
Japan	9.0	1.1	14.9	0.0	1.5	0.5	0.8	1.2	11.4	1.7	0.1	0.5	42.7
Australia & NZ	2.2	0.2	1.5	3.4	0.9	0.2	0.1	0.5	2.8	0.5	0.1	0.3	12.6
DME Africa	17.0	0.2	4.9	0.8	0.0	2.0	0.3	0.9	0.8	0.2	0.6	0.5	28.2
DME America	10.0	0.8	15.5	2.5	0.3	0.6	4.5	0.8	1.5	0.9	0.5	0.5	38.2
Mid. East	8.3	0.2	3.5	6.2	0.3	1.0	1.2	4.1	4.4	1.2	0.5	0.7	31.7
DME Asia	30.4	3.4	48.3	27.3	4.0	3.1	1.3	4.9	42.5	8.4	0.7	2.7	177.0
CPE Asia	7.4	0.6	6.1	11.6	0.6	0.9	0.3	1.6	37.8	0.0	2.2	2.6	71.6
Eastern Europe	22.4	0.2	1.2	0.7	0.1	2.2	0.6	3.1	0.9	1.7	14.2	25.1	72.4
Former USSR	11.9	0.0	0.4	1.1	0.0	0.8	0.1	0.8	4.5	1.1	26.2	0.0	47.0
Total	400.4	38.0	156.1	79.5	14.8	23.9	28.2	32.0	138.6	21.8	50.5	39.7	1,023.5

Table 2
Breakdown of the embodied energy balance in 1989 (million TOE)

Regions	Direct Input E	Net Domestic Input $E^{(Z)}$	Export $E^{(e)}$	Import $E^{(i)}$	Net Import $E^{(i)}-E^{(e)}$	Ratio (%) $\frac{E^{(e)}}{E^{(Z)}}$	$\frac{E^{(i)}}{E^{(Z)}}$	$\frac{E^{(i)}-E^{(e)}}{E^{(Z)}}$
Western Europe	1.402	1.469	334	400	66.8	22.7	27.3	4.5
Canada	221	214	45	38	-7.5	21.3	17.8	-3.5
USA	1.942	1.975	123	156	33.0	6.2	7.9	1.7
Japan	406	443	43	80	36.9	9.6	18.0	8.3
Australia & NZ	97	99	13	15	2.2	12.7	15.0	2.2
DME Africa	200	196	28	24	-4.3	14.4	12.2	-2.2
DME America	360	350	38	28	-10.0	10.9	8.1	-2.8
Mid. East	235	235	32	32	0.3	13.5	13.6	0.1
DME Asia	495	457	177	139	-38.4	38.8	30.4	-8.4
CPE Asia	645	595	72	22	-49.8	12.0	3.7	-8.4
Eastern Europe	368	346	72	51	-21.9	20.9	14.6	-6.3
Former USSR	1.372	1.365	47	40	-7.2	3.4	2.9	-0.5
World Total	7.743	7.743	1.023	1.023	0.0	13.2	13.2	0.0

Table 3
Breakdown of the embodied energy balance per capita in 1989 (TOE / capita)

Regions	Direct Input	Net Domestic Input	Export	Import	Net Import
Western Europe	3.25	3.41	0.77	0.93	0.15
Canada	8.50	8.21	1.75	1.46	-0.29
USA	7.83	7.96	0.50	0.63	0.13
Japan	3.30	3.60	0.35	0.65	0.30
Australia & NZ	4.85	4.96	0.63	0.74	0.11
DME Africa	0.40	0.40	0.06	0.05	-0.01
DME America	0.92	0.89	0.10	0.07	-0.03
Mid. East	1.87	1.87	0.25	0.25	0.00
DME Asia	0.31	0.29	0.11	0.09	-0.02
CPE Asia	0.58	0.53	0.06	0.02	-0.04
Eastern Europe	2.97	2.79	0.58	0.41	-0.18
Former USSR	4.75	4.72	0.16	0.14	-0.03
World Average	1.55	1.55	0.21	0.21	0.00

Table 4
Breakdown of the embodied energy balance per GDP in 1989 (TOE / million $US)

Regions	Direct Input	Net Domestic Input	Export	Import	Net Import
Western Europe	273	286	65	78	13.0
Canada	437	422	90	75	-14.7
USA	401	408	25	32	6.8
Japan	153	167	16	30	13.9
Australia & NZ	323	330	42	49	7.4
DME Africa	597	584	84	71	-13.0
DME America	438	426	46	34	-12.1
Mid. East	357	357	48	49	0.4
DME Asia	587	541	210	164	-45.6
CPE Asia	1,655	1,527	184	56	-127.8
Eastern Europe	1,061	998	209	146	-63.1
Former USSR	560	557	19	16	-3.0
World Average	402	402	53	53	0.0

With respect to the balance of import and export, Northern regions including the USA, Western Europe and Japan are net importers, while Southern regions including DME (Developing Market Economies) and CPE (Centrally Planned Economies) in Asia, DME in Africa, DME in America are net exporters. In terms of per capita net import and/or export, Japan imports 0.30 TOE per capita, while Eastern Europe exports 0.18 TOE per capita. By and large, there already exists a large gap in direct energy use per capita between industrialized and developing countries. This gap is further expanded if the indirect input of energy by trade is taken into account. Thus the present world economic structure tends to increase the disparity between North and South in terms of resource consumption and the responsibilities for the impact on the global environment.

The largest flow of embodied energy to Japan is accounted for from DME in Asia which include Dynamic Asian Economies such as South Korea and the ASEAN countries, and then comes the USA, then CPE in Asia including China. With respect to the flow from Japan, the largest are the USA and DME in Asia. As a net balance, Japan recorded an import surplus of 36.9 million TOE.

Figure 1 shows the export from various regions to Japan for different years between 1975 and 1989. The export from DME in Asia to Japan almost doubled over the period from 1975 to 1980, and then the rate of its increase was stabilized. Since mid-1980s, however, it has begun to increase sharply again. Export from CPE in Asia to Japan showed a constant increase over the whole period, reflecting closer economic relation between China and Japan.

Figure 2 shows exports from Japan to various regions. The largest counterpart here is the USA, then comes DME in Asia, and then Western Europe. Exports to the USA showed a rapid increase over the period from 1975 to 1986, but then turned to decline. Exports to DME in Asia and Western Europe also showed a trend of increase until 1986, but then stabilized.

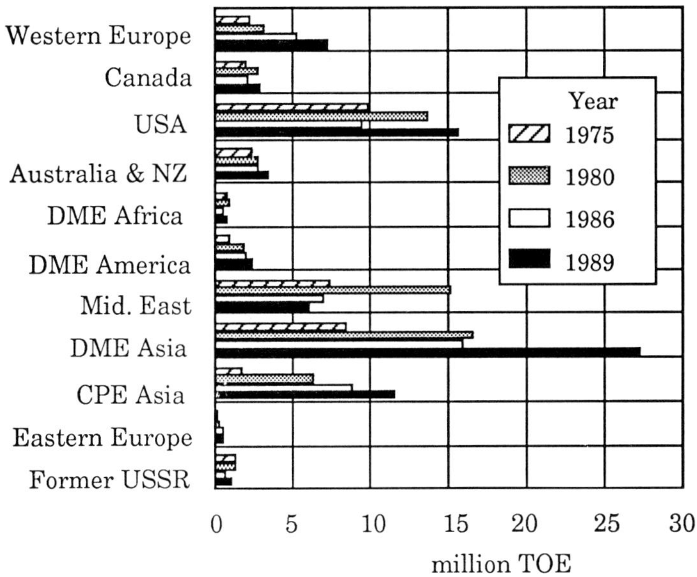

Figure 1. Flow of the embodied energy from various regions to Japan

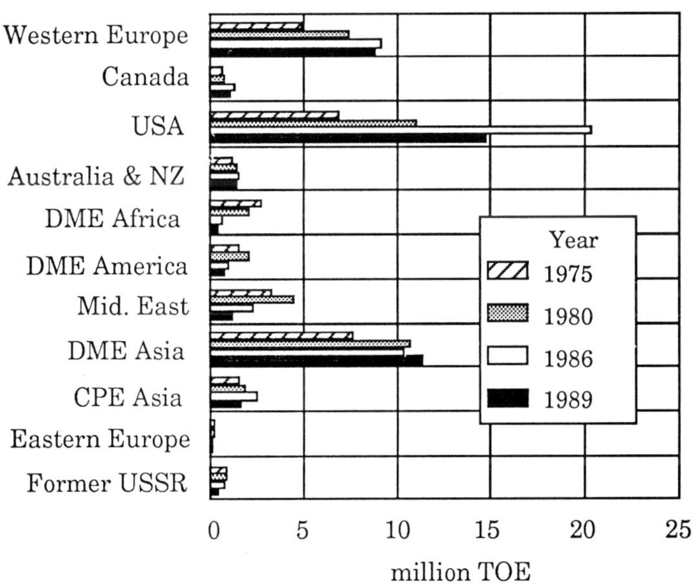

Figure 2. Flow of the embodied energy from Japan to various regions

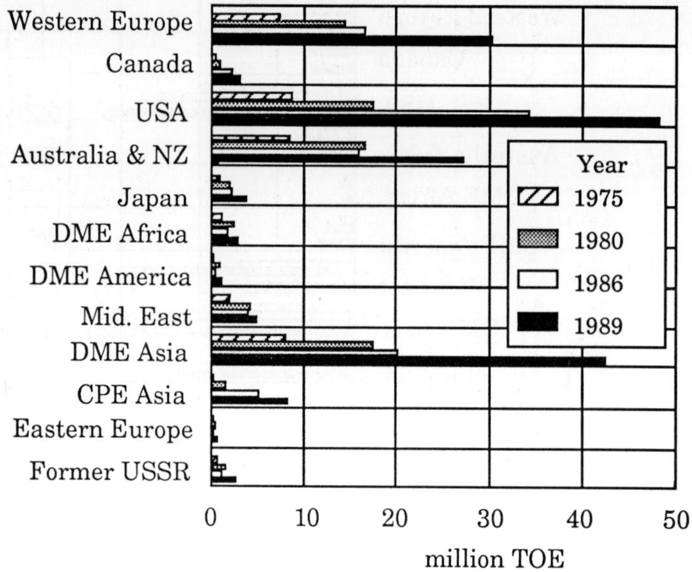

Figure 3. Embodied energy flow from developing market economies in Asia

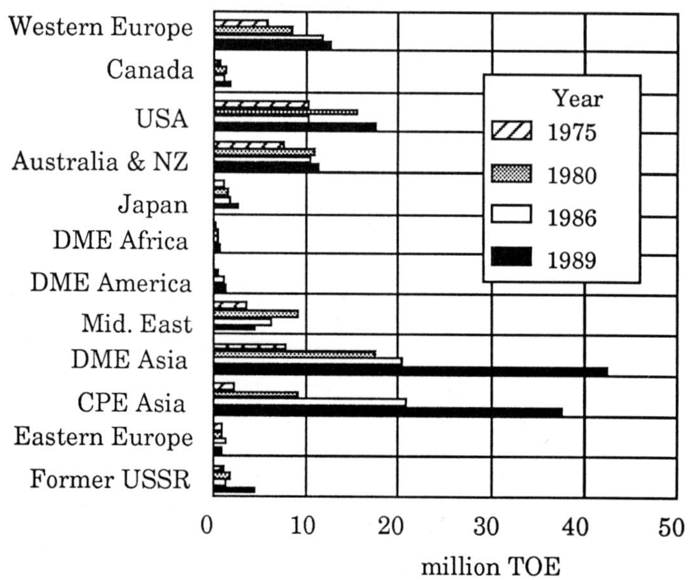

Figure 4. Embodied energy flow to developing market economies in Asia

Table 5
Trade balance of the embodied CO_2 flow between regions in 1989 (million tC)

Export from \ Export to	Western Europe	Canada	USA	Japan	Australia & NZ	DME Africa	DME America	Mid. East	DME Asia	CPE Asia	Eastern Europe	Former USSR	Total
Western Europe	168.5	2.6	18.0	5.0	2.2	7.0	3.1	6.9	8.9	1.9	3.4	3.5	231.0
Canada	2.9	0.0	21.2	1.8	0.3	0.2	0.4	0.3	1.1	0.3	0.0	0.1	28.5
USA	25.1	20.3	0.0	11.6	2.5	1.6	10.2	2.8	12.8	2.1	0.3	1.2	90.4
Japan	6.4	0.8	10.7	0.0	1.0	0.4	0.6	0.9	8.2	1.2	0.1	0.4	30.6
Australia & NZ	1.8	0.2	1.2	2.8	0.8	0.1	0.1	0.4	2.2	0.4	0.1	0.2	10.2
DME Africa	13.8	0.1	4.0	0.7	0.0	1.6	0.2	0.7	0.6	0.2	0.5	0.4	23.0
DME America	6.9	0.5	10.7	1.7	0.2	0.4	3.1	0.5	1.0	0.6	0.4	0.4	26.5
Mid. East	6.0	0.2	2.5	4.5	0.2	0.8	0.9	2.9	3.2	0.9	0.4	0.5	23.0
DME Asia	24.3	2.7	38.6	21.8	3.2	2.4	1.1	3.9	34.0	6.7	0.6	2.2	141.6
CPE Asia	6.9	0.5	5.7	10.8	0.6	0.8	0.3	1.4	35.2	0.0	2.0	2.4	66.7
Eastern Europe	18.1	0.2	1.0	0.5	0.1	1.7	0.5	2.5	0.7	1.4	11.5	20.3	58.5
Former USSR	8.5	0.0	0.3	0.8	0.0	0.6	0.1	0.6	3.2	0.8	18.6	0.0	33.4
Total	289.1	28.1	114.0	62.1	11.2	17.7	20.5	23.9	111.2	16.5	37.9	31.5	763.5

Asian countries including ASEAN and China have recorded rapid economic growth since the 1980s, and they are expected to become a growth center of the world in the next century. With this economic growth, exports from and to DME and CPE in Asia showed a remarkable expansion in the 1980s, as demonstrated in Figures 3 and 4.

3.2. Import-export balance of embodied CO_2 among regions

The above results obtained for the embodied energy flow can be translated into the flow of embodied CO_2 by using the emission coefficients of fossil fuels. This is shown in Tables 5 and 6 corresponding to Tables 1 and 3.

Since the principal energy source in most countries is fossil fuels, major characteristics observed for the results of embodied CO_2 are very similar to those of embodied energy.

Table 6
Breakdown of the embodied CO_2 balance per capita in 1989 (tC / capita)

Regions	Direct Input	Net Domestic Input	Export	Import	Net Import
Western Europe	2.22	2.36	0.54	0.67	0.13
Canada	5.17	5.15	1.10	1.08	-0.02
USA	5.76	5.85	0.36	0.46	0.09
Japan	2.33	2.58	0.25	0.50	0.26
Australia & NZ	3.98	4.02	0.51	0.56	0.05
DME Africa	0.33	0.32	0.05	0.04	-0.01
DME America	0.63	0.62	0.07	0.05	-0.02
Mid. East	1.35	1.36	0.18	0.19	0.01
DME Asia	0.25	0.23	0.09	0.07	-0.02
CPE Asia	0.54	0.50	0.06	0.01	-0.04
Eastern Europe	2.42	2.26	0.47	0.31	-0.17
Former USSR	3.37	3.36	0.12	0.11	-0.01
World Total	1.15	1.15	0.15	0.15	0.00

3.3. Economic development and trade

As a general trend, total (direct and indirect) energy requirement per capita increases with increasing income per capita and a rising standard of living. Figures 5, 6 and 7 show the relationship between CO_2 emission and per capita GDP (in 1985 US dollars) for various regions over a period.

Figure 5 shows the amount of per capita embodied CO_2 pertinent to net domestic consumption and investment. The graph suggests the existence of two groups: the regions of the first group, comprising the USA, Canada, Australia & New Zealand, the former USSR and Eastern Europe, lie in the upper part of the graph, while the regions belonging to the second group such as Japan, Western Europe, the Middle East and all developing economies are situated in the lower part. Furthermore, the first group is divided into two sub-groups: one to the right part (the USA, Canada, and Australia and New Zealand) and the other to the left (the former USSR and Eastern Europe). Within each group, there is a linear relationship between per capita CO_2 and per capita GDP.

Then, Figure 6 shows the amount of per capita net import (i.e., import minus export) of embodied CO_2. Japan and Western Europe have been net importers, while Eastern Europe has been a net exporter throughout the period under consideration. The USA changed its position from exporter to importer in the mid 1980s, while Canada became an exporter after 1980.

Figure 7 focuses on the situation in developing regions. The most notable feature here is that the net export from DME in Asia has expanded in accordance with their increasing per capita GDP. The export from CPE in Asia has also increased with their economic development, although the change in their per capita GDP is not clearly demonstrated due to the statistical problems in estimating GDP in dollar values. The result shows that these regions have achieved economic growth by taking over the responsibilities for the environmental loads which are induced by the demand for goods and services in industrialized nations.

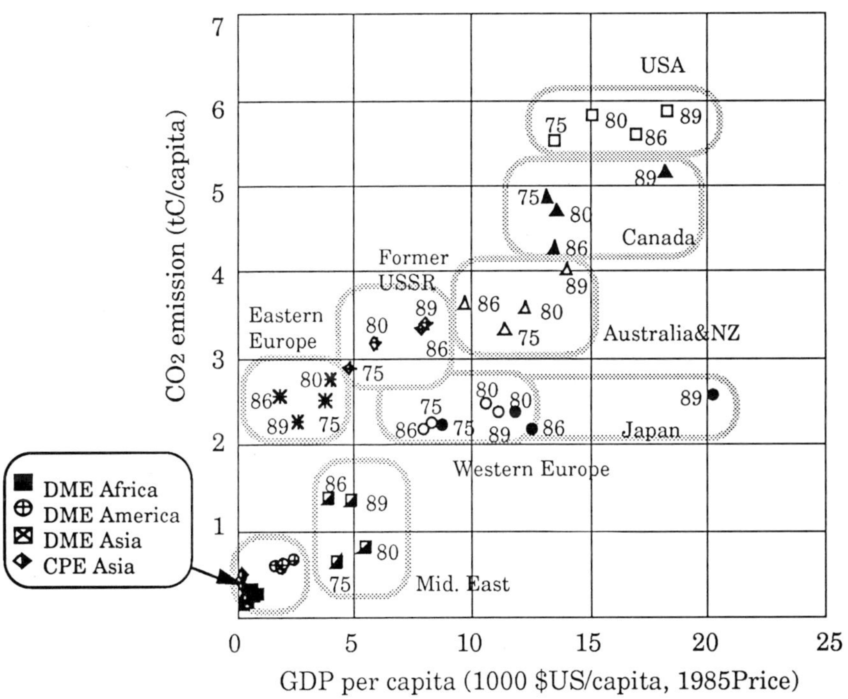

Figure 5. Per capita GDP and CO_2 emissions for domestic input

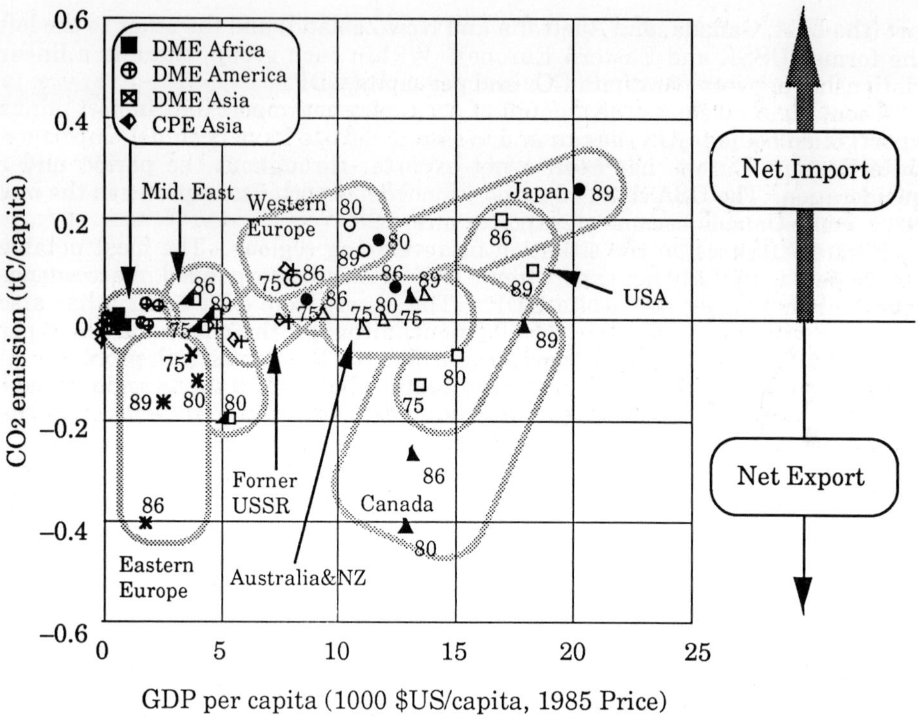

Figure 6. Per capita GDP and net import of CO2 (whole world)

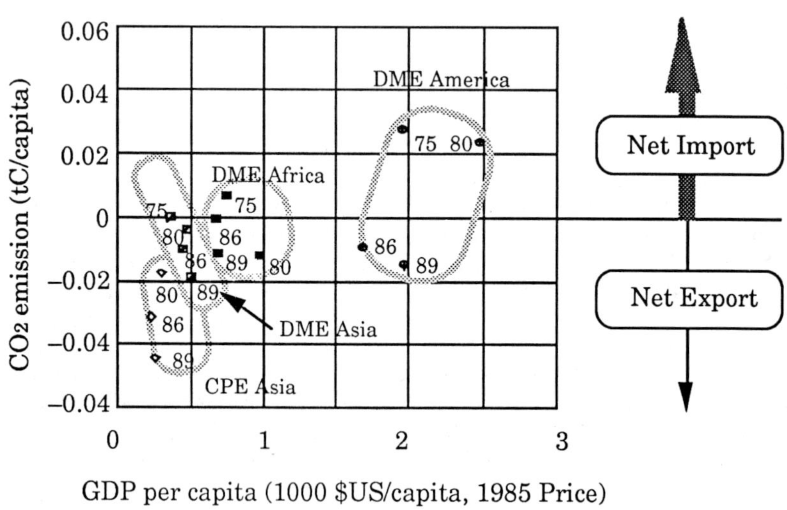

Figure 7. Per capita GDP and net import of CO_2 (developing regions)

4. ECO-BALANCE IN THE ASIA-PACIFIC REGION

In the analysis presented in the previous chapter, each country or region is represented by a single industry. This, of course, is an over-simplification.

Tiwaree and Imura [11] adapted the standard input-output model in order to analyze carbon dioxide emission in ten countries in the Asia-Pacific region: Indonesia, Malaysia, Philippines, Singapore, Thailand, China, Taiwan, South Korea, Japan and the USA. Here, the region's trade with the rest of the world was neglected and the region was treated as if it was a closed economy. The analysis was made by combining the multinational input-output table with energy statistics available for each country.

Firstly, the "Asian International Input-Output Table 1985" of the Institute of Developing Economies [12] was reduced to 200 x 200 sectors input-output table for ten countries. Three types of energy sources (i.e., coal, oil and natural gas) were considered, and their consumption data for energy purposes were obtained from published international sources [3, 13, 14] .

Table 7 compares the total (direct plus indirect) embodied CO_2 intensities of a unit monetary value demand (i.e., in terms of tC/1000 $US) of differing industry sectors in ten countries. The intensities differ greatly between industry sectors in a country as well as between the same sectors in different countries. In most countries, energy related sectors have the highest CO_2 intensities. In Indonesia, Malaysia, Philippines, Thailand, Taiwan and South Korea, the agricultural sector exhibits the lowest value of CO_2 intensity, while in Singapore, Japan and the USA, the services sector has the lowest value; and in China, the and livestock and then agriculture sectors have the lowest value. Among the countries, China has notably high total carbon dioxide intensities in almost all industry sectors.

The emission intensity of a country depends upon the country's industrial structure (e.g., the relative importance of the unprocessed, semi-processed or finished goods sectors in the total national economy), and the efficiency of production technology, especially energy-related technology. In addition however it depends upon the industrial structure and production technology of other economies from which each sector receives goods and services for its own output. Therefore, there is an interdependence between sectors within an economy and between different sectors of other economies with which trade in goods and services takes place.

Table 8 is an estimate of the trade in embodied energy associated with the transaction of goods and services between the above-mentioned ten countries. The country with the largest export is China, followed by the USA, Indonesia and Japan, whereas the largest importing country is Japan and then the USA, Singapore and South Korea. Countries such as Singapore and Japan depend largely for their energy consumption and carbon dioxide emissions upon other countries. Oil-producing countries such as Indonesia and Malaysia exhibit a peculiar feature here as their economies rely heavily upon the export of oil products: the energy consumption for its exports is larger than that for its domestic consumption and investment.

Table 7
Comparison of the total embodied CO_2 intensities of differing industry sectors of ten Asia-Pacific countries in 1985 (tC/1000 $US)

Industry Sectors / Country	Indonesia	Malaysia	Philippines	Singapore	Thailand	China	Taiwan	S. Korea	Japan	USA
Agriculture	0.05	0.08	0.04	0.40	0.14	0.47	0.16	0.11	0.12	0.48
Livestock	0.09	0.19	0.07	0.08	0.23	0.32	0.29	0.29	0.14	0.42
Forestry & Products	0.19	0.34	0.26	0.15	0.24	1.07	0.32	0.19	0.15	0.28
Fishery	0.15	0.18	0.30	0.79	0.83	0.49	0.52	0.49	0.38	0.79
Oil & Natural gas	2.68	3.66	2.96	0.64	4.88	14.26	3.16	3.80	3.73	5.21
	(1.45)	(2.05)	(0.14)	(0.05)	(0.56)	(7.05)	(0.22)	(1.13)	(0.30)	(2.20)
Mining (including coal)	0.46	0.95	2.01	0.62	2.22	40.74	12.88	15.28	10.32	12.71
	(0.20)	(0.49)	(0.53)	(0.51)	(0.48)	(2.93)	(1.2)	(0.44)	(0.56)	(1.98)
Food, Beverages & Tobacco	0.10	0.30	0.12	0.14	0.26	0.69	0.28	0.23	0.16	0.33
Textile, Leather & Products	0.23	0.46	0.15	0.08	0.25	1.12	0.37	0.40	0.20	0.39
Pulp,Paper & Printing	0.37	0.22	0.36	0.08	0.38	1.50	0.43	0.41	0.24	0.36
Chemical Products	0.73	0.46	0.31	0.44	0.33	3.86	0.50	0.79	0.45	1.15
Rubber Products	0.26	0.40	0.20	0.36	0.43	1.25	0.35	0.34	0.24	0.47
Non-metallic Mineral Products	0.64	0.66	0.78	0.21	1.22	4.90	1.75	2.43	1.33	1.39
Metal Products	0.57	0.55	1.59	0.09	0.69	4.88	0.73	0.79	0.41	0.77
Machinery	0.12	0.24	0.44	0.04	0.29	2.01	0.37	0.36	0.19	0.24
Transport Equipment	0.25	0.07	0.36	0.07	0.26	1.96	0.44	0.34	0.20	0.26
Other Manufacturing	0.14	0.41	0.42	0.13	0.17	1.60	0.40	0.48	0.21	0.31
Electricity, Steam & Hot Water	7.33	4.41	2.84	0.67	2.29	11.52	2.66	1.88	1.13	2.42
	(6.69)	(3.8)	(1.09)	(0.67)	(1.87)	(9.55)	(0.78)	(0.99)	(0.47)	(1.88)
Construction	0.21	0.25	0.50	0.10	0.60	3.02	1.07	0.81	0.43	0.43
Trade & Transport	0.25	0.19	0.43	0.14	0.46	1.72	0.40	0.51	0.18	0.30
Services	0.13	0.15	0.18	0.04	0.24	1.10	0.21	0.19	0.11	0.10
Overall Average	0.38	1.90	0.35	0.21	0.39	1.96	0.46	0.45	0.25	0.42

Note: Figures for energy industries in parentheses show the direct energy requirement.

Table 8
Balance of the embodied energy flow between Asia-Pacific countries in 1985 (million TOE)

from \ to	Indonesia	Malaysia	Philippines	Singapore	Thailand	China	Taiwan	South Korea	Japan	USA	Total Export	Direct Input	Net Import (a)	Net Input (b)
Indonesia	10.9	0.1	0.1	0.7	0.1	0.1	0.3	1.0	12.4	5.1	19.9	30.8	−17.1	13.7
Malaysia	0.1	4.4	0.2	1.1	0.8	0.1	0.2	1.1	4.3	0.3	8.1	12.5	−4.7	7.8
Philippines	0.0	0.0	10.0	0.1	0.0	0.1	0.0	0.1	0.6	0.3	1.3	11.3	2.0	13.3
Singapore	0.2	0.5	0.0	3.7	0.2	0.0	0.0	0.1	0.6	0.2	1.8	5.5	11.8	17.3
Thailand	0.0	0.1	0.1	0.1	14.8	0.1	0.1	0.0	0.3	0.5	1.1	15.9	2.4	18.4
China	0.8	1.5	0.9	9.8	1.4	544.3	0.2	0.4	27.8	8.6	51.4	595.7	−46.7	549.0
Taiwan	0.1	0.1	0.1	0.3	0.1	0.3	27.1	0.2	1.0	2.3	4.6	31.6	0.9	32.6
S. Korea	0.1	0.1	0.1	0.1	0.1	0.0	0.1	47.6	3.3	1.9	5.9	53.0	4.8	57.8
Japan	0.7	0.5	0.2	0.4	0.4	2.2	1.0	2.2	338.6	6.4	14.0	352.6	59.8	412.4
USA	0.9	0.5	1.6	1.1	0.4	1.9	3.5	5.7	23.3	1,682.0	38.8	1,720.8	−13.2	1,707.6
Total Import	2.8	3.4	3.2	13.6	3.5	4.7	5.5	10.7	73.8	25.6	147.0	2,829.7	0.0	2,829.8

Notes: (a) Net import = Import – Export
(b) Net Input = Direct Input + Import – Export

5. ECO-BALANCE OF REGIONS WITHIN A COUNTRY

Many developed countries have set national targets for stabilizing or reducing the emission of carbon dioxide. In most cases however these national targets are mere guidelines rather than mandatory standards. In order to achieve the targets, voluntary action by the public and business should be promoted. At the levels of government closest to the people, the initiatives of local authorities are especially vital in support of voluntary action. Accordingly, many prefectural and municipal authorities in Japan have prepared a Local Action Plan to reduce emissions of greenhouse gases, especially CO_2 emissions from fossil fuel use. These Local Action Plans generally contain a profile of baseline CO_2 emissions associated with energy use within their jurisdiction, and an inventory of measures and technologies appropriate to reduce or stabilize CO_2 emissions.

The collected Japanese data suggest that disparities in CO_2 emission per capita and per GDP among regions within a country are even greater than those among countries. Direct CO_2 emission in the Tokyo metropolis [15] is as small as 0.83tC per capita, less than one-third of the national average, since its people consume goods and services provided from outside. On the other hand, emissions in prefectures and municipalities with heavy industry and power plant are several times larger than the national average, as demonstrated in Table 9 [16]. A typical example of the indirect use of fossil fuels is electricity consumption. There is an argument about whether CO_2 emission from electricity generation should be allocated to original producers or final users.

Table 9
Estimate of the CO_2 emissions in some municipalities and prefectures in Japan for 1990 (1000 tC)

	Tokyo	Yokohama City (1989)		Fukuoka City	Yamaguchi Pref.		National Total
Sectors	B	A	B	B	A	B	A
Energy Conversion	0	2811	-	0	3341	237	105600
Industry	3350	1885	2681	101	6764	7759	107300
Residential	1990	608	1359	439	544	231	18200
Commercial	830	179	754	490	428	125	16200
Transportation	3560	1136	1226	566	573	606	56200
Waste	-	-	-	135	290	293	12000
Cement production	-	-	-	-	2308	2308	-
Total	9860	6619	6020	1688	13622	12158	318000
per capita (tC)	0.83	2.10	1.91	1.36	8.63	7.70	2.57

Note: Two estimates A and B are shown above. Estimate A allocates the responsibility for CO_2 emission in electricity generation to original producers, while estimate B allocates it to final users.

Therefore, responsibilities of local communities for reducing CO_2 emissions should be discussed both in terms of direct and indirect consumption of fossil fuels.

6. TRADE IMPLICATIONS OF LIFE CYCLE ASSESSMENT (LCA)

Life cycle assessment is an integrative review of the entire life cycle of a product "from the cradle to the grave." It aims to quantify the cumulative impact of a product from the point where materials are extracted from the earth up to either a certain point in the product's life cycle or, in the most complete case, to the final disposal of the remaining wastes back into the earth [17]. Thus LCA is an information base that can be applied to define general policy strategies and also to regulate processes or products by means of product certification or eco-labeling, etc. LCA is also a tool used to identify and measure both "direct" and "indirect" environmental, energy, and resource use associated with a product or process.

Recently, application of LCA by manufacturers has been actively discussed in ISO (International Standardization Organization) and other international forums. The limitations of the currently available methodologies of LCA, however, is that the boundary of the system's analysis is not well-defined, and the input of materials from other countries is not explicitly taken into account. International input-output analysis can supplement this deficiency, although it is an aggregated approach and can not deal with microscopic differences in materials.

Moreover, LCA has been mostly applied to consumer goods and it pays little attention to capital goods such as buildings and infrastructures or services. Input-output models can circumvent such difficulties. Ikeda and Imura [18] analyzed the total energy induced by construction, and discussed the cumulative energy embodied in the stock of buildings, roads and other urban infrastructures.

7. CONCLUDING REMARKS

Embodied energy and CO_2 flows accompanying transactions in goods and services at different levels is discussed. The responsibility of countries and sectors for causing environmental loads is treated on the basis of model calculations. There already exists a large gap in direct energy use per capita between industrialized and developing countries. This gap is further widened if the indirect input of energy associated with trade is taken into account. It means that the present world economic structure tends to increase the disparity between North and South in terms of resource consumption, and responsibility for impacts on the global environment.

The net export of embodied CO_2 from Developing Market Economies and Centrally Planned Economies in Asia has expanded in accordance with their economic growth. It has been demonstrated that the economic growth in these regions has been achieved by taking over the environmental loads induced by the demand for goods and services in industrialized countries.

International harmonization of environmental standards, such as the setting of an equal emission allowance per unit of particular output across countries, could be a solution for reconciling trade and the environment [19]. However, there are no valid unambiguous criteria for comparing physical emissions in two different countries. Here, embodied energy and CO_2 can serve as good indicators to assess the environmental impact of goods and services required in a country and those traded between countries. The trade balance of embodied energy and CO_2 between countries and regions would provide a rational basis for allocating the true environmental costs of trade, and for discussing the responsibilities of respective countries and regions in protecting the global environment.

The authors would like to thank Mr. H. Shiratsuchi for his help in carrying out model calculations.

8. REFERENCES

1 UN Conference on Environment and Development, The Rio Declaration on Environment and Development, Note by the Secretary-General of the Conference, Rio de Janeiro, 1992.
2 IEA/OECD, Climate Change Policy Initiatives, Paris, 1992
3 The Energy Data and Modeling Center (ed.), Energy and Environment Statistics '94, Energy Conservation Center, Tokyo, 1994.
4 R. E. Miller and Peter D. Blair, Input-Output Analysis, Foundations and Extensions, Prentice-Hall, New Jersey, 1985.
5 R. Costanza, Science, 210 (1980) 1219.
6 P. W. Gay and J. L. R. Proops, Applied Energy, 44 (1993) 113.
7 Y. Moriguchi, Y. Kondo and H. Shimizu, Energy and Resources, 14 (1993) 32.
8 Y. Moriguchi, Y. Kondo and H. Imura, Proceedings of the Annual Meeting of the Society of Environmental Science (1993) 15 (in Japanese).
9 H. Imura, Y. Moriguchi, H. Shiratsuchi and T. Sakai, Environmental Science, 7 (1994) 225 (in Japanese).
10 United Nations, The World Trade Statistics, New York, 1994.
11 R. S. Tiwaree and H. Imura, Environmental Systems Research, 22 (1994) 376.
12 Institute of Developing Economies (ed.), Asian International Input-Output Table, I. D. E. Statistical Data Series No. 60, I. D. E., Tokyo, 1992.
13 IEA, Energy Statistics of OECD Countries, Paris, 1990.
14 IEA, Energy Statistics and Balance of Non-OECD Countries, Paris, 1990
15 Tokyo Metropolitan Government, Action Plan for the Protection of the Global Environment, Tokyo, 1992.
16 H. Imura, Environmental Research Quarterly, 94 (1994) 56 (in Japanese).
17 R. Heijungs, G. Huppes and H. A. Hales, LCA in Environmental Decision Making, Centre of Environmental Science, Leiden, 1994.
18 H. Ikeda and H. Imura, Environmental Systems Research, 21 (1994) 192 (in Japanese).
19 K. Steininger, Ecological Economics, 9 (1994) 23.

TOWARD GLOBAL PLANNING OF SUSTAINABLE USE OF THE EARTH
Development of Global Eco-Engineering
S. Murai (Editor)

Thoughts on the Conference "Breakthrough" and Comments on: Seawater Irrigation of Desert Coasts as a Global Eco-Technology for Food, Bio-Fuel Production and Community Nuclei, While Balancing the Atmospheric Carbon Cycle for Sustained Economic Growth

Carl N. Hodges

Chief Executive Officer, Planetary Design Corporation, Two Arizona Center, Suite 1555, Phoenix, Arizona 85004

Abstract

This paper searches for a vocabulary to title a goal for "global eco-engineering" that is beyond "sustainable." A goal of sustainability is not adequate to inspire the effort that must be made to solve the planet's environmental problems. A vision must be articulated that conveys to the people of the Earth the possibility of a future that is continually enhanced...a future that is better and better instead of one that "continues the present." Many people on the Earth are so disadvantaged that a goal of staying the same is not a goal.

The paper looks at seawater irrigation of desert coasts as an example of one new technology that is "enhancing." This technology is presented in the context of the Biosphere 2 Project. Continual improvements are being made there as a result of Biosphere 2 being energetically open (solar energy for biological photosynthetic processes) and informationally open for human intelligence input (technological and social improvements). The Earth (Biosphere 1) can be viewed in the same context.

There are 40,000 kilometers of desert seacoasts in the world, and if a significant percentage of these coasts were developed in ecologically sensitive ways, the "greening" of large areas at the edge of the sea could significantly contribute to and possibly be the major contributor to balancing the Earth's atmospheric carbon cycle. And, importantly, this solution could be executed as a highly successful economic growth venture. There are now technologies which make it possible to use seawater for irrigation of large-scale agricultural farms for food, fiber and the biochemical inputs for a blended petrochemical/biochemical future. Seawater may be used for domestic and commercial sewage systems, and the subsequent effluent used for ornamental landscaping with salt-tolerant plants.

Future cities supplied with seawater will provide habitats for the expanding world population and reduce the pressure on existing communities. If correctly designed, these communities will provide a continually increasing stream of wealth...a goal worthy of defining.

1. PRESENTATION TEXT

I am sure that all of us went through a process of evolution in our thinking about our participation for this conference. It started in the fall of 1993, when we accepted, intensified when we actually sat down to write a contribution, and has continued, I am sure, since arriving at the TOYOTA MOTOR CORPORATION MIKKABI CREATIVE CENTER. How can we individually, and as a group, contribute to success and at the end of the conference, say that we have produced something valuable as a result of our efforts?

When I drafted the above abstract, I had planned to write a relatively conventional scientific paper to review the work of the Environmental Research Laboratory of the University of Arizona, the Planetary Design Corporation, and the Biosphere 2 Project. I assembled previous papers I had written and ones by others relative to the topic. Then, I re-read the invitation letter of December 24, 1993 from Professor Shunji Murai and the July 4, 1994 letter from Secretary General Masato Kimura with the program and wondered about our time together at Toyota's "Creative Center."

The third paragraph of Professor Murai's description of the conference said:

> "Through intensive discussion and exchange of information among experts in a wide variety of fields...the conference will contribute to formulation of a framework for further research. <u>A breakthrough is expected</u>* toward the development of global eco-engineering, which seeks continuous human habitation, the improvement of human welfare...."
> *Underline added.

So, here we are at a world-class "creative center" with a program that offers "an extensive exchange of information," "...the opportunity for intensive discussion," and an expectation of a "BREAKTHROUGH." Would a traditional paper from me contribute much?

Our program has our presentations broken into 12 sections, plus opening and closing remarks. I thought ahead to our "closing remarks." Surely they would include an eloquent description of our "Breakthrough." Our breakthrough would be a wonderful thing, something that by means of the new technology of "global eco-engineering" would provide "continuous human habitation and improvement of human welfare," based on the *sustainable* utilization of global environmental resources and *preservation* of global eco-systems.

I liked our expectations. I am a sincere believer in the phenomenon of "self-fulfilled prophecies." In 1976, I became involved with the Walt Disney Company, when I was asked to participate in the design of Walt Disney's dream of an "Experimental Prototype Community of Tomorrow (EPCOT)." Walt Disney believed that we needed a vision of positive futures, if we were to invent a future that would lead to happiness, for us individually and as a society. I shared that philosophy and received great satisfaction and enjoyment from my work for the scientific design of the most successful pavilion at EPCOT, The Land Pavilion, which is the story of agriculture and offers the guest a spectrum of visions of ways that the world can feed itself in the future.

> *I would like to say at this point that, although I work with Disney, I do not consider myself a "Pollyanna." In fact, I find great empathy for Dr. Dennis Meadows' paper, "It Is Too Late to Achieve Sustainable Development, Now Let Us Strive for Survivable Development." I agree with him that the world population is already beyond an optimal desirable level. I would rather have an equal number of human beings enjoy the earth over many millennia instead of the numbers we have at present, and even less desirable, the numbers that are projected before our population levels, hopefully, by intentional means, or as Dr. Meadows says, "...a population dieback which takes the population back down to sustainable levels." If we can create a true "breakthrough" at this conference, we will impact the population issue.*

As I drafted my paper for the conference, I felt some satisfaction in that the "eco-technologies" that I describe are tools that can be part of a positive prophecy of the future. I became concerned, however, as I tried to fit what I wanted to say with the vocabulary used to describe the conference. It was clear to me that a "usual" scientific paper would not make an adequate contribution to the "intensive discussion and exchange of information" we would need for our "breakthrough."

I decided to unbind my thinking and visualize a situation where we were a group of friends, with few constraints, gathered together at a center for creativity. The first thing I would offer to you would be a gift of one of my favorite, if not my absolute favorite, books, *The Grace of Great Things (Creativity and Innovation)* by Robert Grudin.(1) I hope you have had a chance to glance at yours. I have highlighted my two favorite quotes for our thinking together about a breakthrough. And, I brought copies of an article not unlike what I would have presented here, if I had not decided to try to go a bit beyond...that is, a reprint from *AMBIO Magazine*, The Journal of the Human Environment, published by the Royal Swedish Academy of Sciences.(2) (Please see page 490 of this reference for a general list of references for this paper.)

What I would like to do most would be to sit in one of the beautiful Japanese gardens and read through *The Grace of Great Things* together and discuss creativity.

> "The creative mind is able to redefine itself and reality at large...thus generating whole new sets of alternatives." "...creative achievement...is characteristically an offering to society...this, in turn, suggests that individual-society interaction, in which society's past or predicted reception of creative achievement, becomes one of the directing factors in a creator's

> work." "The redefinition of problems is a creative act that...involves an aggressive rebuilding of reality...."

Further, Grudin points out that a truly creative solution to a problem is often

> "...based on the seemingly absurd, but often viable premise that, if one feels uncomfortable with a major problem in life, one may simply trash it, saying something like, I don't like this problem. I want a bigger, bolder problem, or a fatter, happier one. In some cases, creative redefinition is so radical as to constitute a kind of paradox."

I would propose that we have a grand paradox vis-a-vis our planetary environmental problems. We are not, in my opinion, trying to solve an adequate problem when we set our goal as "sustainability." I believe we need a "bigger, bolder problem...a happier one." It is my belief that we will not be able to muster the creativity necessary for "sustainable development," and again, with respect for Dr. Meadows' view, even less likely that the world will have the creativity to be successful for a goal of only "survivable development." I would propose that part of our breakthrough might be the proposing of a goal of "continually enhancing development," one that is significantly beyond "sustainable development."

My proposal on such a goal in no way is meant to devalue the incredibly important work and contribution of those defining the seriousness of our environmental problems and alerting us to the catastrophic deterioration of our planetary environment. But, the condition of so many people on the earth is at such a low level that a goal of "sustainability" will not inspire their participation in solutions. A goal must be for a better future. And I think there should be components of it that include unlimited growth, not in population, but in wealth, in the intellectualism of life, in our method for valuing biological and cultural diversity, etc.

The latest reading I have done is the work of The World Resources Intitute, where Dr. Robert Repetto says, "Three factors...economic development, availability of contraceptives, and the improved status of women are all necessary for rapid demographic transition."(3) My contribution here is to the first of those three, which, in turn, will help the other two.

I have a second gift, a small book, *Toward a New Economic System in Decline or Transformation*, by Willis Harman.(4) It is a beautiful piece of thinking from the President of the Institute of Noetic Sciences:

> "We are witnessing a national and global system in decline...." "This is not bad news any more than is the decline of the caterpillar in the process of becoming a butterfly. Aware persons in the business and professional worlds can perceive a constructive factor in the present metamorphose, but only if we are willing to 'face the good news' (the butterfly possibility) and to share responsibility for the whole in a way they never taught us in Biz School." "The transformation...is the real challenge of our times."

Willis Harman does a good job of defining how that transformation might come about. I recommend his writings for our joint creative discussions.

I have difference with Harman, but it is not with the magnitude of the problem he asks us to address. Transformation is a big, fat, juicy problem. My disagreement is with Harman's acceptance of the premise that "...in the end the world is a closed system." That is the premise of many environmentalists, perhaps best exemplified by Jeremy Rifkin in his book, *Entropy: Welcome to the Greenhouse World.*(5) From my perspective, the earth is certainly NOT a closed system. Admittedly, it is essentially materially closed, with the exception of satellites we send into space and meteorites that impinge on the earth. But, more importantly, the earth is energetically and informationally open. We receive every day from outside the earth via sunshine the principal energy source that powers our biosphere. Every day we receive from our satellites (as has been reported here) information about our earth. And, every day we send information into space seeking information in return. Most importantly, the earth is not entropic because of life processes. Photosynthesis is a major organizing force using the external energy source of sunlight, and our intelligence is used to constantly increase our store of knowledge.

I will briefly, mainly for later informational discussion, introduce a tool to help us with the accumulation of new knowledge. Biosphere 2 is located between Phoenix and Tucson, Arizona. I had the pleasure of heading up the largest scientific consulting activity. Biosphere 2 is the first system complex enough to support human life independent of the earth's living systems. It would take more than my allotted time just to adequately introduce the project. Let me say, in summary, that it is a complicated project, one which, like activities in Biosphere 1, has had perturbations, management changes, etc. But the concept of the project is magnificent. A principal lesson of Biosphere 2 is that there is no "away." Things cannot be thrown "away" in Biosphere 2. Everything is part of cycles. And in Biosphere 2, many of the cycles run much faster than in Biosphere 1.

In Biosphere 2, we designed a system to maintain a certain carbon dioxide level, and to do that, we had to increase the green area. It is my belief that we must define the levels at which we want Biosphere 1 to operate...population, CO_2, etc., and develop biospheric systems to provide these levels. In the *AMBIO* paper, I discuss the amount of area of earth that should be greened with direct seawater irrigation of plants and how that could be one component of balancing atmospheric carbon levels, even with increasing fossil fuel use, and how ultimately these new seawater-irrigated plantations could be contributors to a "transformation" to a blended petrochemical bio-chemical future. And, most importantly, that transformation relies on a resource that, in a certain context, has been overlooked...or at least not creatively utilized.

Figure 1 is a presentation on the world's water supply. On the earth, only three-tenths of one percent is fresh water that is renewable. There are many publications on the water shortages the world now faces. There are predictions that wars will be fought over water. Those wars will not be necessary. Fresh water on the earth is insignificant compared to the total water supply. In fact, an interesting metaphor is, "If all the world's water fit into a bathtub, the portion of it that could be used sustainably in any given year would barely fill a teaspoon." Most of that teaspoon of water is used in irrigated agriculture; ...of the global water use...73%. And in many cases, the irrigation of land is solely for livestock food production. In California, one of the United States' major agricultural states, that is the largest single use of water, and yet the State of California is one of the areas of the United States where water is of great concern. Rationing has occurred in major cities.

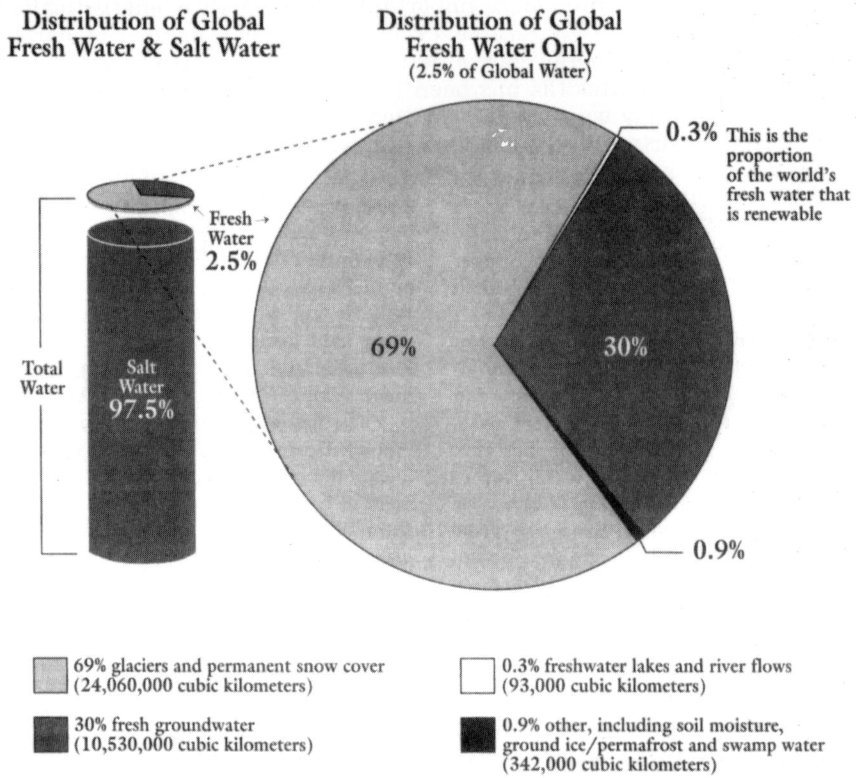

Figure 1. The World's Water.

(Note: Percentage figures do not add up to 100% due to rounding.)

(Source: Igor Shiklomanov, "World Fresh Water Resources" in Peter H. Gleick, ed., *Water in Crisis: A Guide to the World's Fresh Water Resources*, 1993.)

Let me present to you, in a creative way, I hope, how CO_2 and seawater are related. (Video of Marlon Brando and the Seawater-Irrigated Farm.)

On our earth today, 50% of the population lives within 50 kilometers of the sea. Within 50 years, the population of the earth could double, and at that time, 50% of the people would still live within 50 kilometers of the sea. So in many of our lifetimes, and certainly in our children's lifetime, as many people as there are on earth today could live at the interface of the sea and the land. To accommodate that growth, we would have to build what would be the equivalent of four new Cairos every year. Obviously we will not build four new mega-cities every year. But I do believe that, right now, however, the nuclei for major new cities are being established on the coasts of the earth. One of them is Puerto Peñasco, Sonora, Mexico, where a new community, "Los Ambientes," is being developed using seawater for sewage, agriculture, landscaping, and biochemical feedstock production. Another is in Jubail, Saudi Arabia. (Jubail is a $21-billion industrial city of Saudi Arabia.) In the *AMBIO* article, there are concepts of those communities.

I would propose that the "breakthrough" that we might develop would have two parts:

(1) We should define a goal that is far greater than sustainability. If that goal were defined and espoused by a group such as this, it would be looked upon as something worthy of consideration.

(2) We should design "intellectual catalysts" that demonstrate our goal is achievable. One might be a prototype community of tomorrow (PCOT) that would be environmentally enhancing. Every time a square meter was added to PCOT, the entire planet would be better off. The automobiles would be efficient and would run on a blend of petro and biochemical fuel. Every mile that one was driven, more carbon would be taken from the atmosphere than put in. Every time a park was greened or a food supply field enlarged, the atmosphere of the world would be improved. And, most importantly, if every citizen would continually be an example of the anti-entropic action (intelligence), they would continually contribute. Another intellectual catalyst might be to define the field of Urban Biopsherics, as in the following table.

Table 1
Urban Biospherics, key to planetary design

- As agriculture becomes more efficient, rural populations gravitate to cities which act as magnets.
- Rural lands are then free for other purposes, including ecological restoration and support and systems for a great diversity of non-human species.
- But older cities lack the elasticity to readily accommodate a vast influx of new people, which strains hygiene facilities, sewers, water lines, air quality, garbage disposal, transportation, housing, education, health care, jobs, etc., to their breaking points.

Table 1
Urban Biospherics, key to planetary design

- Effects on older cities include: degraded environment, impoverishment of municipal structures, crime, flight of affluent, abandonment of real property, etc.
- Therefore, there is a need to: (1) recreate older cities and their suburbs to make them more elastic, i.e., more livable for larger populations; (2) create supplemental cities to relieve population pressure and provide amenities; and (3) invent new cities to take advantage of new ecosystems, new industries, new population resources.
- Growth is the key to make the planetary environment hospitable to human and all other species.
- The world economic pie is too small, i.e., there is not enough value (wealth) to provide for the growing population.
- We need to create vast amounts of new wealth to provide all our citizens with a degree of affluence, decent quality of life, jobs and education. Education and jobs are key elements in overcoming society's problems rooted in over-population, and manifested in the demise of cities and the environment, pockets of terrorist activities, disease epicenters, etc.
- That means new industries must be created, plus infrastructures to support them--thus creating new wealth, new jobs, new markets, new consumers.
- But the history of industrialization, until now, has also been a history of environmental degradation; hence development has been seen as the antithesis of environmental preservation.
- It need not be, if the need to balance and speed up natural cycles is appreciated. For example, development to avoid and remove toxic materials could be a positive environmental contribution.
- This calls for new bio-industrialization (eco-industrialization?) in which all development has an environment-enhancing component largely based upon intelligent harnessing of biological processes.
- This is the real urban frontier!
- The organizations that demonstrate their ability to pioneer and manage this process will enjoy a massive competitive edge in the future, and the more experienced they are, the more credible they will be.

Perhaps our new intellectual catalyst communities might be as described by Mr. Brando.

A Message from Marlon Brando, First Resident of Los Ambientes:

THE PROMISE OF LOS AMBIENTES, THE WORLD'S FIRST SEAWATER-BASED RESIDENTIAL, AGRICULTURAL, AND INDUSTRIAL COMMUNITY

Los Ambientes is a new kind of environment-enhancing community. A prototype for the seawater-based cities that will transform areas of the world's desert seacoasts in the early years of the twenty-first century. It will be built in Sonora, Mexico, near the shrimping town of Puerto Peñasco.

What makes Los Ambientes Different?

The land--gardens, parks and farms, public spaces, putting greens and other recreation areas--will be irrigated with seawater drawn from deep wells dug into the coastal sands. The seawater from these wells will be a constant seventy-four degrees Fahrenheit--cool enough to moderate the heat from the bright sun and warm enough to swim in comfortably. Pools, brooks, canals, fountains and waterfalls will all be seawater--as will the water that flushes the toilets and washes the streets.

How is it possible to irrigate gardens, parks and farms with seawater? Do we plan to desalinate first? No, we do not. The gardens and farms will grow plants called halophytes, plants that naturally grow in salt marshes where the sea and the land touch. There are many such plants. One of them, an oil-seed plant called *Salicornia*, will provide the principal crop for our farms and the basis of much of our industry.

A Crop to Restore the Earth.

Salicornia is one of nature's miracles, a crop so useful it is almost too good to be true. It is the highest yielding oil-seed crop in commercial production. The oil it produces makes a marvelous salad oil, an excellent cooking oil, a fine domestic oil. Mixed with diesel fuel, you can drive your car on it.

This alone would make *Salicornia* an attractive commercial crop, but this is just the beginning.

Because *Salicornia* is literally irrigated with seawater, it marks the first time in the history of human agriculture that farmers can actually reverse the flow of nutrients from the land to the sea. After ten thousand years of washing our soils and their nutrients into the oceans, *Salicornia* farming gives us a chance to put them back onto the land and to rebuild the soils that time, farming and erosion have destroyed.

Salicornia culture can and will rebuild desert sands into soil. It will also fix an unusual amount of carbon, holding out the promise that we can balance the atmospheric carbon cycle without compromising the lifestyle of the developed world, and without, in fact, giving up our cars and our electric power plants.

In addition, *Salicornia* meal and *Salicornia* straw can serve as fodder for domestic animals, whose hooves and droppings also play a key role in building soils. After we feed our cows and goats at Los Ambientes, we will package some of our leftover meal and straw to be used as high-quality, high-protein animal feed.

If you are wondering how we know *Salicornia* can be grown economically in commercial quantities, we know because it has already been done. For some years, there has been a *Salicornia* seed farm at Puerto

Peñasco. The seed has been going to Ras Al Zawr in Saudi Arabia to help establish a large-scale oil-seed farm.

In time, this Saudi *Salicornia* farm will become the basis for a major seawater-based city on the Kingdom's desert coastline . The scale of the Saudi enterprise will be larger than Los Ambientes, but many of the same techniques will be applied. A requirement for the proposed Saudi seawater community (as for Los Ambientes) is that, in its total activities, it removes more carbon from the atmosphere than it adds through its combined use of fossil and biofuels.

Aquaculture as well as Agriculture.

The sand-filtered seawater that comes from our deep on-shore wells is ideal for aquaculture as well as agriculture.

There is, of course, a direct connection between the aquaculture and the agriculture. The water in which the shrimp and fish are raised, further enriched by their droppings, will be recycled to irrigate the *Salicornia.*

In addition to high-quality edible oils, cosmetic and pharmaceutical oils and specialty lubricants, high-protein animal feeds, sea-salt and gourmet food items like shrimp and tilapia filets, we plan to establish a series of other specialty activities. Among them are entrepreneurial enterprises making baked goods, goat cheese, gourmet shrimp paste, tilapia leather for crafts, even specialty writing papers, wallpaper and building materials produced from *Salicornia* straw.

Why seawater cities?

As everyone knows, the world's population is growing at an astounding rate. What is equally obvious is the fact that people, wherever they are, prefer to live in cities. (Well over fifty percent of humanity are already city-dwellers.) To accommodate these new inhabitants, the world is going to need the equivalent of hundreds of new cities.

Without forethought and planning, these new cities will very likely grow up around, adjacent to and within cities that already exist. These existing cities have culture, industry, and amenities that act as magnets to attract new immigrants from rural areas.

While the great cities of the world have long absorbed huge influxes of new citizens over time, most are now close to or past the breaking point. Their housing is inadequate, their water, sewer, and school systems are strained or dysfunctional, their air is polluted, their social services unable to cope, their transportation systems are grid-locked and their ability to provide new jobs is outstripped by the number of eager applicants.

The world's great cities are almost all endangered species. These urban environments need to be rebuilt, restored, recreated. In some cases, as Tokyo is already doing, supplemental cities will have to be built to provide recreational facilities and other amenities for the expanded population. But there will also be a need for completely new cities. Where are these new cities to be built?

An ideal place, it seems to us, is the world's desert seacoasts.

There are over 40,000 kilometers of desert seacoast in the world. These are places where there is little or no development, little or no agriculture, little or no settlement--places previously regarded as inhospitable and barely inhabitable, simply because they are not easily adaptable to old-fashioned freshwater agriculture. These same places, however, are ideal for seawater-based agriculture and aquaculture. They are, in short, ideal for seawater-based communities. We have only to demonstrate that such communities can be created and thrive and grow; and that as they grow, they can enhance the environment of our whole planet.

That is the mission of Los Ambientes--to provide a model for the new world of environmentally-enhancing seawater cities to come. We invite you to join us in this great adventure.

Who is behind Los Ambientes?

The Environmental Research Laboratory (ERL), of the University of Arizona, conducted most of the original research that led to the development of the concepts we have outlined here. In some of this they had aid and cooperation from the University of Sonora in Mexico. The Planetary Design Corporation (PDC), a technology transfer company which benefits the work of the University of Arizona, has done most of the work necessary to turn these concepts into commercial applications. (The principals in PDC include environmentalist business leaders from the United States, Mexico, Sweden, England, Saudi Arabia and Japan.) Its associated companies, GENESIS in Mexico, and Halophyte Industries, Inc. in Saudi Arabia, have been responsible for actual commercial operations.

2. POSTSCRIPT

This "postscript" is being written on an airplane from Tokyo to Honolulu as thoughts run through my mind as to the pleasantness and degree of success of the Eighth Toyota Conference. I have met the wonderful people who attended the conference and gotten to know my host. With hindsight now, particularly relevant to me again are Professor Shunji Marai's comments that I quoted at the beginning of this paper: "Through extensive discussion and exchange of information among experts in a wide variety of fields...the conference will contribute to formulation of a framework for further research. A breakthrough is expected toward the development of global eco-engineering which seeks continuous human habitation, the improvement of human welfare...."

Certainly there were "intensive discussion and exchange of information among experts in a wide variety of fields...." That interaction took place in the most pleasant of surroundings. The conference was well-organized, sessions began promptly, and the food was delicious. At the conclusion of the conference, there was a recommendation "for further research" and a follow-up conference.

I think of my airplane reading from a previous flight, i.e., John Kenneth Galbraith's, *The Culture of Contentment*.(6) His book addresses what happens when the majority of participants of an event (in his analysis, a culture) are "content."

Galbraith suggests that "...there will be contrary voices. These will be heard, and often with respect, but not to the point of action...the policy avowed by the contented...does not deny the problem...rather it delays action. Notably, it proposes more research."

Perhaps our conference was too pleasant. Being "contented," the easiest of recommendations would be more research and another conference.

I do not believe we achieved a "breakthrough." Perhaps that expectation was simply too high. I do not think so, but....

So, as a postscript, I recommend another reading, i.e., Galbraith's book. I will also agree with the contented in recommending another conference. I would propose, however, that the future conference be a little less pleasant. We should have less formal sessions. We should all work harder. We should have read all of the papers before we gather. We should start with the content of a typical conference as a given.

Most of all, we should rise to a level of creativity as described in *The Grace of Great Things.* I propose that we work on a "bigger, bolder, fatter, happier problem..." than simply more research and a future conference. We should create a breakthrough by defining a global eco-engineering task, and we should implement it.

My thanks to the Toyota Motor Company, to Professor Murai, to Masata Kimura, to the participants, and to all the others who did such a wonderful job of contributing to our contentment.

3. REFERENCES

1 R. Grudin, The Grace of Great Things, Ticknor & Fields, New York, 1990.

2 C. N. Hodges, T. L. Thompson, J. J. Riley and E. P. Glenn, Reversing the Flow: Water and Nutrients from the Sea to the Land, AMBIO, J. of the Human Environment, Royal Swedish Acad of Sci, 22 (7) (1993), 483-490.

3 R. Repetto, The Second India Revisited, World Resources Inst, Washington DC, 1994.

4 W. W. Harman, Toward a New Eco-nomics, Inst of Noetic Sci, Sausalito, California, 1994.

5 J. Rifkin, Entropy, Into the Greenhouse World, Bantam Books, New York, 1981.

6 J. K. Galbraith, The Culture of Contentment, Houghton Mifflin Co., Boston,1992.

TOWARD GLOBAL PLANNING OF SUSTAINABLE USE OF THE EARTH
Development of Global Eco-Engineering
S. Murai (Editor)
1995 Elsevier Science B.V.

ENVIRONMENTAL MONITORING AND CONSERVATION BY WISE USE

Prof.Dr.Dr.h.c.mult. **Paul Müller**
Director of the Institute of Biogeography
University Saarbrücken, Germany

On the closure of the **Rio Summit** (1992), the following appeal was signed by 425 members of the scientific and intellectual community:

We want to make our full contribution to the reservation of our common heritage, the Earth. We are however worried, at the dawn of the 21st century, at the emergence of an irrational ideology which is opposed to scientific and industrial progress and impedes economic and social development. We contend that a Natural State, sometimes idealized by movements with a tendency to look toward the past, does not exist and has probably never existed since man's first appearance in the biosphere, insofar as humanity has always progressed by increasingly harnessing Nature to its need and not the reverse.

We fully subscribe to the objectives of a scientific ecology for a universe whose resources must be taken stock of, monitored and preserved. But, we herewith demand that this stock-taking, monitoring and preservation be founded on scientific criteria and not on irrational preconceptions. We stress that many essential human activities are carried out either by manipulating hazadous substances or in their proximity, and that progress and development has always involved increasing control over hostile forces, to the benefit of mankind. We therefore consider that scientific ecology is no more than an extension of this continual progress toward the improved life of future generations.

The greatest evils which stalk our Earth are ignorance and oppression, and not Science, Technology, and Industry whose instruments, when adequately managed, are indispensable tools of a future shaped by Humanity, by itself and for itself, overcoming major problems like overpopulation, starvation, and worldwide diseases.

In our world with more than 5 billion people, **environmental monitoring** and **realistic environmental conservation strategies** are two of the most important scientific ecological targets. The intensity of land use has increased dramatically during the last two decades. More than 100.000 chemicals are listed in the European Inventory of Existing Substances (EINECS). Even if only 10.000 of these listed chemicals may be produced on a larger scale their potential risk for man and environment is not well known. Therefore, the risk assessment of existing chemicals is one of the most urgent problems the industrialized countries have to solve.

1. ENVIRONMENTAL MONITORING AND ENVIRONMENTAL RISK ASSESSMENT

The procedure for the **environmental risk assessment** of a compound consists of comparing its concentration in the environment with its highest concentration at which **no effects on organisms or ecosystems** are expected to occur (cf. Technical Guidance on Environmental Risk Assessment of Existing Chemicals in Accordance with the Requirements of Council Regulation EEC No. 793/93; July 1994; cf. also ASTM 1991, BURTON 1992, DE GREEF and DE NIJS 1990, EMANS et al. 1993, EPA 1990, EUROPEAN COMMUNITIES 1993, HOESTRA and EWIJK 1993, LOKKE 1994, OECD 1993, OKKERMAN et al. 1992, PACK 1993, PEDERSEN et al. 1994, RÖMBKE et al 1993, WAGNER and LOKKE 1991). For the environment the protection goals are up to now limited to the following:

- aquatic ecosystem
- terrestrial ecosystem
- top predators
- micro-organisms in sewage treatment systems
- atmosphere

Measured or estimated environmental concentrations (PEC) are compared with concentrations, which, taking into account adequate safety factors, are regarded as a level below which the balance of **probability is that an unacceptable effect will not occur (PNEC)**. PNEC values are usually based on results from monospecies laboratory tests or in very few cases on model ecosystem tests.

The general objectives of environmental monitoring are to recognize environmental pollution problems at as early a stage as possible, to analyze the type, source, and path of contamination and effects and to develop recommendations as to solve the problem before serious effects become manifest. Anoher important task is to control the effectiveness of remedial measures. These tasks require sophisticated strategies and methods for selecting and collecting environmental samples (ELLENBERG, FRÄNZLE and MÜLLER 1978, ERNST 1994, LEWIS 1987, MARKERT 1993, MÜLLER 1979, MÜLLER and WAGNER 1985, STOEPPLER 1994, WAGNER 1993a,c).

The task of a monitoring project may be, for example, to characterize and evaluate the burdening of a certain area by environmental chemicals. The concentration of SO_2 in the ambient air or the content of cadmium in the dustfall measured over a certain period of time at a somehow selected spot alone will give only poor information about the real objective, because the real SO_2 burden depends not only on its concentration in the air, but mainly on its impact on the organisms and materials present in the area, which is influenced by the specific sensitivity of the different species and their stages of age and development, their exposiion, the wind velocity, temporal changes and spatial differentiation of SO_2 concentrations and the concentrations of interfering compounds such as, e.g. NO_x and O_3. There is no fairly simple model available which would allow a realistic calculation of the real effects resulting from the total concentration of a pollutant analyzed in a single environmental medium. This means that a spectrum of samples has to be taken which is able to represent the totality of

General procedure for the environmental risk assessment of a substance in a given compartment*

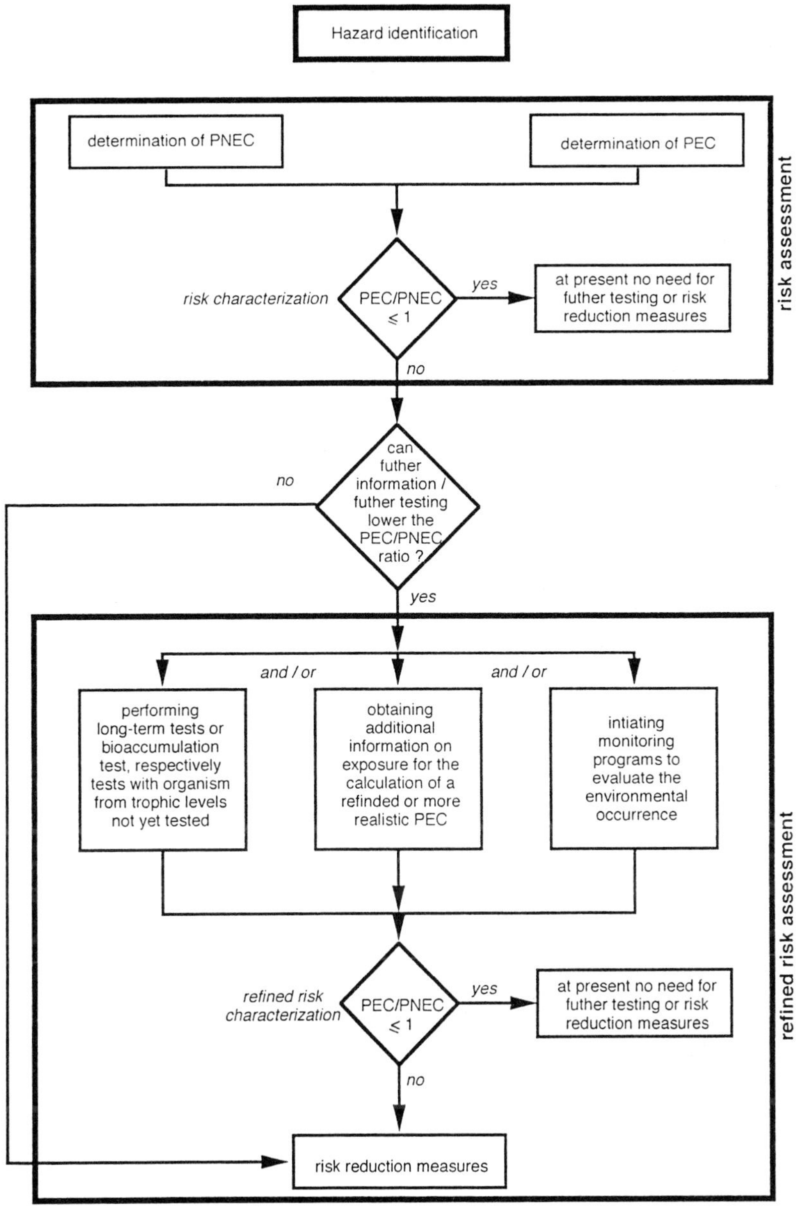

* (aquatic, terrestrial, air)

Testing strategy for bioaccumulation and risk charakterization for secondary poisoning

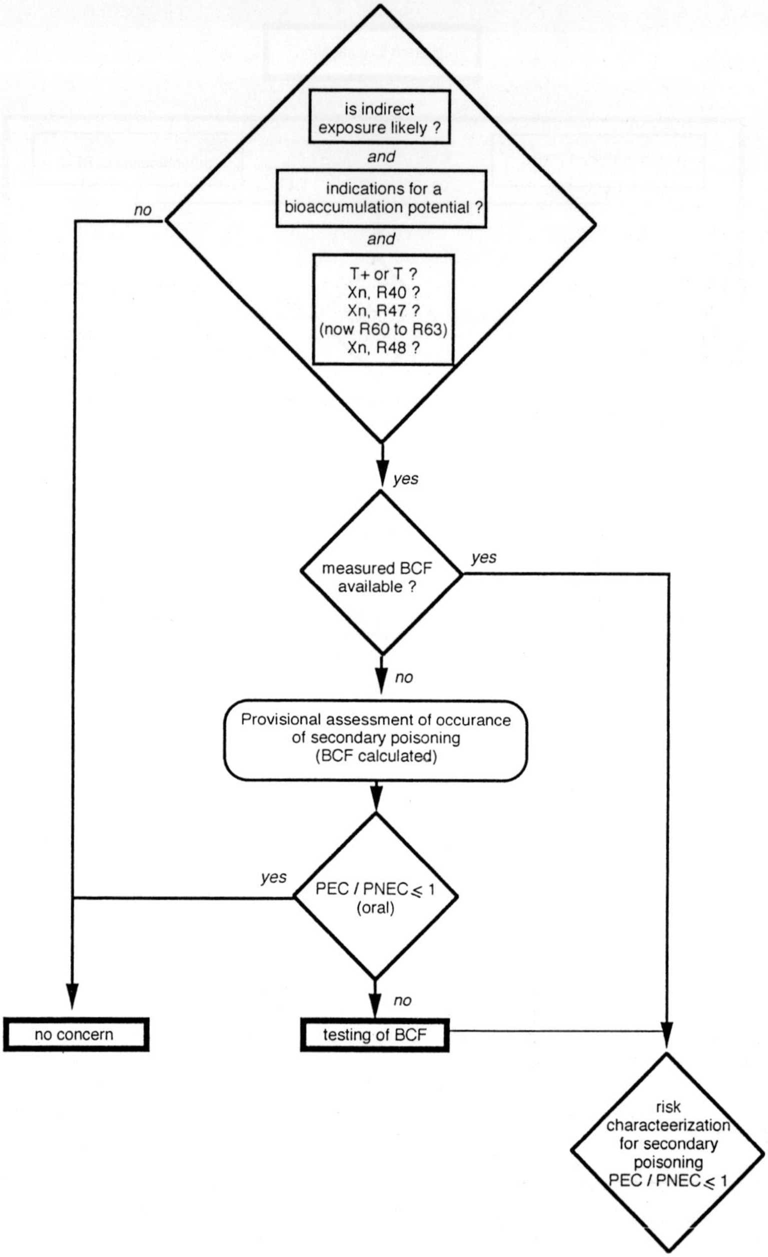

possible or intended targets present in the area to be studied. One possible way to reduce the effort of such an inventory to a reasonble extent is to use suitable **biological indicators** (ARNDT et al. 1987, MÜLLER 1980, 1981, WAGNER 1993a,b, 1994a).

Moreover, the objectives of environmental monitoring studies seldom relate to the properties or pollution impact of a specific point or spot in the landscape. Rather a more or less extended and carefully delimited area is to be characterized. Nevertheless, most technical measurements or sampling procedures for air, water, or other environmental media are carried out on discrete spots, which are expected to be more or less "representative" for the area. The data obtained are then taken for the whole study area or extrapolized to "the surroundings" of this point, but it is seldom examined whether the data really represent the quality of the area to be investigated. Thus the internal spatial heterogeneity of the area and with that the variability of the data remain unknown. **Sampling is the first and the least corrigible step in environmental monitoring and analysis.** While methodology and techniques for physico-chemical measurements of emission and the impact of pollutants are well developed and standardized, the measurement of effects, including the accumulation of pollutants in the habitats, food and bodies of living organisms is still in full progress. This is the reason why concentrates, in this paper, our attention on approved strategies for sampling environmental specimens such as animal tissues, plants, and soil as matrices for chemical analysis on nutrients, trace elements, and pollutants for purposes of environmental monitoring. Special emphasis is placed on possibilities, approaches, and methods for quality assurance and quality control in planning and field work.

2. REPRESENTATIVENESS AND ACCURACY OF ENVIRONMENTAL SAMPLES

The term representative is a keyword in environmental monitoring. However, this frequently used term is seldom exactly defined or specified. Violation of representativeness may cause systematic errors of high magnitudes (see WAGNER 1994b for more details). In effect it refers therefore to the accuracy of analytical results in the broadest sense. Neglecting the several different aspects of representativeness is the main reason for bad quality and a costly accumulation of often rather useless data.

As we deal with areas and spatially distributed phenomena, the geographical or spatial aspect of representativeness is the first to be discussed. **Geographical representativeness** means that the whole area is characterized by a set of samples or data. This is often claimed by describing the quality of more or less extended area by the central tendency and variation of the relevant parameters. However, in the case of significant heteogeneity of the area, the mean is not an adequate measure. In this case, the distribution pattern has to be described by spatially definition of gradients or by deviding the area into homogeneous subareas from the results of preliminary sampling and ascertaining mean and variability for each of them by measurements or samples from an adequate number of spots randomly or systematically distributed over the area.

Ecologically representative means that the samples should represent the total targets population or ecosystem. This requirement may be relatively easy to fulfil in the case of area dominated by one or few species, as in the case of agriculture and forestry. "There is no better indicator for the health of a spruce forest, than Norway spruce itself, nothing better than beech trees in a beech forest ..." (KNABE 1982). In diverse ecosystems like tropical rain forests it is extremely difficult and laborious if not impossible to obtain ecologically representative samples of the whole vegetation, because the different species may have different accumulation behaviour and resulting concentrations of ingredients and pollutants. In such cases only stratified random sampling strategies can lead to useful results.

Standardized biological indicator systems react with known specivity and sensitivity to specific pollutants and have the ability to randomize or integrate pollutant inputs spatially and temporally. Such **indicator systems** can be used vicariously as representatives instead of the spatially, physiologically and temporally often extremely variable total population. Of course the accumulation behaviour of this indicator system has to be similar or comparable to that of the majority of the total population of the ecosystem to be characterized. In other words, it has to be biologically and physiologically representative.

Genetical representativeness may be a problem in genetically very heterogeneous populations or in comparing areas with genetically different populations. This is particularly the case for example in spruce forests which have been planted without considering the origin or provenance of the trees, as in many spruce or pine forests planted after the first and the second world war. Differences in the growth rates and phenotypical properties caused by different physiological adaptation of the genotypes can influence the accumulation behaviour and resulting chemical composition. However, this seems to be of relatively minor importance compared with other factors of natural variability.

Another problem is linked with the adaptation of populations to specific contamination or long-term pollution of their habitats. There are different possible adaptation strategies (ERNST 1974, 1994) but as an effect often the accumulation behaviour and/or the distribution of the pollutant in the body is changed which may impair the comparability of the resulting concentrations of these contaminants in the biological tissues.

In order to establish temporal trends, e.g. of atmospheric pollution, strictly quantitative comparability of consecutive samples of a series is the first priority. Optimizing repeatability means reducing the influence of natural (spatial, biological, temporal) variability of the sampled population in order to yield maximum precision and repeatability of the samples. This demand can only be achieved by stratifying the population to be sampled, standardizing the sampling procedure, and defining a very homogeneous sampling area. However, most of the measures claimed here to improve repeatability and precision generally reduce the spatial representativeness of the samples. The more the size of a sampling area is reduced in order to get homogeneous conditions, the more effort has to be spent to protect it against undeliberate changes, because the continuation of the trend monitoring is very sensitive to accidental changes in the case of a small sampling area or population.

Another demand on the results of environmental monitoring is the widespread desire to compare the burdening of different, sometimes far apart areas by quantitatively comparing monitoring data. This seems to be easy with, for example, the concentrations of chemicals in water or ambient air. The problems become directly obvious for soil data, because there are no two spots in the world with two identical soils and without additional information about the composition and chemical properties of the soil the concentration of, for example, cadmium in a soil sample can say almost nothing about its possible effects on biota, ground water, etc. Concentrations of chemicals in biological tissues used as biomonitors are also influenced by the local complex of environmental factors and spatial differences in the genotypical and phenotypical composition of the populations. On the other hand, also th potential and real effects of identical concentrations of a substance like sulfur dioxide in the ambient air in different areas can be very different as well, depending on environmental factors like moisture, insulation, and the semultaneous presence of other additives like ozone. In the light of these relations the "biologically standardized" reactions and accumulation behaviour of well-known and genetically comparable monitor organisms have a great potential to improve the spatial comparability of pollution monitoring data related to their effects. Preconditions for using this potential are carefully selection and stratification and standardized sampling methods.

For supra-regional or large-scale comparisons across different climatic zones, the basic should on principle not be restricted to the level of simple concentrations of desired and undesired substances in environmental media and commercial goods like food and feed products, but extended to their specific importance, hazards, and effects on human health as well as on the functioning and structure of prevailing ecosystems. The geographical basis for monitoring and comparison on an ecosystem level are preferably small, well-defined ecosystems with a minimum of diffuse input-output relations like small watersheds. This approach is thoroughly described in the guidelines for the "Integrated monitoring of the Nordic countries" (BRAKENHIELM 1990, LEWIS et al. 1989, PAULUS et al. 1990, NORTHERN COUNCIL OF MINISTERS 1988).

There is a broad spectrum of biological methods available for monitoring and evaluating accumulation as well as toxicological and ecotoxicological effects of pollutants. The borders of biological monitoring methodology are marked on the passive side by indicator organisms showing directly observable effects of a certain influence on their habitat, like specific growth form and coloring of plants growing on soils with high contents of heavy metals. Passive biological monitoring depends on analyzing organisms which are sampled where they grew up (natural habitat or normal cultivation). These organisms, if adequately selected and sampled, give reliable information directly related to the ecosystem they belong to. As far as direct influence or contamination can be excluded and quality assurance criteria are observed, such samples possess high probative force, but often precision may be rather poor because of effects of various sources of natural variability.

The active or experimental side of this spectrum is characterized by standardized cultivation of well selected and genetically homogeneous organisms which are exposed to the environmental influences of the location to be monotored

2.1 SAMPLING STRATEGIES FOR PLANTS (PASSIVE BIOMONITORING)

The examples of sampling strategies explained in this chapter are intended as a collection of approved methods and special recommendations rather than a systematic overview of existing procedures.

Sampling by chance

The roots of bioindication and effect monitoring are undoubtedly located in the observation of pollution impacts and sampling of affected tissues by chance. In the meantime, information, knowledge, awareness, and methodology have improved but sampling by chance is still an important source of information about manifest, hitherto unknown or unexplained effects like, for example, suddenly appearing, unexplored diseases in human populations, unaccountable damage to vegetation, fish deceases in various water bodies without obvious reason, seabird deceases, or the seal decease some years ago in the European North Sea (e.g., ELLIOTT 1984, MÜLLER 1988). Beyond the general problem of only subsequent, reactive research on already manifest damage, the samples received by this kind of sampling are often already severely changed by autolysis, decay, and other influences (LEWIS and KLEIN 1990, LILLESTOLEN et al. 1993), so that in such cases only coarse or incomplete information about the possible cause of the event can be received from such samples.

Selected sampling

Preliminary investigations or initial research into a new problem often require the analysis of samples selected for special purposes. In the case of air pollution effects on plants, selected sampling is often used in the form of comparing pairs of affected and not affected plants or plant organs, e.g., ill and healthy-looking trees, twigs or leaves with and without chlorosis, necrosis, aphid infestation, and so on (ZIMMER 1993).

For screenings, samples are often selected according to gradients or factors present in the area or population, which are potentially correlated with the concentration of a sought-after substance e.g. health status of the organisms, elevation, exposition, edaphic factors, human influences etc., in order to get information about sources of variation or the potential spans of concentrations of substances in the population.

The distribution of substance in different organs or tissues of an animal or plant (organotropy) is also often analyzed using accidentally available or especially selected individuals. This knowledge is a precondition for any attempt to reduce variability or to increase the predicative power of the samples by stratification and/or standardization of the sampling (e.g., KRIVAN and SCHALDACH 1986, LIETH and MARKERT 1990, MÜLLER and WAGNER 1985, WAGNER 1987, WYTTENBACH et al. 1988, ZIMMERMANN and PLANKENHORN 1986).

Random sampling

The main criterion for differentiating random from selected sampling designs is the requirement that in the first case spots or individuals to be sampled are not selected by subjective criteria in the field. "In completely random sampling designs, each potential sampling unit must have an equal chance of being collected" (LEWIS 1987). Random sampling strategies offer the advantage that subjective selection criteria are eliminated and most of the common statistical methods are suitable for analyzing and describing the results (GREEN 1979). From the aspect of quality assurance, true random sampling designs have some disadvantages in practical work:

(1) the designated sampling points in the plan are often difficult to localize exactly in the field or the coordinates of the randomly chosen points are difficult to determine exactly, respectively;
(2) spatial patterns or gradients may not be recognized because of uneven density of sampling points;
(3) it is often not possible to control whether the realization in the field is in full harmony with the principles of the random sampling design. In other words, random sampling is often postulated or pretended, but not always executed in the strict sense of the word in the practical field work. With respect to the difficulties of (1) and (3) improvement is promising thanks to the increasing availability and accuracy of Geographical Position Systems (GPS) combined with the use of Geographical Information System (GIS). These tools provide new possibilities and chances for an increased use of random sampling designs as well as quality assurance and quality control in environmental monitoring in the future (WAGNER 1994b).

Systematic sampling (transects, grids)

The most essential aspect common to both random and systematic sampling designs is the fact that they exclude subjective decisions and selection criteria in the individual case during the sampling. Systematic sampling is generally preferred when spatial structures or patterns of a parameter have to be analyzed, e.g., for large-scale screenings, surveys, or inventories (e.g., BAU 1992, HODGESON 1985, RÜHLING and TYLER 1969, SCHOLZ et al. 1993, STEINNES 1982).

The most simple form of systematic sampling is to place a transect following a known or supposed gradient, for example from a point source of pollution in the main wind direction, and to take samples on this line in equally or in some other way preselected dis tances from the sources (e.g., ERNST and LELOUP 1987). Other common systematic sampling designs are, for example, more or less sophisticated polar grids for source-related monitoring and regular, rectangular, triangular, or hexagonal grids to cover a given area. The optimal density of the grid has to be determined by preliminary sampling if no specific experience is available. For geostatistical analysis of the results the grid should be at least so dense that, generally spoken, the concentrations of neighbouring points lie nearer together than those of distant points (ENGLUND and SPARKS 1988, SCHERELIS and BLÜMEL 1988, WAGNER 1993d). The sampling points have to be previously determined in the sampling protocol and then identified in the field as accurately as possible without looking for any additional criteria. This of course is plain theory, because nobody, for example, can collect a meaningful soil sample from a preselected spot if a house, a street, or a trunk of a tree covers the surface. If such places are excluded and, for example, a minimum distance of 10 m from buildings and roads is set up, we will come automatically to a stratified sampling design. This means that the results of our investigation will not be valid for the total area but only for that part of the area which is more than 10 m away from buildings and roads.

This example indicates that in practice true unrestricted, systematic or random sampling is the exception rather than the rule. Decisive requirements are therefore to define clearly
(1) the area to be investigated with all exceptions and exclusion criteria (and to consider these exceptions in the interpretation and evaluation of the data) and
(2) whether such points are totally excluded or how they may be replaced (in order to prevent any subjective solutions on the spot).

The critical point is not **how** these determinations are made but that they are made **previously** in the sampling plan and things are not left to chance, which means to subjective decisions in the particular case. Those subjective decisions can severely bias the results, as GREEN (1979) says: "The assumption of independence of errors is the only one in most statistical methods for which violation is both serious and impossible to cure after the data have been collected".

Stratified random sampling

Stratification can be used as a strategy to improve reproducibility or to reduce the necessary number of samples in the case of diverse systems on several different levels. The principle possibilities of stratification are explained with the example of a guideline for sampling spruce (*Picea abies*). The first level of stratification is carried out by mapping the area to be analyzed and deviding it into homogeneous subareas. For an area-related stratified random sampling, the subareas covered with spruce trees are overlaid with a regular rectangular grid. The number of crossings in each subarea has been counted and the total number of samples divided in relation to the sizes of the subareas. For each subarea the sampling locations were chosen by determining the respective number of lines crossing with random numbers. These points have been transferred to an aerial photo map and the spruce tree next to such a point in each case has been identified and chosen for sampling.

The next level of stratification depends on the spruce tree itself. The main objective in this case was not to characterize the whole tree itself but to obtain a reproducible and meaningful sample, whose concentrations of several substances should be related to the pollution and health status of the tree. According to previous monitoring programs and based on comprehensive knowledge about the distribution of different substances in the spruce tree (e.g., WYTTENBACH et al. 1988, ZIMMERMANN 1989) the one year-old shoots of the seven year-old branches near the top of the tree have been chosen for sampling.

Further standardization of the sampling protocol in order to improve reproducibility and comparableness of the samples from year to year and between different areas requires for example the definition of exclusion criteria for individual trees (e.g., age >40 years, average health status), the sampling period (between snow thaw and beginning of shooting in early spring), and so on. These measures bring about stratification effects on further levels which have to be considered in the evaluation of the results.

Analogous sampling designs have been developed for the German environmental specimen banking program for pine shoots and for leaves of beech trees (*Fagus sylvatica*) and Lombardy poplars (*Populus nigra 'Italica'*) (UBA 1989).

2.2 EXPERIMENTAL MONITORING WITH PLANTS

Further standardization as well as the use of biological indicators or monitors for assessing places where they cannot be sampled by chance are possible by culturing such plants and exposing them in the field under standardized conditions (active or experimental monitoring, MÜLLER 1977, ARNDT et al. 1987, WAGNER and KRÜGER 1982). The most frequently used plant species for active biomonitoring are grass cultures for assessment of environmental pollution through fluorine and lead (*Lolium multiflorum*, Standardized Grass Culture Method, SCHOLL 1974, VDI 1978-86), exposition of lichens (*Hypogymnia physodes*, ERHARDT 1987, VDI 1991), borecole for passive accumulation of polycyclic aromatic hydrocarbons (PAH's), dioxins, and furans (GRIMMER and HILDEBRAND 1982, PRINZ et al. 1990).

2.3 SAMPLING STRATEGIES FOR SOILS

Soils are one of the most important media for environmental monitoring and risk assessment. Status and changes of pollutant concentrations and other soil parameters in the different soil horizons of a site are valuable indicators of environmental quality, change, and ecosystem integrity. The greatest problem connected with the use of soils is their high structural, morphological, and functional variability in all scales of space and time (BOURGESS et al. 1981, MIEHLICH 1976, SCHERELIS 1988). It is therefore difficult and time-consuming to obtain useful answers from soil analysis for purposes of environmental monitoring and assessment. If soil samples are taken to give reasonable information about a certain area the following general requirements have to be observed:

- the samples have to represent the soils of the area or ecosystem to be characterized and give information about the central tendency and variation of the desired parameters for the whole area or homogeneous subareas;
- the samples must not be contaminated or changed in any way during sampling, sample treatment, transport and storage;
- the samples have to refer to the different soil horizons rather than to sampling depth;
- the sampling method and sample characterization should allow the analytical results to refer not only to sample mass but also to specific fractions like fine soil (<2 mm), clay or organic matter as well as to surface units (e.g. g/m^2) or to volume units (e.g. mg/l) respectively.

There are numerous, more or less sophisticated guidelines and standard procedures for collecting soil samples for various purposes. Most of them only partly correspond to these requirements. The International Standard Organization is now working on a series of standards for soil quality, which will give guidance on sampling strategy, sampling techniques, preservation of samples, safety precautions, and procedures for the investigation of natural, near-natural and cultivated sites (PAETZ and CROESSMANN 1994, ISO/TC 190 1993).

2.4 SAMPLING STRATEGIES FOR ANIMALS

There are some important differences between plants and animals in relation to the indicator function and availability which influence the sampling strategy in an important way.

Indicator function of animals

The **mobility** of many animal species is the most important fact in this context because mobile animals may be regarded as integrators of the spatial burdening of xenobiotic chemicals. This means that spatial representativeness has a different quality and a different importance compared to plants. Therefore, the sampling strategy for mobile animals does not have to based upon spatial representativeness. It has to be emphasized that the efficiency of spatial integration depends not only on the surface of home range but also on the number and density of relations to the site over a period of time. In fact, there is a wide range of efficiency related to the integration of the spatial burdening. For example, a mature earthworm with a small home range represents one limit of this range, which may under certain conditions be comparable to plants, whereas a predator like the red fox, older than two years, shows the other limit by integrating over a large area and over at least one year (MÜLLER and WAGNER 1984).

Furthermore, the **ecological functions** of animals are a special feature. Whereas many (higher) plant species have more or less the same function of primary production, the diversity of functions of animal species is much higher. This means that animals are exposed to the xenobiotic chemicals in a different way to plants and the differences of exposure within animals are normally higher than those within plant species (KLEIN and PAULUS 1995).

A third aspect influencing the indicator function refers to the **toxicokinetic**, which vary considerably between animals and plants. The differences of accumulation are based upon the varying kind of uptake and clearance and concern the parts or organs of the organisms where the pollutants are accumulated.

Overall, the type of information about the environmental pollution stored in plants and animals is often very different. Animals indicate the pollution of environment "closer" to man., i.e., their results are better transferable to human beings. First, suitable animal indicators show the hazards of xenobiotics especially for the final consumers as a result of the phenomena of ecological magnification and secondly they react on the pollutants more similarly to man. In this context it is important to emphasize that the main goal of environmental monitoring is to show the hazards of xenobioticc for man.

Furthermore, another important fact has to be stressed as a consequence of the different kind of information between plants and animals. Plants and animals can never represent each other but they should always be used together in an environmental monitoring program. Therefore a set of suitable specimens should include plants and animals of different trophic levels in the context of their use as accumulation indicators (KLEIN and PAULUS 1994, WAGNER 1993b,c).

Availability of animals

The **availability** (second level) refers to several aspects. First, availability concerns the **number of individuals** which may be collected at one site on one sampling date. A large and stable population forms the basis for this aspect of availability. The final consumers cause the most problems in this context because the population density decreases naturally with the increase in the trophic level. There are only some species that are suitable representatives of this ecological function. Many populations of such species are reduced in number or destroyed by man or have to live in unsuitable habitats.

Secondly, availability means a **sufficient sample quantity** for the chemical analysis that may be sampled from one individual. This meaning of availability expresses itself in two ways. On the one hand, organs like liver and kidneys as target tissues are some times too small for analyzing a high number of different chemical compounds. This is found at the level of consumers like small fish species or small mammals.

On the other hand, individuals themselves are too small for analyzing any interesting pollutant. In addition, they are also too small to be determined with certainty in the field. A lot of reducers like bacteria, fungi or many "worms" (not in a taxonomic sense) are examples of this second meaning. This is the reason why only some species of the genus *Lumbricus* or *Aporrectodea* are suitable indicators at the trophic level of reducers.

Thirdly, availabilty refers to the **sampling methods**. Each method works selectively so that only certain parts of an animal population may be sampled. This is a very important problem in the context of sampling strategy for animal in environmental monitoring because it concerns the *representativeness of samples* related to a population. How can it be possible to obtain representative samples if it is not possible to evaluate the structure of a population as the statistical reference basis? Apart from representativeness, the sampling methods should fulfil a further important requirement. They should guarantee that the specimen type must be readily and reliably available. This requires **standardized sampling methods** that influence the sampled population and the sampled individuals at a very low level. Every capturing method puts animals under stress and this is why a certain unwanted influence on the samples always exists. To assure the quality of animal samples this kind of influence has to be minimized.

Fourthly, availability has a **lawful aspect**. Collections must conform to applicable laws and ethical considerations. The final consumers are the most endangered species and very often protected by law so that again they are hardly available.

Stratified random sampling of less mobile animals

Matured individuals of deep-burrowing anécique earthworm species like *Lumbricus terrestris* or *Aporrectodea longa* (*Lumbricidae, Oligochaeta*), which have a tendency to K-selection in relation to their population growth (WALLWORK 1983), are very loyal to their habitat but do not really integrate the burdening in a spatial dimension any differently to plant species. The sampling strategy has to take this fact into account if spatially representative specimens are to be sampled.

Furthermore, matured individuals of both species could be accepted as being almost ramdomly distributed in suitable habitats outside the breeding season (EDWARDS and LOFTY 1977). This is important because it makes it possible to select the collection sites randomly but it should not be mistaken for existing habitat preferences. Matured individuals are not distributed randomly in all habitats in a given area. These facts should be the starting point for developing a suitable sampling strategy.

The first step is to look for large-scale environmental pattern in the area to be sampled. An environmental pattern results from the combination of all parameters that are able to characterize an area, such as vegetation types, soil types, land use or meso-climatic types.

The next step is a screening to study the distribution and abundance of the indicator species in the area (PAULUS et al. 1994). For this purpose each subdivision is overlayed with a grid. Each unit of the grid has the same surface (size) depending on the capturing method (KLEIN 1993), which has to be the same for the actual sampling of specimens. The unit to be sampled must be selected randomly. The size of sample in each subdivision has to be proportional to its part of the surface of the area and must be very high.

The result will in most cases be that the species is not distributed and not abundant in all subareas of the environmental pattern. If the indicator species occurs abundantly in one dominant subdivision it would be able to characterize the burdening of this part of the area. It has to be emphasized that it is the wrong method to try to use the species in subdivisions where it is not well distributed and abundant because there will be no basis for representative and standardized sampling. The right method is to use other indicator species for characterizing the whole area. This is an additional argument for a set of species.

Screening is the basis for obtaining representative samples with the actual sampling. The procedure of the actual sampling corresponds in principle to the screening but the spatial reference basis is different. The suitable subdivision(s) of the environmental pattern is (are) overlayed by a grid and units to be sampled are selected randomly as mentioned above. The number of selected units depends on the surface and homogeneity of the subdivision(s), on the one hand, and on the size of each sample and the time spent on sampling, on the other hand.

This procedure is not a proper random sampling nor a proper stratified random sampling related to the whole area but it may be regarded as a combination of both. The area is divided into strata, at least one stratum is selected according to the distribution and abundance of the target species and a random sampling is carried out within the selected stratum. Therefore, this selection could be regarded as subjective but it is the only possibility to obtain samples providing correct information on the burdening of xenobiotic chemicals (see above). In relation to subdivision, this sampling strategy is a real random sampling and therefore it is suitable for pooling all samples from one subdivision if the quantity and quality of samples are equal. In the case of *Lumbricus terrestris* or *Aporrectodea longa*, the limitation to matured individuals and to one sampling date in autumn assure the comparable quality of samples.

The presentation of the sampling strategy up to here is only the basis. Potential modifications depend on the criteria for defining subdivisions and/or on the number of suitable subdivisions according to the distribution and abundance of indicator species. Only two of the possibility are to be mentioned here.

If the spatial distribution of pollution within an area is interesting, the definition of the environmental pattern should include the prior knowledge or assumptions concerning this spatial burdening. For example, if a plant emitting pollutants is dominant in or nearby the area to be sampled, the predominant wind direction is one important criterion for defining subdivisions of the environmental pattern. This point is of minor interest or rather should not be a criterion if the burdening of the area in comparison to others is studied.

If there are a lot of suitable subdivisions in an area for one species, the sampling strategy can be modified in such a way that a random sampling is realized within each spatial stratum. If this is done, the differences between the subdivisions become clearer and the samples may be pooled under the above-mentioned requirements with regard to quantity and quality of samples. The advantage of this elevated sampling in relation to the sample size is the reduction in the number of chemical analyses and finally in the costs.

Stratefied random sampling of mobile animals

Roe deer (*Capreolus capreolus*) and fish species like bream (*Abramis brama*) and roach (*Rutilus rutilus*) are suitable accumulation indicators (EDGREN et al. 1981, HECHT 1991, KRÜGER and KRUSE 1984, TATARUCH 1991). Due to their mobility they are able to integrate the pollutants over a certain area. This is the reason why spatial stratification does not have to be carried out and the sampling itself may be regarded as random in relation to space and individuals. Nevertheless, this sampling strategy must be based upon a necessary standardization at the population level to assure the quality of samples.

The big problem with the representation of population is that the statistical reference basis is not really known. The conclusion is very simple: representativeness of samples referring to population could not be realized from a statistical point of view. The consequence could be either to realize it as well as possible according to the used sampling techniques or to do sampling without it. The decision is based on the question of whether it is necessary to represent a population in order to use indicator species.

What we want to know concerns the information about the environmental pollution stored in organisms, i.e., what is interesting is the ability to indicate. We do not want to know the variation of the contamination concentration in a population. Therefore, the right question is: "How is it possible to get representative samples in relation to their ability to correctly indicate the environmental pollutions?" To avoid misunderstandings: the knowledge of accumulation behaviour within a population is very important for studying the **suitability** of an animal species as indicator if constitutes the basis for its **use** in monitoring programmes.

So far the decision should be to do sampling without the need to represent a population. If we do not have to and if we are not able to represent a population, the possibility exists to standardize the samples. Reproducibility and representativeness join at this point. This means that standardization does not reduce the representativeness of samples related to the information content of organisms. The principle of stratification is suitable for obtaining the "best" stratum of the population. There are several criteria for stratifying a population, such as age, sex, weight, length and genotype (LEWIS 1987). Age is the most important criterion for selecting individuals of freshwater fish species and roe deer because accumulation is very often age-dependent (TATARUCH 1991, NIIMI 1983, DISSER et al. 1992, BIGNERT et al. 1993).

A suitable stratum in *roe deer* population is relatively quickly found using age as a stratifying criterion. The age can only be determined with certainty if the individuals are not older than one year. A reliable age determination is very important in the context of quality assurance. The selection of suitable specimens within the group of individuals not older than one year is based upon the consideration of the advantage and disadvantage of fawns and exactly one-year-old individuals. Fawns are only suitable up to an age of about seven to eight months because younger fawns are suckled. The consequence is that lipophilic chemicals are overrepresented whereas metals like cadmium are underrepresented in suckled fawns because they do not pass the placenta (TATARUCH 1991) and are normally accumulated during a long period without reaching a saturation concentration. The problem with older fawns is their availability is not equally satisfactory in all hunting areas because of the fact that in some areas fawns are already hunted in September and October. One-year-old individuals are mostly hunted between mid May and the end of June directly after the opening of the hunting season. Sampling one or two months earlier would be desirable because up until that time the young males (not only the females) are also loyal to their habitat, where they have lived since their birth. Later in the year, the probability is higher that they are driven away by older males.

The above-mentioned fish species cause more problems when it comes to finding a suitable age class. This has several reasons. First, age cannot be determined with certainty in the field during sampling. It causes problems even in the laboratory (BAGENAL 1974). Weight and length are the indicators of age during sampling but they are specific for each waterbody. Therefore a screening to study the relation between age and weight and age and length has to be carried out if prior information is not available. Generally, eight to twelve year-old individuals of the bream and four to seven year-old individuals of the roach seem to be suitable because of their availability in different water bodies and the fact that age-dependent accumulation is not recorded within these groups.

A random sampling may be carried out within the selected age classes and is generally assured by the used sampling techniques (KLEIN and NENTWICH 1993, KLEIN and PAULUS 1993). In conclusion, sampling strategy for mobile animals like freshwater fish and roe deer may be regarded as a modification of the stratified random sampling because one stratum is selected subjectively. It has to be emphasized that the selected stratum is assumed to correctly represent the information content of the indicator species.

Zebra mussels as examples for sampling under semi-artificial conditions

In contrast to the previous sections in which sampling strategy is strictly based upon sampling of specimens under natural conditions this strategy is founded on the manipulation of specimens in such a way that the samples are more readily available with the disadvantage that the specimens are living on an artificial substratum. The sessile zebra mussel (*Dreissena polymorpha*), which has been well studied as an accumulation indicator (BUSCH 1991, KLEIN and ALTMEYER 1992, KLEIN and WAGNER 1993, NALEPA and SCHLOESSER 1993), is an example for sampling under semi-artificial conditions.

The manipulation is realized by using artificial substrate for colonization with mussels. Uncolonized plates are exposed in order to sample specimens originating from the water body to be studied. The plates are made from additive-free polyethylene (PE) and have a surface size of 900 cm^2. Seven such plates are screwed together at a distance of 7 cm each to form a so-called plate tower or rack. The spatial steel screw connections are concealed by spacers made from polytetrafluorethylene (PTFE). These plate racks can be installed in the water either individually or in a group of several racks connected with each other. It is important that all used materials have no influence on the concentrations of pollutants in the mussels.

The advantage of this method is that individuals of zebra mussels from one specific suitable age or length class are more readily available with regard to a sufficient number of individuals, easy handling during collection of the specimens and using zebra mussels as indicators even in water bodies where they do not occur naturally. This is realized by colonizing the plates with zebra mussels in a water body where they are abundant and replacing the colonized plates in another water body to be monitored where mussels link.

It has to be stressed that this method is only semi-artificial because the mussels settle down themselves. This fact does not change even if the mussels on the plates are replaced. The difference to natural conditions is minimized by exposing the plates as near to the ground of water body as possible. It is well known that the neighborhood to sediment is an important factor in the context of accumulation (DEKOCK and BOWMER 1993).

Similar techniques for sampling under semi-artificial conditions are used for other species like feral pigeons (ALTMEYER 1993) and honey bees (BROMENSHENK 1978, MÜLLER and WAGNER 1985).

2.5 ENVIRONMENTAL MONITORING AND ENVIRONMENTAL SPECIMEN BANKING

Environmental Specimen Banking is the systematic and long-term storage of selected environmental materials for deferred analysis and evaluation. Banked specimens such as soils, sediments, plants, animals, and human tissues **offer the chance to analyse retrospectively** the concentrations and effects of potentially dangerous substances, mixtures, and pathogens in the environment which are hitherto unknown, undetectable or unsuspicious. There is an international cooperation of Environmental Specimen Bank in the Federal Republic of Germany, the USA, Canada, Denmark, Sweden and Japan (WISE at al. 1990, STOEPPLER and ZEISLER 1993).

"A Specimen Bank is the systematic collection, characterization, and secure storage of samples for deferred examination (analysis or other evaluation). These samples must be stored under conditions which will minimize the occurance of any changes in the sample (i.e.. physical, chemical, or biological). The examination of the samples is generally deferred for a period of years, even decades, following collection, so that samples can be studied in relation to newer knowledge, new substances of interest, and new and improved analytical techniques. ..."(BMU/U.S.EPA 1989).

The Environmental Specimen Banking Programme of the Federal Republic of Germany was started in 1985. The conception, built up on the result of an extensive pilot phase and a series of international workshops, is presented in the following main parts:

Selection of research areas

Sampling areas have been chosen as to form a national Network of Ecological Assessment Parks coordinating Environmental Specimen Banking with long-term ecological research and environmental monitoring. The list of at present fourteen areas comprises the major ecosystems and habitat types that occur within the Federal Republic of Germany including two main rivers, two urban industrial ecosystems, and three coastal areas.

Selection of specimen types:

As a (minimal) set of specimen types to represent main functions in these ecosystems the following species have been selected:

Atmospheric essence (Filters) and wet deposition
Soil: Litter/Humus Layer and Topsoil

Terrestrial plants:
- Norway spruce *Picea abies*
- Scots Pine *Pinus sylvestris*
- Beech *Fagus sylvatica*
- Lombardy Poplar *Populus nigra 'Italica'*

Terrestrial animals:
   Earthworms *Lumbricus terristris/ Aporrectodea longa*
   Urban pigeons (Eggs) *Columba livia f. domestica*
   Roe deer (Liver) *Capreolus capreolus*

From Fresh Water ecosystems:
   Lake and River sediments
   Zebra Mussel *Dreissena polymorpha*
   Bream *Abramis brama*

From Wadden Sea:
   Brown Algae *Fucus vesiculosus*
   Common Mussel *Mytilus edulis*
   Lugworm *Arenicola marina*
   Eelpout *Zoarces viviparus*
   Herring Gull *Larus argentatus* (Eggs)

Human specimens:
   Whole blood, urine, hair, adipose tissue and others

Standard operating procedures (SOP):

The SOPs are the basis for the comparability, reliability, and repeatability of the banked samples. They contain detailed instructions for the

- selection of sampling sites and specimens,
- sampling,
- providing cover for repeatability of sampling,
- area and sample characterization,
- sample treatment and long-term storage,
- documentation of sampling and storage conditions,
- chemical anlysis,
- data processing and evaluation,
- quality assurance.

3. ENVIRONMENTAL CHANGES AND CONSERVATION STRATEGIES

Every species has a scientifc value. But from the majority of species we know only the scientific names. Every eradicated species is first a destruction of largely undocumented knowledge. Species diversity and species distribution pattern have strong interrelationships with area size and land use dynamics. Environmental monitoring describes also the fluctuation of population in space and time. That is the basic knowledge for a realistic conservation strategy. The intensity of land use has increased dramatically during the last two decades. Today some of the European and Asian countries have more than 240 inhabitants/km^2. The "losers" of the changes in the land use have to crowd together in the last areas remaining for wildlife. Biosphere, Reserves, National Parks or Protected Areas have become the classical means of nature conservation. The "philosophy" of conservation, which is, in spite of opposite affirmations, the official nature protection strategy of many countries, ignores global interlinkage among all parts as well as the fact, that there are no untouched "islands" where nat

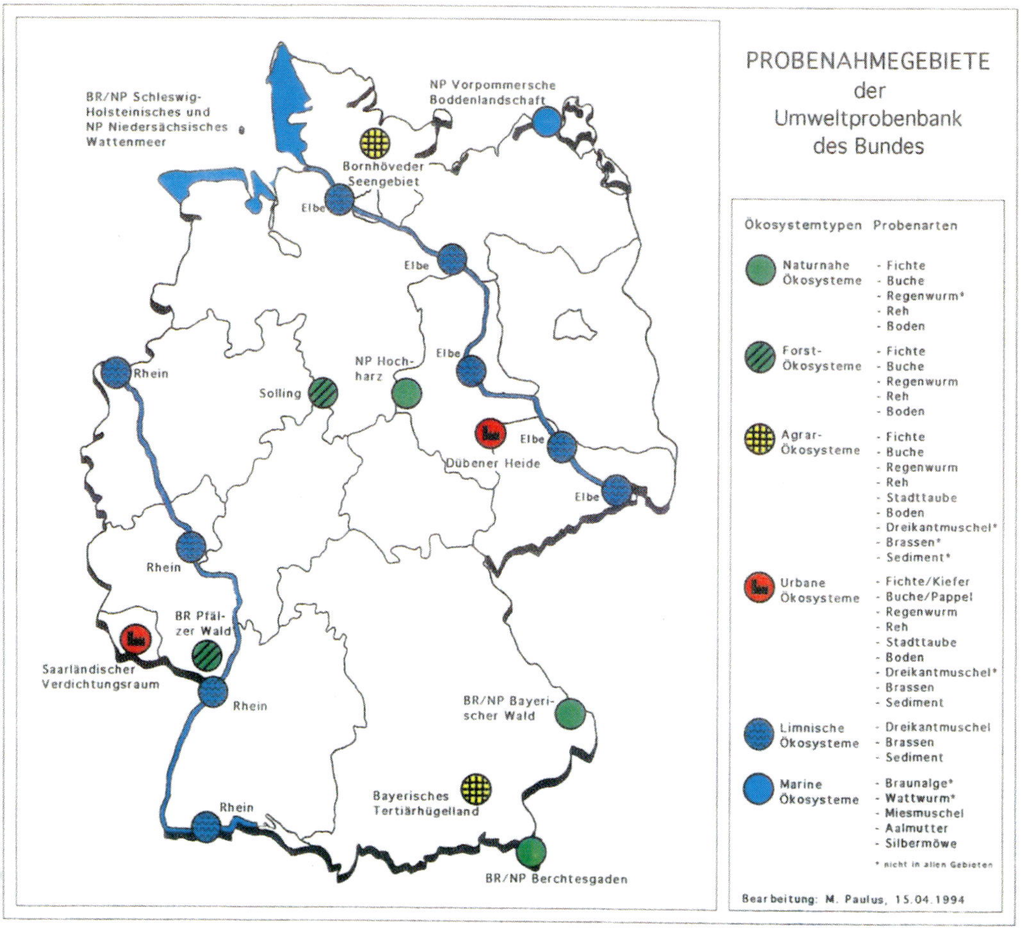

PROBENAHMEGEBIETE
der
Umweltprobenbank
des Bundes
BR/NP Schleswig-Holsteinisches und NP Niedersächsisches Wattenmeer
NP Vorpommersche Boddenlandschaft
Bornhöveder Seengebiet
Elbe
Elbe
Elbe
Elbe
Elbe
Rhein
Rhein
Rhein
Rhein
Solling
NP Hoch-harz
Dübener Heide
BR Pfälzer Wald
Saarländischer Verdichtungsraum
BR/NP Bayerischer Wald
Bayerisches Tertiärhügelland
BR/NP Berchtesgaden
Ökosystemtypen
Probenarten
Naturnahe Ökosysteme
- Fichte
- Buche
- Regenwurm*
- Reh
- Boden
Forst-Ökosysteme
- Fichte
- Buche
- Regenwurm
- Reh
- Boden
Agrar-Ökosysteme
- Fichte
- Buche
- Regenwurm
- Reh
- Stadttaube
- Boden
- Dreikantmuschel*
- Brassen*
- Sediment*
Urbane Ökosysteme
- Fichte/Kiefer
- Buche/Pappel
- Regenwurm
- Reh
- Stadttaube
- Boden
- Dreikantmuschel*
- Brassen
- Sediment
Limnische Ökosysteme
- Dreikantmuschel
- Brassen
- Sediment
Marine Ökosysteme
- Braunalge*
- Wattwurm*
- Miesmuschel
- Aalmutter
- Silbermöwe
* nicht in allen Gebieten
Bearbeitung: M. Paulus, 15.04.1994

ural processes take place independently of the global changes/of land use. At the same time the question arises whether a conservation strategy based on "equilibrium hypotheses" and confidence in regeneration of natural systems, ignores the real nature of both - nature and humans (REMMERT 1988). Scientific arguments mostly are replaced by psydo-ethical principles. The land user is considered as an "ecosinner", land use is seen as contrary to nature conservation and the fact is overlooked that the right way of use is of relevance.

In our world with more than 5 billion people, nature protection strategies can only be successful if the second component of conservation, the conservation through wise use, is implemented into laws and directives. "Winners" and "losers" of the changing land use cannot be treated equally. With respect to the continuously changing habitats, a static conservation strategy which considers organisms and all types of ecosystems will be a failure (HASSEL 1978).

The current conservation policy is based on the idea of an equilibrium with respect to both nature and culture. This may be referred to as "preserving nature conservation". Losers and winners receive the same status of protection. Especially with regard to reasons for species protection this is a wrong strategy.

Another problem is the fact that most of the conservation hypotheses are based on the idea of nature being in a state of equilibrium (BERRYMAN 1991, TULJAPURKAR 1990, WOLDA 1991). But at different times different equilibria are recorded for the same space (REMMERT 1989, 1991).

All ecosystems are characterized by a typical periodic cycle. This cycle is determined by the replacement of different keystone species, at roughly constant rates in a given area. There is no equilibrium in such a system, but the different successional stages are adjacent like in a misaic. The size of mosaic elements varies from the size of a single canopy to many square kilometers. There are strong interrelations between mosaic cycles and stochastic dynamic processes. In this changing environment all species and populations are tested for their competitive capacity. As a result of this we have "winners" and "losers".

Nevertheless, in search of an optimal solution to many questions we frequently loose sight of the patient. From a viewpoint of a satisfied European it is easy to argue about the moral of a hungry African. Often we realize too late that nature, if considered of value, should be protected, even under conditions of poverty.

3.1 THE ELEPHANT PROBLEM: IVORY BAN AND/OR CONSERVATION BY WISE USE

The protection of the African elephant and its habitat has been the key reason behind the creation of some of Africa's finest national parks (BOSMAN and HALL-MARTIN 1989). But elephants, like man, can modify the landscape by changing composition and structure of the natural vegetation and ecosystems. The effect of elephants on their habitat has been one of the most rigorously debated topics in African wildlife conservation (MÜLLER 1990).

Several scientists have explained the "elephant problem" in terms of biological cycles (cf. the Mosaic-Cycle Concept of Ecosystems; REMMERT 1991). Some hold that elephant numbers will build up in an area, woodland will be destroyed, the elephants will die and the woodland will recover. Evidence for such cycles has been found in the age distribution of tree species in some areas. The Mosaic-Cycle Concept was developed in 1938 by AUBREVILLE, using as an example the pristine forests in Western Africa. In the succeeding years the concept was seldom referred to, with a few rare exceptions, until it was revived and expanded by REMMERT (1988, 1989, 1991). But today the national park areas are too small and the management or manipulation of African elephant populations is usually an attempt to solve a conflict between elephants, ecosystems and man.

In addition to control elephant movements by barriers, fences and shooting, other actions taken specifically to "manage" elephants include the provision of water supplies, attempts to control poaching and the regulation of professional ivory hunting, subsistence hunting and trophy hunting.

But with the appearance of "elephant problems" in national parks in the last two decades, the culling of excess numbers of elephants has become a further management procedure. Elephant culling was initiated in the Murchison Falls National Park (Uganda) where contractors removed a specified number of animals. To minimize the disturbance, entire family groups were shot and the operations were carried out quickly and clean. The same technique was later used in Kenya and Tanzania for single operations, and in the Luangwa Valley (Zambia) for a few years. This method is still used in culling operations in Zimbabwe.

Today elephants are found in only 35 countries in Africa (MÜLLER 1994). In all countries other than Botswana, Equatorial Guinea, Gabon, Malawi, South Africa, Tanzania and Zimbabwe, elephant numbers are reported to be declining.

Elephant conservation is a question of managing natural resources for the mutual benefit of humans and wildlife. For the Zimbabwean Wildlife Society was the ban of ivory trade a "cynical exercise in high-powered politics" (see Zimbabwe Wildlife, September 1989).

The ban of the sale of ivory by CITES has a profound effect in Zimbabwe. The cornerstone of all African wildlife conservation has been to make wildlife a valuable and usable asset on a sustainable basis. When game on private land become the property of the land-owner and not of the state, there was an immediate improvement in the way in which it was looked after. Recently major efforts have been made to convince rural communities that wildlife is as valuable or even more valuable a crop than any other.

With this policy suddenly reversed, ivory will have only a value for the poacher and it is unlikely that rural people would remain on the side of law and order.

Ivory bans will kill more elephants than they save.
The best way for the elephant population is protection by wise use. Welfarists and conservationists must work together. Aiming at ensuring survival of the elephants.

3.2 THE GREY PARTRIDGE BETWEEN HUNTERS AND ANTI-HUNTERS

There are totally different relationships concerning the grey partridge. Hunting has no influence on total mortality of the populations. This has been shown by comparative studies in Luxemburg, Lorraine (Lothringen), Saarland and Rheinland-Pfalz in 1990, which have been carried out in my institute.

In the **Saarland**, 268 hunting grounds (70%) were analyzed. An increasing population has been recorded for 103 of them (40%). In spring 1990, 1923 breeding pairs (BP), and in the following autumn 1931 "chains" (with an average chain composition of 9.7 partridges) have been reported. Winter mortality was 65% of which hunting caused less that 1%.

In **Luxemburg** 67% of all hunting-grounds were registered (KRIER 1987). The density of partridges, which is comparable to those of the Saarland, varies depending on natural landscape units and land use. Winter mortality is also 65%, **although partridges are not hunted**.

In **Rheinland-Pfalz** information from 873 hunting-grounds was obtained until 18 March 1991. Density of breeding pairs is 2.3/100 ha cultivated land (for the Saarland: 1.7 BP/100 ha; Department Moselle:1.4 BP/100 ha). But also in Rheinland-Pfalz there are natural landscape units with much higher densities of breeding pairs (amongst others the Nördliche Oberrheinniederung, Kaiserslauterner Senke, Mittelrheinisches Becken, Moseleifel).

In **Lorraine** 461 hunting-grounds were registered. For 1986 and 1987 BP values were 4.3, for 1988 and 1989 6.8, in 1990 it was 8.3. Winter mortality is 65% like in Saarland, in Rheinland-Pfalz and in Luxemburg. 20% on were due to hunting.

The studies show that the type of distribution of the partridge is influenced not only by vegetation structure and land use but also by climate.

As with all steppe species, also intrinsic factors contribute to population fluctuations, which have been overlooked in most previous studies. The high winter mortality in most parts surveyed (upon which hunting has almost no influence, except for Lorraine) shows that other factors have a decisive influence on population dynamics.

Important factors regulating density of breeding pairs in Spring are:

- structural diversity
- diversity of land use on a small scale
- frequency of lines with vegetation cover
- frequent changes between small "Ruderal"-sites and areas with high vegetation cover and wild herbs
- abundance of insects during the period of rearing and abundance of seeds during winter
- application of chemicals for plant protection
- climate and microclimate

Additional ecological constraints for partridges are:

- heavy rains during the rearing period
- retarded late winters
- early ploughing of stubble fields
- destruction of contact with covered population density
- lack of "blinds"
- loss by predators

This loss by predators - at least in Germany - is much underestimated (POTTS et al. 1980).

Change in our cultivated landscape is often slow, and in many cases hidden. This is the reason why the consequences for plant and animal species become evident only years later. The winners of these changes are the fox (by killing adult birds), crows (by destroying the nests) or ermines (by killing the chicken). In areas where the habitats of the partridge being destroyed, where grassland areas and small fields are missing, even the last partridge will fall victim to them.

The chain-composition of partridge during autumn (up to 10 young adults) shows clearly, that the dangers to both nests and chicken are numerous, even under climatic favorable conditions during the rearing period. In cleared land, hedgerows are missing which provide cover against crows specialized in robbing nests. Young partridges are taken by all animals of prey.

British scientists could prove by experiments under natural conditions that the influence by animals of prey in cleared land is of great inportance (TAPPER et al. 1990, 1991).

The results of the European Wildlife Research Institute (EWI) and of Game Conservancy show that a guarantee for the survival of the populations is the maintenance of the partridge by the hunters and their interest in the possibility of sustained use by hunting. The results show as well that investment for the maintenance of the partridge also is of benefit to a variety of other plant and animal species of agricultural landscape in Europe.

3.3 THE QUELEA -STORY: CONTROL THROUGH A LAND USE POLICE

African elephant (*Loxodonta africana*) and partridge are used as game animals, but nevertheless there is a number of animal species which primarily are not utilizable or which are not used, but the management of which is urgently necessary, to prevent damage to nature or particular types of human land use. Among the latter category is, for instance, the red billed quelea (*Quelea quelea*) from the savanna areas of Africa or some corvids from Europe.

The red billed weaver bird (*Quelea quelea*) which occurs in extremely dense flocks in the Sahel Zone and which causes tremendous damage in certain areas under cultivation has been fought againt by the use of avizides etc. Millions of birds were annually exterminated in Somalia, Sudan or zambia without population intensity having been markedly reduced. Reasons for this has been shown by PEVELING (1990) in our institute.

PEVELING showed that a particular form of land use can significantly reduce natural nestling mortality. A specific land use mosaic (short grass savanna through overgrazing, scattered sorghum fields, permanent irrigation or water supply for cattle, remaining stands of Acacia trees) is in favour of the red billed weaver bird. This bird is a winner of changes in land use; present land use "creates" high-population densities with concomitant damage to agriculture.

Similar has been described by other colleagues in our institute for *Passer hispaniolensis* in Marocco (RISS 1988) and for *Passer luteus* (KLEIN 1988) in Niger.

These examples clearly illustrate that changed land use changes conditions for particular species populations. PEVELING (1990) demonstrated that, owing to the evolution of coadaptive mechanisms between grasses and cattle (herbivores) grazing directly improves the feed potential of pastures for granivores birds and causes the population to concentrate in the vicinity of susceptible grain crops. Natural regulation sets in only after grains cultivated are destroyed.

As, given conditions of poverty and hunger, nobody agrees with such a mechanism, killing of millions of birds is continued. The solution to this problem lies in a land use policy. As long as land use policy is not changed the winner of the change in land use has to be controlled.

4. PROTECTION AND/OR CONSERVATION BY WISE USE

With regard to nature conservation policy the examples shown in section 3 have different information content.

The example of the elephant clearly shows that big herbivores (but also carnivores) have, with increasing land use conflicts (rural population, vegetation, carrying capacity of systems, islands effect, etc.) to be sensibly managed to protect the environment. In the case of roe deer and red deer this has been accepted by all interest groups.

Both Europeans and Americans should realize that there are only a few people who would like to have an elephant in their rose garden. Even Africa is becoming smaller for the elephants. Under conditions of poverty they can survive only when they are considered of value. Consequently wise use should also be applied for CITES. For the elephants it is high time to develop a rational and country-specific conservation concept before it is too late (MÜLLER 1990, 1994).

The example of the partridge contains the message that the hunting together with the interests of the hunters may guarantee free conservation for this species which is also threatened by our land use system. Hunting impact does not increase total mortality, it is a good example for compensatory mortality. Preservation methods for the partridge are of advantage for many animal and plant species in an agricultural landscape.

Wise use of partridges becomes a conservation strategy for other threatened organisms but also becomes an indicator showing that private nature conservation activities can sometimes be more efficient than protection methods supervised by the State.

The example "Quelea" shows that land use policy is the key factor for an understanding of population dynamics of many species. Land use determines which species are winners and which are losers.

For many cases the evolved ratios between predators and prey change in favor of the predator. On a long-term basis in such cases only a change in land use policy will be of help. In case this is not possible, as for "Quelea", winners have to be reduced by means of well-defined management strategies.

All examples show that nature conservation can no longer be successful if it relies only on natural regeneration capacity in remaining habitat islands.

This "cheese-cover-concept", completely isolated and as the only valid nature conservation strategy, will lead to increased extinction. It is important to create both useful conservation and development strategies for our different types of ecosystems.

Wise use as the right way of use should not rely on species only but should be related to the habitat, the natural ecosystems or ecosystems which have been degraded to various extents. There should be a closer link between wise use and conservation of systems (organisms, communities, ecosystems). But the fundemental basic for the environmental risk assessment and for our conservation strategies is the scientific environmental monitoring.

5. LITERATURE

ALTMEYER, M. (1993): Biomonitoring mit Stadttaubeneiern zur Erfassung von Chemikalien und deren Wirkungen in Verdichtungsräumen. Dissertation, Universität des Saarlandes, Saarbrücken.

ARNDT, U., NOBEL, W. and SCHWEIZER, B. (1987): Bioindikatoren, Möglichkeiten, Grenzen und neue Erkenntnisse. Ulmer, Stuttgart.

ASTM-AMERICAN SOCIETY FOR TESTING AND MATERIALS (1991): Standard Guide for Conducting Sediment Toxicity Tests with Freshwater Invertebrates. ASTM Standard E 1383-90 (Marcia Nelson; 314/875-5399). American Society for Testing and Materials, Philadelphia, PA: 1-20.

AUBREVILLE, A. (1938): La forêt coloniale: les forêts de l'Afrique occidentale française. Ann. Acad. Sci Colon. 9: 1-245, Paris.

BAGENAL, T.B. (1974): Aging of fish. Surrey: Unwin Brothers Limited.

BAHADIR, M. (1991): Ökologische Chemie. Chemie in unserer Zeit 25 (5), 239-248.

BAU, H. (1992): Der Einsatz pflanzlicher Bioindikatoren in der Bundesrepublik Deutschland. VDI-Berichte Nr. 901, 37-72.

BERRYMAN, A. (1991): Stabilization or regulation: what it all means! Oecologia 86: 140-143.

BIGNERT, A., GÖTHBERG, A., JENSEN, S., LITZEN, K., ODSJÖ, T., OLSSON, M. and REUTERGARDH, L. (1993): The need for adequate biological sampling in ecotoxicological investigations: a retrospective study of twenty years pollution monitoring. The Science of the Total Environment 128: 121-139.

BOSMAN, P. and HALL-MARTIN, A. (19989): Elephants of Africa. Safari Press, Long Beach.

BOURGEES, T.M., WEBSTER, R. and MCBRATNEY, A.B. (1981): Optimal Interpolation and Isarithmic Mapping of Soil Properties. VI. Sampling Stratgeies. J. Soil Sci 32, 643-659.

BRAKENHJELM, S. (1990): The National Swedish Environmental Monitoring Programme (PMK). Swedish Environmental Protection Agency, Uppsala.

BROMENSHENK, J.J. (1978): Investigations of the impact of coal-fired power plant emissions upon insects: Entomological studies in the vicinity of Colstrip, Montana. In: E.M., LEWIS, R.A. (Eds.): The Bioenvironmental Impact of a Coal-fired Power Plant: Preston, third Interim Report, Colstrip, Montana, EPA 600/3-78-02. U.S. Envrionmental Protection Agency.

BURTON, G.A. (1992): Biological Test Method: Acute Test for Sediment Toxicity using Marine or Estuarine Amphipods. Report, EPS 1/RM/26, Environment Canada, Ottawa, Ontario K1A 0H3, Canada.

BUSCH, D. (1991): Entwicklung und Erprobung von Methoden für einen Einsatz der Süßwassermusxchel Dreissena polymorpha (Pallas) für ein Biomonitoring von Schwermetallen für das Ökosystem Wasser. Dissertation, Universität Bremen.

DE GREEF, J. and DE NIJS, A.C.M. (1990): Risk assessment of new chemical substances. Dilution of effluents in the Netherlands. RIVM report No. 670208001.

DE KOCK, W.CHR., BROUMER, C.T. (1993): Bioaccumulation, biological effects and food chain transfer of contaminants in the Zebra Mussel (Dreissena polymorpha). In: NALEPA, T.F., SCHLOESSER, D.W. (Eds.): Zebra Mussel: Biology, Impacts, and Control. Boca Raton, Ann Arbor, London, Tokyo: Lewis Publishers, pp. 503-536.

DISSER, J., BRUNN, H., NAGEL, A. and PRINZINGER, R. (1992): Untersuchungen zur Belastung von Vogeleiern mit Chlorkohlenwasserstoffen unter besonderer Berücksichtigung der PCBs. Ökol. Vögel (Ecol. Birds), 14: 173-209.

EDGREN, M., OLSSON, M. and REUTERGARDH; L. (1981): A one year study of the seasonal variations of DDT and PCB levels in fish from heated and unheated areas near a nuclear power plant. Chemosphere, 10 (5), 447-452.

EDWARDS, C.A. and LOFTY, J.R. (1977): Biology of Earthworms. Chapman and Hall, London.

ELLENBERG, H., FRÄNZLE, O. and MÜLLER, P. (1978): Ökosystemforschung im Hinblick auf Umweltpolitk und Entwicklungsplanung. BMI Bonn.

ELLIOTT, J.E. (1984): Collecting and archiving specimens in Canada. In: LEWIS, R.A., STEIN, N. and LEWIS, C:W: (Eds.): Envrionmental specimen banking and monitoring as related to banking. Martinus Nijhoff, The Hague.

EMANS, H.J.B., PLASSCHE, E.J. et al. (1993): Validation of some extrapolation methods used for effect assessment. Environ. Toxicol. Chem. 12: 2139-2154.

ENGLUND, E. and SPARKS, A. (1988): GEO-EAS (Geostatistical Envrionmental Assessment Software), Programs and user's guide. Envrionmental Monitroing Systems Laboratory, U.S. Envrionmental Protection Agency, Las Vegas, N.Y.

EPA (1990): Methods for Assessing Exposure of Chemical Substances 11: Methodology for Estimating the Migration of Additives and Impurities from Polymeric Materials. EPA 560/5-85-015, USA.

ERHARDT, W. (1987): Die Bedeutung von Akklimatisation und Referenzexposition für die Auswertung von Flechtenexpositionen. VDI-bericht 609, 701-714.

ERNST, W.H.O. (1974): Schwermetallvegetation der Erde. Fischer Verlag, Stuttgart.

ERNST, W.H.O. and LELOUP, S. (1987): Perennial herbs as monitor for moderate levels of metal fall-out. Chemosphere 16, 233-238.

ERNST, W.H.O. (1994): Sampling of plants for environmental trace analysis in terrestrial, semi terrestrial and aquatic environments. In: MARKERT, B. (ed.): Sampling of envrionmental materials for trace analysis. VCH-Publisher, Weinheim, New York, Tokyo (in press).

EUROPEAN COMMUNITIES (1993): Technical Guidance Documents in Support of the Risk Assessment Directive (93/67/EEC) for New Substances Notified in Accordance with the Requirements of Council Directive 67/548/EEC, Brussels.

FISCHER, P. (1991): Statistische Überprüfung und Optimierung von Beprobungsplänen für die Umweltprobenbank am beispile von Fichte (Picea abies) und Buche (Fagus sylvatica L.). Unpublish. thesis, Biogeography, Saarbrücken.

FISCHER, P. and WAGNER, G. (1993): Definition homogener Probenahmeflächen innerhalb eines flächenrepräsentativen passiven Biomonitorings. Ecoinforma '92, Band 4, S. 221-228, Bayreuth.

FISCHER, P. and WAGNER, G. (1994): Sampling design for qualified area-representative data in long-term passive environmental monitoring, SPRUCE Conference 1992 in Lisabon (in press).

GREEN, R.H. (1979): Sampling design and Statistical Methods for Environmental Biologists. John Wiley & Sons. New York, 257 pp.

GRIMMER, G. and HILDEBRAND, A. (1965): Der gehalt polycyklischer Kohlenwasserstoffe in verschiedenen Gemüsesorten und Salaten. Dt. Lebensmittel-Rundschau 8, 237-239.

HASSEL, M. (1979): The dynamics of Arthropod Predator-Prey Systems. Pribnceton University Press, New Yersey.

HECHT, H. (1991): Rehwild - ein Bioindikator zur Erfassung des Ausmaßes von Umweltbelastungen durch Störfälle von kerntechnischen Anlagen. VDI-Berichte, Nr. 901, 937-954.

HEINISCH, E. (1992): Umweltbelastung in Ostdeutschland. Fallbeispile: Chlorierte Kohlenwasserstoffe. Wiss. Buchgesellschaft Darmstadt.

HENLE, K. und KAULE, G. (1991): Arten- und Biotopschutzforschung für deutschland. Forschungszentrum Jülich, Jülich.

HERZIG, R. (1993): Multi-Residue Analysis with Passive Biomonitoring: A New Approach for Volatile Multi-Element Contents, Heavy Metals and Polycyclic Aromatic Hydrocarbons with Lichens in Switzerland and the Principality of Liechtenstein. In: MARKERT, B. (1993): Plants as Biomonitors - Indicators for Heavy Metals in the Terrestrial Envrionment: 285-328. VCH Weinheim.

HODGESON, J.M. (1985): Soil Survey Field handbook: Describing and Sampling Soil Profiles. Bartholomew Press, Dorking: Harpenden.

HOEKSTRA, J.A. and EWIJK, P.H.v. (1993): Alternatives for the no-observed-effect level. Environ. Toxicol. Chem. 12: 187-194.

ISO-TC 190 (International Standard Organization) (1993): ISO/CD 10381-1-4, Soil Quality - Sampling - Part 1: Guidance on the design of sampling programmes, 47 pp. Part 2: Guidance on sampling techniques, 46 pp. Part 3: Guidance on safety, 49 pp. Part 4: Guidance on the procedure for the investigation of natural, near-natural and cultivated sites, 28 pp.

KAULE, G. (1991): Arten- und Biotopschutz. 2. Aufl., Ulmer-Verlag, Stuttgart.

KEITH, L.H. (1987): Principles of Envrionmental Sampling. American Chemical Society, ACS Professional refernce Book.

KLEIN, R. (1988): Zur Ökologie des Goldsperlings (Passer luteus Licht). in der Bundesrepublik Niger unter besonderer Berücksichtigung seiner Schadwirkungen in Kulturlandschaften. Diss. Biogeographie, Universität Saarbrücken.

KLEIN, R. (1993): The animal specimens of terrestrial and limnetic ecosystems in the Envrionmental Specimen Banking Programme of Germany. In: STOEPPLER, M. and ZEISLER, R. (eds.): Biological Envrionmental Specimen Banking. The Science of the Total Envrionment 139/140: 203-212.

KLEIN, R. and ALTMEYER, M. (1992): The Zebra Mussel Dreissena polymorpha: In: NEUMANN, D. and JENNER, H.A. (Eds): Limnologie Aktuell Vol. 4, 255-265, Gustav Fischer Verlag, jena.

KLEIN, R. and PAULUS, M. (1992): Umweltprobenbank als Bestandteil ökosystemarer Monitoringprogramme. Ecoinforma 4: 49-58.

KLEIN, R. and NETWICH, K. (1993): Richtlinie zur Probenahme und Probenbearbeitung Reh (Capreolus capreolus). In: BMU (Bundesministerium für Umwelt, Naturschutz und Reaktorsicherheit), Referat N ii 2 - 71022/1, Bonn.

KLEIN, R. and PAULUS, M. (1993): Richtlinie zur Probenahme und Probenbearbeitung Stadttaube (Columa livia f. domestica). In: BMU, referat N ii 2 - 71022/1, Bonn.

KLEIN, R. and WAGNER, G. (1993): Richtlinie zur Probenahme und Probenbearbeitung Reh (Capreolus capreolus). In: BMU, Referat N ii 2 - 71022/1, Bonn.

KLEIN, R. and PAULUS, M. (1995): Umweltproben für die Schadstoffanalytik im Biomonitoring: Standards zur Qaulitätssicherung bis zum Laboreingang. Gustav Fischer Verlag, Jena, 1995, in preparation.

KLEIN, R. and WAGNER, G. (1993): Strategien für die Umweltprobenbank der Bundesrepublik Deutschland. In: ERDMANN, K.-H. and NAUBER, J. (eds.): MAB-Mitteilungen, 37, pp. 117-122.

KNABE, W. (1982): Monitroing of air pollutants by wild life plants and plant exposure: suitable bioindicators for different immission types: In: STEUBING, L. and JÄGER, H.-J. (eds.): Monitoring of Air Pollutants by Plants, 59-72, Junk, The Hague.

KNABE, W. (1983): Immissionsökologische Waldzustandserfassung in Nordrhein-Westfalen (IWE 1979). Forschung und Beratung, Reihe C 37, Landwirtschaftsverlag Münster.

KNABE, W. (1984): Merkblatt zur Entnahme von Blatt- und Nadelproben für chemische Analysen. AFZ 39, 447-484.

KRIVAN, V. and SCHALDACH, G. (1986): Fresenius' zeitschrift für analytische Chemie, 324, (1986), pp. 158-167.

KRÜGER, K.-E. and KRUSE, R. (1984): Fische als Bioindikatoren für anorganische und organische Umweltkontamination in Seen und Flüssen unterschiedlicher Ökosysteme. F+E-Report, Umweltbundesamt Berlin, Cuxhaven.

LEWIS, R.A., STEIN, N. and LEWIS, C.W. (1984): Environmental Specimen Banking and Monitoring as Related to Banking. Martinus Nijhof Publ., The Hague.

LEWIS, R.A. (1987): Guidelines for Environmental Specimen Banking with Special Reference to the Federal Republic of Germany: Ecological and Managerial Aspects. U.S. Department of the Interior, Natiobnal Park service, U.S. MAB Report No. 12.

LEWIS, R.A., PAULUS, M. HORRAS, C. and KLEIN, B. (1989): Auswahl und Empfehlung von ökologischen Umweltbeobachtungsgebieten in der Bundesrepublik Deutschland. MAB-Mitteilungen 29, Bonn.

LEWIS, R.A. and KLEIN, B. (1990): A brief History of Specimen Banking: Storage, Institutions and Applications. Toxicological and Envrionmental Chemistry 27, 251-266.

LIETH, H. and MARKERT, B. (1990): Element Concentration Cadasters in Ecosystems. Methods of Assessment and Evaluation. Weinheim.

LILLESTOLEN, T.I., FORSTER, N. and WISE, S.A. (1993): Development of the National Marine Mammal Tissue Bank. In: STOEPPLER, M. and ZEISLER, R. (eds.): Biological Envrionmental Specimen banking. The Science of the Total Envrionment 139/140, 97-107.

LIMA, ST. and VALONE, TH. (1991): Predators and aviasn community organisation: an experiment in a semi-desert grassland. Oecologia 86: 105-112.

LOKKE, H. (1994): Ecotoxicological extrapolation: tool or toy? In: DOKERS et al. (eds.): Ecotoxicology of soil organisms. SETAC Special Publ. Ser., Lewis Publishers.

LUEPKE, N.P. (1979): Monitroing Environmental Materials and Specimen Banking. Martinus Nijhof Publ., The Hague.

MARKERT, B. (1993): Plants as Biomonitors - Indicators for Heavy Metals in the terrestrial Envrionment: 285-328. VCH Weinheim.

MARKERT, B. (1994): Sampling of environmental materials for trace anlysis. VCH Weinheim.

MIEHLICH, G. (1976): Homogenität, Inhomogenität und Gleichheit von Bodenkörpern. Z. Pflanzenern. Bodenk. 139: 597-609.

MÜLLER, P. (1977): Biogeographie und Raumbewertung. Wissenschaftliche Buchgesellschaft Darmstadt.

MÜLLER, P. (1979): The Need for and Possibilities of Employment of an Envrionmental Specimen Bank in the federal Republic of Germany. F+E Report Umweltbundesamt Berlin, Saarbrücken.

MÜLLER, P. (1980): Biogeographie. UTB, Stuttgart.

MÜLLER, P. (1981): Arealsysteme und Biogeographie. pp. 704, Ulmer Verl., Stuttgart.

MÜLLER, P. (1987): Ecological side effects of Dieldrin, Endosulfan, Deltamethrin and Cypermethrin application against Tsetse-flies in Adamaoua, Cameroon. Publ. by GTZ (Eschborn) and World Bank (Washington).

MÜLLER, P. (1987): Eine Niederlage für den Artenschutz. Wild und Hund 8: 4-6.

MÜLLER, P. (1988): Phoca vitulina, Chrysochromulina polylepis und andere ökosystemare S.O.S.-Signale der Nordsee. In: Umweltdiskussion: Krise des Ökosystems Nordsee. Zeitschrift für Angewnadte Umweltforschung 1988, 209-216.

MÜLLER, P. (1990): Götterdämmerung für die grauen Riesen Afrikas. Hilft der Stopp des Elfenbeinhandels den Elefanten? Wild und Hund 1: 78-81.

MÜLLER, P. (1990): Naturschutz durch Nutzung tropischer Regenwälder? Zeitschrift für angewandte Umweltforschung 3 (2): 194-199.

MÜLLER, P. (1991): Bestandssituation des Rebhuhnes im Saar-Lor-Lux-Raum und Rheinland-Pfalz. Saarjäger 43.

MÜLLER, P. and WAGNER, G. (1986): Probenahme und genetische Vergleichbarkeit von repräsentativen Umweltproben im Rahmen des Umweltprobenbank-Pilotprojekts. F+E-Report T 86-040, Bundesministerium für Forschung und Technologie, Bonn.

MÜLLER, P. and WAGNER, G. (1985): Untersuchung von Probenarten und Entwicklung von Probenahmerichtlinien für Biomonitoring im Rahmen der Umweltprobenbank. F+E-Report UBA Berlin, Saarbr.

NALEPA, T.F. and SCHLOESSER, D.W. (1993): Zebra Mussel: Biology, Impacts, and Control. Lewis Publishers, Boca Raton, Ann Arbor.

NIIMI, A.J. (1983), in: NRIAGU, J.O. (Ed.): Aquatic Toxicity: pp. 207-246. John Wiley and Sons, New York, Chichester, Brisbane, Toronto, Singapore.

OECD (1993): Application of Structure - Activity Relationships to the Estimation of Properties Important in Exposure Assessment, Envir. Monographs 67.

OKKERMAN, P.C. et al. (1992): Validation of some extrapolation methods with toxicity data derived from multiple species experiments. Ecotox. Envrion. Sof. 25: 341-359.

OLSSON, M. (1976): Mercury level as a function of size and age in Northern Pike, one and five years after ther mercury ban in Sweden. Ambio, 5(2): 73-76.

PACK, S. (1993): OECD-Draft: A review of statistical data anlysis and experimental design in OECD aquatic toxicology test guidelines. Shell Research, Sittingbourne, UK.

PAULUS, M., HORRAS, C., KLEIN, B. and LEWIS, R.A. (1990): Vertiefte Auswahl von Probenahmeregionen für die Umweltprobenbank und ökologische Beratung zu ihrem betrieb. Final report UBA 90/108 08001.

PAULUS, M. and KLEIN, R. (1994): Umweltprobenbanken als Instrumente zur umweltchemischen Beweissicherung und retrospektiven Bioindikation. In: GUNKEL, G. (ed.): Bioindikation in aquatischen Ökosystemen. Fischer-verlag jena (in press).

PAULUS, M., ALTMEYER, M., KLEIN, R., HILDEBRANDT, A., OSTAPCZUK, P., OXYNOS, K. (1994): Aufbau flächenrepräsentativer Probenahmen von Umweltproben zur Schadstoffanalytik am Beispiel der Regenwürmer in landwirtschaftlich genutzten Räumen. Umweltwissenschaften und Schdstoff-Forschung, UWSF-Z. Umweltchem. Ökotox. 6(5), in press.

PEDERSEN, F. et al. (1994): Ecotoxicological Evaluation of Industrial Waste Water Miljoprojekt Nr. 254, Denmark.

PEVELING, R. (1990): Möglichkeiten der Reduktion von Schäden in der Getreideproduktion der Demokratischen Republik Somalia durch den Blutschnabelwebervogel (Quelea quelea L.) unter besonderer Berücksichtigung der ökologischen Grundlagen. Dissertation, Biogeographie, Universität Saarbrücken.

POTTS, R. (1980): The effects of modern Agriculture. Nest predation and Game Management on the Population Ecology of Partridges (Perdix perdix and Alectoris rufa). Advances in Ecological research 11: 1-79.

PRINZ, B., KRAUSE, G.H.M. and RADERMACHER, L. (1990): Polychlorierte Dibenzodioxine und Dibenzofurane, Untersuchungen zur Belastung von Gartenböden und Nahrungspflanzen. Staub - Reinhalt. Luft 50: 377-381.

REMMERT, H. (1988): Naturschutz. Springer, Heidelberg.

REMMERT, H. (1989): Ökologie. 4. Aufl., Springer, Heidelberg.

REMMERT, H. (1991): The Mosaic-Cycle concept of Ecosystems. Ecol. Studies 85, Springer, Heidelberg.

RICHARDS, P.W. (1981): The tropical rain forest. Cambridge Univ. Pr.

RISS, B. (1988): Zur Populationsökologie und -dynamik des Weidensperlings (Passer hispaniolensis TEMM) in Marokko. Dissertation, Biogeographie, Universität Saarbrücken.

RÖMBKE, J. et al. (1993): Grundlagen für die Beurteilung des ökologischen Gefährdungspotentials von Altstoffen im Medium Boden; Entwicklung einer Teststrategie. F+E Nr. 10604103, Batelle Europe, Frankfurt.

ROSSBACH, M., SCHLADOT, J.-D. and OSTAPCZUK, P. (1992): Specimen Banking. Springer Verlag.

RÜHLING, A. and TYLER, G. (1969): Ecology of Heavy Metal - a Regional and Historical Study. Bot. Notiser 122: 248-259.

SCHERELIS, G. (1988): Untersuchungen zur profildifferenzierten Variabilität der Schwermetalle Chrom, Mangan, Eisen, Nickel, Kupfer, Zink und Blei in rezenten und fossilen Parabraunerden Baden-Württembergs. Stuttgarter Geographische Studien, Bd. 112, Universität Stuttgart.

SCHERELIS, G. and BLÜMEL, W.D. (1988): Geostatistik und ihre Anwendungsperspektiven in der Geoökologie am Beispiel des Krieging-Verfahrens. Karlsuher Manuskripte zur Mathematischen und Theoretischen Wirtschafts- und Sozialgeographie, Heft 92, Universität Karlsruhe.

SCHLADOT, J.D. and BACKHAUS, F. (1988): Preparation of Sample Material for Envrionmental Specimen Banking Purposes - Milling and Homogenization at Cryogenic Temperatures, in: WISE, S.A., ZEISLER, R. and GOLDSTEIN, G.M. (1988): Progress in Envrionmental Specimen Banking, NBS Special Publication 740: 184-194.

SCHOLL, G. (1971): Ein biologisches Verfahren zur Bestimmung der Herkunft und Verbreitung von Fluor-Verbindungen in der Luft. Landwirtsch. Forsch. 26: 29-35.

SCHOLZ, R.W., NOTHBAUM, N. and MAY, T.W. (1993): Starre und hypothesengeleitete Probenahmepläne. ECOINFORMA 4: 229-240.

SPRENGART, J. and WAGNER, G. (1993): Richtlinie zur Probenahme und Probenbearbeitung Boden. In: Bundesministerium für Umwelt, Naturschutz und Reaktorsicherheit. Referat N II 2 - 71022/1, Bonn.

STEINNES, E. (1982): Monitoring of Trace Element Deposition by means of Mosses. In: NATO Air pollution Pilot Study Assessment Methodology and Modelling 1975-1979, Suppl. to the 3rd follow-up report, No 134, 1-8.

STÖCKER, G. (1993): Analytical and biological data of banked samples - a requirement to interpret accumulation and leaching of pollutants in biological specimens. In: STOEPPLER, M. and ZEISLER, R. (eds.): Biological Environmental Specimen Banking. The Science of the Total Environment 139/140: 491-505.

STOEPPLER, M. (1994): Probenahme und Aufschluß - Basis der Spurenanalytik. Springer-Verlag (in press).

STOEPPLER, M. and ZEISLER, R. (1993): Biological Environmental Specimen Banking. The Science of the Total Environment. BESB special issue, in press.

TAPPER, ST., POTTS, D., REYNOLDS, J., STOATE, CH. and BROCKLESS, M. (1990): The Salisbury plain experiment - year six. The Game Conservancy Review of 1989: 42-47, Fordingbridge.

TAPPER, ST., BROCKLESS, M. and POTTS, D. (1991): The Salisbury plain predation experiment: The conclusion. The Game Conservancy Review of 1990: 87-91, Fordingbridge.

TATARUCH, F. (1991): Freilebende Wildtiere als Bioindikatoren der Schwermetallkonzentration. VDI-Berichte Nr. 901, 925-936.

TULJAPURKAR, S. (1990): Population Dynamics in variable environment. Springer, Heidelberg.

VDI (1978): VDI Bl. 1-3: Messen der Wirkdosis; Verfahren der standardisierten Graskultur. VDI-Handbuch Reinhaltung der Luft, Düsseldorf.

VDI (1991): VDI 3799 Bl. 2: Messen von Immissionswirkungen; Ermittlung und Beurteilung phytotoxischer Wirkungen von Immissionen mit Flechten; Verfahren der standardisierten Flechtenexposition. VDI-Handbuch Reinhaltung der Luft, Düsseldorf.

WAGNER, C. and LOKKE, H. (1991): Estimation of ecotoxicological protection levels from NOEC toxicity data. Wat. Res. 25: 1237-1242.

WAGNER, G. (1981): Monitoring of airborne heavy metal pollution using selected accumulation indicators. Intern. Conf. Heavy Metals on the Envrionment, Amsterdam, pp. 421-424. CEP-Consultants, Edinburgh, GB.

WAGNER, G. (1987): Entwicklung einer Methode zur großräumigen Überwachung der Umweltkontamination mittels standardisierter Pappelblattproben von Pyramidenpappeln (Populus nigra 'Italica') am Beispiel von Blei, Cadmium und Zink. In: STOEPPLER, M. and DÜRBECK, H.W. (eds.): Beiträge zur Umweltprobenbank Nr. 5, Jül. Spez. 412, KFA Jülich.

WAGNER, G. (1993): Plants and soils as specimen types from terrestrial ecosystems in the environmental specimen banking program of the Federal Republic of Germany. In: STOEPPLER, M. and ZEISLER, R. (eds.): Biological Envrionmental Specimen Baning. The Science of the Total Environment 139/140: 213-224.

WAGNER, G. (1993): Einsatzstrategien und Meßnetze für die Bioindikation im Umweltmonitoring. Ecoinforma '92, Band 4, S. 23-31, Bayreuth.

WAGNER, G. (1993): Umweltprobenbanken - neue Instrumente für Umweltforschung, -analytik und -planung. Ecoinforma 4, 71-80, Bayreuth.

WAGNER, G. (1993): Large-Scale Screening of Heavy Metal Burdens in Higher Plants. Chapter 5.6.7. In: MARKERT, B. (Ed.): Plants as Biomonitors for Heavy Metal Pollution of Terrestrial Environment, S. 425-434. VCH-Publishers, Weinheim/New York.

WAGNER, G. and KRÜGER, J. (1982): Experimentelles Monitoring als Risikoanalyse - Wirkungs- und Trendkataster für die Bewertung von Städten. Das Gartenamt 31: 516-524.

WAGNER, G., BAGSCHICK, U., BUROW, M., MOHL, C., OSTAPCZUK, P. and STOEPPLER, M. (1985): Spruce needles as indicators for heavy metal pollution: A comparative study for the Envrionmental Specimen Bank in differently polluted areas. In: International Conference. Heavy Metals in the Environment. Athens, pp. 515-518.

WAGNER, G., ALTMEYER, M., KLEIN, R., PAULUS, M. and SPRENGART, J. (1993): Richtlinie zur Probenahme und Probenbearbeitung Fichte (Picea abies) und Kiefer (Pinus sylvestris). In: Bundesministerium für Umwelt, Naturschutz und Reaktorsicherheit, Referat N II 2 - 71022/1, Bonn.

WAGNER, G. (1994): Biologische Umweltproben. In: STOEPPLER, M. (Ed.): Probenahme und Aufschluß in der Spurenanalyse, pp. 71-83, Springer-Verlag, Berlin/Heidelberg/New York.

WAGNER, G. (1994): Basic approaches and methods for quality assurance and quality control in sample collection and storage for environmental monitoring. The Science of the Total Environment (in Prep.).

WALLWORK, J.A. (1983): Earthworm Biology. Southampton, Cametot Press Ltd.

WINTER, A., MÜLLER, P. and WAGNER, G. (1992): Versuche zur Verbesserung der Nachweisgrenzen für Schwermetalle bei standardisierten Graskulturen durch alternative Wuchssubstrate. VDI-Berichte 901, 729-748, VDI-Verlag, Düsseldorf.

WISE, S.A., ZEISLER, R. and GOLDSTEIN, G.M. (1988): Progress in Environmental Specimen Banking. NBS Special Publication 740, U.S. Dep. of commerce, U.S. Government Printing Office, Washington.

WOLDA, H. (1991): The usefulness of the equilibrium concept in population dynamics. Oecologia 86: 144-145.

WYTTENBACH, A., BAJO, S., TOBLER, L. and KELLER, T. (1985): Major and trace element concentrations in needles of Picea abies; levels, distribution functions, correlations and environmental influences. Plant and Soil 85: 313-325.

WYTTENBACH, A., TOBLER, L. and BAJO, S. (1988): Major and trace elements in the twig axes of Norway spruce and the relationship between twig axis and needles. Trees 1988, 204-212.

ZIMMER, M. (1993): Auswirkungen von Insektenbefall an Fichtentrieben und -nadeln auf die Gehalte der Makronährstoffe Calcium, Kalium und Magnesium sowie auf die Schwermetalle Cadmium und Blei. Unpubl. thesis, Universität des Saarlandes, Saarbrücken.

ZIMMERMANN, R.-D. (1989): Schadstoff- und Nährelementverteilung im Baumkronenbereich der Rotbuche. AFZ 11, 4-7.

ZIMMERMANN, R.-D. and PLANKENHORN, W.E. (1986): Methodik der Blattprobenentnahme an der Rotbuche unter immissionsökologischem Aspekt. Allgemeine Forst Zeitschrift 41, 33-35.

TOWARD GLOBAL PLANNING OF SUSTAINABLE USE OF THE EARTH
Development of Global Eco-Engineering
S. Murai (Editor)
1995 Elsevier Science B.V.

Restoration of biodiversity in Urban and Peri-Urban environments with native forests

Akira Miyawaki

Japanese Center for International Studies in Ecology (JISE)
Shonan Village Center, Hayama-machi,
Kanagawa, 240-01, Japan

Abstract

Monotonous biota and standardized artificial structures, invested with non-biological materials and petrochemical fuels, characterize the typical urban sceneries today. The most desirable and reliable method to restore biodiversity of the urban and peri-urban environments is to create native forests by native trees based on the study of the potential natural vegetation.

The Miyawaki team has succeeded in restoring biodiversity by extending man-made forests to cities, industrial and housing complexes over 340 locations throughout Japan since 1973. The revegetation technique is the integration of the knowledge of Japan's indigenous Chinju-no-mori forest and modern vegetation ecology. The technique has been successfully applied to commercial sites in Malaysia and Thailand as well.

1. Introduction

The glories of the human civilzations in the past 6000 years are great assets of mankind. The studies of ancient cities in Mesopotamia, Egypt, Greece, and the Roman Empire reveal, however, that impoverishment and loss of diversity, and eventual extinction of their native forests triggered the fall of the civilizations. The surrounding areas of these historical sites still remain dry and barren today.

In Japan, forests have also been cleared in the past 2000 years to make villages, towns and cities; however, the sites sensitive to human impact, such as ridges, steep slopes, resevoirs and water fronts, were kept intact. Whenever new villages and cities were made, native forests with native trees were enshrined and preserved by evoking reverence for the sites in people's mind. The native forests were sanctuaries. After the Second World War, rapid industrialization and urbanization as well as loss of religious awe of nature started and accelarated the extinction of these native forests.

Kanagawa Prefecture is one of the good examples. It is adjacent to Tokyo, occupying 1/200 of Japan's total land area with the population of 8.3 million. If concentration of population were an indication of urban advancement, it is highly advanced. However, most of the 2850 Chinju-no-mori forests identified in a previous survey have disappeared; only 40 sites retain the natural level equivalent to natural monuments (Miyawaki, Tohma and & Suzuki 1979).

We integrated the theories of vegetation science and plant ecology of Europe, and the knowledge of Japanese Chinju-no-mori forests, and have applied the technique to the creation of native forests since early 1970's (Miyawaki 1975).

The efforts have been successful, with the cooperation of industries, local municipalities and some Ministries of the Japanese Government, in most of the vegetation zones from the northern Island of Hokkaido to the subtropical Ishigaki Island of Okinawa Prefecture, covering a distance of 3000km (Miyawaki 1989, Miyawaki, Fujiwara, Nakamura & Kimura 1983, Miyawaki, Fujiwara & Ozawa 1993, Miyawaki & Golley 1993). The conventional succession theory has it that it takes 300 years to restore original native forests once they are destroyed. Our restoration technique based on the study of potential natural vegetation, however, has successfully created native forests in about 1/10 of that time span (Tuxen 1956). The seedlings planted 22 years ago form well-developed crown of 20m now. The soil fauna, an important element of ecosystems, steadily develops along with the trees (Aoki & Harada 1985, '89). In about 15 to 20 years, a diverse forest ecosystem which is approximately equivalent to natural native forests may be restored (Fig. 1).

This process was claimed to take over 200 years by Clements in his Succession Theory (1916).

This technique has been successfully extended to commercial areas in Malaysia, to the rain forest in Borneo, Belem in Amazon/Brasil and to *Nothofagus* forest in

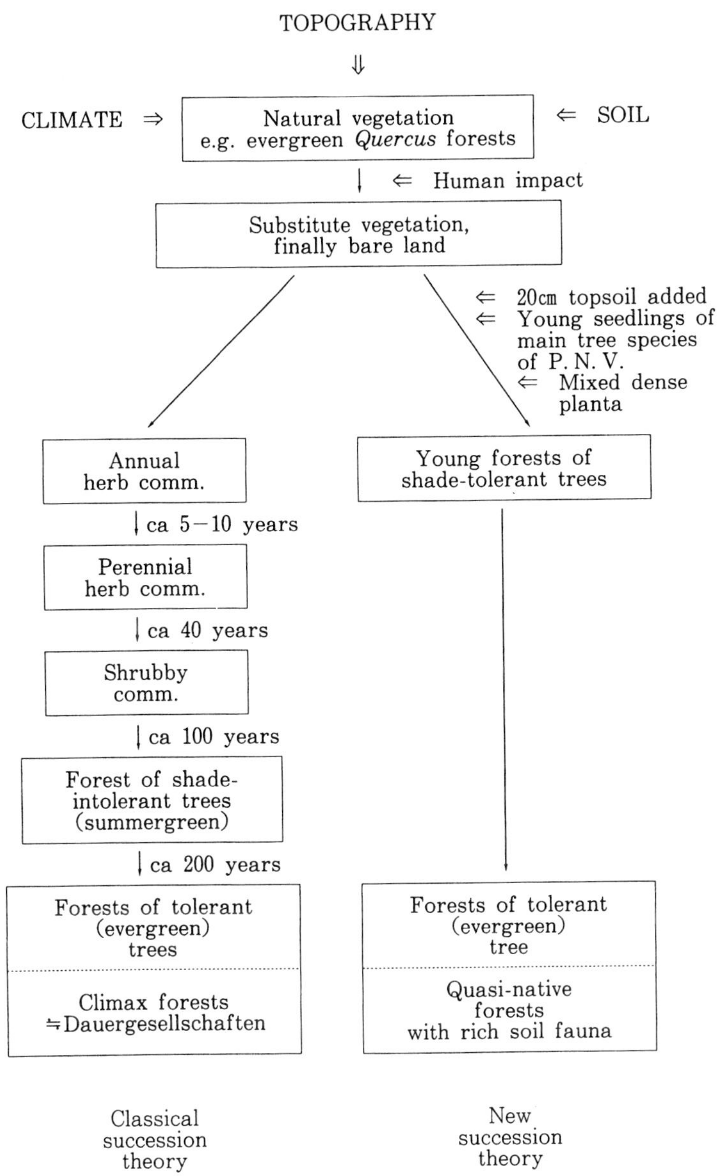

Figure 1. Comparison between new succession and classical theory (Miyawaki, 1992).

Concepcion in Chile (Miyawaki, 1993 a, b.).

2. Methodology

The initial stage of the vegetation-ecological process of restoration of native forests is phytosociological field study (Braun-Blanquet, 1928, '51, '64) to obtain information on the actual vegetation of the site. Remnants of house forests and Chinju-no-mori forests, land use, topography and soil profiles in the vicinity are also surveyed. Comprehensive study of these elements is made to detect the potential natural vegetation, a concept developed by R. Tuxen (1956). Then maps of the actual vegetation and of the potential natural vegetation are prepared. Our team has conducted intense field study and completed the maps of actual vegetation and potential natural vegetation, though at a small scale, covering the total land area of Japan. (Miyawaki ed. 1980-1989 Vegetation Japan Vol. 1-10 Appendix Actual and Potential Natural Vegetation Map 1:500000)

The process of the phytosociological field study is illustrated in Fig. 2.

It is most important to select, for planting, the main tree species, or at least one of the main tree species with potential to form a forest. In Japan's laurel forest zone, for example, the main tree species in lowland and alluvial areas are *Persia thunbergii* and *Castanopsis cuspidata* var, *sieboldii* in shallow soil, and evergreen *Quercus* species i.e. *Q. myrsinaefolia, Q.* glauca etc. in inland areas (Miyawaki, 1976, '79, 1980-1989). Since the deep, straight roots of these species make transplanting difficult, the seeds are grown in containers for about 2 years until the seedlings are 30 to 50cm high and the containers are filled with the root systems.

Drainage is important in areas where the annual rainfall exceeds 1000-1500mm such as on the Japanese Archipelago. Low mounds of loose soil are created, therefore, along streets, boundaries between public facilities and industrial sites and railways to improve air and water drainage even if the width of the mound is only 1m. Steep slopes of 30-45# along highways, for example, must be reinforced with bamboo or wood to make beds for seedlings. Mixed and dense

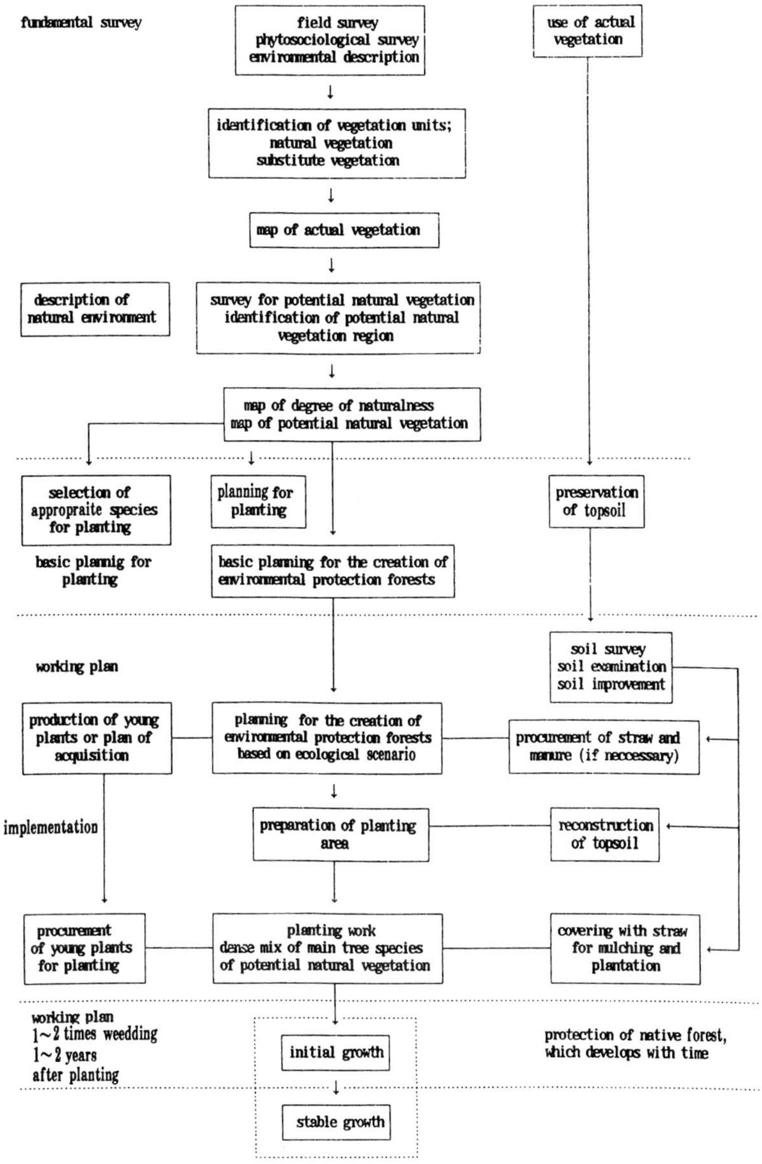

Figure 2. Flow chart for the restoration and creation of native forests.

planting of different species is the best way to restore natural diversity in urban areas. The number of seedlings is 2-3/ m^2 in normal stands and 5-6/ m^2 in beachfront and windy sites, to encourage a density effect. After the first winter, the seedlings planted in this manner are most likely to grow 1m a year. In the first 2-3 years, weeds must be eliminated and then placed on the mound as extra mulch. After planting, mulching is provided, in Japan, by placing 4kg/ m^2 of rice straw around seedlings. The mulching is effective in preventing weeds, protecting the seedlings from cold and dryness. With mulching, erosion does not take place on slopes even under heavy rainfall.

Decomposed straw contributes to the improvement of the soil. In three years the canopy develops and shields the forest bed, preventing the growth of weeds. After this point, it may be safely said that "no management is the best management."

3. Objective and Result

Our method and technique has been successful throughout Japan from the Summergreen broad-leaved forest zone, partly Subalpine zone to the Subtropical Zone, and from the beachfront lowland to the Summergreen broad-leaved forest zone with *Fagus crenata, Quercus mongolica* var, *grosseserrata, Q. dentata* etc. at the altitude of 1000m. On manmade islands and reclaimed land, along urban and industrial complexes, railways, highways and around public and educational facilities, efforts were made to create native forests by native trees. The northernmost man-made native forest is at Sempaku induscrial park in Hokkaido and the southernmost stand is at the Second thermal power plant in Ishigakijima Island in Okinawa Prefecture. It is always the citizens and school children who play the important role of planting seedlings according to the ecological scenarios (Miyawaki, Harada & Ude 1979, Miyawaki, Fujiwara and Ozawa 1993, Miyawaki & Golley 1993). The efforts have been also successful overseas, in Kuala Lumpur and Malacca in Malaysia and in Bangkok. In Bangkok, for example, the seedlings planted around a new Jusco shopping center have grown to be 4m in 20 months (Miyawaki 1994).

Restoration of biodiversity has been successfully achieved in native forests by native trees throughout Japan since 1970. The achievements are highly evaluated in Japan as well as abroad (Miyawaki 1993a,b., Miyawaki and Golley 1963 and others).

In subtropical Malaysia, an indigenous green environment and biodiversity has been restored around a shopping center. Low native flowering shrubs are planted as an edge community along roads and alleys so that citizens can enjoy the blossoms as well as the verdure. In case of emergency, the native forests function as shields against typhoons, fires and earthquakes.

4. Conclusion and further development

Since 1990, more than 300,000 potted seedings of 33 species of Dipterocarpaceae, the main tree species of the potential natural vegetation of Bintulu, Sarawak (Borneo) have been grown there.Currently, under the supervision of the our group and University of Agriculture Malaysia, the seedlings are steadily growing and covering the part of the university campus (800 ha), which were once burned for farming purposes.

In Belem, in the lowland around the estuary of the Amazon River, the Miyawaki team and University of Agriculture Para formed a joint study group and created 7-m high native forests in 20 months. 200,000 seedlings of *Virola* and 92 other species of main native trees have been grown in a year with the cooperation of the Mitsubishi Trading Company and EDB. The planting has passed the experimental stage. The native trees are growing steadily to recover the biodiversity of that part of the globe.

Simultaneously in Concepcion in Chile,*Nothofagus* forests have replaced the conventional species of Australian *Eucalyptus* and American *Pinus taeda* with the support of the Chilean Government. *Nothofagus* trees were planted in areas which were burnt for the purpose of farming or by accident. Recovery of biodiversity in native forests by native trees has started and so far been successful in the southern hemisphere as well.

In all of these instances, the government, local municipalities and the private sector are the producers and stay behind the scene. It is the citizens who play the main role of planting seedlings. The grass-roots movement for the recovery of biodiversity and green environments has started throughout the world (Miyawaki, Fujiwara & Box 1987).

6. References

1 Aoki, J. & Harada, H. 1985. Formation of environmental protection forests and changes in soil fauna, especially *Oribatid* mites.Bull. Inst. Environ. Sci. Technol. Yokohama Nat. Univ. 12:125-135.

2 Aoki, J. & Harada, H. 1989. A Soil-Zoological Sutudy of Urban Ecosystem around Tokyo. Research Institution of Washington, Washington, DC.

3 Braun-Blanquet, J. 1928. Pflanzensozlologie:Grundzuge der Vegetationskunde.330pp, Berlin. 3. Auflage 1951 and 1964, 865pp, Wien, New York.

4 Clements, F. 1916. Plant Succession: An Analysis of the Development of Vegetation. 512pp. Carnegie Institution of Washington, Washington, DC.

5 Miyawaki, A. 1975.Entwicklung der Umweltschutz-Pflanzungen und Ansaaten in Japan. Sukzessionsforschung fur Vegetationskunde, Vaduz, Cramer, pp. 237-254.

6 Miyawaki, A. 1976. Die Umwandlung immergruner in sommergrune Laubwalder in Japan.Werden und Vergehen von Pflanzengesellechaften. Bericht uber das Internationale Symposium der Internationalen Vereinigung fur Vegetationskunde, Vadus, Cramer, pp. 367-381.

7 Miyawaki, A. 1979. Vegetation und Vegetationskarten auf den Japanischen Inseln. Bull. Yokohama Phytosociol. Soc. Japan (Pestschrift fur prof. r. Tuxen),Yokohama, 16:49-70, mit 2 farb, Vegetationskarte.

8 Miyawaki, A. 1981. Energy policy and green environment on the basis of ecology. Beyond the energy crisis opportunitu and challenge. In: Fazzolage,R,A,& Smith, C,B,(eds), pp. 581-587. Oxford & New York.

9 Miyawaki, A. 1982a. Umweltsschutz in Japan auf vegetationsokologischer Grundlage. Bull. Inst. Environ. Sci. and Technol. Yokohama. Yokohama National University, 11:107-120(German with English Synopsis).

10 Miyawaki, A. 1989. Restoration of evergreen broad-leaved forest("laurel forest") in Japan. In:Academy, Ch. The World Community in Post-Industrial Society, 5.The human encounter with nature: Destruction and reconstruction.

pp.130-147.
11 Miyawaki, A. (ed.),1980-1989. Vegetation of Japan vols. 1-10(each vol. 400-70pp with color vegetation maps and tables). Tokyo Shibundo. (Japanese with German and/or English summary).
12 Miyawaki, A. 1993a. Restoration of native forests from Japan to Malaysia. H. Lieth and M. Lobmann(eds.), Restoration of Tropical Forest Ecosystems, p.5-24. Kluwer Academic Publishers.
13 Miyawaki, A. 1993b. Global perspective of green environments-restoration of native forests from Japan to Malaysia and South America, based on an ecological scenario. IGARSS'93,Better Understanding of Earth Environment. vol.1, p.6-8. Tokyo.
14 Miyawaki, A. 1994. Toward restoration of global green environments -from Japan to SB Asia and South America -. Japanese Scientific Monthly. 47(5):6-17.(Japanese), JSPS. Tokyo.
15 Miyawaki, A., K.Fujiwara, & E. O. Box. 1987. Toward harmonious green urban environments in Japan and other countries. Bull. Inst. Environ. Sci. Technol. Yokohama Natn. Univ. 14:67-82, Yokohama.
16 Miyawaki, A., Fujiwara, K., Nakamura, Y. & Kimura, M. 1983. Okologische und Vegetationskundliche Untersuchungen zur Schaffung von Umweltschutzwaldern in den Industrie-Gebieten Japans. Bull. Yokohama Phytosociol. Soc. 22, Ⅰ. 84pp, Ⅱ.151pp, with color vegetation maps (Japanese with German and English summaries)Yokohama.
17 Miyawaki, A., Fujiwara, K. & Ozawa, M. 1993. Native forest by naive trees-Restoration of Indigenous forest ecosystem-(Reconstruction of environmental protection forest by Prof. Miyawaki'smethod). Bull. Inst. Environ. Sci.Technol. Yokohama Natn. Univ. 15(1):95-102. (Japanese and English). Yokohama.
18 Miyawaki, A. & Golley F. B. 1993. Forest reconstruction as ecological engineering. Ecological Engineering 2:333-345. Elsevir Amsterdam.
19 Miyawaki, A., Harada, H. & Ube, H. 1979. Vegetationskundliche Untersuchung zur Schaffung von Umweltschutzwaldern um Industrie-Anlagen,erlautert am Beispiel der 11 Fabriken der Tore-Industrie-AG. Bulletin of Yokohama Phytosociology Society 8, 50pp, with 2 color vegetation

maps (Japanese with German summary), Yokohama.

20 Miyawaki, A., Tohma, H. & Suzuki, K. 1979. Pflanzensoziologische Untersuchungen der Shinto-Schrein- und Buddistischen Tempelwalder in der Prafektur Kanagawa (Hauptstadt Yokohama). 180pp, (Japanese with German summary),The Board of Education of Kanagawa Prefecture, Yokohama.

21 Tuxen, R. 1956. Die heutige potentielle naurliche Vegetation als Gegenstand der Vegetationskartierung, Angewandte Pflanzensoziologie, Stolzenau/Wes. 13:5-42.

TOWARD GLOBAL PLANNING OF SUSTAINABLE USE OF THE EARTH
Development of Global Eco-Engineering
S. Murai (Editor)

RELIEF OF IMPACTS OF URBAN ACTIVITIES ON GLOBAL ENVIRONMENT

Keisuke Hanaki and Toshiaki Ichinose

Research Center for Advanced Science and Technology,
The University of Tokyo, 4-6-1, Komaba, Meguro, Tokyo 153, Japan

Abstract

Energy consumption in urban area depends both on socio-economical factors and urban structure. The former are determined by circumstances of the country, whereas the latter can be controlled by the urban planning. Share of commercial, office and service sectors in energy consumption is large in typical Japanese large cities. Growth rate of energy consumption in this sector has been much higher than population growth and higher than GDP growth. Per capita energy consumption in Japanese large cities are smaller than the national average. Relatively high amount of energy is used in large cities in developing country such as Thailand and China. Per capita energy consumption in these cities are almost comparable to Japanese city and much higher than the average of the country they belong to. The energy consumption in Japanese cities is much more sensitive against temperature than Chinese cities. Heat recovery from solid waste incineration or sewage can provide energy in urban area although the recovered amount is not very high. Heat recovery system in urban area should be well planned. Life cycle assessment is necessary to evaluate the true effectiveness of technology or city with less energy consumption.

1. IMPACT OF CITIES ON GLOBAL ENVIRONMENT

1.1. Global ecosystem

Global environmental issues have become significant factors influencing the future development of humanity. The global warming issue will be a particularly difficult problem to overcome, since it results from activities that humankind has pursued since the industrial revolution. Our daily lives result in the production of excessive amounts of greenhouse gases (GHGs), primarily because of the consumption of fossil fuels.

Although global warming and other global environmental issues were the subject of academic debate among natural scientists, there was a lack of awareness on the part of the general population, including engineers. Traditionally environmental problems were the focus of concern on a local scale, such as within cities or in their environs. Examples of these problems include the deterioration of air quality due to atmospheric pollution in cities, and water pollution affecting rivers and lakes in areas around cities.

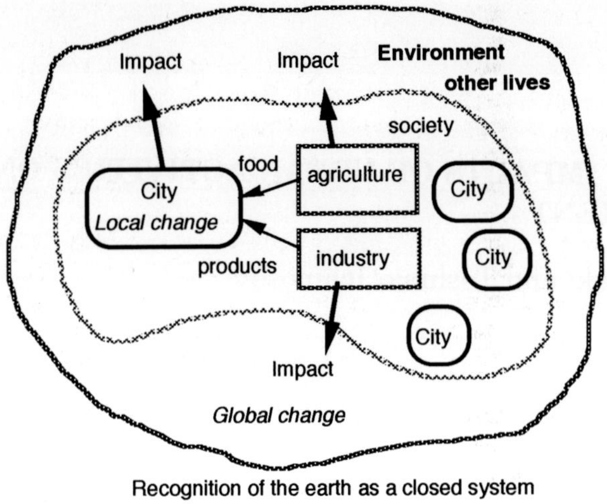

Figure 1. Global and local change of environment surrounding cities.

As shown in Figure 1, cities function in human society as concentrations of human residence and activity. The diagram also shows that cities are sustained by agricultural and industrial activity, and that the external environment is influenced by the existence of cities and agricultural and industrial activity. Various substances are emitted into the external environment, which includes ecosystems other than humanity. Except where extremely toxic substances (such as PCBs) were emitted, however, it was assumed that this situation was acceptable unless local environmental problems arose. We now recognize that this is not true, and we have been forced to learn the obvious fact that the Earth is a closed system. Changes that occur globally are referred to as "global changes" as distinct from local changes. Various forms of anthropogenic activity contribute to these changes. The contribution from cities is significant and expected to increase especially in developing countries of which population continues to increase until the middle of the 21st century. Relieve of impacts from urban areas are essential to prevent global environment from serious deterioration.

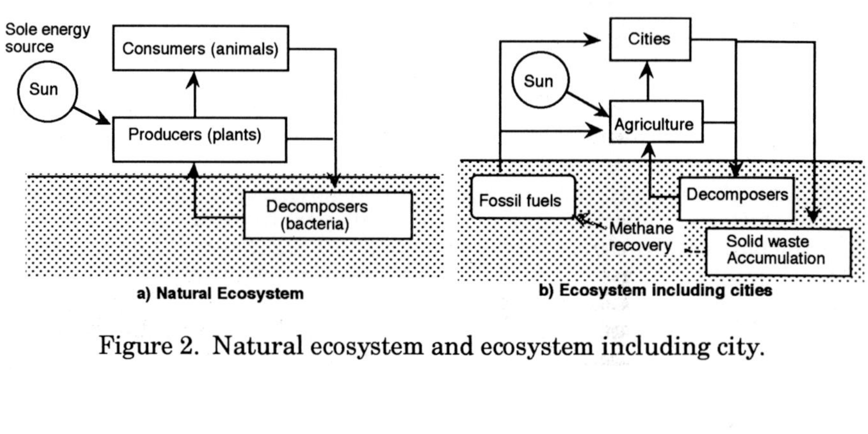

Figure 2. Natural ecosystem and ecosystem including city.

Figure 2 shows a modern artificial ecosystem, which encompasses the strong impact of anthropogenic activity, modeled on the concept of a conventional ecosystem. In the natural global ecosystem (Figure 2a), the sun is the only source of energy. The producers (plants) provide usable energy for the ecosystem by using the sun's energy to synthesize cells from carbon dioxide through photosynthesis. Animals generally survive by eating plants or plant-eating animals. Plant and animal remain and waste products are broken down by decomposers, such as bacteria, and return to carbon dioxide.

Clearly this natural ecosystem is being disrupted by anthropogenic activity (Figure 2b). Humans have added energy to the system in the form of fossil fuels. Of course, fossil fuels themselves derived from substances that once belonged to the ecosystem. Problems arise because they are being consumed by humans over an extremely short period of time, compared with the time required for their formation.

Agriculture is fundamentally a means of producing organic products in the form of products by means of energy provided by the sun. In practice, however, even agriculture relies on the use of fossil fuels, and our cities are totally dependent on them. The extreme expansion of urban activity and agricultural activity through dependence on fossil fuels has led to the production of vast volumes of waste products. Since the quantities of waste exceed the capacity of decomposers, the surplus is accumulated through the formation of landfills. If a route that would return these waste products to fossil fuels can be activated, the loop would be closed. In reality, however, this is merely realized by extracting methane gas at limited site of landfills. As a result, ecosystems that include modern cities are based on one-way flows of substances from fossil fuels to wastes.

1.2. Greenhouse gases emission

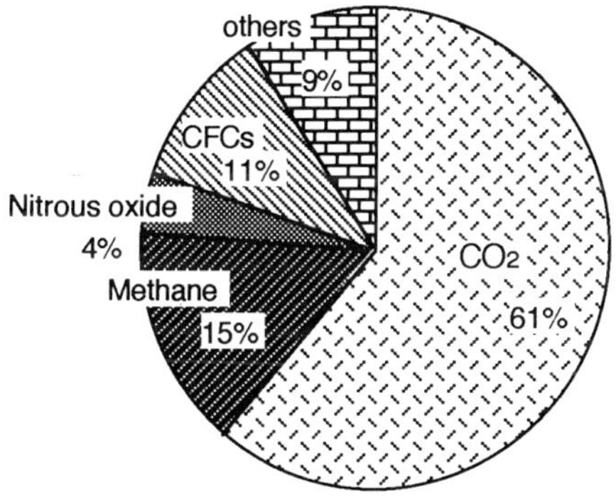

Figure 3. Contribution of each greenhouse gas emitted in 1990 with time horizon of 100 years.

Figure 3 shows the estimated global contribution of each greenhouse gas emitted in 1990 with the time horizon of 100 years (Intergovernmental Panel on Climate Change (IPCC) [1]). Carbon dioxide is the largest contributor, followed by methane and nitrous oxide. Actual measures to deal with greenhouse gases obviously include efforts to reduce the emission of carbon dioxide, which is strongly linked to consumption of fossil fuels. However, it will not be easy to achieve satisfactory reduction of greenhouse gases solely through efforts on carbon dioxide, and it will also be necessary to develop countermeasures against methane and nitrous oxide emissions, even though their overall contribution is apparently small. Sources of methane and individual greenhouse gases are discussed in detail in the IPCC's 1990 report [1].

Table 1 lists the main sources of greenhouse gases and their relationships to anthropogenic activity, particularly urban activity. Fossil fuel consumption in the context of urban activity is a major source of carbon dioxide. Cities also play a role in deforestation caused by land use changes and logging. In addition, cities account for a substantial percentage of demand for the cement industry, which is the third source of carbon dioxide.

Table 1
Contribution of urban activity to anthropogenic emission of greenhouse gases.

GHG	Source	Anthropogenic contribution	
		General	Urban area
Carbon dioxide	Fossil fuel	industrialized society	high consumption density urbanization = energy consumption increase (developing countries)
	Deforestation	land use change commercial logging	logging for paper, wood product
	Cement industry		construction of infrastructure
Methane	Rice cultivation	increasing food demand	
	Livestock	increasing food demand	
	Solid waste landfill		solid waste generation exceeding capacity of land
Nitrous oxide	Fertilizer	anthropogenic fixation of nitrogen gas	
	Combustion	biomass burning	automobiles, solid waste and sewage sludge incineration
	Wastewater		loading exceeding natural capacity
Ozone	Nitrogen oxide & hydrocarbon		air pollution (automobiles, industry)
CFCs	Industrial products	industrialized society	

Anthropogenic activities, including cities, are sources of methane emissions, since methane production resulting from rice or livestock farming can be attributed to population growth and the resultant rise in food production to feed growing populations. The emission of methane from solid waste in landfills is a phenomenon peculiar to cities. In areas that are free of the high-density human occupation that occurs in cities, waste products are broken down within the natural decomposition capacity, and the methane production does not occur.

In the case of nitrous oxide, there is a strong correlation between fertilizer-related emissions and food production. Relatively large urban sources include automobile exhaust gas, sewage sludge and solid waste incineration [2], and wastewater [3, 4]. In addition, cities are the specific source of tropospheric ozone, which is formed from nitrogen oxides emitted by automobiles and other incineration sources.

Cities in developed countries are expected to lead the effort to reduce GHG emission levels in the context of future countermeasures against global warming. However, GHG emissions in the cities of developing countries will be an important factor, since these are expected to increase to a greater or lesser degree due to the explosive growth of energy consumption in those cities.

1.3. Material Flows in Urban Areas

Urban activity is analogous to metabolism. The word "metabolism" belongs to the terminology of biochemistry. A typical example of metabolism is the way in which human beings obtain energy and sustain growth by eating food.

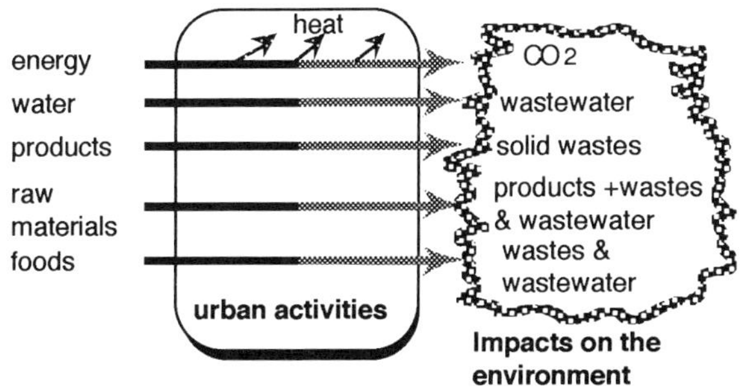

Figure 4. Transformation of materials through urban activities.

Figure 4 shows the typical material changes that occur as a consequence of urban activity. From the perspective of their environmental impact, the key materials that support urban activity, such as food, commodities and energy, are sources of waste products, wastewater and carbon dioxide. Where their use is accompanied by the consumption of energy, heat is also generated.

A number of approaches can be used to reduce the impact of cities on the environment. The first approach is based on "end-of-pipe technology," which is the technology used to manage waste immediately prior to its release into the environment, such as through wastewater treatment. Such technology has traditionally been developed in the field of environmental engineering. However, reduction of environmental impact, especially global impact, by such methods is limited. This is apparent when one considers such problems as the difficulty of finding sufficient waste disposal sites to meet demand. The second approach involves the use of technology to reduce emissions of substances that impact on the environment, such as the development of energy-saving equipment. A related concept that has emerged in recent years calls for the development of manufacturing processes that reduce the amount of waste produced. The third approach is the raising of public awareness about the environment. The fourth approach, and the one that has the potential to make a contribution in the context of urban planning and management, is the improvement of the urban metabolic system. Possible areas of improvement include the recycling of materials, the effective utilization of heat within cities, and the reduction of city energy consumption through appropriate urban planning.

2. ENERGY CONSUMPTION IN URBAN AREA

2.1. Factors affecting energy consumption

Energy is consumed in city for very wide range of purposes from household use to commercial or industrial use. The amount of energy consumption in urban area is determined by both socio-economical factors and structure of urban area as shown in Figure 5. The former includes industrial development, per capita GDP, which represents the standard of life, energy price, and life style of the residents. These factors mostly depend on rather circumstances of the country than the situation of the urban area itself. The energy consumption is also influenced by the differences between urban area and rural area in development or life standard, which are large in many of the developing countries, whereas such difference is small in the developed countries. These socio-economical factors are hard to be controlled.

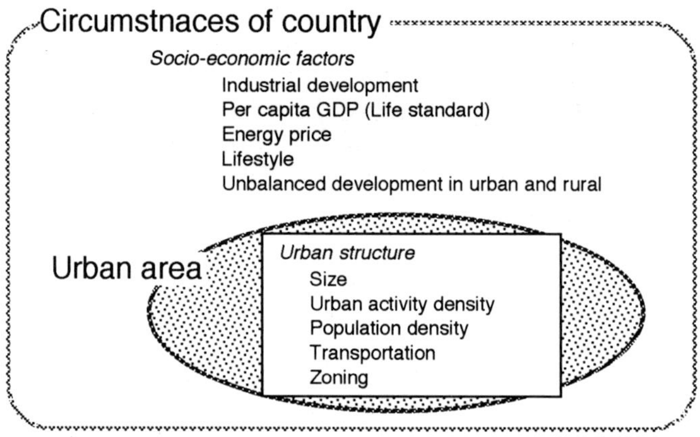

Figure 5 Socio-economic factors and urban structures governing energy consumption in urban area.

The energy consumption also depends on physical structure of city, such as city size, the density of various urban activities as well population density, transportation system and zoning policy. City with concentrated activity and population has more or less smaller energy consumption than that with low density. Transportation system is obviously important factor which influences urban energy consumption. Individual automobile transportation consumes much higher energy than mass-transit system. Zoning policy currently taken tries to apart residential area far from business or commercial district to ensure good local environment. However, such discrete zoning contributes to higher energy consumption for commuting. Introducing mixed zoning with ensuring satisfactory environment is expected to decrease energy consumption in the transportation sector. Urban planning can control these factors on the physical structure of city. However, magnitudes of the effects of these factors on energy consumption in urban area are yet to be estimated.

2.2. Urban energy consumption in Japan

Energy consumption in seven large Japanese cities, namely Sapporo, Tokyo, Yokohama, Nagoya, Osaka, Kobe, and Fukuoka, are analyzed here. The energy consumption is categorized here into four sectors, namely, household; commercial, office and service; industry; and transportation sector based on the purpose of energy use. Various statistics both in user side, such as household expenditure survey, and supplier side such as electricity supply statistics were used to estimate the energy consumption. Figure 6 illustrates the contribution of each sector to the energy consumption (end-use basis) in these seven cities. The share of industry sector was almost 50% in Osaka, while Sapporo and Fukuoka had industrial sector with only less than 10%. Fukuoka showed low value because adjacent Kitakyushu City had high accumulation of industry. Commercial, office and service sector, which is typical in cities, was as high as 50% in Fukuoka followed by Tokyo (23 wards).

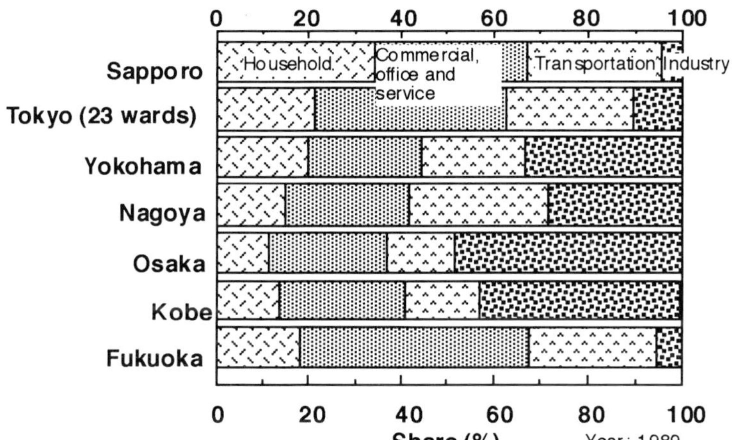

Figure 6. Share of each sector in energy consumption in seven large Japanese cities.

Average value of annual growth rate in these seven cities in energy consumption in each sector was compared with the population and GDP growth (Figure 7). Only energy consumption in the industry sector grew slower than the population. Other sectors showed much higher growth rate than the population in these cities. The highest growth rate was observed in commercial, office and service sector, which are typically representing various activities in urban area. The annual increase of energy consumption in this sector is even greater than GDP growth and almost ten times greater than growth rate of population. Controlling the growth of energy consumption in this sector in urban area is essential in developed countries such as Japan.

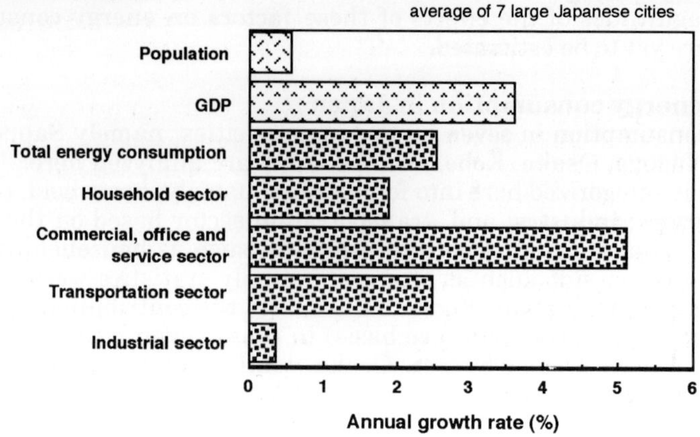

Figure 7. Annual growth rate of energy consumption in each sector and population in the period between 1980 and 1988 (growth rate values in seven cities, namely Sapporo, Tokyo (23 wards), Yokohama, Nagoya, Osaka, Kobe and Fukuoka were averaged).

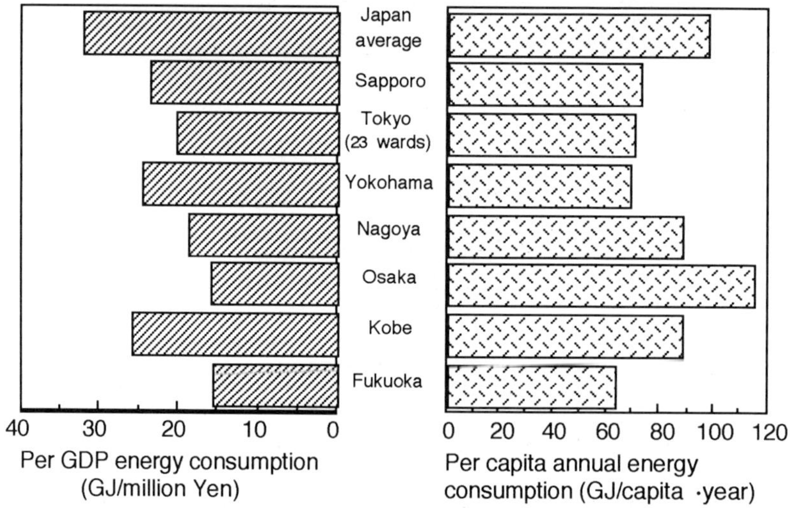

Figure 8. Per capita and per GDP annual energy consumption (end-use basis) in Japan.

Energy consumption per person and that per GDP in the seven cities are compared with national average values in Figure 8. Per capita energy consumption in these cities was smaller than the national average value except for Osaka which has high industrial activities as previously shown in Figure 6. These comparisons indicate that the urban area generally consumes smaller amount of energy per person than non-urban area because the main activities are more commercial or service oriented in the urban area. Per GDP energy consumption are obviously lower in all of the seven cities than the non-urban area. This is probably because there are no lifestyle differences between urban and non-urban areas in developed countries, because high population densities result in higher energy efficiency through small housing floor space and effective transportation such as railway in urban areas, and because commercial activity is more prevalent than industrial activity.

Lower per capita energy consumption in large cities than non-urban area is probably common tendency in developed countries. It should be, however, noted that these facts do not directly mean that nation-wide progress in urbanization can reduce national energy consumption in total. Many of the urban activities are supported by industrial products which have caused significant energy consumption outside of that urban area.

2.3. Energy consumption in cities in developing countries.

Energy consumption was estimated in Thailand and China, both of which are developing countries. Figure 9 shows Bangkok region's shares of various key indicators evaluated [6] based on statistics of energy consumption [7]. With just 2% of Thailand's area and 16% of the country's population, Bangkok region (Bangkok Metropolitan area and Bangkok vicinity) consumes approximately 60% of national energy consumption while earning one-half of GDP. This means that energy consumption and GDP per capita in Bangkok region are extremely higher than the national averages.

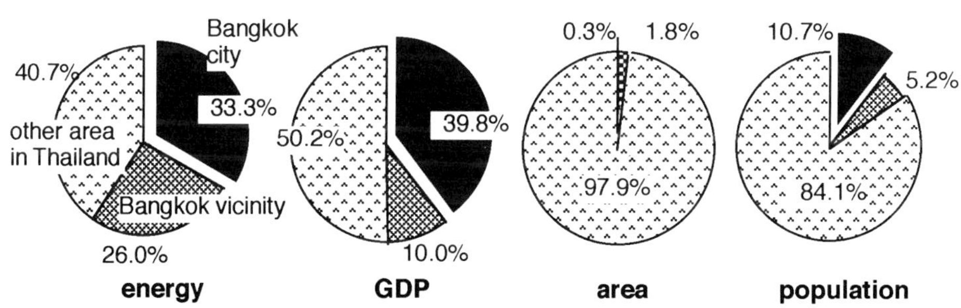

Figure 9. Share of Bangkok or Bangkok vicinity in Thailand.

Per capita energy consumption and per GSP (Gross Social Product) energy consumption for provinces and large cities in China are plotted in Figure 10 [6]. Beijing, Shanghai and Tianjin, which are developed area in China, show higher per capita consumption than other provinces. However, these cities showed relatively lower per GSP energy consumption. It indicates that development has increased the energy consumption with urbanization in these three cities, but energy is not wasted there. Contribution of these cities to GSP is very great.

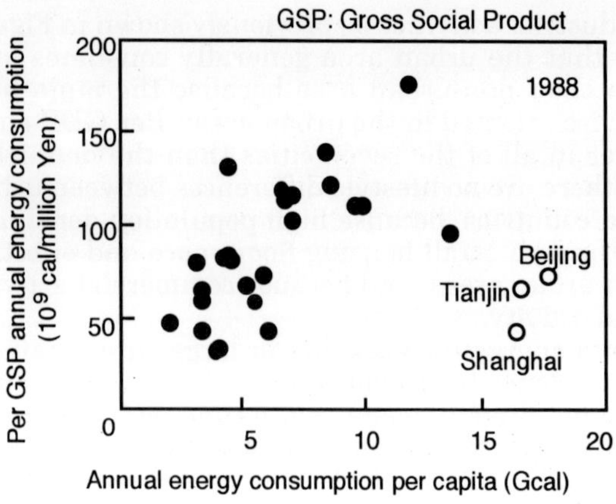

Figure 10. Per capita and per GDP energy consumption in China.

Higher per capita energy consumption in urban area than non-urban area seems common in developing countries. Development of urban area is normally accompanied by industrial activities, and lifestyle of people in cities is different from that in non-urban areas. People move from rural area to urban area, and at the same time they begin to consume more energy than before.

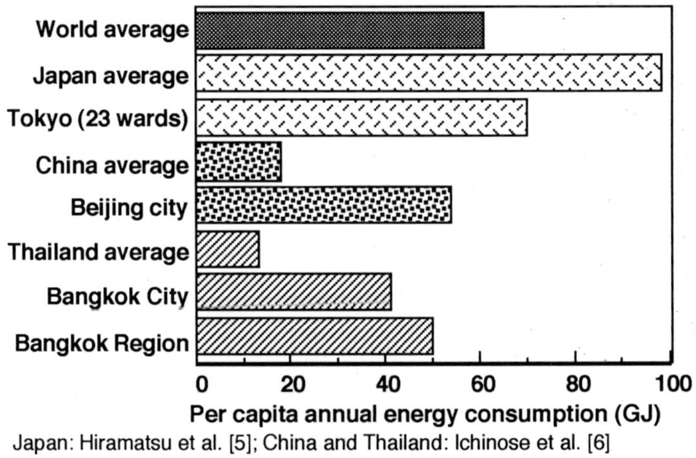

Figure 11. Comparison of per capita energy consumption between large city and national average in developed and developing countries.

Figure 11 shows the comparisons of per capita energy consumption between national average and large city in Japan, China and Thailand. The values of Beijing city, Bangkok city or Bangkok region showed almost comparable value to

Tokyo (23 wards). However, these values were much higher than the national average within each country.

Urbanization in the developing country causes great increase in per capita energy consumption. Increase both in population and in per capita energy consumption remarkably contribute to the explosive energy demand increase in the developing country. Such an increase in energy consumption is undesirable from the perspective of global warming. From the viewpoint of improving living standards in developing countries, however, it is a change that should not be denied.

2.4. Effect of climate on energy consumption

Energy consumption in household sector is influenced by climate. Sensitivity of energy consumption against climate is often discussed by using index of degree-hour, degree-day, or degree-month which represent both magnitudes of coldness or hotness and duration. Figure 12 shows the dependency of annual consumption of kerosene which is almost absolutely used for heating in 47 prefecture capitals in Japan on degree-month (standard 18°C) [6]. There seems a threshold point beyond which people use fuel very intensively.

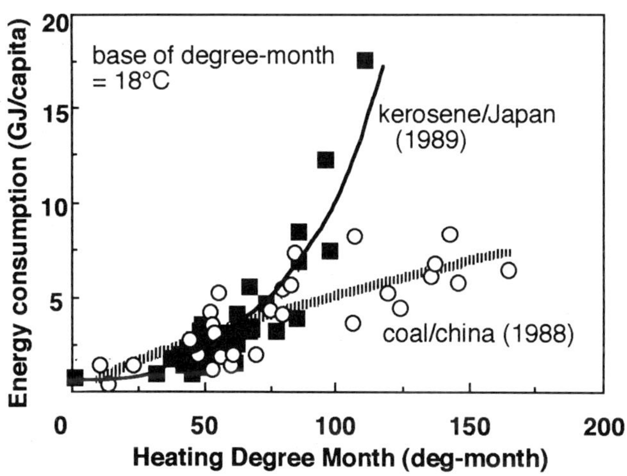

Figure 12. Dependency of kerosene consumption in Japan and coal consumption in China on heating degree month [6].

Similar analysis was attempted for Chinese provinces and cities [6]. Chinese coal consumption, which showed highest correlation with the heating degree month, is far less sensitive to temperature as shown in Figure 12. Such low sensitivity in China might be caused by centrally planned and enforced energy allocation. Future development of the country may cause the increase in temperature sensitivity of energy consumption.

3. HEAT AND ENERGY RECOVERY IN URBAN AREA

3.1. Potential of heat and energy recovery

In recent years effort has been channeled into efforts to find effective uses for the unused waste heat and natural heat that exist in cities. Potential heat or energy recovery from solid waste (garbage) or sewage and their characteristics are summarized in Table 2.

Table 2. Possible methods of energy recovery from solid waste and wastewater.

Resources	Method of energy recovery	Potential energy recovery per capita [a] (GJ/person • year)	Advantages	Disadvantages
Solid waste	Electricity and/or heat recovery at incineration plant	3.7	High quality of heat energy obtained	Location of incineration plant is often far from CBD (Central Business District)
	Methane recovery from landfill	0.9	Greenhouse gas (methane) emisison also reduced	Low recovery percentage
Sewage	Heat recovery from sewage	1.9	Large flexibility in site of heat recovery	Low quality of heat, heating purpose only
	Methane recovery from sewage sludge	0.2	Established process, good fuel obtained	Low energy recovery

a: See the text for assumption for calculation.

The amount of heat produced by solid waste incineration is around 10 MJ (megajoules) per kilogram of solid waste (wet basis). Since urban solid waste is produced at the rate of about one kilogram per day per person in Japan, it would be possible to produce 3.7 GJ of energy per person-year. This is equivalent to about 5% of per capita energy demand, which amounts to 70 GJ per annum (Figure 11). However, in view of the percentage of energy demand that relates to heating and cooling, and the fact that energy produced from solid waste collected over a wide area would be supplied over a limited area, it appears that this would be an effective source of heat in supply areas. As the quality of heat obtained from the incineration is high, using it for heat only is not wise. Generating electricity together with recovering the waste heat through co-generation is realistic way.

Another option of energy recovery is that through methane fermentation process which is conducted by bacteria working without oxygen. Treatment of solid waste in a designed reactor is possible, but more feasible way is to recover methane gas from solid waste landfill site in which microbiological methane production takes place for many years. Part of the calorific value of solid waste is converted to that of methane gas in the landfill site. A rough estimation if half of the original calorific value is converted to methane and that half of the produced methane gas in the landfill site is recovered, the recovered methane accounts for 0.9 GJ/year per person, which is much smaller than the energy obtained in the incineration. Recovery and utilization of methane from the landfill site are very important in a different aspect. It can reduce the emission of methane gas, which is greenhouse gas. Because methane from the solid waste landfill is estimated to be an important source of methane [1], its recovery is desirable.

Sewage can be also important energy source. Energy is recovered from sewage, which has higher temperature than atmosphere, by using heat pump for heating through district heating system. By assuming if for 5°C temperature difference is recovered from sewage of which discharge per person is 250 l/d, 1.91 GJ/year per person is recovered from sewage. A disadvantage here is that the quality of the recovered heat is poor and using it with heat pump for space heating and hot water supply are the almost only way to utilize this energy. On the other hand, great advantage is that this heat can be recovered anywhere along sewer pipe with simple equipment such as heat pump.

The other way of energy recovery from sewage is again through methane production. Direct methane conversion from sewage is, however, not feasible in the developed country like Japan because of poor effluent from the sewage treatment plant. Instead, methane production from sewage sludge by the process known as anaerobic digestion, can be applied. By assuming if each person discharges sewage containing 100g of COD (Chemical Oxygen Demand), and that only half of this organic matter is converted to methane in the anaerobic digestion of sewage sludge, 0.2 GJ/year is recovered. This value is smaller than the energy recovered by other source, but this technology has been well established through its history for more than 50 years.

3.2. System of heat recovery in urban area

Technology of energy recovery from solid waste, sewage, or river water is established to some extent as described in the previous section. More important and practical issue is its application to actual urban area which has uneven distribution of residential, commercial, or business area. High density of heat demand is preferable in the district heating system from the viewpoint of efficiency especially pipe construction cost. Central business district, or high-rising apartment house seems suitable site to apply district heating system. On the other hand, energy source such as solid waste incineration plant, sewage treatment plant is normally located in suburb or fringe zone of urban area. Such discrepancy should be taken into consideration in the planning of energy recovery.

Figure 13 illustrates currently applied system and possible options. Many of the solid incineration plants in Japan recover energy, and the typical case of heat utilization is supply of heat to swimming pool adjacent to the plant (Figure a)), because business district is not located near the plant. Energy from solid waste is of course utilized in this case, but it does not contribute to the reduction of fossil fuel consumption in urban area. A simple way to meet energy source with demand is to transport heat to the central business district (CBD) in which district heating/ cooling plant is constructed (Figure b)). Heat loss and high construction cost of pipe are the problems to be solved in this case. Another plan accompanying substantial change of urban planning is relocation of solid waste incineration plant to CBD (Figure c)). The plant can be constructed underground and is expected to function as energy plant.

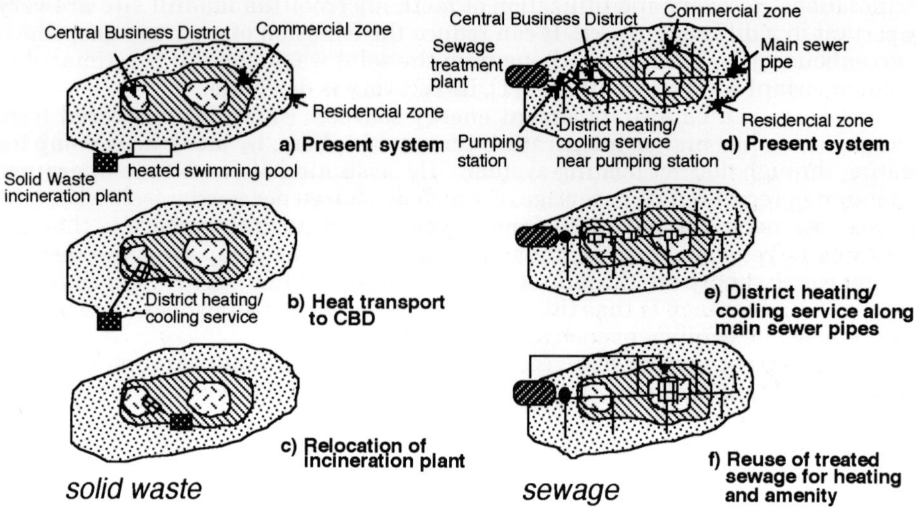

Figure 13. Heat recovery system from solid waste and sewage.

Utilization of sewage heat also has several options. The current system is recovering heat of sewage at sewage treatment plant or existing pumping station (Figure d)). However, location of this facility does not fit to the area of high energy demand. A possible option is to construct district heating/ cooling plant(s) along the main sewer pipes which usually run through CBD. Potential of recoverable energy increases with increase of sewage flow. Location of heating/ cooling plant is determined by calculating the recoverable energy based on sewage flow data and energy demand which is estimated based on land use. Another option is a combination of energy recovery and reuse of treated sewage. Treated sewage is pumped up again to CBD and let the water flow through urban area after the recovery of heat. Such a stream can create pleasant environment in CBD.

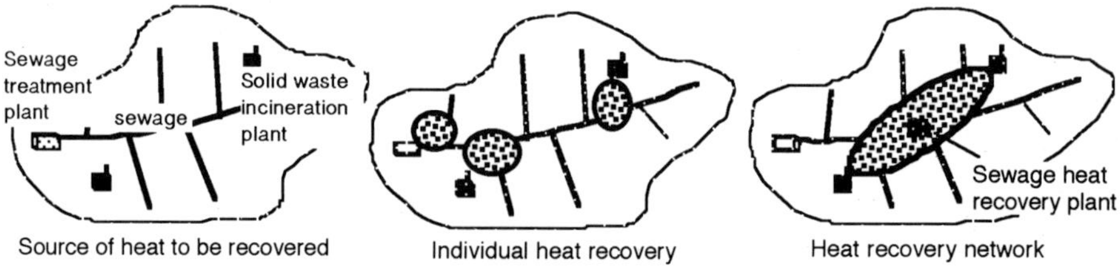

Figure 14. Heat recovery network in urban area

Most of the urban area has more than one source of heat for recovery (Figure 14). Besides introducing individual heat recovery to each source, formation of a network for heat recovery is also a possible option. Different kinds of heat, such as sewage heat and solid waste heat are used in such a network which has large service area.

4. TOWARD THE ENVIRONMENTALLY SOUND CITY

The concept of the environmentally sound city represents a desirable future goal for cities that currently face circumstances such as those described in Section 1. An environmentally sound city is a city that (1) imposes a minimal impact on the external environment, (2) has a high degree of resource self-sufficiency, and (3) maintains a high quality of urban amenity.

A particularly important issue in relation to the first of these characteristics is the impact imposed on the global environment. In order to evaluate the global environmental impact, it is necessary to determine the impact imposed through the production, use and disposal of a facility or product by means of a life cycle assessment (LCA).

Method of LCA was developed to analyze overall environmental impacts of products. All steps beginning from obtaining raw materials, manufacturing, and up to the disposal step are analyzed and cumulative impacts are evaluated. One can judge how one particular product gives overall impact on the environment by knowing the cumulative impacts.

Application of the method of LCA is being expanded to technology or even to city. Comprehensive analysis becomes necessary in such cases. Some of the technologies for reducing energy consumption sometimes require more facility or building, of which construction needs more energy. Intelligent building is an example of this. Energy can be saved during the operation of such building, but its construction needs more energy than ordinary building. LCA is really necessary to judge whether such intelligent building truly contribute to the reduction of environmental impact.

The achievement of amenity frequently conflicts with the minimization of the environmental burden or the achievement resource self-sufficiency. The word "amenity" encompasses a wide spectrum of meanings. In developing countries it may refer to the availability of basic human needs, such as safety, education and medical services. In developed countries the focus is tending to shift from material amenity to mental amenity. The concept of urban amenity in the mental sense encompasses such questions as how people living in cities can be provided with mental satisfaction, and how culture can be maintained. The improvement of mental amenity need not be incompatible with the reduction of the environmental impact, and it is here that we can find a path to the goal of the environmentally sound city.

REFERENCES

1 Intergovernmental Panel on Climate Change (1990), "IPCC Working Group I Report (Scientific Assessment)"

2 K. Hanaki, Y. Hirose and T. Matsuo (1993), "Estimation of anthropogenic emission of nitrous oxide gas in regional scale." Proc. 1st JSCE Symposium on Global Environment, 240-245. (in Japanese)

3 K. Hanaki, H. Zheng and T. Matsuo (1992), "Production of nitrous oxide gas during denitrification of wastewater." Water Science and Technology, Vol. 26, 5/6, 1027-1036
4 H. Zheng, K. Hanaki and T. Matsuo (1994) : "Production of nitrous oxide gas during nitrification of wastewater." Water Science and Technology, to be published.
5 N. Hiramatsu, K. Hanaki and T. Matsuo (1992), "Comparison of energy consumption among seven major cities in Japan" Environmental Systems Research, Japan Soc. Civil Engineer, Vol. 20, 252-261 (in Japanese).
6 T. Ichinose, K. Hanaki and T. Matsuo (1993) "International comparison of energy consumption in urban area." Proc. Environmental Engineering Research, Japan Society of Civil Engineer, Vol. 30, 371-381 (in Japanese).
7 Thailand Development and Research Institute (1990) "Industrializing Thailand and its impact on the environment."

TOWARD GLOBAL PLANNING OF SUSTAINABLE USE OF THE EARTH
Development of Global Eco-Engineering
S. Murai (Editor)

Eco-sound Technologies for Urban Planning and Management.

Hiroshi Shimizu

Regional Environment Research Division,
The National Institute for Environmental Studies,
16-2, Onogawa Tsukuba Ibaraki 305, Japan

Abstract

Contemporary society is reviewed from a perspective of world history. It is argued that a new style of urban planning and management will be introduced in the near future in Asian countries, especially in Japan. It is also pointed out that environmental problems such as climate change are the limiting condition in the creating of a new style in present society. This means that technologies to decrease CO_2 emissions considerably to cope with climate change are required in the introduction of a new urban style. The concepts of eco-sound technologies and stepwise technologies are introduced to achieve the needed reductions in CO_2 emissions. Eco-sound technologies are defined to have the properties of both environmentally friendliness as well as other merits such as ease of use, comfortableness, and high efficiency. A stepwise technology is one that operates on a different principle from conventional technology. As an example of eco-sound technologies, several examples are presented and analyzed. The first one is an energy technology, the solar cell. Solar heating and the insulation of buildings are other possibilities which applications in houses and urban architecture. In the traffic field, an ultra-low emission and energy-saving automobile, a multilayered road system, and a highly efficient railway freight system are offered. The way to widely introduce these technologies into society is also discussed.

1. INTRODUCTION

Global environmental problems are among the most serious threats challenging the world today. Among these problems, climate change is by far the most important, both because its effects are likely to be widespread and the measures to deal with it difficult to implement. In this paper, the problem of climate change is discussed from the viewpoint of world history. Policies to cope with global warming are discussed along with possible technologies to alleviate it. In the discussion of technologies, the concept of "eco-sound technologies" will be introduced.

2. HISTORICAL INTERPRETATION OF CLIMATE CHANGE

The history of human beings can be viewed as the history of the development of the world's large urban communities such as Rome, Florence, Paris, London, and New York. When, why, how and by whom were these cities built? The processes of the development of these great cities seem to have a common pattern. Typically, there is a major war or revolution after which new social and political systems are established. An emphasis is then placed on the development of primary industries and agriculture because of the shortages and hunger that prevailed during the war or revolution. Around 10 to 20 years afterwards, agricultural production becomes sufficient. At this point, the emphasis shifts to the development of secondary industries with much of the technology imported from more developed countries. Around 50 years after the war or the revolution, as economic wealth grows, the development or reconstruction of large urban communities begins. At the same time there is an emphasis on fundamental learning and scholarship. After about one generation, the fruits of this investment in research are realized. About 100 years later, an emphasis on the arts and culture flourishes.

During the periods of urban construction and rebuilding, new styles are introduced. These new styles are influenced by several important factors including social needs, the availability of capital, certain critical technologies to support specific styles of construction, and various limiting factors. Table 1 show the characteristics of several important periods.

The Japanese style was established during the Edo era. The typical architectural styles were the Shoin Zukuri (a rather formal style for official buildings such as temples and castles) and Sukiya Zukuri (a less formal style suitable, for example, for the homes of wealthy merchants). The social needs were the movement of people to big cities and the Danka system which was introduced to curb the spread of Christianity. The supporting technology was tools made from iron. The limiting factors were the protection of the cities from enemies and the policy of self-sufficiency reflected in the closing of Japan to outsiders.

Now, fifty years after the second world war, a new period is dawning. The accumulation of wealth is progressing, especially in Japan. Many more people are moving to big cities. As these trends continue, we can expect a new style to develop. What will be the limiting factor

Table.1 Characteristics of Stytles

Name of Style	Typical Archtecture	Era	Typical Place	Social Requirement	Economic Source	Background Technology	Limiting Factor
Japanese	Shoin-zukuri Sukiya-zukuri	Edo	Edo Kyoto	Movement of Chomin to Cities Danka System	Growth of Chomin Class	Iron Tools	Enemies Sakoku Sytem
Circus Crescet	Circus Crescet	After The Glorious Revolution	London Perse	Movement of Middle Class to Cities	Overseas Activity	Blick	Big Fires
Modern	Skyscraper	After 1920	America	Big Business	New Industries	Steel Cement Elevator Automobile	Efficiency of Economy
New Japanese		After 2000	Japan	Change of Industrial System	Electronics		Global Environmental Problem

of this new style? The answer is global environmental problems, especially climate change. In the past, as styles were created, they were affected by the limiting factor, and the factor had to be dealt with. For example, in the Circus–Crescent style in England, large fires were the limiting factor. This limit was dealt with by using bricks to build their structures. Brick buildings were more expensive but England was rich enough to afford them because of the wealth arising from overseas British activities.

3. THE DIRECTION OF TECHNOLOGY TO COPE WITH CLIMATE CHANGE

Many technologies have been developed and used widely to decrease pollution. These technologies have two common characteristics. One is that they do not change the main principle or purpose of the polluting underlying technology. The second is that these technologies had only one purpose: to decrease pollution. For example, technologies to reduce emissions of sulfur or nitrogen oxides from power plants did not change the purpose or principle of power generation. Characteristics of the technologies to reduce pollution are that they cost money and consume energy. It is for these reasons that in the past there have been conflicts between polluting industries and those that are affected by the pollution. When we consider climate change, the conflicts between polluters and those at risk from climate change are likely to be less. This is so because everyone will be affected by climate change, both polluters and the affected public. Moreover the problem is so large that it will require everyone working together to solve it.

Solving the problem of climate change will require the introduction of two new concepts. One is that of "eco–sound technologies," and the other is of "stepwise technologies."

a) Eco–sound Technologies.

As technologies develop, they typically become more useful, more efficient, less polluting, and more effective. For example, locomotives changed from steam power to electric power. The reasons for the change were higher speed, elimination of smoke, reduction in noise, and higher energy efficiency. All of these benefits were realized with the replacement of steam locomotives with electrically powered ones.

Now we must solve the problem of climate change. This problem cannot be solved by actions on the part of only one sector of society. Rather, it will require the concerted cooperation of all segments. The technologies to decrease the risks of climate change will have to be acceptable to all important segments of society. To achieve this acceptance, technologies should be both effective in decreasing the risks of climate change and attractive to those who will use them. We call such technologies "eco–sound technologies."

Eco–sound technologies have two properties. The first is that they are effective environmentally. The second is that they provide benefits for the user compared with conventional technologies. The concept of an eco–sound technology is shown in Figure 1. The horizontal scale reflects the merits such as usefulness, comfortableness, and efficiency. The vertical scale reflects the environmental benefits. Conventional technologies to cope with pollution belong in the upper left–hand quadrant. Most technologies that are widely used today belong in the lower right–hand quadrant. Eco–sound technologies would belong in the upper right–hand quadrant.

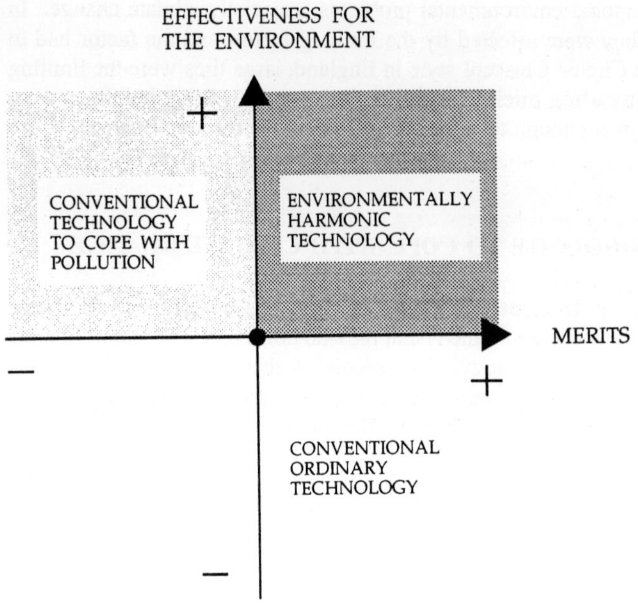

Fig.1 CONCEPT OF ECO−SOUND TECHNOLOGY

(b) Stepwise technology

The difficulty in coping with climate change is that the target to solve the problem is very high. Several targets have been proposed to decrease CO_2 emissions, a list of which is shown in Table 2. The first one is the recommendation from the IPCC in 1990 and indicates that stabilizing CO_2 atmospheric *concentrations* at twice the preindustrial level would require a reduction in CO_2 *emissions* of 50 percent below those of 1985. The second and the third targets are the actions proposed by the Japanese Government and the United Nations Framework Convention on Climate Change, respectively. The target of

Table 2 TARGETS TO DECREASE CO_2 EMISSION

1. THE IPCC SCIENTIFIC ASSESSMENT (AUG. 1990)

IF CARBON DIOXIDE EMISSIONS ARE REDUCED TO 50% OF 1985 LEVELS, CONCENTRATIONS ARE STABILIZED AT TWICE THE LEVEL OF THE PREINDUSTRIAL AGE

2. THE JAPANESE GOVERNMENT ARREST GLOBAL WARMING (OCT. 1990)

THE EMISSIONS OF CO_2 SHOULD BE STABILIZED ON A PER CAPITA BASIS IN THE YEAR 2000 AND BEYOND AT ABOUT AS IN 1900

3. UNITED NATIONS FRAMEWORK CONVENTION ON CLIMATE CHANGE (JUN. 1992)

RETURN THE ANTHROPOGENIC EMISSIONS OF CARBON DIOXIDE TO EARLIER LEVELS BY THE END OF THE PRESENT DECADE

the action program and the Framework Convention are almost the same. Their primary goal is to prevent an increase in emissions; achieving these targets would *not* stabilize atmospheric CO_2 concentrations. The recommendation of the IPCC, which is based on scientific understanding, is more strict and would lead to a stabilization of the atmospheric CO_2 concentrations.

Now, since the quality of life and the level of GDP are almost proportional to the emission rate of CO_2, decreasing CO_2 emissions could lead to reductions in the quality of life and economic activity. And achieving the target of IPCC would require reducing the quality of life to half its present level. This target would be very hard to reach in any simple way.

If we should attempt to achieve a 50–percent reduction in emissions by applying new technology, the effectiveness of decreasing CO_2 emissions would have to be amazingly high. Table 3 shows the CO_2 emission rate from each sector. To decrease the overall emission rate to half its present level, the emission rate in each sector should be decreased by half.

When we apply new technologies to solve problems, we first consider modifying conventional technologies. We will call this way incremental improvement. Incremental improvements have several advantages. The effect of the improvement can be easily predicted, the lead time to get the improvement is not long, and the cost to realize the improvement is not high. By using incremental improvements, some targets such as those expressed in the Japanese action program, can be achieved. But it will be very difficult to achieve the goal of the IPCC recommendation. To meet this target we

Table 3 JAPANESE EMISSIONS BY SECTOR OF CARBON DIOXIDE FOR 1990 (TOTAL AMOUNT OF EMISSION : 318Mt - C)

SOURCE	DIRECT EMISSION (%)	DIRECT INDIRECT (%) EMISSION
ENERGY CONVERSION	33.3	—
ELECTRIC UTILITIES	24.6	—
INDUSTRY	33.8	51.4
AGRICULTURE	2.8	3.1
CONSTRUCTION	1.3	1.4
FOOD	0.7	1.6
PAPER & PULP	2.1	3.4
CHEMISTRY	3.0	5.7
CEMENT & CERAMICS	6.4	7.4
IRON & STEEL	12.2	16.5
RESIDENTIAL & COMMERCIAL	10.9	25.2
RESIDENTIAL	5.7	13.2
COMMERCIAL	5.1	11.7
TRANSPORT	17.7	19.4
CAR	15.4	16.0
WASTE INCINERATION	3.9	3.9

must adopt new technologies whose principles are different from conventional ones and whose effects on the environment are small. We call this approach "stepwise improvement." The concepts of incremental improvements and the stepwise improvements are shown in Figure 2. Examples will be discussed below.

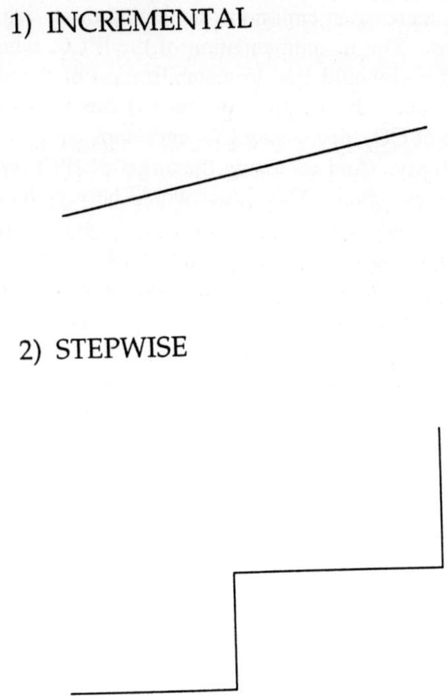

Fig.2 CONCEPT OF THE INCREMENTAL AND STEPWISE WAYS

4. EXAMPLES OF ECO–SOUND TECHNOLOGIES

There are numerous sources of carbon dioxide. CO_2 is emitted from all industries, businesses, vehicles, and public activities. This fact means that there are a large number of technologies that can be used to reduce CO_2 emissions. Among these technologies, several ones are very effective and can be applied in urban planning and management. Several of these will be introduced and discussed as examples of technologies that are eco–sound and stepwise improvements.

(a)Solar cell

Solar cells generate electricity from the energy of sunlight. The energy of sunlight is about 1 kW per square meter when the sunlight is incident to the perpendicular plane to the direction of the sun. If the efficiency of the solar cell is 10 percent, 100 watts (W) of electric energy can be generated from 1 square meter of solar cells. If half of the housing area all over Japan were used to support solar cells, the amount of power generated wold be the same as the total electric generation in Japan.

The energy payback time of solar cells has been calculated at about 2 to 3 years. As the lifetime of solar cells is 20 to 30 years, the energy payback time is short enough compared with the life of solar cells. As a result, solar cells represent an important means of mitigating climate change.

vehicles. The third is as a corridor for pipes carrying water, waste water, and natural gas. The fourth is for cables carrying electricity and telecommunications. Lastly, under the road, subways can be constructed. Roads can also serve recreational functions, for example, for festivals and parades and -- where traffic is low -- as playgrounds for children. If all of these functions were met by a single-layer road, the road would have to be very wide to avoid congestion. To provide for these multiple purposes, a new concept is offered to make roads multilayered. In a multilayered road, the road has several layers and each layer would have a different function. In Figure 4, an example of a multilayered road is shown. In this case, the ground surface is used for automobiles, the second layer is for public transportation and public services such as water supply and electricity. The top layer is for pedestrians. There are several advantages of a multilayered road. First, land is used more efficiently. Second, construction costs are relatively low. And third, there are distinct safety and pollution benefits.

The concept of a multilayered road can be applied to many types of roads. In a narrow road, it can be used to increase the capacity of traffic. In a wide road, the highest layer can be used as a park or public plaza. An expressway can also be reconstructed using this concept, where the first layer is for heavy trucks and buses and the second and the third layer are for passenger cars or lightweight vehicles.

The multilayered road can increase traffic capacity. It is an attractive means of cutting CO_2 emissions and pollutants because traffic congestion can be decreased if the density of automobile traffic does not changed. Since a multilayered road has a variety of merits, shown above, it can be said to be an eco-sound technology.

Fig. 4 An example of a multilayered road

MW per year in Japan. It is very important therefore to reduce the cost of solar cells. This capacity would be about 30 times larger than the yearly production of solar cells in 1990. So, if this application is realized, it would be a very useful means of cutting the price of solar cells.

(b) Solar heating

About 10 percent of CO_2 is emitted in heating water in residential and commercial buildings in Japan. From a technical point of view, the easiest way to decrease the emissions from water heating is to use solar hot water heaters. Solar heating uses solar panels which collect the sunlight and heat water by a very simple means. By the use of solar heating, about 50 to 60 percent of the energy can be saved and the energy payback time is only 1 year. But the use of water heaters has not spread widely, partly because the purpose of solar heaters is only to save energy and there are no other advantages. If the use of solar water heaters is to spread, some additional benefit will have to be added. One such benefit would arise with the combination of solar water heating and the 24–hour bath system, now under testing. In the 24–hour bath system, the used hot bath–water is cleaned and purified by circulating it through a series of filters. Typically, an electric heater is used to maintain the temperature of the water. In the new system, the water temperature is kept hot using solar water heaters. Thus, if the merits of both systems are combined, they become eco–sound technologies.

(c) Insulation of buildings

About 10 percent of CO_2 is emitted from the heating of residential and commercial buildings in Japan. Insulating homes and buildings is a very effective way to decrease the energy needed for heating and air conditioning. If 10 cm thick insulation is put on walls, roofs, and floors, and double glazing is used in windows, the heating energy is reduced to about one third that of the non–insulated house or building. The temperature difference between the window side and the central part of the room becomes small when insulation and double glazing are used. This makes the room more comfortable. Windows and walls do not have dew condensation in an insulated house. As the insulation provides both energy savings and comfort, it can be said to be eco–sound. Although insulating a building is expensive now, if all houses and buildings were insulated, the costs would decrease quickly.

(d) Ultra low emission and energy saving automobiles

About 16 percent of CO_2 is emitted from automobiles in Japan. There are five major ways to decrease the emission of CO_2 from automobiles. One is to make vehicles smaller and to use high–efficiency technology. As the energy consumption of automobiles is almost proportional to their weight, downsizing is very effective. High–efficiency engines, low rolling–drag tires, low air–drag bodies and light–weight body materials are all important technologies for highly efficient automobiles. The second is through the management of traffic flow. Traffic signal control and the construction of truck terminals are examples of this category. The third is changing life styles. The fourth is a modal shift, and the fifth is to use so–called low emission cars, such as electric vehicles and hybrid cars.

(e) A multilayered road system

As a major example of the management of traffic flow, a multilayered road system is considered in an urban area. A road has many purposes in the lives of people, especially in urban areas. The first purpose is as a place for walking. The second is a place to drive motor

It is said that the biggest problem in using solar cells is the price. Because of the high initial cost, electricity from solar cells is about 3 times more expensive than the price of conventionally generated electricity. But the price of solar cells has decreased to about one twentieth of what it was 10 years ago. It is expected that the price will drop by another two thirds within 10 years.

Are solar cells an eco-sound technology? If they are used for the generation of electricity instead of using power plants, they are not always eco-sound, because there are no other benefits than the energy savings for the user of electricity. In this situation, solar cells are located at point A in Figure 3.

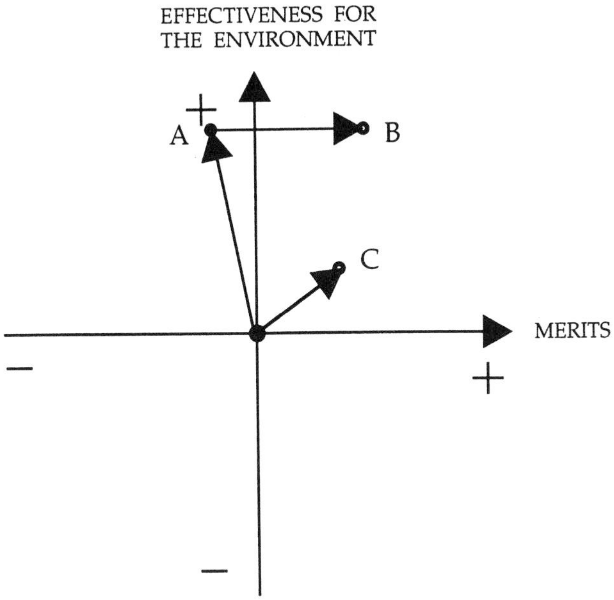

Fig.3 EVALUATION OF SOLAR CELLS
A : PRESENT SOLAR CELL
B : SOLAR CELL WITH AN ADDITIONAL VALUE
C : SOLAR CELL USED IN ANOTHER PURPOSE

If solar cells have other merits, they will become an eco-sound technology, like B point in Figure 3. The effective way to achieve this merit is to find other applications. As one such application, consider the ventilation system of an automobile. The air conditioner of an automobile is very powerful now because it must cool down the car from very high temperatures reached when it is parked under the sun in the summer season. If the air in the automobile is ventilated while it is parked, the inside temperature is almost the same as that of the outside. Then less heat needs to be removed from the car, the capacity of the air conditioner can be reduced, and the driver will not feel so hot when he or she gets into the car, even if the automobile is parked in the hot sun. The ventilation can be achieved by the use of solar cells which are part of the roof of the automobile.

The price of the solar cells could be offset by the lower cost of the air conditioner the capacity of which can be decreased. And overall energy consumption is reduced. When solar cells are used for such an application, they become an eco-sound technology, and they are plotted at point B in Figure 3. Suppose all new passenger cars were equipped with solar cells with 30 W output power. The total amount of production of solar cells would reach at least 300

(f) A highly organized railway

Trains are more efficient than road vehicles for several reasons. The rolling resistance of a steel wheel on a rail is much smaller than that of a tire on a road, air drag of each car in a train is much smaller than that of automobiles, and the energy efficiency of a train is much higher than that of an internal-combustion engine car. But trains are not used as much compared with road motor vehicles, especially in the transportation of freight. The reason is that the trains are not as convenient for users because freight must be carried from the origin of the trip to the station and from the final station to the freight's destination. Moreover, freight trains stop at stations for a very long time. Some ideas are offered here to deal with these problems. One is what we call "short-cut piggy back." The other is a land ferry system.

The short-cut piggy back is a variation of a piggy back system. Turntables are put on the freight car of the train and enable the truck to get on and off at any place of the train. A picture of the turntable is shown in Figure 5. When the train is moving, the direction of the table is parallel with that of the rail. When trucks get on or off, the turntable turns as shown in the figure, and the trucks move onto or off the turntable by themselves. A

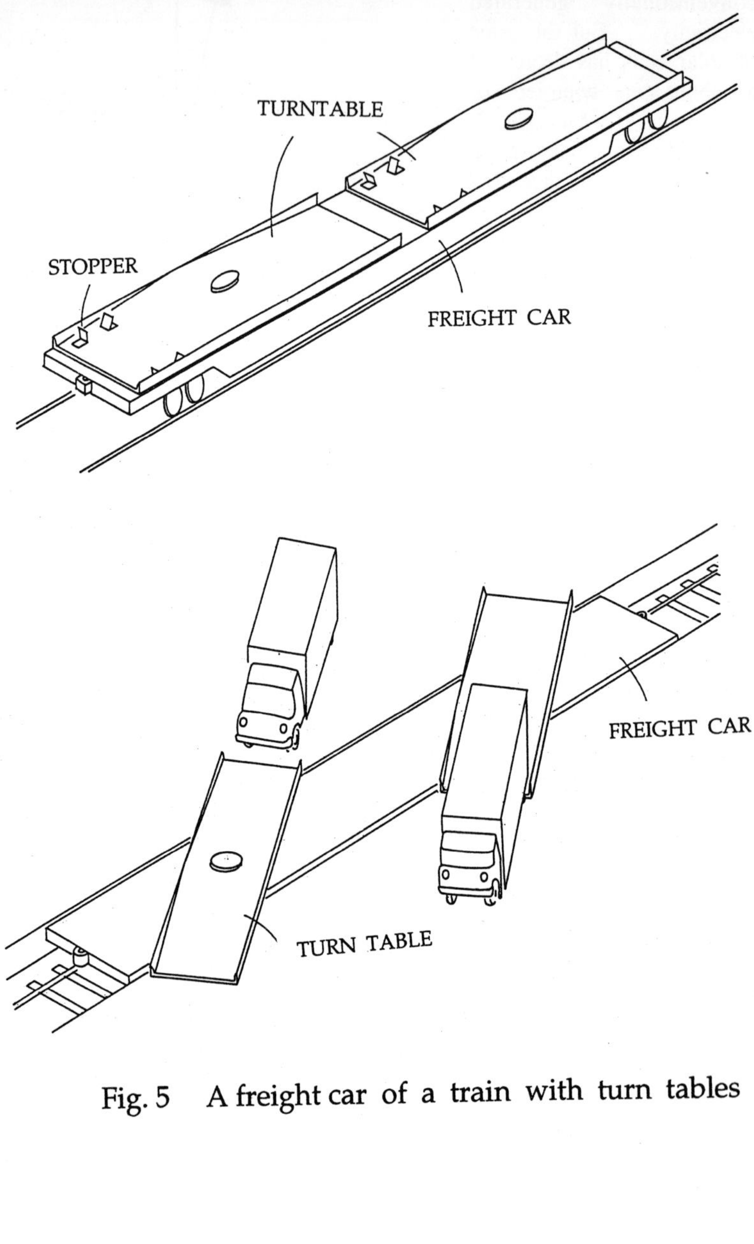

Fig. 5 A freight car of a train with turn tables

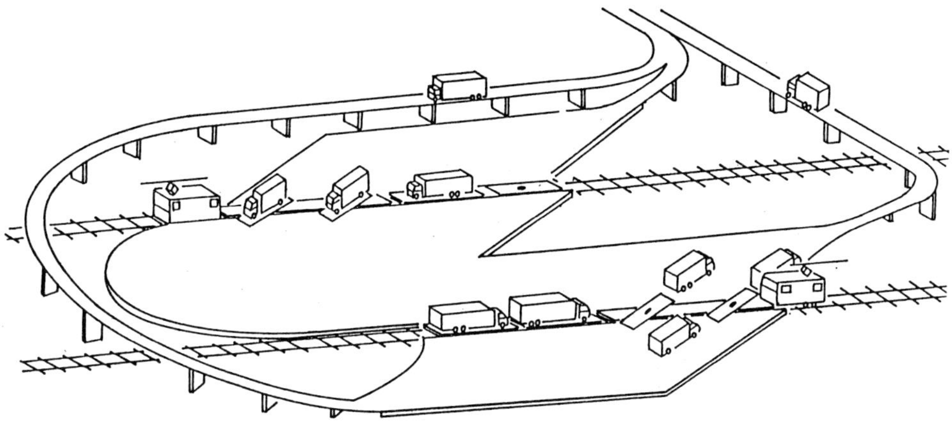

Fig. 6 A conceptual drawing of truck parking

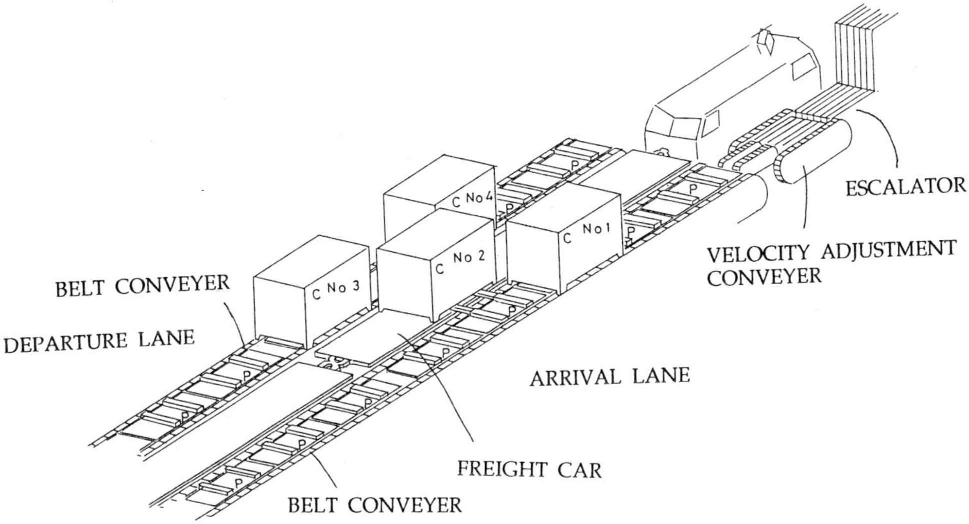

Fig. 7 A schematic drawing of an automatic container handling system

special truck parking area would be constructed at the train station to facilitate boarding and to shorten the waiting time. A conceptual drawing of the truck parking facility is shown in Figure 6. The short–cut piggy back would be useful to reduce truck traffic in a big city.

The land ferry is an effective container freight transportation system. The main technology is an automatic container handling system which would be installed at the station. This system would permit handling a large number of containers very quickly. Figure 7 shows a schematic drawing of the automatic container handling system. Containers are loaded and unloaded quickly and automatically from the freight train. As result of this automation, the land–ferry system allows the freight train to travel as fast as a passenger train. To allow the rapid handling of containers, a facility to store the containers is also required. A schematic drawing of the total system including the container handling station is shown in Figure 8. The containers that have been off–loaded or that are to be carried by the train are stored in the building behind the automatic container handling system. By using this system, freight can be carried quickly and economically, allowing on–time delivery even when the traveling distance is long.

If the short–cut piggy back and the land ferry are realized, the train system for freight becomes more efficient and convenient compared with transportation by trucks. The energy

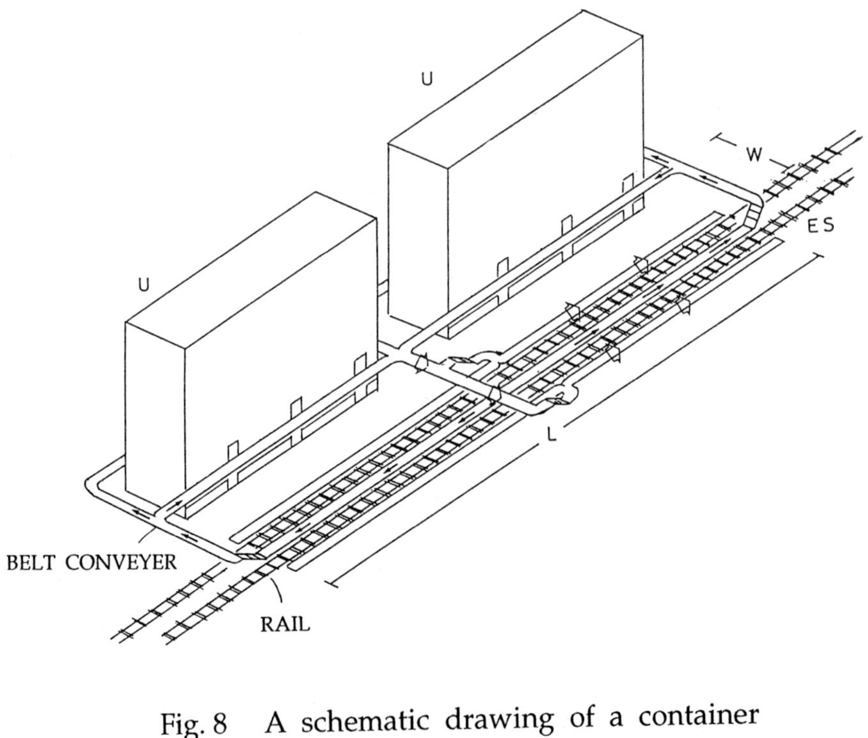

Fig. 8 A schematic drawing of a container handling station

efficiency is good as a conventional railway. With these characteristics, the train system would become an eco-sound technology.

5. WAYS TO ACCELERATE THE USE OF ECO-SOUND TECHNOLOGIES

Several technologies have been discussed from the stand point of eco-sound and stepwise technology. Among those discussed, solar cells, ultra-low emission and energy saving automobiles, the multilayered road, and the highly organized railway are examples of stepwise technologies. The insulation of buildings, the multilayered road, and the highly organized railway are eco-sound technologies by themselves. And if other ideas are incorporated, solar cells, solar heating, and ultra-low emission and energy saving automobiles will also become eco-sound technologies.

If these technologies were widely introduced into society, the consumption of fossil fuels would decrease considerably. But these technologies, as they are, are not in wide use. Because present technologies are so widely used, the introduction of new systems requires a large-scale change in the present system. For example, if solar cells were widely used, the electric power generation system would change from a centralized one dependent on large power plants to one dependent on decentralized sources. The technology and know-how for generating and distributing power would change completely. Such a change could lead to considerable anxiety among power companies, anxiety which should be avoided if at all possible.

There are two reasons to adopt new technologies, one negative and the other positive. The negative one is climate change which will occur unless some counter measures are adopted. The positive motivation is that, historically, the time is approaching for developing a new style. In this new era, technologies will be accepted without difficulty because society is expecting positive changes. This means that the acceptance of eco-sound technologies can be understood from a historical perspective. Many people are pessimistic about the future. They are afraid of major changes because of the risk of failure. But actually, there is every reason to be optimistic about the future. Major changes will give us the opportunity for great success, especially in Japan. If people were to adopt more optimistic outlooks, eco-sound and stepwise technologies -- which will bring big changes to society -- will be accepted and used successfully.

6. CONCLUSIONS

In this paper, the application of new technologies has been discussed that could be used in innovative urban planning and management. The overall goal would be a major reduction in the emission rate of CO_2 to cope with climate change. It was also pointed out that the adoption of a new style will occur in the near future. As a result, the present era is a good time to begin the large-scale restructuring of urban areas. Two concepts of technologies are proposed. One is that of eco-sound technologies and the other is stepwise technologies. Examples of technologies which can be categorized by each concept were presented. The way to introduce these technologies widely was also discussed. To achieve sustainable development, the attitudes of people, the directions of companies, and the goals of society will have to change, worldwide. As we consider such changes, it is important that we approach the future with an optimistic viewpoint. Then the final result will be satisfactory.

efficiency is good as a conventional railway. With these characteristics, the train system would become an eco-sound technology.

3. WAYS TO ACCELERATE THE USE OF ECO-SOUND TECHNOLOGIES

Several technologies have been discussed from the stand point of eco-sound and appropriate technology. Among those discussed, solar cells using low emission and energy saving automobiles, the elevated road, and the highly organized railway are examples of eco-sound technologies. The insulation of buildings, the elevated road, and the highly organized railway are eco-sound technologies in this sense. Wind, solar cells, solar heating, and ... automobiles ... are also eco-sound technologies.

TOWARD GLOBAL PLANNING OF SUSTAINABLE USE OF THE EARTH
Development of Global Eco-Engineering
S. Murai (Editor)

Global Planning for the Sustainable Use of the Earth Based on Land Suitability Analysis

Ryosuke SHIBASAKI*
Shunji MURAI**
Takao ENDO*
Bai XUEMEI***
*Institute of Industrial Science, The University of Tokyo
7-22-1 Roppongi, Minato-ku, Tokyo 106, JAPAN
fax:+81-3-5411-0441
Email:shiba@shunji.iis.u-tokyo.ac.jp
**Asian Institute of Technology
Division of Computer Science
AIT GPO Box. 2754, Bangkok 10501, THAILAND
*** International Center for Ecology
1560-39, Kami-yamaguchi, Hayama, Kanagawa, 240-01, JAPAN

ABSTRACT:

To secure a sound basis for continuous human habitation of the Earth and improvement of human welfare, it is necessary to establish sustainable use of global environmental resources. Authors propose a concept of global landuse planning which aims at sustainable use of the Earth and mitigation of global changes. Under a proposed framework of global landuse planning, land suitability analysis for forest conservation/reforestation and agricultural food production is conducted on a global scale . Suitability of forest conservation/reforestation is evaluated for each grid with 10 min. by 10 min. size with respect to efficiency of carbon storage in forest biomass, while suitability for agricultural land is assessed for the same size of grids with respect to food productivity in terms of energy. In evaluating suitability, NPP(Net Primary Productivity) estimated from remote sensing data and soil erosion possibility data estimated from climate and land condition data are integrated.

The analysis revealed the trade-off relation between the amount of carbon storage by reforestation and that of food production due to the limited resource of land. Land suitability analysis for other aspects of natural renewable resources such as timber/fuel wood production and agroforestry is left for further researches. More importantly, distribution of demands (population) to products from renewable resources and transportation should be taken into account in evaluating suitability of lands for resource utilization.

1.Introduction

Explosive expansion of human activities causes environmental changes on the global scale which may threaten basis for continuous human habitation of the Earth. For example, increase of energy and resource consumption is leading to exhausting non-renewable resources and to increasing discharge of toxic wastes. Excess development of agricultural land and over-grazing to meet the increasing food demand may accelerate degradation of renewable resources, such as soil.

Countermeasures have been proposed, such as legal regulation on the use of fluoro carbons, taxations on the carbon dioxide discharge and technological development to fix carbon dioxide or to remove toxic wastes. These aim at gradually improving efficiency and decreasing discharge rate of socio-economic and technological systems.

On the other hand, we can seek a sustainable use of global environmental resources which allow to fulfill basic requirements for continuous human habitation of the Earth. Sustainable use of global environmental resources will require discussion on the spatial patterns and distribution of resource use based on quantitative and homogeneous data, because quantity and characteristics of global environmental resources have a extremely wide spatial variety.

In forecasting human impacts on global environmental systems, patterns of resource use, that can be called "land use" for terrestrial environmental resources, play important roles. Emission of green house gas (GHG) through land use changes such as deforestation are comparable to that from fossil fuel consumption. Modeling mechanisms of land use changes has crucial importance not only in forecasting GHG emission but also in formulating policies to direct land development/management activities. Even in forecasting and assessing impacts due to global changes, such as changes in crop productivity, increase of risk of natural disasters, spatial patters of population density and economic activities, which are reflected in land use, are important factors.

We can extend a concept of management of global environmental resource use for continuous human habitation of the Earth from "natural" spaces to "artificial spaces" , that are urban spaces, by including urban infrastructures, such as sewerage systems and railway networks in global environmental resources. Extreme degradation of environmental quality and increasing impacts on global environment of Mega-cities explosively growing in developing countries are the results of excess concentration of population in areas with very poor man-made environmental resource base.

Focusing terrestrial resource uses, we can propose a concept of "global land use planning", where more desirable land uses, that is, spatial patterns of human activities and uses of environmental resources, and strategies for guiding human activities are discussed to fulfill basic requirements for continuous human habitation of the Earth and improvement of human welfare.

To effectively support the process of promoting researches and policy formulation for global planning, a computer environment have to be established to allow experts and policy makers to make an access to global environmental data, analyze them to identify and model underlying mechanisms, and to make quantitative evaluation of cost and effectiveness of countermeasures. Global GIS integrated with remote sensing for global monitoring is a key component of such a computer environment, but in this paper these topics are not discussed due to space limitations.

The objectives of this paper are as follows:
1) to propose a conceptual framework of global land use planning,
2) to conduct land suitability analysis for carbon stock in forest biomass and agricultural food production as an experimental study for making "a master plan of the Earth".

2. A Framework of Global Land Use Planning

2.1 Goals of Global Land Use Planning

The final goal of global land use planning, "securing a basis for continuous human habitation of the Earth" can be disaggregated into several sub-goals as follows.

1) Sustainable use of global environmental resources.
2) Decrease of environmental impacts due to land use activities.
3) Mitigation of impacts from global changes.

Sustainable use of renewable resources requires conservation of reproduction basis of the resources and to keep the exploitation rate within reproduction rate. As for non-renewable resources, lowering consumption rate and heightening recycling ratio is necessary to minimize consumption rate of virgin resources. For this purpose, global land use planing should be complemented by economic measures such as taxation on non-renewable resources.

Decrease of environmental impacts due to land use activities and mitigation of impacts from global changes is requested to decrease risks for future human society.

2.2 Policies for Implementing a Global Land Use Plan

In global land use planning, implementing policies aim at directing and controlling resource uses (land uses) to promote sustainable uses and to mitigate environmental impacts due to land use activities and so forth. Policies include direct measures which control land uses directly such as zoning and environmental impact assessment (EIA) of development projects, and indirect measures which affect and direct land use activities. Land use activities may be changed or redirected by changing costs and benefits by direct economic measures such as taxation and subsidies. Technology improvement/transfer and education may shift directly cost functions. Technological improvement and transfer projects will have to be accompanied by establishment of global funding systems incorporating taxation systems. Modification of legislation system for land holdings and utilization can alter a framework of behavior of individual land owners. In addition, long-term indirect measures may include educations to emphasize the importance of equity between human generations.

Global efforts coordinated under a framework of a land use plan are indispensable, to secure necessary fund, technological knowledge and human resources and to effectively implement these policies.

2.3 Possibility of Global Cooperation for Sustainable Use of the Earth

Usual planning procedure is conducted at the regional to national level, not exceeding the boundaries of nations. How about the possibility of obtaining consensus for global planning and it's implementation? Global land use planning has several strengths pushing up the possibility of realization.

1) Goal of global land use planning, "securing basis for continuous human habitation of the Earth is so basic that it may effectively drive different nations and regions to make better coordination as members of a global community.

2) By detecting changes of land uses and land covers, most human impacts on global environmental systems, such as deforestation and desertification are easily monitored from space. Through monitoring from space, global community can not only grasp present situations of global environment on the real-time basis, but also watch how each member follows agreements of global land use planning, which can be a strong incentive to realize global land use plans.

3) Transparency and rationality of the planning process should be considered as the most important factors determining the feasibility of a global plan. Homogeneous and reliable global data which can be obtained with remote sensing and other advanced ground-measurement techniques offer a sound basis for rationality of the planning process. And integrating and visualizing global data and different models/knowledge in a GIS environment, transparencies can be improved in the process, which may result in quicker and more rational agreements among global community members

2.4 A Planning Procedure for Global Land Use Plans

A process of global land use planning comprises three major componets (**Fig.1**).

1) Global monitoring and development of global databases to assess quantity and characteristics of global environmental resources.

To effectively support the process of promoting researches and policy formulation for global planning and to improve transparency and rationality of the planning process, a computer environment have to be established to allow scientists, experts and policy makers to make an access to global environmental data, analyze them to identify and model underlying mechanisms, and to make quantitative evaluation of cost and effectiveness of countermeasures. Global GIS integrated with remote sensing for global monitoring is a key component of such a computer environment. To generate consistent and quality global databases using a wide variety of data with different spatio-temporal resolutions and accuracies, integration techniques will be of crucial importance.

2) Analysis and forecast of land use changes due to expansion of human activities and their impacts on global environment.

To forecast global land use changes and their impacts on global environment systems, global models (global landuse - environment-interaction models) have to be developed using the

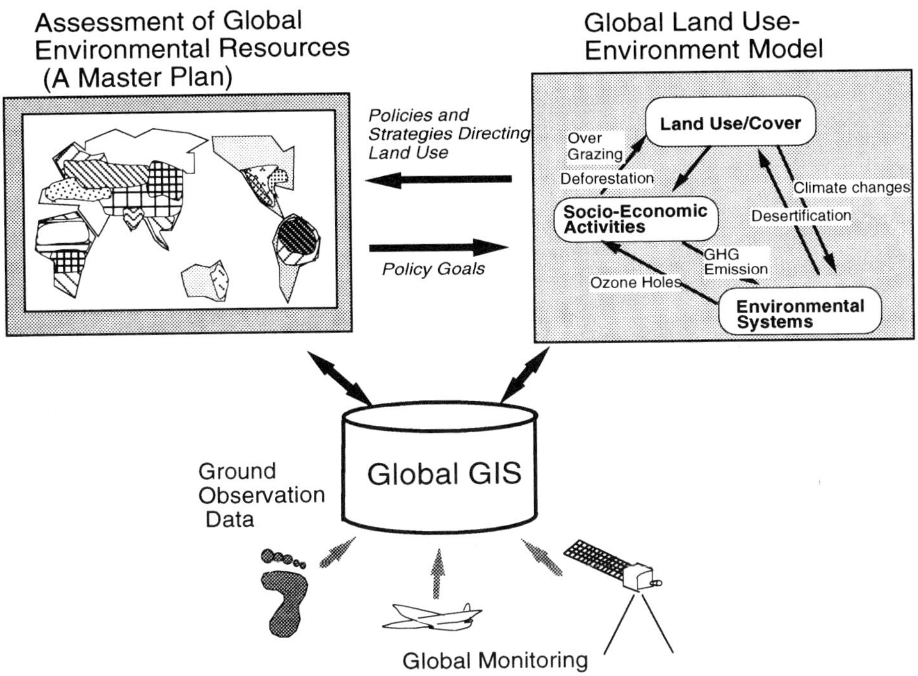

Fig.1 Major Components in Global Land Use Planning

global environmental databases. Three major components are human activities(socio-economic systems), land use/land cover and environmental systems. Development of global models of climate systems are already being attempted while development of other more foot-tight environmental systems such as vegetation ecosystems seems to relatively insufficient. Development of dynamic global models of land use/land cover linked with socio-economic systems and terrestrial ecosystems.have recently started by the initiative of IGBP.

3) Assessment of global environmental resources

Global environmental resources have such a large geographic variety. that creates a demand to assess global environmental resources with respect to reproduction rate, vulnerability to global changes and so forth. Based on the assessment, we can estimate the carrying capacity of the Earth quantitatively and set up the target of global land use planning under several assumption and estimates on future population, technological development, GHG discharge from socio-economic activities and so forth. Policies directing land use activities towards the goal of global land use plans will be tested using global landuse - environmental models. with respect to cost and effectiveness (benefits) among generations.

3. A Case Study of Global Planning based on Land Suitability Analysis

Under the framework of global land use planning, a case study for estimating the carrying capacity of the Earth based on land suitability assessment at the global scale. Among global environmental resources, forest and agricultural area are major components in terms of area. Forest is very important not only for conservation of natural ecosystems and hydrological circulation but also for sequestering carbon to mitigate global warming, while agricultural lands are essential for food production for exploding global population. It is often the case, however, that lands suitable for agricultural food production are also suitable for reforestation. Agriculture and reforestation are conflicting with each other for limited resource base.In a case study, focusing carbon sequestration and storage by reforestation and forest conservation and food production by agriculture, land suitability assessment are conducted to identify the trade-off relations quantitatively for estimating carrying capacity of the Earth.

3.1 Land Suitability Assessment for Forest Conservation and Reforestation

Land suitability analysis is conducted to identify areas for forest conservation and reforestation with respect to carbon storage. A procedure of evaluating suitability for forest conservation/reforestation and global data used is summarized in **Fig.2**.

Suitability for forest conservation and reforestation is evaluated with respect to the followings:

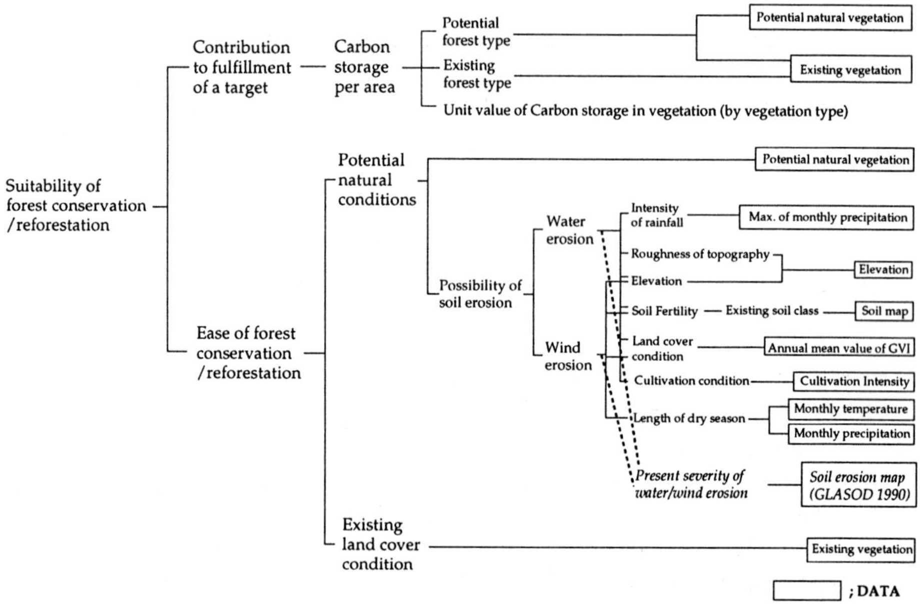

Fig.2 Procedure of Suitability Evaluation for Forest Conservation/Reforestation

1) contribution to carbon storage by forest biomass, and
2) ease of forest conservation / reforestation.

Contribution to carbon dioxide fixation can be evaluated as the amount of carbon stored in forest which are temporally averaged over the period to 2050. For conserved forest, amount of carbon stock is assumed to remain constant under the assumption that biomass of matured forest remain constant. For reforested area, the amount of carbon of biomass is assumed to increases linearly to 2050.

The concept of "ease" corresponds with the level of cost of conservation and reforestation. Sustainability of planted forest affects the level of ease through the cost of forest conservation and reforestation. Level of ease of forest conservation and reforestation are affected by several factors, such as natural conditions, utilization levels of forest resources and level of silviculture technologies. Natural conditions mainly comprise climate and soil conditions. Utilization levels of forest resource may affect cost and difficulties in conserving forest and promoting reforestation. Availability of seeds and young plants, skilled labors and local technologies are included in the technological conditions. In this paper, however, only natural conditions are evaluated due to the lack of global data for evaluating the other aspects.

In evaluating natural conditions, potential vegetation and the possibility of soil erosion are major components. Potential vegetation data are from (Box, 1994). Areas where potential vegetation is forest are given higher scores for their potential for forest conservation and reforestation. The possibility of soil erosion is determined using intensity of rainfall, roughness of terrain surfaces, length of dry season, cultivation intensity (see **Table 1**) and so forth.

Important factors are identified and relative weights are determined in the evaluation process based on the statistical analysis on spatial patterns of severity of water/wind erosion reported by global data set on soil erosions ("GLASOD"; UNEP/GRID, 1990).

Fig.3 a) shows the result of the regression analysis on water erosion factors. Almost all items have reasonable tendency and statistical significance. Concerning cultivation intensity, the item with the widest range, non-cultivated land corresponds with the lowest severity and more intensive cultivation causes higher severity. GVI (Global Vegetation Index), the item with the second widest range, because lower GVI value which means low density of vegetation, lower value corresponds with higher severity and vice versa.

Fig.3 b) shows the result of the regression analysis on wind erosion factors. Concerning the length of dry season which has the widest range, longer length causes higher severity. Lower value of GVI corresponds with higher severity and vice versa.

The possibility of water/wind erosion (**Fig.4**) is estimated by substituting the lowest GVI value and the highest cultivation intensity value to the formula for water/wind erosion estimation, because we assume deforestation and/or agricultural development may cause the lowest GVI value and the highest cultivation intensity value.

Fig.5 shows the results of suitability evaluation for forest conservation and reforestation.

Cultivation Intensity in this study [10min.×10min./pixel]		Olson World Ecosystem Classes Ver. 1.4D [10min.×10min./pixel]	
		Non-cultivated Land	Cultivated Land
Matthews Cultivation Intensity (the areal extent of cultivated land) [1deg.×1deg./pixel]	0-50 %	1	2
	50-100%	1	3

Table 1 Categorizing Method of Cultivation Intensity

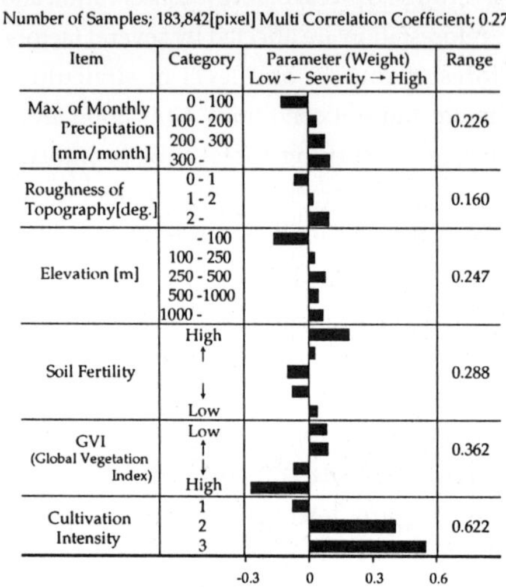

Fig.3 a) The Result of an Analysis on Water Erosion Factors

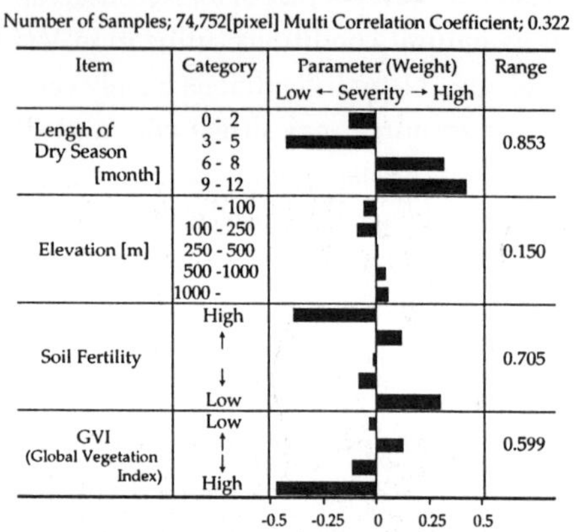

Fig.3 b) The Result of an Analysis on Wind Erosion Factors

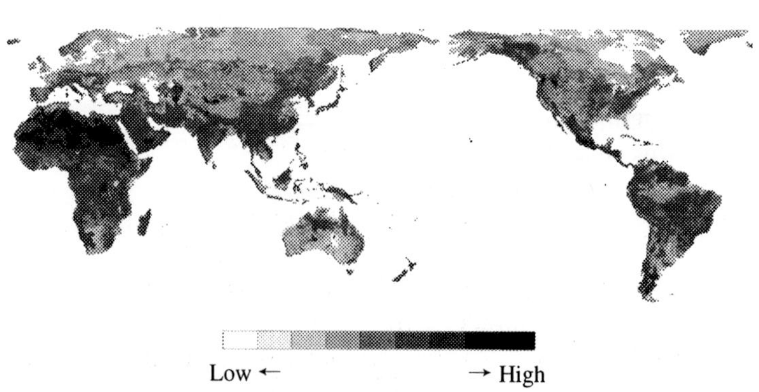

Fig.4 A Global Map of Estimated Soil Erosion Possibility

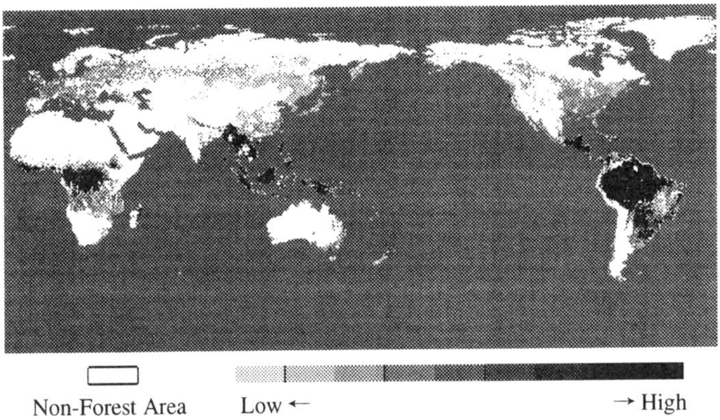

Fig.5 a) A Global Map of Land Suitability of Forest Conservation

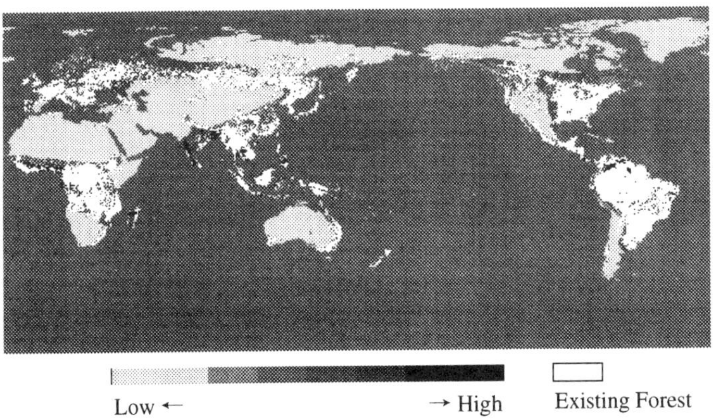

Fig.5 b) A Global Map of Land Suitability of Reforestation

$$\text{Crop productivity} = e^{5.031} \cdot \text{GVI} \cdot \text{Prep}^{0.409} \cdot \text{Fert}^{0.063} \cdot \text{GNPpc}^{0.554}$$

t value (8.6) (1.2) (14.4)

Correlation coef. $=0.82$ (Number of samples $=128$)

Crop productivity: Unit: calorie/ha/year
Prep: Annual average precipitation (mm/year)
Fert.: Rank of fertilizer application per unit area based on a table below
GNPpc: Rank of GNP per capita based on a table below

Rank value	1	2	3	4	5
Fert.	<90kg/ha	<200kg.ha	<300kg/ha	<400kg/ha	>=400kg/ha
GNPpc	<$2500	<$5000	<$7500	<$10,000	>=$10,000

3.3 Land Suitability Analysis for Agricultural Food Production

Land suitability of agricultural production is assessed based on potential agricultural productivity in terms of energy, and "soil stability". Soil stability, reflecting major part of sustainability of agricultural production, is a evaluated based on possibility of soil erosion and soil fertility in the same manner as in the analysis for forest conservation/reforestation.

Potential agricultural productivity is estimated by applying a GVI-based model (Murai et.al,1993) for estimating actual agricultural productivity.

Fig.6 shows actual agricultural food productivity estimated with the above formula and a GVI-based NPP(Net Primary Productivity) model for live stock productivity in grass lands.

Under the assumption that GVI value of newly developed agricultural area equal to a country-based average value of GVI, country-based GVI is substituted to estimate potential productivity of newly developed agricultural areas. Fertilizer application and GNP per capita have a common value for each country by assuming almost the same level of agricultural inputs in a country. Spatial variations in a country solely depends on spatial patterns of precipitation.

Potential productivity of grazing land is estimated by converting net primary productivity estimated from GVI to live stock yield (Murai et.al.1993). Areas which can be used for crop production are allocated to crop production, because crop production has higher productivity than grazing in terms of energy production. **Fig.7** shows a global map of agricultural suitability.

3.4 Identifying Trade-off Relation between Carbon Storage Capacity in Bio-mass and Agricultural Food Production

By allocating each land unit to use with the highest suitability, "better" land use patterns can

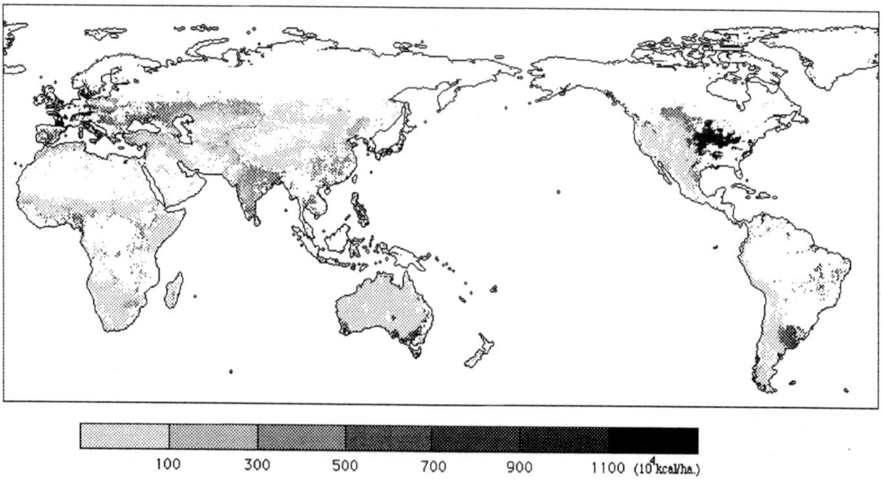

Fig.6 A Global Map of Agricultural Food Productivity

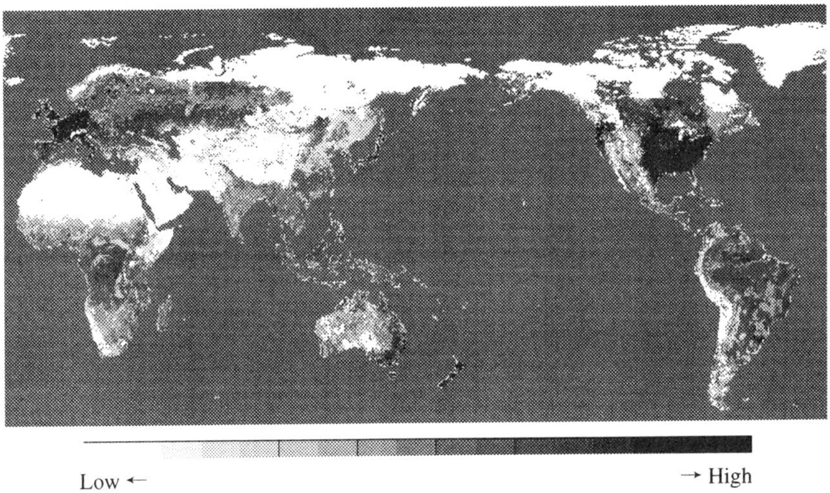

Fig.7 A Global Map of Land Suitability of Agricultural Food Production

be estimated in terms of total potential agricultural production and carbon storage in forest on the sustainable basis. In order to compare the level of suitability for carbon storage and food productivity, it is necessary to value carbon in biomass against food from agricultural production. If the relative value of carbon become higher, more lands will be allocated to forest and vice versa.

By computing sets of total potential food production and carbon storage in forest biomass for each relative value of carbon over food, trade-off relations due to limited resource base of the Earth can be plotted. **Fig.8** shows an estimated trade-off relation curve.

The trade-off curve can be called a "frontier curve", because, on the curve, every land unit with any potential of agricultural food production and/or carbon storage is allocated to agriculture land or forest. We cannot go beyond the frontier curve.

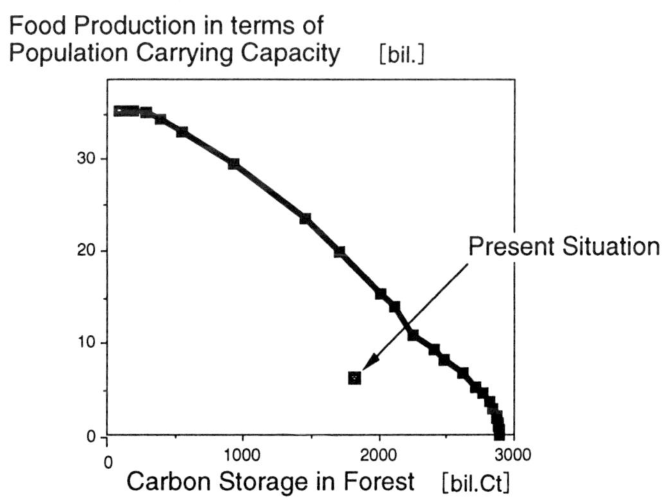

Fig.8 A Trade-off Relation between Agricultural Food Production and Carbon Storage in Forest

As relative value of

food from agricultural production over carbon stock in forest increases, a point corresponding a land use pattern moves along the frontier curve from right to left.

In land use allocation, however, there is no minimum allowable level of "ease" of forest conservation/reforestation or "sustainability" of agricultural production. By setting a threshold value of "ease" or "sustainability", a more "sustainable" frontier curve can be drawn inside of the frontier curve in **Fig.8**.

4. Conclusions and Future Prospects

This paper proposes a conceptual framework of global planning for sustainable use of the Earth. And a case study based on land suitability assessment are conducted to estimate a frontier curve showing the limited resource base of the Earth. To support the global planning process ranging from global data acquisition to evaluation of impacts of a proposed landuse policies, a computer environment integrating remote sensing and global GIS is indispensable.For example, land suitability assessment as shown in this study is based on trial-and-error process requiring a wide variety of knowledge in many research fields and quality global datasets with different levels of spatial and temporal resolution and reliability. Global GIS plays a core role in supporting such analysis and planning process as well as in managing global datasets.

5. References

1) Abe,I., Shibasaki,R. et al. (1992): A Global Map for Forest Conservation, Proc. of 13th Asian Conference on Remote Sensing.

2) Box,E.O. (1994): Global Potential Natural Vegetation: Dynamic Benchmark in the Era of Disruption, Preprints for the Eighth TOYOTA Conference, pp.6-1−6-19.

3) Endo,T., Shibasaki,R. (1994): An Analysis on Soil Degradation Factors at the Global Scale, Proc. of JSPRS Spring Convention, pp.155-158 (in Japanese).

4) Burrough (1986): Principles of Geographical Information System for Land Resources Assessment, Claredon Press, London.

5) Goto,S. (1991): A Study on Forecast of Global Changes in Biosphere and Estimation of Human Habitability of the Earth using Remote Sensing Data, Doctoral Dissertation, University of Tokyo (in Japanese).

6) Murai,S., Honda,Y. (1992): Global Ecoclimate Mapping and Assessment of Human Activities, Asian-Pacific Remote Sensing Journal, vol.4, no.2, pp.39-46.

7) Murai S., Honda,Y. (1991): Global Change Monitoring using the NOAA Global Vegetation Index and Geo-information, Asian-Pacific Remote Sensing Journal, vol.4, no.1, pp.63-73.

8) UNEP/GRID (1991): GLASOD: A Users Guide to the Global Digital Database, Document file in GLASOD.

9) UNEP (1992): World Atlas of Desertification , Edward Arnold.

10) Murai,S. et.al. (1993): Integrating Remote Sensing and GIS for Global Planning of Sustainable Use of the Earth, Proc. of Beijing GIS'93.

A Global Map of Land Suitability of Reforestation

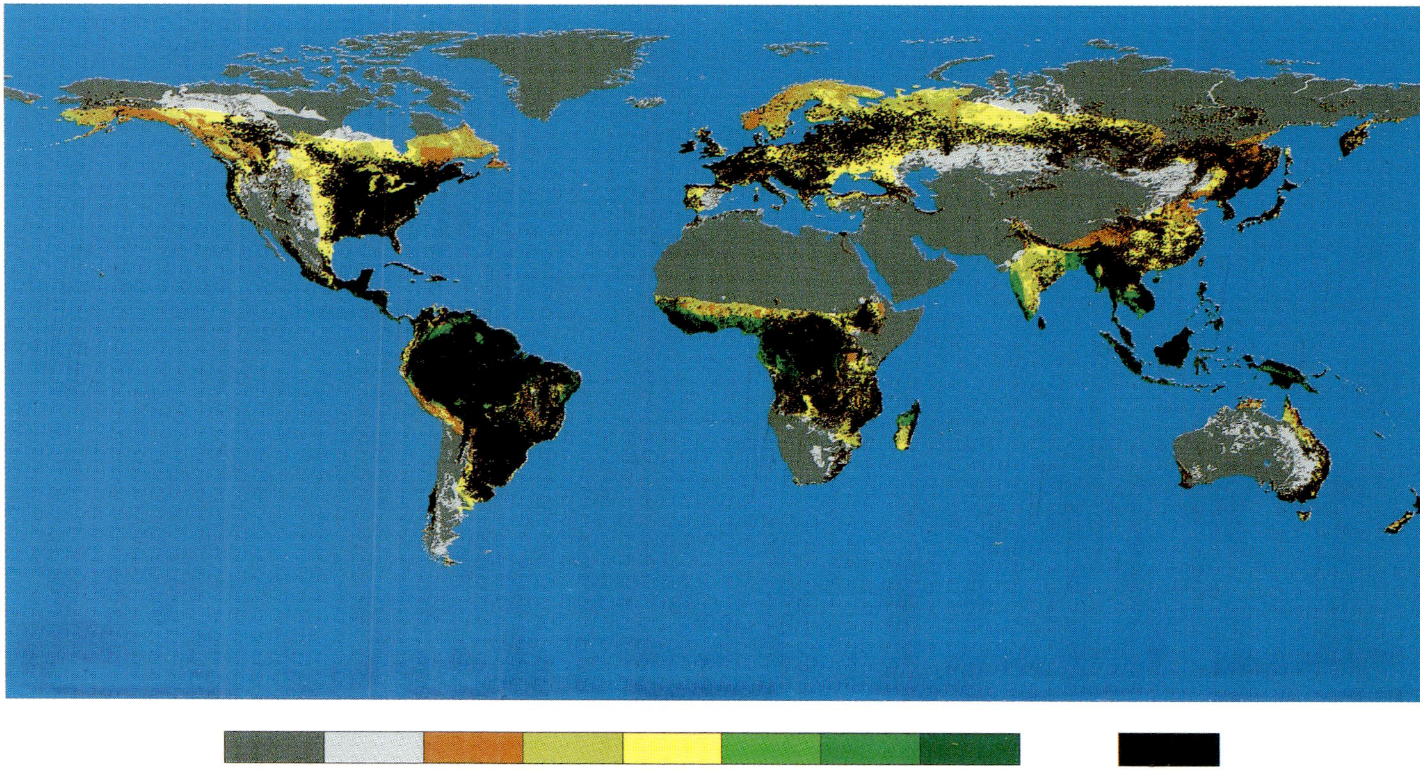

A Global Map of Land Suitability of Forest Conservation

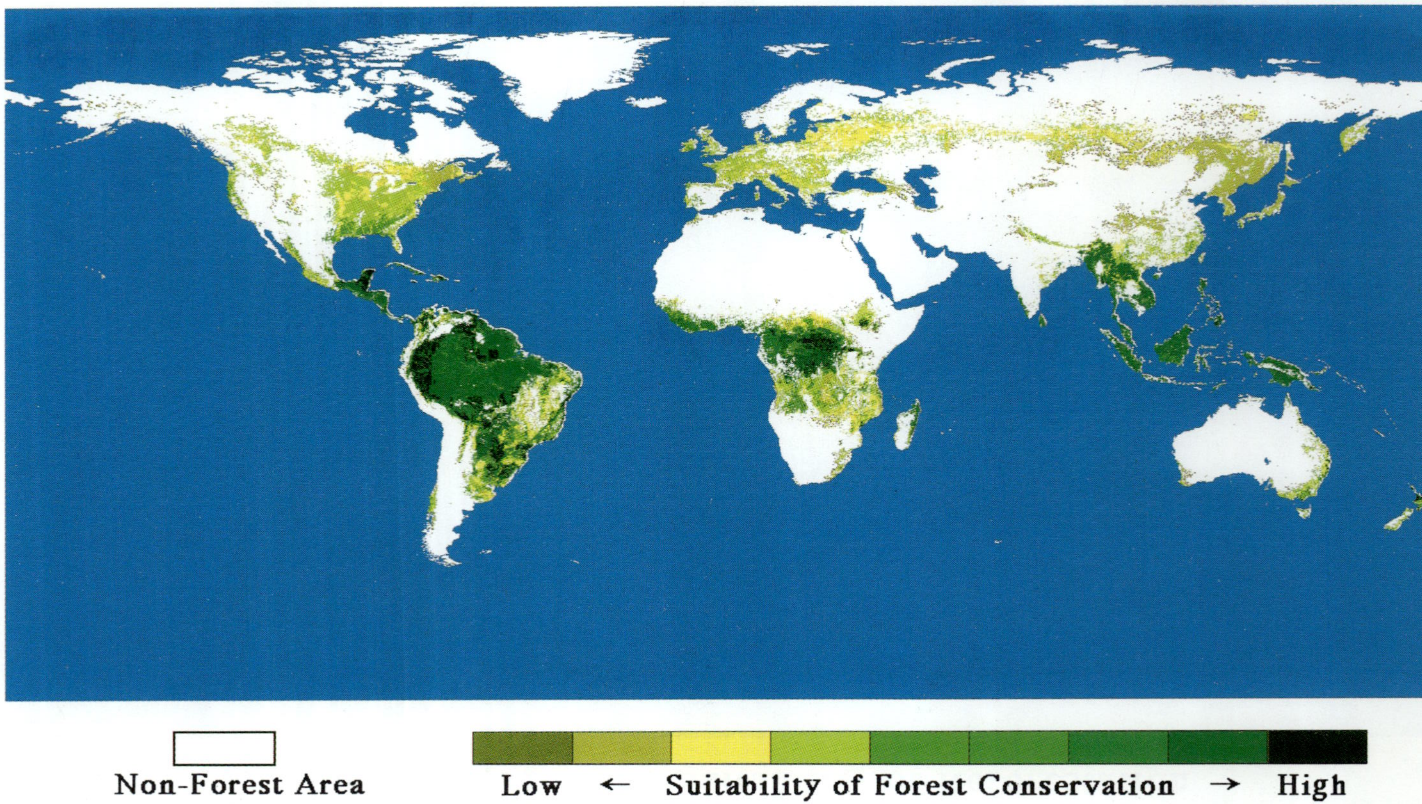

TOWARD GLOBAL PLANNING OF SUSTAINABLE USE OF THE EARTH
Development of Global Eco-Engineering
S. Murai (Editor)

Environmental Ethics Might Change the Ego-society to Eco-sound-society

Naito Masaaki
National Institute for Environmental Studies
Japan Environment Agency
16-2 Onogawa, Tsukuba, Ibaraki 305 Japan

ABSTRACT

The latest prediction of the global environment tell us that the situation is highly critical. To cope with the situation, drastic change of the social paradign must be inevitable.

The paradigm shift reguires us to sacrifice our material wealth to a large extent. Therefore, environmental ethics is only a motive force to move the society.

"Sustainability", "independence", and "co-existence" will be three essential norms evolved from the environmental ethics. These norms will alter the present ego-society to eco-sound society.

It is not easy to figure out this eco-sound society, yet in which resources are almost totally cycled and energy basically relies on natural sources. All the industrial, urban and institutional structure will be largely different from the existing one.

1. INTRODUCTION

"Why we must preserve the environment or nature" is a crucial yet difficult question continuously raised in the human history. An argument about this question has again emerged today in company with the fear to the crisis of the global environment. Thus several reviews of past discussion on "the rights of nature" or "eco- ethics" have recently appeared mainly in the U.S. and also in Japan. An outstanding monograph among others is by Bryan G. Norton who made extensive review of the conventional studies on rationale to preserve natural variety and did organization of them. Another one is an excellent monograph by Roderick F. Nash who made thorough review of the arguments on the environmental ethics back to the fifth century. Similar book but focused just on the modern American environmental movements is summarized by Riley E. Dunlap and Angela G. Mertig.

In Japan quite a number of publications have also appeared on the thought or phylosophical aspects of environmental issues just recently. Y. Shizume has written about a little broader topics "Toward a Science on the Relation of Mind, Material, and Gods". Y. Fujiwara's monograph entitled "Structure of Natural View and Environmental Ethics" is also outstanding one regarding historical view of nature. "Ecology and Christianity" which is a collected dialogues between well-known environmentalists and theologians in Japan is also interesting. One of the environmentalists, J. Takagi, has written a book "How We View Nature To day". He developed an introductory yet unique argument on view of nature in it.

2. Background Situation of the World Problems

Now the world is facing a number of serious problems such as population explosion, starvation, south-north problem besides environmental detrioration. They all closely interact and acceralate each other, so that they are often called a visious cycle composed of three "P" of "population", "poverty" and "pollution" as in Fig-1.

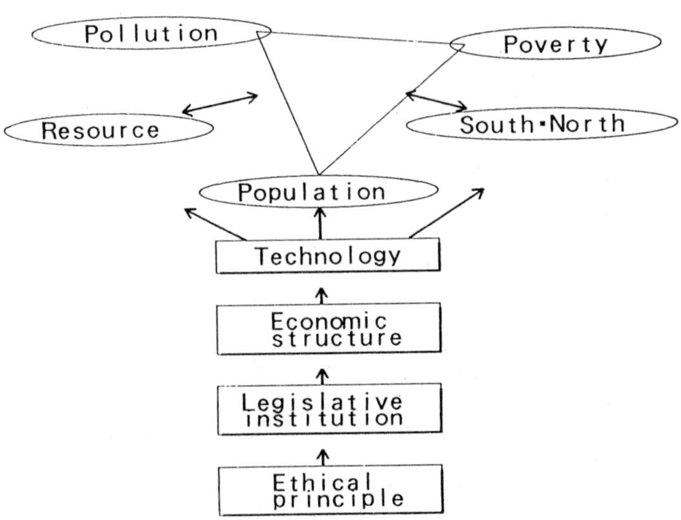

Fig 1 Vicious Cycle of the World Problems

Refering to the structure of the problems seen in the figure, the framework of the discussion should encompass such wide area of disciplines as phylosophy, socio-economy, technology as a whole. This is shown in Fig-2 whose details will be discussed in the following chapters.

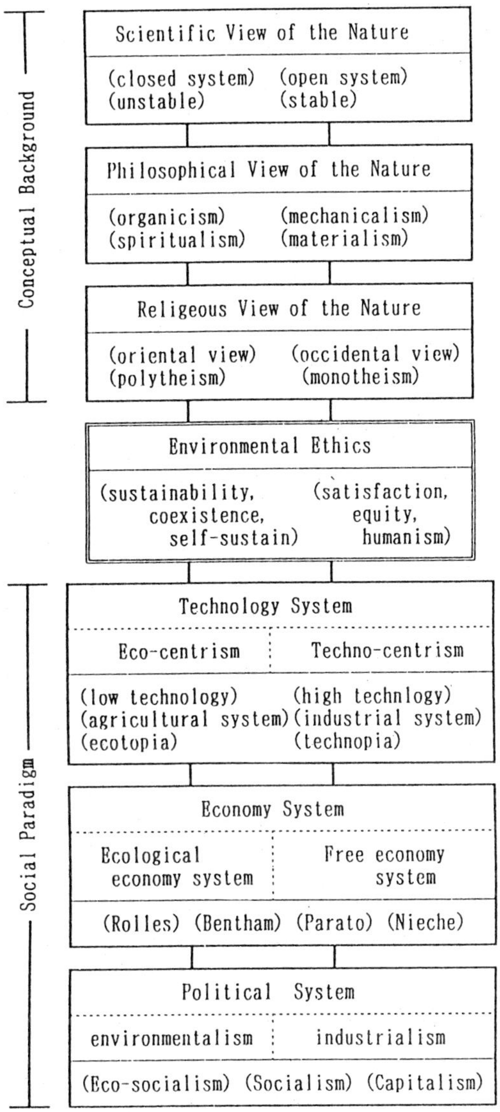

Fig. 2 Various aspects of inter- related issues relevant to social paradigm

Among the above world problems, the global environmental crisis is the one to be focused in this work. To evolve a resolution, first needed is to understand the cause of the problem. As shown in Fig-3, there are major three causes. They are so expressed as the "technological debt", "Socio-economic debt", and "mankinds debt" which are mainly transfered to "future generations", "developing countries", and "non-human creatures", respectively.

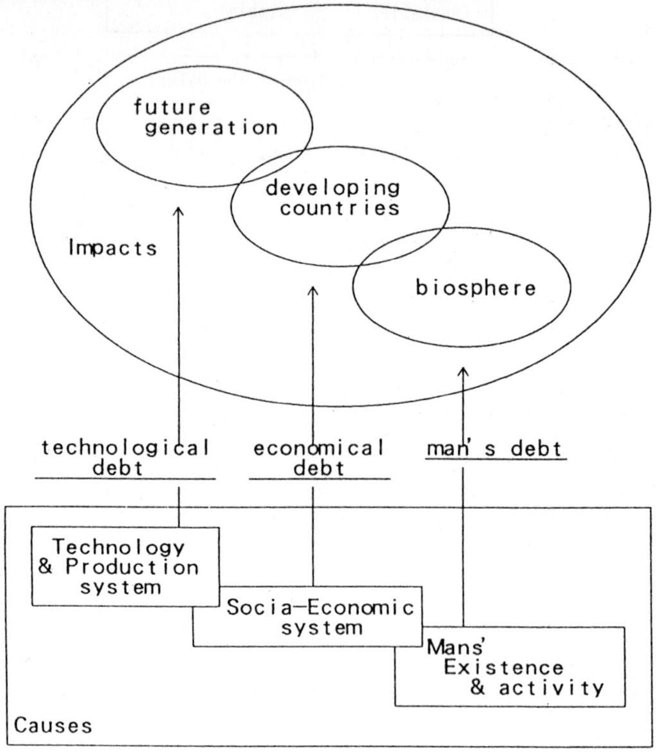

Fig.3 Causes and Results of the World Problems

3. Pursuing a New Ethical Principle

3.1 Concept embraced in the new "Environmental Basic Law" of Japan

In November 1993, the new environmental basic law was enacted to respond the current difficult and manifold environmental problems. The new law embraces a new concept which is quite different from the conventional one. This is clealy expressed by the several essential keywords appeared in the law. They could be categorized into three couples of op posing notion as summarized in Table-1. The background idea behind these three couples of keywords can be interpreted as follows.

Table-1 Three couples of opposing notion involved in the basic law

	(A)		(B)	Interpretation
①	Welfare of the Nation	vs.	Welfare of the Mankind (Article-1)	→ Not transfer the debt of Domestic Activities to Developing countries
②	Enjoy Blessings of the Envir.	vs.	Inherit of the Blessing (Article-3, 4)	→ Present generation to Future Generation
③	Contact to the Nature	vs.	Conservation of Bio-diversity & Species (Article-14・2)	→ Human to Nonhuman creatures

The opposing three couples imply that we are now forced to pursue hard and essentially contradicting objectives. The group A designates the objectives which we have ever pursued for "ourselves". Whereas the group B denotes the objectives we have to attain for "others" through carefully avoiding the side-effects caused by pursuing A.

The side-effects are all directed toward "others" such as,

① economically weak peoples
② future generations
③ non-human creatures

who are all weak existences in this world. The intrinsic feature of the global environmental problems is nothing more than this.

The concept above is illustrated in Fig-4. What this means is simply stated that there exists two stages of conflicts. First level conflict lies between "Our Economy" and "Our Environment" and the second level one does between "Our Economy and Environment" and "Survive and Welfare of Future Generation, Developing Nations, and Other Creatures or Nature". Now our hard task is how to manage this delicate double stage balance composed of three sectors.

To do this, needed are the idea on

① how to weigh each sector.
② how to find strategies toward eco-sound society with assigned weight and assigned level of welfare of each sector.

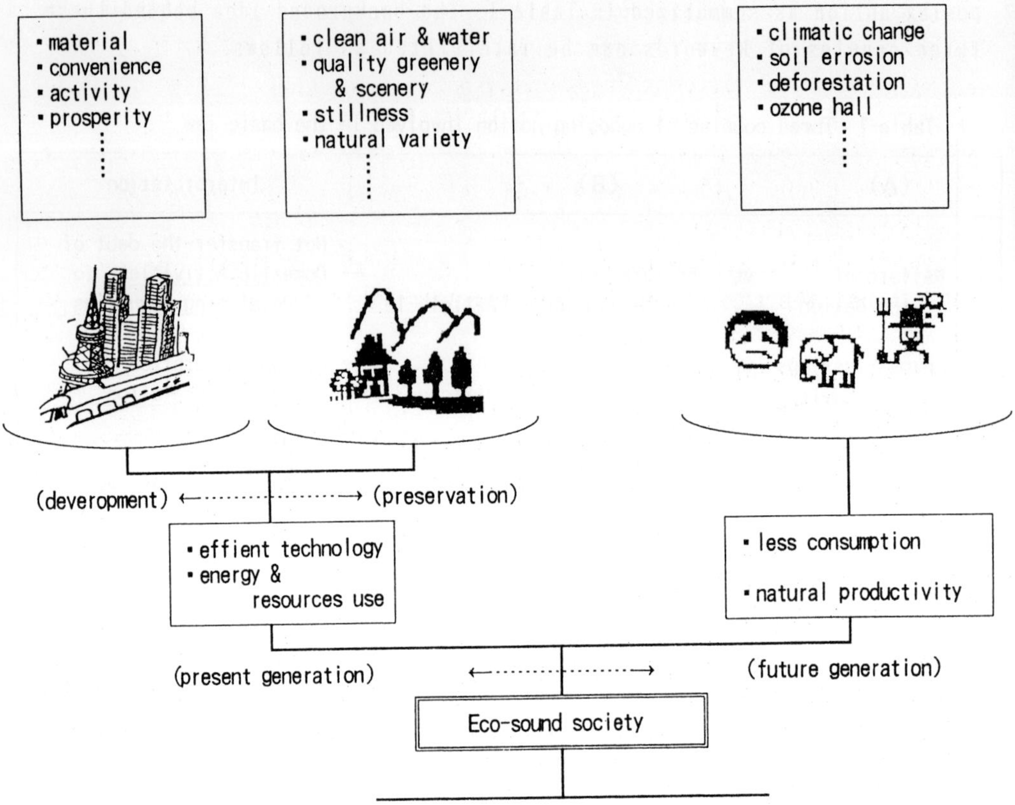

Fig. 4 Touchy Balance of Three Objects

3.2 Where to find reasons to preserve the "others"

It is not easy to find the reason why we have to preserve the global environment or "others", that is, nature, future generation, other peoples. To find this reviewing the history of the environmental administration in Japan will be of great use. The first phase of the problems was serious pollution episode. There was no discussion in this phase to protect human life or health. The well known "non-economic approach" was adopted at that time in which economy was almost disregarded in order to control the environmental pollution.

At the second phase when the serious situation improved, it had become difficult to find out a ground to preserve the environment. Effort has been made to derive a ground. One is "economic valuation" of various functions of the environment, such as recreational use, cultivation of water resources or mitigation of climatic variation. An approach usually employed for quantitative evaluation is to measure "willingness to pay" of individuals by means of "hedmic pricing" or "contingent valuation" methods.

The fault of economic valuation is that proper evaluation of the function of the environment is difficult and also the economic value is usually not high enough to compensate fruits brought about by developing or economic use of the environment.

Another way of valuating the environment is made based on "esthetic" aspect like historical or cultual value, which are uncountable in monetory term. Such sort of value is not too much appreciated in comparison with economic development particularly in Japan. Even the precious historical and sinic beauty of Ohara in Kyoto, for ex ample, is now in a critical situation by a resort development.

The last and final ground to assure preserving the environment is "Ethics". This is of course almost nothing or little to do with "utility" of the environment for the present generation. Yet without this ethical view, there are no reasons to preserve the global environment so far as it does not benefit our wealth but does sacrifice it more or less.

Summarizing the above description, let me conclude that, as the ground or reason to protect the environment, "E" principles (Economy, Esthetic, Ethics) are observed along with the historical change of the Japanese environmental administration.

More regorous argument about such subject is seen in numerous ever published monographs, such as by Bryan G. Norton. Refering to such deep phylosophical review, valuation in terms of "Economy" and a part of "Esthetic" correspond to anthropocentrism in which the environment is ap preciated as "instrumental value" for the utility of the present generation. This valuation could be extended to the future generation to some extent as "transformatine value". "Ethics" is deduced from non-anthropogenic stand point in which "intrinsic value" of the environment is appreciated. This is eventually comes into Leopold's "land ethics" or Gaianism that claims "species exist as ends in themselves". "Shallow ecology" appreciates "instrumental value" of the environment and "deep ecology" does "instrinsic value".

An interpretation of environmental ethics is depicted in Fig-5. In the figue, herizontal exis implies temporal herizon from the past (ancestor) to the future (future generation) and vertical axis indicates contempoary existans. "Me" lies on the cross of the two axes and the distance from the center implies the degree of inverse of intimacy from me.

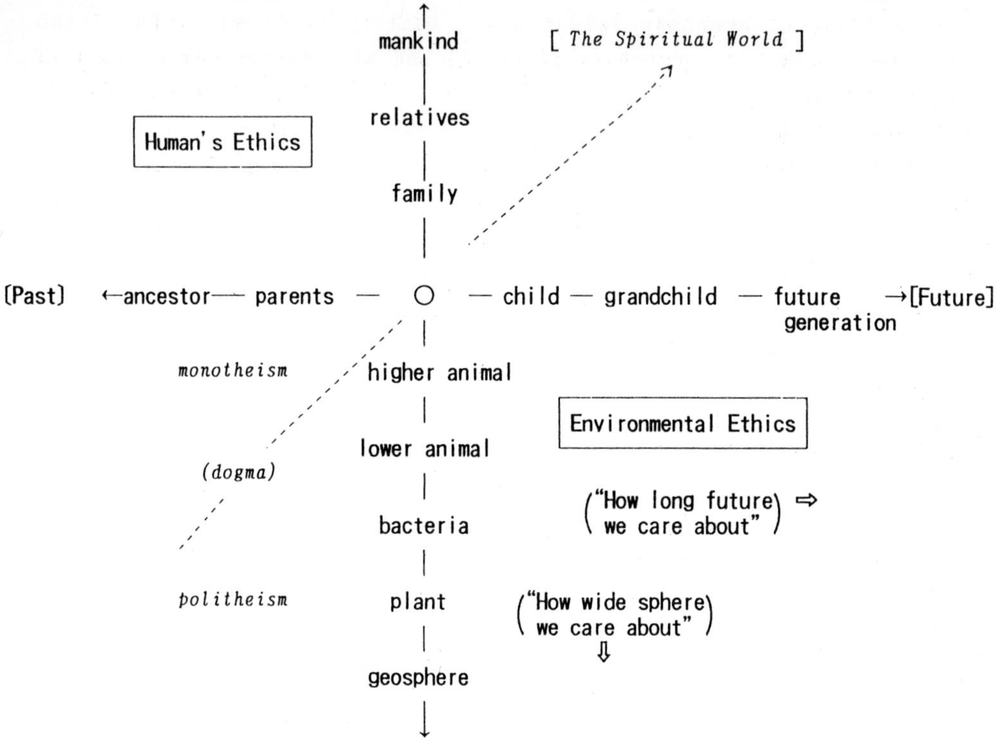

Fig. 5 The Objects Where Environmental Ethics Targets

Environmental ethics, contrasting with conventional "(Human) Ethics", might be identified that how long future and how wide sphere "me" cares about. The argument on this domain is called "-ism", whereas "dogma" does not to this real world but does to the spritual world. But somewhat related to the environmental ethics is in a thought of nature on monotheism versus politheism.

The author's profession is far from such phylosophical argument. The purpose of the above description is, therefore, not to propose any of new ethical view or claim but to roughly review previous theories.

So that new guiding principle why and practical method how to attain "Eco-sound Society" may hopefully be derived lead.

4. How to Achieve the Eco-sound Society

4.1 How technology and industrial development can do

There is a strong assertion that technological development can overcome the current world problems. On the contrary, another sect of people insists that technology and the modern industrial society itself has caused such critical problems, not merely as a side-effect from misusing technology. The former is called "techno-centrism" and the latter is "eco-centrism". It is almost impossible to give definite judgement of which is true. Therefore, they are called "-ism".

One evidence at this moment is derived from the data recently announced by Japan Environment Agency. The data implies that the CO_2 control plan of Japanese Government based on the agreement at Rio-Summit already be broken down, because of economic growth and the limited effect of technological development. As a matter, it is well known fact that the target of Rio-Summit Agenda is far from sufficient to curtail increase in CO_2 concentration in the global atmosphere. This information may at least reveals that no matter how the current possible technological progress may occur, it can hardly improve the situation unless some drastic breakthrough takes place such as realization of"nuclear fussion".

4.2 Way to the eco-sound society

So far as the technological resolution is very difficult, another way is to shift the paradigm of existing society. The goal toward eco-sound society is figured out in Fig-6. The essence of the chart means that three kinds of debts generated from the modern society should be solved by virtue of remodeling "technological system", "socio-economic system", and "lifestyle" based on the new social norm of "sustain", "independence" and "co-existence", respectively. The crucial issue herewith is to identify how remodel the paradigm. It is often termed as "eco" or "environmentally friendly" in a word. Yet, the detail is not clearly defined but is of very confuse.

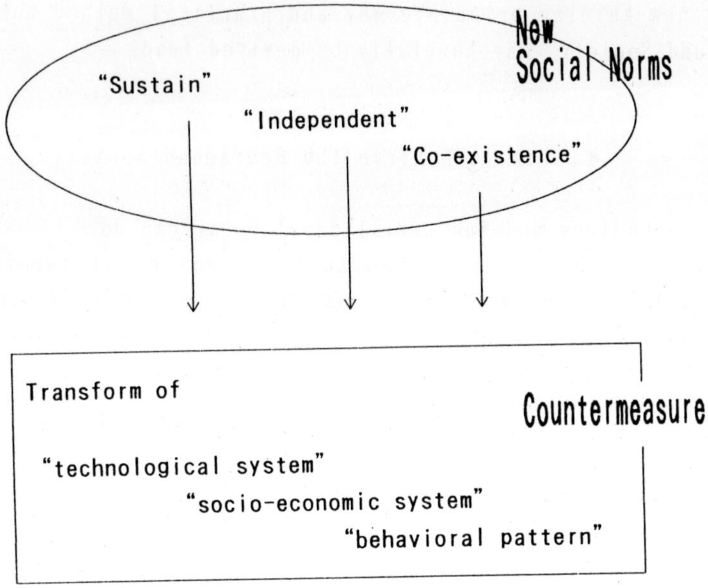

Fig. 6 Transformation of the Social Structure Based on New Social Norms

Here proposed is a society which consists of total material cycle and natural energy. The essential portion of the society is shown in Fig-7. Key issues lie in how to compose total network and how to restruct the proper ratio of primary, secondary, and tirtial industries. The current ratio is too unbalanced in Japan. Then, recent tendency of industrial restructuring is desirable, such as big motor company own and manage farm land. There must be a number of hardles to overcome until realizing the final figure. Among the numerous difficulties, the shift of "heartware", that is, moral or ethical aspect is vital rather than hardware (say, developing energy system and material cycle) and software (say, remodelling economic system, changing GNP). But unless otherwise, human being will have little possibility to survive or at least to maintain civilization. After all, achievement of eco-sound society is entirely depending upon how much we can control our "ego" for the sake of other people and creatures.

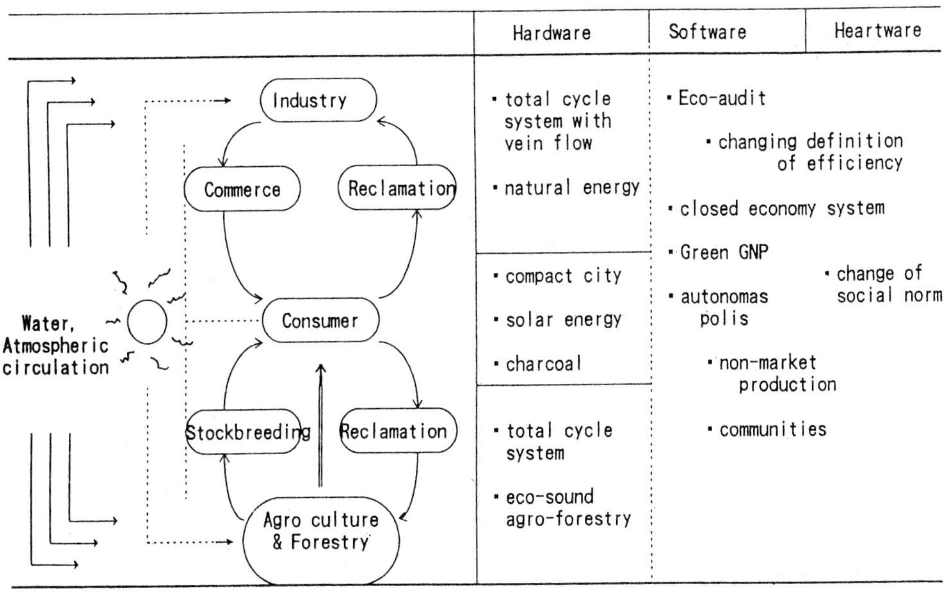

Fig. 7 A Concept of New Society with Eco-sound Metabolism

4.3 A real sample of the eco-sound society

In recent years, a number of samples seem to be similar to the here proposed eco-sound society are realizing though they are still quite small in scale.

The biggest one among them is yet the size of 5,000 population who costitute 50 communities in Japan and 5 in oversea countries. Their production is based on agro-forestry, stockbreedings, and food processing. One of the major features of their system is that the whole production sectors are totally linked so that materials are properly cycled within the sectors as in the natural ecological system.

Another far more important aspect of the society lies in their behavioral or moral principle whose central norm is set in achieving "my happiness" through realization of other's welfare and happiness. This sounds like "eutopian dreams" which have often appeared in the history whenever some barrier emerges ahead of mankind. No one can defintely tell whether the current examples are different from the conventional ones or not. So far as the above introduced sample is concerned, based on the survey the author have made last couple years, the community seems to be largely different from the conventional examples because it is carefully lerning from the past failures. The present "eutopian dream" could save the catastrophic destruction of the global environment from the "Americal dream" which symbolizes material life based on enomous consumption of the environment. This must be the last chance. Promissing future will be expected only when this sort of ethical community collaborates with TOYOTA.

TOWARD GLOBAL PLANNING OF SUSTAINABLE USE OF THE EARTH
Development of Global Eco-Engineering
S. Murai (Editor)

The role of traditional management systems and local knowledge in planning environmental and natural resources management in the global age

Kenneth Ruddle

International Resources Management Institute, c/o Matsugaoka-cho 11-20, Nishinomiya, Hyogo 662, Japan

Abstract

Familiar financial, human and technological resources available to help solve global environmental and natural resource allocation problems should be complemented by "unconventional" human resources, to inform the design and implementation of natural resource and environmental management systems for sustainable development. Principal among them are traditional resource management systems and their associated local (traditional) knowledge bases.

Traditional, community-based resource management systems survive because they enable communities to meet basic human needs, spread the risks inherent in resource uncertainty and scarcity, institutionalize area management, access limitations, and technology and allocation regulations, and are adapted temporally to fluctuating resource availability, thereby contributing to local community stability.

Local knowledge sustains the operation of traditional management systems. The sets of rules that comprise a management system derive directly from local knowledge and concepts of the resources on which the system is based. Thus local knowledge of the environment and resources used, as well as of the society within which the resultant goods and benefits are distributed, is fundamental to the continuity of sound community-based management practises and to the design of new systems of sustainable resource management.

The practical usefulness of traditional management systems and associated bodies of local knowledge is examined here with reference to tropical coastal-marine environments and fisheries. Such systems are a practical antithesis to conventional models, which assume that fishing strategies evolve as a consequence of competition. As such they embody fundamental concepts relevant to fisheries management worldwide.

1. INTRODUCTION

The complex inter-connected problem of overpopulation, inappropriate development and environmental destruction, and the impoverishment of both global environment and society that this causes, is well-recognized. What is less widely recognized, but equally obvious if we pause to consider it, is that the problem is of such a vast scale that existing financial, human and technical resources are inadequate to tackle it, if *conventional* means are used and *conventional* solutions sought.

The present reality is that given the various enormous and competing demands on conventional resources, environments and societies in the poorer countries of Africa, Asia, Latin America, and the Pacific Basin will be the real losers. Some of the world's richest nations are already switching resources from those regions to assist Eastern Europe and the former USSR. In other words, geopolitical and strategic realities now rank higher than humanitarianism in decisions to allocate scarce and shrinking development resources.

This implies that if the poorest and most threatened areas environmentally and socially in the Third World are to be assisted, *non-conventional* resources must be utilized as either a complement to or replacement for conventional ones. In other words, much development effort in such places will have to be based on transforming existing systems of natural resource use and management, and building on what human resources already exist to create and manage appropriate environmental technologies for sustainable development [1]. This is reinforced by a growing awareness that "scientific" solutions to problems of renewable natural resources management do not necessarily contribute to their sustainable development and continued use and management [2].

Two of the principal non-conventional human resources in poorer countries are traditional ways of using and managing natural resources and environments in a sustainable manner - most societies do or did have traditional environmental technologies and community-based management systems to regulate them - and local (traditional) knowledge that underpins any management system. Such traditional management systems and their associated knowledge bases are two inseparable, non-conventional resources that must be exploited if the environmental crisis is to be halted, or at least mitigated, and poor societies given a chance to launch themselves on a path toward sustainable development.

In this contribution I demonstrate the importance and potential role of traditional management systems and local knowledge, using examples from tropical coastal-marine environments and coastal multi-species fisheries, and particularly from the Asia-Pacific Region, since most of my more recent work has focused on those topics in that area [3].

2. TRADITIONAL FISHERIES MANAGEMENT SYSTEMS

Numerous reviews of contemporary, "scientific" fisheries management have documented serious, inherent weaknesses. Most occur in the fisheries of developed countries [4]. Nevertheless the same biological and economic models responsible for those problems are still generally recommended for fisheries development and management in Third World contexts. This is ironical, "... first because in many Third World societies there already exist sophisticated fisheries management systems well-adapted for local use; second, because similar systems also exist (although often extra-legally) in many parts of the First World; and third, because many such Third World systems might be readily adaptable for managing fisheries in the First World" [5].

Traditional systems of fishing rights and tenurial relationships of small-scale fishers to resources and resource areas have been documented throughout the world. Many traditional management systems - often common property regimes to which access to a particular territory was limited to a defined user group, in which operational rules were specified and where control resided in traditional local authorities - were either deliberately or inadvertently weakened or destroyed by colonial administrations, and replaced by centralized fisheries institutions nominally responsible for all aspects of fisheries management, from policy formulation through enforcement [6].

Traditional systems were not usually replaced by more effective central or regional management institutions. The cumulative effect of restructuring fisheries management has been the impoverishment of both tropical fisheries resources and the fishing communities dependent on them. Compared to the scope of the problems facing government agencies attempting to manage inshore fisheries, their administrative and technical abilities are generally relatively weak. As a result, devolution of resource management and allocation decisions to local communities, within the framework of co-management, is increasingly seen as an alternative to ineffective management undertaken by distant, understaffed and underfunded government agencies [7].

Alternative systems of fisheries management, such as traditional, community-based systems, are a practical antithesis to conventional models, and embody fundamental concepts relevant to fisheries management worldwide. They have survived so widely for several reasons. First, they enable communities to meet basic human needs. Second, inclusion rights spread the risks inherent in resource uncertainty and scarcity. Third, because they institutionalize area management, access limitations, and regulations for overcoming gear externalities and allocation problems, and are adapted temporally to fluctuating resource availability, they contribute to local community stability. This is in striking contrast to the assumption that fishing strategies evolve as a consequence of competition. Since the resource(s) is managed and controlled by the community, the regime locally devised and upheld, and because individuals are taught to comply with it early in their socialization, transaction costs (a major cost under other types of management regime) are low [8].

Common property regimes are often tightly bound within and inseparable from the shared cultural matrix of traditional communities. As such they constitute an element of a lifestyle rather than of just an economic sector or occupation. Compared to their "scientific" counterparts, they are easily, inexpensively and locally enforceable, based on local authority [9].

Unlike conventional fisheries management, traditional community-based systems of marine resources management are focused on resolving *gear externalities* and *allocation problems.* In contrast too, traditional systems are implemented based on *defined geographical areas* and *controlled access,* are self-monitored by the local fishers, and are enforced by local moral and political authority. Gear externalities are of two types, attracting fish away from other gear and direct physical interference with other gear. Allocation problems arise through excessive competition for prime spots, or when there are too few spots to sustain the efforts of the number of fishing units present. This has both social and economic costs. Such problems arise because of the uneven spatial and temporal distribution of fish; all sea areas do not contain productive fishing spots, and not all spots are of equal productivity. These are the great strengths of such systems, and these are what they have to contribute in modern management designs [10].

Both the problems of gear externalities and assignment are overcome in traditional community-based management systems at the first level by (1) control of a fishing area, as a *property,* and (2) defining exactly who has access *rights* to that area. At the second level *rules of operational behavior* then specify assignments of time and place within that space and group having access. The first level is sustained by rights of exclusion, or limited access, that maintain the private area of a community of local fishers against outsiders. The second level, intra-group operational rules, is sustained by local authority that has the power to invoke sanctions on offenders [11].

Thus under well-functioning traditional systems of community-based resource management rent dissipation is overcome since access is controlled. Further, in a great many systems sustainable harvesting practises are enforced, thereby leading to resource conservation.

3. LOCAL KNOWLEDGE SYSTEMS

Local knowledge is an important human resource that guides and sustains the operation of traditional management systems; knowing where, when, and how to fish governs a great many of the fishing decisions made by small-scale fishers. Further, the sets of rules that comprise a fisheries management system derive directly from local knowledge and concepts of the resources on which the fishery is based. Thus local knowledge of the environment and resources used, as well as of the society within which the resultant goods and benefits are distributed, is fundamental to the continuity of sound community-based management practises and to the design of new systems of sustainable resource management.

In coastal societies, bodies of local knowledge are empirically-based and practically-oriented, and some are complex and highly organized. Most such knowledge combines empirical information on fish behavior, marine physical environments and fish habitats, and the interactions among the components of ecosystems into comprehensive and often complex fish taxonomies, directed toward ensuring regularly successful catches, and, often, longterm sustainment of aquatic resources. In some instances, explicitly conservationist objectives are evident [12].

But throughout the world local knowledge systems are changing, sometimes rapidly, owing to the pressures of modernization, urbanization, commercialization, and monetization, and to the elitist values often engendered by non-traditional education. In a great many instances, local knowledge is thus becoming hybridized with extra-local elements. As a reaction, however, and partly related to the resurgence of ethnic pride and indigenous rights movements worldwide, there has been a welcome burgeoning of interest in the subject [13].

3.1. Common Characteristics of Local Coastal-Marine Knowledge

There are important commonalities that characterize corpuses of local knowledge of coastal-marine environments and resources in many widely separated parts of the world [14]. The principal ones are that:

(1) they are based on longterm, empirical, local observation, that is adapted specifically to local conditions, embraces local variation, and is often extremely detailed;

(2) they are practical and behavior-oriented, focusing on important resource types and species;

(3) they are structured, which makes them somewhat compatible with Western biological and ecological concepts through a clear awareness of ecological links and notions of resource conservation. However, local logic is essentially inductive, with conclusions based on many prior observations and hypotheses not always either verified or refuted by systematic observation. Rather, theory is developed by producing new *post hoc* explanations, or by dropping current explanations; and

(4) they are often dynamic systems capable of incorporating an awareness of ecological perturbations and of merging this awareness with an local core of knowledge.

4. COMPONENTS OF LOCAL MARINE KNOWLEDGE

4.1. Fish behavior

The main component of local ecological knowledge of marine resources concerns the behavior of fish and other animals, particularly species of principal economic and ritual importance. Fish behave differently in different places at different times. Such information, often combined with knowledge about astronomical cycles, general climatological processes, sea conditions, and terrestrial resources, is a principal key to successful fishing. Whereas much of this knowledge deals with habitual aggregations of important food species, details of general fish behavior are known also, including the relationships among different reef species. Special behavioral characteristics are often embodied in fish names [15].

Fish behavior varies temporally according to fixed diel, monthly and longer cycles. Thus a knowledge of the causes and characteristics of different types of fish aggregation behavior by species is fundamental to successful fishing. And the relative importance of the types of such behavior will vary according to the types of gear employed. Widespread is high predictability accorded to aggregations associated with reproduction, since this form of behavior correlates closely with lunar phase. Lunar phase is the major indicator of predictable events on fishing grounds, since many fishes and other marine animals form large aggregations for reproduction in known locations during certain months and lunar phases. Throughout the tropics "prime spots", where predictable aggregations of important food species occur, are known to local fishers. For such spots knowledge of the timing and location of fish aggregations is capture-oriented, highly prized and often jealously guarded [16].

Most reef fish aggregations occur at specific times of the year during a few days around new or full moon, although some species have more extended aggregation periods. Predictable aggregations are marked not only with reference to where and on what days of the lunar month they occur, but also pinpointed by hour and by such diel change of location as might result from tidal movements, for example [17]. This knowledge is geared specifically to the efficient and guaranteed capture of large numbers of preferred food fish. Thus fishing for specific target species becomes predictable, as has been verified by field observations in Okinawa [18] and Solomon Islands [19], among many other places.

4.2. Marine Physical Environments and Fish Habitats

Knowledge of the marine physical environment is of major importance to both the fishing technologies employed and to the targeting of fish by species. As among the Torres Strait Islanders, of Northern Territory, Australia, for example, physical environments and fish habitats are also denoted by often finely sub-divided taxonomies that are fitted together in a coherent, interactive framework that is a once both elegantly sophisticated and eminently practical. For them, "... the sea is knowable, useable and predictable. Marine plants and animals and sea and weather conditions are not thought of as being 'out there' in some sort of undefinable, geographic mush, undifferentiated with respect to species, season or site"; the timing, location and conditions of the occurrence of marine phenomena is regarded as entirely predictable. As but one example of the richness and sophistication of this knowledge base, in their decision-making Torres Strait Islanders refer to a taxonomy of more than 80 distinct terms to describe different tides and tidal conditions, using a "...lunar-based system that keeps track of the four daily tides, changes in height and current speed, time of occurrence and duration, seasonal shift (due to changes in earth, sun and moon orbits), water clarity, surface conditions and associated movements of fish, sea turtles and dugongs" [20].

The fishers of Jambudwip, India, coordinate the complex variables of seabed topography, sea water conditions and sequences of tide and ebb with fish behavior, to both ensure successful catches and their own safety at sea. In their selection of the appropriate seabed over which to conduct their activities these fishers are "... like the agriculturalists who tend to classify the soil according to its relative fertility and the types of crops grown" [21]. The "soil" of the seabed is classified by its capacity to support the net poles and its fertility regarding the types and quantity of fish in the waters above it. Undulations in the seabed affect patterns of water flow and the color of surface waters, which, in turn, are used as indicators of the nature of the seabed.

For example, *juni-bhanga jal* (lit. "water which glows like fire flies") indicates an abundance of fish. Water smell and an oily appearance to the surface are additional diagnostic characteristics. A fish shoal emits a fishy smell that enables fishers to judge its direction of movement. The types of oil released vary among species, and can be distinguished by master fishers. That of the hilsa (*Hilsa ilisha*), for example, is pinkish, and, together with the large area of the shoal and characteristic disturbance of the water, distinguishes the movement of this species [22].

4.3. Local knowledge and ecosystems concepts

In some societies bodies of local knowledge of coastal-marine systems demonstrate a sophisticated understanding of local marine ecology, the behavior of marine fauna, and of the interrelationships among the organisms involved and elements of the physical environment [23]. A "conservation ethic" may exist, expressing a perception of resources being finite and depletable. (The term "conservation ethic" is defined here "... as an awareness of people's ability to deplete or otherwise damage natural resources, coupled with a commitment to reduce or eliminate the consequences" [24]). It is not intended by this to distinguish between conservation based on a desire to protect resources owing to their practical value to people from that based on their perceived intrinsic value.) Certainly not all coastal societies possess a marine conservation ethic. There appears to be none among Torres Strait Islanders [24] or the Ponam Islanders of Manus Province, Papua New Guinea [26], for example, and it has been speculated that a conservation ethic is less likely to develop among people inhabiting a functionally unlimited environment [27]. Torres Strait islanders, unlike many Pacific peoples, have always had access to marine resources greatly in excess of their needs or abilities to harvest. But where they exist, traditional conservation ethics and management practices provide a locally sanctioned code of behavior that can be harnessed to further the objectives of modern marine resource management.

The evidence concerning conservation by traditional fishers is mixed. Several types of marine resource conservation measure may have been consciously employed by traditional societies in the Asia-Pacific Region to conserve stocks and ensure sustained yields (see below, p.8).

In other societies the thinking may be less sophisticated, longterm ecological implications of various types of resource use not always perceived, and target species seldom seen as integral parts of a general marine ecosystem. In some places, as in many parts of the South Pacific, this might be attributable to the many decades of strong mission influence, that has eroded the local knowledge base through "... the concept of an omnipotent and omniscient but inscrutable God" [28].

Although ideas of general ecosystems may not be common in local knowledge, a rudimentary awareness of various marine ecological linkages is commonplace in small-scale fishing communities. General resource depletion, manifested in either increasing scarcity or decreasing mean size of individuals is widely commented on as being "bad" but inevitable. This illustrates a typical problem in attempting to assess the scientific validity of local knowledge, since, although intuitively appealing, this perception is incorrect. Decreasing mean size does not necessarily indicate overfishing, since mean size can be expected to fall when effort is increasing. When a fish population has become exploited to a biologically maximum sustainable level, mean sizes will be significantly lower than those when exploitation was at a low level. Whereas fishers may perceive this to indicate overfishing, in reality the fishery is approaching its most productive state. It is only when mean size falls below a certain limit that overfishing occurs. It is impossible without detailed scientific studies to determine where this point is. Thus biologists claim correctly that such studies are needed because fishers' intuitive feelings can be very misleading. The problem is that the necessary research is so time-consuming that the costs dwarf the benefits, except in large fisheries [29].

The traditional idea of a "rest" period for stock recovery is a resource management practice widespread in the Asia-Pacific Region. In many parts of the world certain fishing methods, especially dynamite-fishing, are perceived of as destructive to the marine environment and resource base, experienced fishers commonly observing that dynamiting not only kills fish indiscriminately but also destroys the coral habitat, and therefore precludes future "good fishing". Local legitimation of the widespread taboos on the use of explosives demonstrates a clear perception of causal linkages.

Wide-scale linkages might also be perceived among ecosystems. In Marovo Lagoon, Solomon Islands, for example, such an understanding extends to an appreciation of the possible wider consequences of large-scale commercial logging, mining and fishing. Logging operations and the possibility of upland mining led locally to an increased awareness of the problem of erosion-induced lagoonal sedimentation, with its consequences for fishing. Commercial fishery operations have also elicited environmentally-aware responses. In Marovo the idea of food chains exists, as expressed by the awareness that baitfishing by the tuna industry depletes the prey of important food fish. Tuna bait boats have been operating there for several years, and some fishers and leaders are increasingly opposed to these activities, on the grounds that if the bait species are overfished food for the larger fish will become scarce [30].

But the interactions between baitfish stocks and other reef and lagoon fish populations are complex and still not well-understood phenomena. Further investigation is required to verify or reject those local perceptions, although research conducted to date tends to reject them [31].

5. THE MODERN PRACTICAL USEFULNESS OF LOCAL KNOWLEDGE

Local knowledge is of continuing importance to present-day subsistence fishing. Acquiring, using and transmitting it is still extremely relevant to livelihoods, particularly where marine resources are still relatively abundant. But in addition, the practical, behavior-oriented corpuses of local knowledge, which focus on the most important species and their habitats, are of immediate value to fisheries planning, monitoring and management, and the enforcement of regulations.

Besides their specific fisheries knowledge, local fishers often know much about a wide range of other significant biological events. This could be important in siting and managing protected areas, for example, especially where recorded scientific knowledge is inadequate [32]. This is especially important when trying to establish reserves where people traditionally depend heavily on the local natural resources.

The principal areas in which modern science is deficient and where traditional fishers can contribute knowledge to development and management planning are examined below.

5.1. Traditional Management Methods

Traditional common property systems of inshore fisheries management occur throughout the world. In many cases they ensure or ensured equitable access and included management measures to ensure the sustainability of coastal fisheries. Many such management strategies include limited entry, seasonal, spatial, gear, size, or species restrictions, prior appropriation rights and the concept of community-based sole ownership, among others. Especially important for sustainable development is an understanding of traditional methods of entry limitation [33].

5.2. Conservation

Many such traditional management systems included deliberate conservation measures. Particularly important are the widespread use of closed seasons and areas, especially as they relate to spawning, and harvestable size limitations. The local knowledge basis for such conservation measures could be of great assistance in the planning of marine reserves, as well as in routine marine resources management [34].

5.3. Stock Assessments

Fishers' knowledge can often provide a useful basis for understanding local fish stocks and their population dynamics. Particularly important in this respect is local knowledge concerning the timing, location and behavior of spawning aggregations of reef and lagoon fishes. Populations of individual such fish are usually thinly and unevenly distributed, making censusing almost impossible. But when these fish spawn they aggregate at specific locations during certain seasons, and often on particular lunar phases [35]. This provides excellent opportunities to monitor stocks, as well as a useful focus for management because exceptionally large catches are often made from them. For instance, there is no conventional scientific information available on West Samoan mullet fisheries. Local knowledge of reef fishers would be a good substitute for this, and could be obtained at a fraction of the cost of conventional methods [36].

5.4. Environmental Impact Assessment

Given local knowledge of spawning migrations and spawning aggregation sites, fishers can offer valuable insights into of the likely impact of coastal infrastructural changes, such as causeway, harbor, landfill, or other constructions [37].

5.5. Local hydrography

This is well-understood in local knowledge. For example, Palauan fishers are familiar with an island wake effect that had never been described by physical oceanographers, and which has an impact on the distribution of tuna and other species. Fisheries biologists unaware of such phenomena are handicapped in their work [38].

5.6. Mapping

Local knowledge also includes an intimate spatial familiarity with the physical environment, including local currents, seabed conditions and other such phenomena [39]. It also often includes a finely detailed mental mapping system [40]. This submarine cartographic knowledge of fishers is of immediate practical importance, especially along rapidly prograding coastlines where conditions can change dramatically in just a few years, since it can be applied to update old marine charts, as in India [41].

5.7. Fishing Methods and Technologies

The local knowledge expressed in fishing methods and technologies is an alternative to high technology development approaches. Further, the knowledge of fish behavior on which traditional methods of fishing are predicated can be adapted to the use of modern technology. For example, high catches are experienced around Western Samoan Fish Aggregation Devises (FADs). At the beginning of one externally-funded development project, FADs were placed more or less randomly, and so attracted few tuna. Later, when they were placed at locations suggested by local fishers, catches improved considerably. Local fishers understand and can predict the location and timing of the daily migrations of yellowfin and skipjack tuna [42]. As Samoans say, the terminus of this migration is "where the tuna sleep". And this is where the FADs work well. This is an excellent example of the knowledge that fishers possess and that scientists lack, and demonstrates an immediate pay-off for employing local knowledge [43].

5.8. Fish Systematics and Biology

Local knowledge of fish names and taxonomies, as well as behavior, can be of immense usefulness in understanding fish systematics and biology.

6. LOCAL KNOWLEDGE TRANSMISSION AND THE LOGIC OF INSTITUTIONS

In addition to the practical aspects of ensuring sustained resource management, the transmission of traditional knowledge is of basic socio-cultural importance to any society. This is because during knowledge transmission over several generations social institutions gradually crystallize and "routine" gradually become "custom".

Further, in the process a transmitter's sense of reality is strengthened, because the rationale underlying "custom", "tradition", normative and actual behavior, and rules and regulations must be systematically provided to "learners" by "teachers" via some form of *educational process* based on *consistent* and *comprehensive legitimating formulae* that are conveyed to *all children*. That leads to the need for social controls to handle deviance and to ensure compliance with social norms. There is a need to *control deviance* by ensuring *compliance* under the threat of *sanctions*.

Thus the process of knowledge transmission leads logically to that of institutionalization. The logic of institutions and that of the linkages among them emerges not from the institutions themselves, but from the way in which they are consciously reflected on them by those that operate within them. This occurs especially during the process of knowledge transmission. When such reflection is common to the various operators, it provides a logical framework for an institution. This logic also emerges from the reciprocity that occurs among operators of different systems, as among fishers and farmers, women and men, and different age sets, for example. Continual acts of reciprocity establish the collective consciousness of a logical framework for linked resource systems and their accompanying institutions. (This can also emerge within a household[s] that practise a mixed economy.) In this way, universes are shared and institutions integrated together [44].

Thus knowledge assumes a pivotal role in any community: integration of an institutional order is understandable only in terms of the "knowledge" that its members have and share of it. It is "... the sum total of 'what everybody knows' about a social world" [45]. At this level, "... every institution has a body of transmitted 'recipe knowledge' [46] ... that supplies the institutionally appropriate rules of conduct" [47].

Such knowledge underlies the dynamics of institutionalized conduct and defines the areas of such conduct, as well as both defining and constructing the roles to be played in the context of such institutions. By definition, then, such knowledge also controls and predicts all such conduct by the operators within a resource system. Since such knowledge comprises a body of generally valid truths about reality, any deviance from the social order is a departure from reality - a deviance that could be variously interpreted as depravity, a symptom of mental illness, ignorance, criminality, wilfulness, or a sign of a power struggle aimed at the eventual usurpation of authority, among other things.

Thus a society's stock of knowledge, when either put into operation or reflected upon, becomes *the local world*; it becomes co-extensive with "the knowable" and provides the framework via which that which is "...not yet known will come to be known in the future" [48], i.e. the acceptance or not of innovation. In these terms knowledge is the key dialectic of society, since knowledge about society both captures everyday social reality and continuously reproduces it.

For example, a body of knowledge develops over generations to refer to the various activities involved in a given resource system, say fishing, takes on a linguistic form. This is extremely important, since:-

(1) local vocabularies define species, habitats, weather patterns, sea conditions, seasons, fish behavior, and the like;
(2) a collection of local "recipes" must be learned in order to fish both correctly and with consistent success;
(3) local knowledge is also a channelling and controlling force that underlies fishing institutions;
(4) in the persistence and crystallization of fishing institutions knowledge becomes the objective description of the local activity/institution; and
(5) a local objective arena/field/ethnoscience of fishing develops parallel with the activity of fishing.

This body of knowledge is transmitted to the next generation as an *objective truth* during socialization, and so is internalized as *subjective reality*. This transmission yields and gives identity to a specific type of person, a local fisher, whose (principal) social universe is constituted by that body of knowledge. As a consequence, to be an active local fisher implies that there exists a local social world defined and controlled by a discrete body of arcane knowledge about fishing.

7. THE POTENTIAL OF TRADITIONAL SYSTEMS AND LOCAL KNOWLEDGE IN MODERN MANAGEMENT

Although eroded or even broken-down in parts of the Asia-Pacific Region, especially because of colonialism or neo-colonialism and the predominance of imported scientific concepts, community-based fisheries management systems and their associated knowledge bases are still used to manage coastal fisheries in a wide range societies. Thus it has frequently been asserted, although usually with scant proof, that such systems can play a potentially major role in the modern world by ensuring equitable access to fisheries, as well as in managing and enforcing conservation measures to ensure the sustainability of coastal fisheries. The thesis generally is that the more the responsibility for the control of local resources can be left to local, traditional users, the fewer will be the social, political, legal, conservation-related and management cost problems that must be addressed by governments [49].

At first sight the adaptation of traditional systems to a modern purpose may appear to invite strong local resistance, since they are often so much a part of the way-of-life. But traditional community-based systems of marine resources management in many parts of the Asia-Pacific Region already incorporate important elements of conventional fisheries management. For example, parallel management strategies include limited entry, seasonal, spatial, gear, size, or species restrictions, prior appropriation rights and the concept of sole ownership, among others [50]. In fact, the use of many such strategies in the Pacific Basin [51] and Japan [52] antedated their adoption in the West. In conventional marine economics terms, sole ownership, limited entry, individual transferable quotas and other such fisheries management schemes are based on the theory of the firm. On the other hand in many societies in the Asia-Pacific Region the local community is the sole owner, and traditions of resource use and management are enforced by community norms that control the behavior of the membership [53]. But this, too, has its parallels in New England and Western European fishing communities, among other places, where socially binding yet unwritten and informal rules carry more weight than official regulations [54].

Most nations in the Asia-Pacific Region face an array of dilemmas in determining rights and delineating responsibilities in marine resources management and development, including what institutions should manage and enforce regulations for subsistence fisheries, legal support for traditional regulation and enforcement, the managerial and developmental role of the central government in small-scale commercial fisheries, the feasibility of centralized management plans versus local decisionmaking, and the nature of the consultative and collaborative process among fishers, local governments and national authorities. Initially these look like local versus central jurisdictional matters, but the underlying issue is one of the policy toward and the means of managing marine resources and of adapting traditional concepts to modern needs and frameworks, such that the range from subsistence fishery to the highly commercialized industrial fishing is served properly [55].

Alternative arrangements can help in overcoming the weaknesses of conventional fisheries management. The most appropriate form of fisheries governance is one in which management authority is decentralized, within a broad policy framework, to enable local governments to fundamentally control local fishing via community-based management systems. Such a system is co-management, whereby decisionmaking is shared between central and/or provincial governments and community-based management authorities. Such arrangements have long-existed *de facto* in many parts of the Asia-Pacific Region. Thus, especially in the archipelagic nations, co-management has, in effective, long been practised [56].

The immediate goals of co-management are more appropriate, more efficient and more equitable management than hitherto. Its broader objectives are community-based development and participatory democracy. In the Asia-Pacific Region co-management of coastal fisheries has the potential to conserve and enhance fish stocks, improve the quality of data and its analysis, reduce excessive and competitive investment by fishers, permit a more equitable and flexible allocation of fishing opportunities, and harmonize the relationship between fishers and government agencies. In particular it holds the promise of greatly reducing the expense and wastage of all aspects of coastal fisheries management [57].

There can be no single system of co-management; such systems must be based on cultural and physical factors, especially resource characteristics and antecedent viable systems of management. Essentially this form of fisheries management will be a syncretic model that blends the complementary domains of conventional management and scientific knowledge with traditional management and local knowledge. In many places there are few practical alternatives to the future management of tropical coastal fisheries being neo-traditional fisheries management [58].

Local knowledge is of great potential value in the modern world. It can provide an important information base for local resources management, especially in the tropics, where conventionally-used data are usually scarce to non-existent, as well as providing a shortcut to pinpoint essential scientific research needs. First, however, it must be systematically collected and organized, and then evaluated and scientifically verified, before being blended with complementary information derived from modern sciences, so as to be useful for resources management.

But peoples' willingness to discuss and share their knowledge varies immensely within and among societies. This often depends on the type of knowledge. It also reflects an understanding of knowledge as power or privilege, especially in socially stratified societies. Further, when market forces intrude, or where individual accumulation is not sanctioned, knowledge is the key to an individual's or community's competitive performance, and thus it is unlikely to be shared widely.

Local knowledge is being rapidly eroded by modernization, urbanization, monetization, and the alienation of youngsters from their traditions as a result of modern formal education [59]. In their future roles as local resource managers, these people will find little guidance in their formal education, unless it has been complemented by local knowledge based on the experience of their elders. Thus educational institutions should offer courses that blend both local knowledge and modern concepts of resources management. These should be made mandatory for students in educational institutions, particularly in the poorer countries.

Tokelau, a small island nation in the South Pacific, provides an example. There the value of local knowledge is undisputed by the residents, as is the need to adapt the traditional framework to the contemporary world, and to establish mechanisms for blending biological information with local knowledge [60]. To counteract the denigration and decline of local knowledge, some islanders believe that the authority of the Council of Elders should be restored, either by restricting it to its traditional roles or by having a Tokelauan with a background in fisheries biology act as an adviser to the Council, as well as by establishing more effective punitive measures for both traditional and modern offenses. There is a recognized need to integrate more fully biological information from stock assessment studies with local knowledge, to modify the age at which traditional education begins, so as to ensure the transmission of knowledge prior to the departure of youngsters for formal education overseas, together with a need to convince the foreign-educated elite of the value of local knowledge.

This is but one small example of a phenomenon occurring throughout the tropics. Given the resurgence of ethnic pride in many areas, as among the

Aborigines of Australia, the Kanaks of New Caledonia, the Maoris of New Zealand, and the First Nation Peoples of Canada, to select but a few, this process is likely to accelerate and become the norm.

9. REFERENCES

1 K. Ruddle and T.B. Grandstaff, Tech. Forecast. Soc. Change 11 (1978) 119-131; K. Ruddle and D.A. Rondinelli, Transforming Natural Resources for Human Development: A Resource Systems Framework for Development Policy. Tokyo, United Nations University Press, Tokyo, 1983.

2 K. Ruddle, Traditional Community-based Marine Resource Management Systems in the Asia-Pacific Region: Status and Potential, International Center for Living Aquatic Resources Management, Manila, 1993.

3 K. Ruddle, Local Knowledge in the Folk Management of Fisheries and Coastal-Marine Environments. In: Dyer, C.L. and J.R. McGoodwin (eds.), Folk Management in the World Fisheries. University of Colorado Press, Niwot, pp. 161-206; K. Ruddle, Nat. and Res. 29 (1994)(in press).

4 see 2.

5 K. Ruddle, E. Hviding and R.E.Johannes, Mar. Res. Econ. 7 (1992) 249-273.

6 see 2; K.Ruddle, A Guide to the Literature on Traditional Community-based Fishery Management in the Asia-Pacific Tropics. Fisheries Circular No. 869, FIPP/C869. Rome, FAO, 1994.

7 see 2.

8 see 2.

9 see 5.

10 see 2.

11 see 2.

12 see 2 and 3.

13 see 2.

14 see 2 and 3.

15 see 2.

16 Ruddle, K. and T. Akimichi, Sea Tenure in Japan and the Southwestern Ryukyus. In: Cordell, J.C. (ed.), A Sea of Small Boats: Customary Law and Territoriality in the World of Inshore Fishing. Report No. 62. Cultural Survival, Cambridge, Mass., 1989, pp. 337-370.

17 R.E. Johannes, Ann. Rev. in Ecol. and System. 9 (1978) 349-364; R.E. Johannes, Using Knowledge of the Reproductive Behavior of Reef and Lagoon Fishes to Improve Fishing Yields. In: J.E. Bardach, J.J. Magnusson, R.C. May and J.M. Reinhart (eds.), Fish Behavior and its Use in the Capture and Culture of Fishes, ICLARM Conference Proceedings 5. ICLARM, Manila, 1980, pp. 247-270.

18 see 16.

19 R.E. Johannes, A Spawning Aggregation of the Grouper *Plectropomus aureolatus* (Rüppel) in the Solomon Islands. In: J.H.Choat, D.J. Barnes, M.A.Borowwitzke, J.C.Coll, P.J.Davies, P.Flood, B.G.Hatcher, D.Hopley, P.A. Hutchings, D.Kinsey, G.R.Orme, M.Pichon, P.F.Sale, P.W.Sammerco, C.C.Wallace, C.R.Wilkinson, E.Wolanski, and O.Bellwood (eds.), Proceedings of Sixth International Coral Reef Symposium, Australian Institute of Marine Science, Townsville, 1989, pp. 751-755.

20 B. Nietschmann, Traditional Sea Territories, Resources and Rights in Torres Strait, In: J.C. Cordell (ed.), A Sea of Small Boats: Customary Law and Territoriality in the World of Inshore Fishing. Report No. 62. Cultural Survival, Cambridge, Mass, 1989, pp. 60-93.

21 see 20, p. 61.

22 B. Raychaudhuri, The Moon and the Net: Study of a Transient Community of Fishermen at Jambudwip. Anthropological Survey of India, Calcutta, 1980.
23 see 2 and 3.
24 R.E. Johannes and K. Ruddle, Human Interactions in Tropical Coastal and Marine Areas: Lessons from Traditional Resource Use, In: A. Price and S. Humphrey (eds.), Applications of the Biosphere Reserve Concept to Coastal Marine Areas. Papers presented at the UNESCO/IUCN San Francisco Conference of 14-20 August 1989. International Union for the Conservation of Nature, Gland, 1993, pp. 19-25.
25 R.E. Johannes and J.W. MacFarlane, Assessing Traditional Fishing Rights Systems in the Context of Marine Resource Management: A Torres Strait Example. In: K. Ruddle and R.E. Johannes (eds.), Traditional Marine Resource Management in the Pacific Basin: An Anthology, UNESCO-ROSTSEA, Jakarta, 1990, pp. 241-261.
26 J.G. Carrier, Marine Tenure and Conservation in Papua New Guinea: Problems of Interpretation, In: B.J. McCay and J.M. Acheson (eds.). The Question of the Commons: the Culture and Ecology of Communal Resources. The University of Arizona Press, Tuscon, 1987, pp. 142-167.
27 M.D. Chapman, Environ. Cons. 12(3) (1985) 217-230.
28 see 2.
29 see 2.
30 E. Hviding, Marine Tenure and Resource Development in Marovo Lagoon, Solomon Islands; Traditional Knowledge, Use and Management of Marine Resources, with Implications for Contemporary Development. FFA Report No. 88/35. Forum Fisheries Agency, Honiara, Solomon Islands, 1988.
31 S.J.M. Blaber, D.A.Milton, N.J.F. Rawlinson, G.Tiroba, and P.V.Nichols. Reef Fish and Fisheries in Solomon Islands and Maldives and their Interactions with Tuna Baitfisheries. In: S.J.M. Blaber and J.W.Copland (eds.), Tuna Baitfish in the Indo-Pacific Region: Proceedings of a Workshop, Honiara, Solomon Islands, 11-13 December 1989. Proceedings No. 30, Australian Center for International Agricultural Research, Canberra, 1990, pp.159-168.
32 see 24.
33 see 2.
34 see 17, Ann. Rev. in Ecol. and System.
35 R.E. Johannes, Words of the Lagoon: Fishing and Marine Lore in the Palau District of Micronesia. University of California Press, Berkeley and Los Angeles; R.E. Johannes, Bull. of Mar. Sci., 31(3) (1981) 673-680; R.E. Johannes, Introduction, In: R.E. Johannes (ed.), Traditional Ecological Knowledge: A Collection of Essays. International Union for the Conservation of Nature, Gland, 1989, pp. 5-8.
36 R.E. Johannes, Ambio 11(5) (1982) 258-261.
37 see 35, Bull. of Mar. Sci.
38 see 35, Johannes Words of the Lagoon....
39 see 38, for example.
40 B. Nietschmann, Torres Strait Islander Sea Resource Management and Sea Rights, In: K. Ruddle and R.E. Johannes (eds.), The Traditional Knowledge and Management of Coastal Systems in Asia and the Pacific. UNESCO-ROSTSEA, Jakarta, 1985, pp. 125-154; see 12; P.T. Sanjeeva Ray, Seafood Export Journal 2 (1990) 1-4.
41 see 40, P.T. Sanjeeva Ray.
42 see 35, Words of the Lagoon....
43 R.E. Johannes, Reef and Lagoon Resource Management in Western Samoa. A Report to the Government of Western Samoa, Apia, and the South Pacific Commission, Noumea, New Caledonia (unpub. ms.), 1982.

44 P. Berger and T. Luckmann, The Social Construction of Reality: a Treatise in the Sociology of Knowledge, Penguin Books, Harmondsworth, 1984.

45 see 44.

46 A. Schutz, Der sinnhafte Aufbau der socialen Welt, Springer, Vienna, 1960.

47 see 44.

48 see 44.

49 K. Ruddle, Mar. Res. Econ., 5(4) (1988) 351-363.

50 R.E. Johannes, So. Pac. Bull. 4 (1978) 31-34.

51 see 50.

52 A. Kalland, Sea Tenure in Tokugawa Japan: the Case of Fukuoka Domain, In: K. Ruddle and T. Akimichi (eds.) Maritime Institutions in the Western Pacific. Senri Ethnological Studies, 17, National Museum of Ethnology, Osaka, 1984, pp. 11-36; A. Kalland, Fishing Villages in Tokugawa Japan: the Case of Fukuoka Domain. Oslo, University of Oslo, Ph.D. diss., 1989; K. Ruddle, The Continuity of Traditional Management Practices: the Case of Japanese Coastal Fisheries, In: K. Ruddle and R.E. Johannes (eds.), The Traditional Knowledge and Management of Coastal Systems in Asia and the Pacific, UNESCO-ROSTSEA, Jakarta, 1985, pp. 157-179; K. Ruddle, Administration and Conflict Management in Japanese Coastal Fisheries, FAO Fisheries Technical Paper No. 273. FAO, Rome, 1987.

53 see 49.

54 J.M. Acheson, The Lobster Fiefs Revisited: Economic and Ecological Effects of Territoriality in the Maine Lobster Fishery, In: B.M. McCay and J.M. Acheson (eds.), The Question of the Commons: the Culture and Ecology of Communal Resources. The University of Arizona Press, Tuscon, 1987, pp. 37-65.

55 see 2.

56 see 2.

57 E. Pinkerton, Introduction: Attaining Better Fisheries Management through Co-Management - Prospects, Problems, and Propositions, In: E. Pinkerton (ed.), Co-operative Management of Local Fisheries: New Dimensions for Improved Management and Community Development. University of British Columbia Press, Vancouver, 1989, pp. 3-36.

58 see 2.

59 K. Ruddle, Mar. Anthro. Stud. 6(1-2) (1994) 1-37.

60 F. Toloa, R. Gillett and M. Pelasio, Traditional Marine Conservation in Tokelau. Can it be Adapted to Meet Today's Situation? In: Anon (ed.), Workshop on People, Society and Pacific Islands Fisheries Development and Management: Selected Papers, Inshore Fisheries Research Project Technical Document No. 5, South Pacific Commission, Noumea, 1993, pp. 19-21.

TOWARD GLOBAL PLANNING OF SUSTAINABLE USE OF THE EARTH
Development of Global Eco-Engineering
S. Murai (Editor)

Beginning the transition to a sustainable global transportation system

James J. MacKenzie

World Resources Institute, 1709 New York Ave., N.W., Washington, D.C. 20006, U.S.A.

Abstract

Worldwide, motor vehicles are major sources of air pollution and greenhouse gas emissions. They are also large consumers of petroleum. Growth in the global fleet has overwhelmed past efforts to reduce overall oil consumption and pollution emissions. Over the next few decades, rapid growth in the global motor vehicle fleet is expected with especially high growth in Asia and Latin America.

For several reasons these trends are not sustainable. Based on past experience, such growth, even with continuously cleaner and more efficient new vehicles, will lead to increases in emissions of carbon dioxide and air pollution. Secondly, the rapid growth in world oil demand, in large part for motor vehicles, will contribute to the peaking of global oil production early in the next century. Replacing crude oil with synthetic oils made from bitumen, tar sands, or coal would be very expensive and would make all the more difficult efforts to reduce global carbon dioxide emissions.

Alternative fuels -- natural gas, ethanol, methanol, electricity, and hydrogen -- are examined to see whether they can solve these linked problems. While offering some benefits, ethanol, methanol, and natural gas are judged to be insufficient to reduce the threats of global warming or air pollution. It is argued that electric vehicles powered by batteries, flywheels, or hydrogen fuel cells, offer the most promising long-term option to reduce emissions of carbon-dioxide and air pollution, with the electricity and hydrogen for these vehicles eventually derived from non-fossil energy sources. The introduction of such vehicles would represent a major step toward a sustainable transportation system.

1. INTRODUCTION

There are about 700 million cars, trucks, and buses in the world today[1] including some 100 million motorcycles. These vehicles undeniably confer many benefits on their users and society at large. Enhanced individual mobility, quicker and improved delivery of commercial and industrial goods and services, and increased economic growth are all facilitated by affordable motor vehicles and the fuel to power them. As anyone living in a major urban city knows, however, there are also some liabilities to extensive motor vehicle use: cars, trucks, buses, and motorcycles are important contributors to serious urban environmental and public-health problems including air pollution, congestion, and accidents. Moreover, they are major contributors to other problems of international concern, including risks to the security of oil-importing nations from depending on potentially insecure sources of oil, and risks stemming from global warming.

As we shall see in this paper, international attempts to reduce the impacts of motor-vehicle use have been only partly successful. For instance, even though the global vehicle fleet continues to become more efficient, global fuel use and carbon-dioxide emissions continue to steadily grow. Similarly, despite major reductions in new-vehicle air pollution emissions, the total amount of pollution, especially in the world's largest cities, continues to increase.

2. NON-SUSTAINABILITY OF PRESENT TRENDS

For a number of reasons, a continuation of present trends in motor-vehicle use is not sustainable. Continued growth in emissions of greenhouse gases from motor vehicles will place the earth's climate in ever greater jeopardy. And even if global warming were not a threat, global oil reserves are finite, so that a shift from oil to other energy sources will be needed as early as the first quarter of the next century. Similarly, continued growth in air pollution emissions -- much of it from motor vehicles -- will make life intolerable in the world's major cities. Finally, without improvements in traffic management and new forms of public transport, vehicle congestion around the world will continue to grow, leading to massive gridlock in many metropolitan areas.

Fortunately, there are technological responses to these linked threats. By changing energy sources, air pollution and greenhouse gases can be controlled without eliminating the use of the personal vehicle. And there are new options under development that will provide more attractive and affordable forms of transport, especially for the very young, elderly, and handicapped. In this paper we will address the fuel-related problems and discuss emerging technologies that offer hope to solve them.

3. AIR POLLUTION

In virtually all major cities around the world air pollution from motor vehicles poses serious problems. The vehicle-related pollutants of most concern are carbon monoxide, nitrogen oxides, volatile organic compounds, sulfur oxides, suspended particulate matter, lead, and ozone.[2] In Mexico City, for example, road transportation accounts for 99 percent of carbon-monoxide emissions, 89 percent of hydrocarbons, and two thirds of nitrogen oxides.[3] Directly or indirectly, these pollutants contribute to photochemical ozone, acid deposition, and global warming. Carbon monoxide -- the result of incomplete fuel combustion -- poses a direct threat to human health and contributes indirectly to the atmospheric buildup of methane, a potent greenhouse gas. Ozone leads to eye and throat irritation, increased risk of asthma attacks, shortness of breath, and greater susceptibility to infection. Ozone also damages crops, trees, and materials. To make matters worse, the nitrogen oxides (NO and NO_2) that participate in ozone formation are removed from the atmosphere as nitric acid, contributing to the problems of acid rain. Acid rain acidifies soils, streams, and lakes, and damages buildings and other structures. Over time, acid rain leads to soil impoverishment and forest declines and the loss of fish. Exposures to lead from motor-vehicle fuel can result in impaired liver and kidney functions, to neurological problems, and to retarded intellectual development.

Gasoline-fueled vehicles are the principal sources of carbon monoxide, nitrogen oxides and lead while diesel engines are major sources of particulates and toxic chemicals. Recent studies of excess deaths in polluted cities of the United States and the UK suggest that small-particle pollution, largely from motor vehicles, is associated with higher rates of lung and heart diseases and premature death.[4] Studies in California suggest that non-smokers living in areas of heavy air pollution experience lung damage similar to that found in cigarette smokers. And recent studies in the UK suggest that episodes of high levels of urban pollution, much of it from vehicles, can lead to a 10 percent increase in mortality from respiratory and cardiovascular diseases.[5] Motor vehicles now constitute the principal source of air pollutants in the majority of cities in industrialized countries.[6] According to the World Bank, the transportation sector accounts for the great bulk of CO emissions (58 to 85 percent of emissions in the OECD countries). Hydrocarbon emissions from transport are estimated to range from 30-87 percent for OECD countries to 37-100 percent for non-OECD nations.

Megacities in the developing world are by no means alone in their high emissions of motor-vehicle pollution. Almost every major city in the world, from Los Angeles to Tokyo, faces similar problems. In the United States, vehicle-related air pollution problems persist despite more than 25 years of regulations aimed at producing cleaner vehicles.

For several reasons efforts to reduce vehicle air pollution have been only partially successful. One reason is that emission controls on gasoline-powered vehicles deteriorate over time. On average, motor vehicles in the United States emit hydrocarbons at three times the rate to which they were certified -- offsetting much of the anticipated clean-air benefits. In addition, some motorists tamper with or remove altogether the pollution devices from their cars. Moreover, many motorists are buying trucks and vans rather than cars, and trucks have less stringent emission limits. There are also important sources of vehicle pollution -- vapor emissions from fuel tanks, for instance -- that are only now being addressed by environmental officials. Finally, there is the relentless growth -- growth in the number of motor vehicles and growth in the number of persons exposed to pollution. Growth in the world fleet worsens air pollution by increasing the number of sources. In the United States, the number of motor vehicles grew from 108 million (10^6) in 1970 to 190 million in 1992, a 75 percent increase. And the number of vehicle-kilometers-traveled (VKT) doubled from 1.86 trillion (10^{12}) to 3.72 trillion over the same period. Worldwide, the number of motor vehicles grew from 246 million in 1970 to 583 in 1990 with a commensurate growth in travel and fuel use.

Growth in the number and use of vehicles is relentless the world over. Figure 1 shows the global growth in motor-vehicles between 1930 and 1991. If present trends continue, there could well be over a billion (10^9) motor vehicles on the road by the year 2010.[7]

The continued growth in the number of sources -- combined with less-than-anticipated reductions from emission controls -- has overwhelmed much of the technological efforts of the past 25 years aimed at reducing motor-vehicle air pollution.

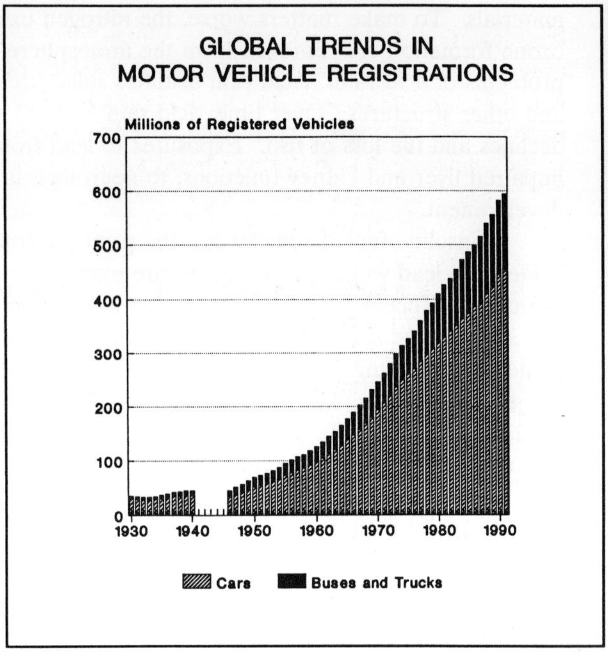

Figure 1

4. THE COMING PEAKING IN GLOBAL OIL PRODUCTION

Today, with world oil prices very low, there is little interest in worrying about when world crude-oil production might peak and start declining. Yet, there is evidence that global oil production may be only a few decades away from beginning its inevitable decline. If this turns out to be true, major new sources of energy will have to be developed since petroleum is used to meet 40 percent of global (commercial) energy needs. The implications for world transportation technology are particularly severe since almost all vehicles are powered by petroleum products. Transportation accounts for about 30 percent of world oil use, 40 percent of oil use in the OECD countries, and about half of oil use in the United States.

I would like to present some simple arguments to the effect that, within a matter of decades -- perhaps by 2010 -- global oil production could begin to decrease because of resource exhaustion. As a result of this decline, new replacement sources for oil will have to be made available, especially for transportation. This kind of analysis was first described in the 1950s by the late U.S. geologist, M. King Hubbert. In analyzing U.S. oil production, Hubbert made three simple assumptions.[8] First, that in a given producing region there was a finite amount of oil that would be ultimately recovered. Second, for a given region, oil production would rise, reach a maximum, and then decline. And third, that annual oil production would be symmetrical about the year of peaking. Hubbert argued that, during the decline phase of production, prices would rise, new technologies would be used to extract the remaining recoverable oil, and that eventually production costs would become too high and new energy sources would be introduced. (In the case of the United States, imported oil turned out to be the cheapest alternative.)

Figure 2 shows the history of oil production in the lower-48 United States. I have fitted these data with a simple curve of the sort used by Hubbert.[a] The total area under the curve of annual production is the total amount of oil that will ever be produced, the so-called "ultimately recoverable resource." In the U.S. case, the curve suggests that ultimate crude-oil production in the lower-48 states will be around 189 billion (10^9) barrels. Of this, 155 billion, 82 percent of the total, had been produced by the end of 1993. The value of 189 billion barrels is in the range of 150-200 billion barrels estimated by U.S. geologists.

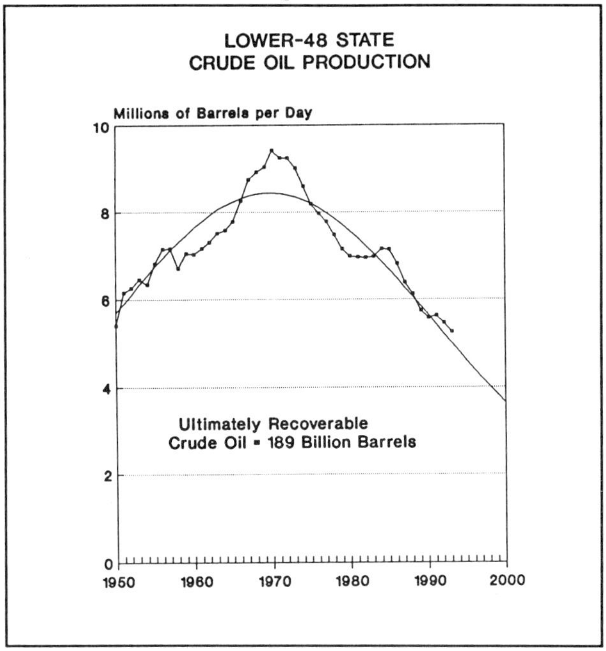

Figure 2

I have performed a similar analysis for world oil production. The U.S. Geological Survey (USGS), in a recent summary of its 10-year World Resources Program, estimated that total recoverable world crude-oil resources are somewhere between 1800 billion barrels and 2477 billion barrels, with a mean value of 2079 billion barrels.[9] Through 1993, the world had consumed about 719 billion barrels of crude oil. Proven reserves are about 1000 billion barrels of oil, of which two-thirds are in the Middle East and three quarters are in OPEC nations. If the USGS is correct, at most, another 758 billion barrels of recoverable oil remain to be discovered.

If we accept the USGS estimates, what can we plausibly expect to happen with global oil production? The above numbers suggest that the world has consumed *at least* 30 percent of the crude oil that will ever be recovered. If the assumption is valid that global oil production will be symmetrical about the year of peaking -- as was the case in

---

[a]The curve is the first derivative of a logistic function.

the United States -- then production will begin to decline when half of the total has been consumed. Under optimistic assumptions -- that the USGS high estimate is correct -- this would occur when a total of 1238 billion barrels (one-half of 2477) were consumed. Since 719 billion have already been consumed, the time of peaking would occur when an additional 519 billion barrels (1238 - 719) have been consumed.

When might this peaking occur? Figure 3 provides a clue. The figure shows world crude-oil production data through 1993 along with two possible scenarios for ultimate production. The area under the lower curve is 2079 billion barrels (the mean USGS estimate); the area under the higher curve is 2477 billion barrels (the high-end USGS estimate). Also shown are two forecasts of future global crude-oil demand for the years 2000, 2005, and 2010; the lower curve is from the U.S. Department of Energy (USDOE), the higher from the International Energy Agency (IEA) of the OECD. In its base case projection, the USDOE estimates that global crude-oil production will reach around 77.8 million barrels per day (MMb/d) by the year 2010.[b] The IEA, in its reference case, forecasts crude-oil demand of 84.5 MMb/d by 2010.

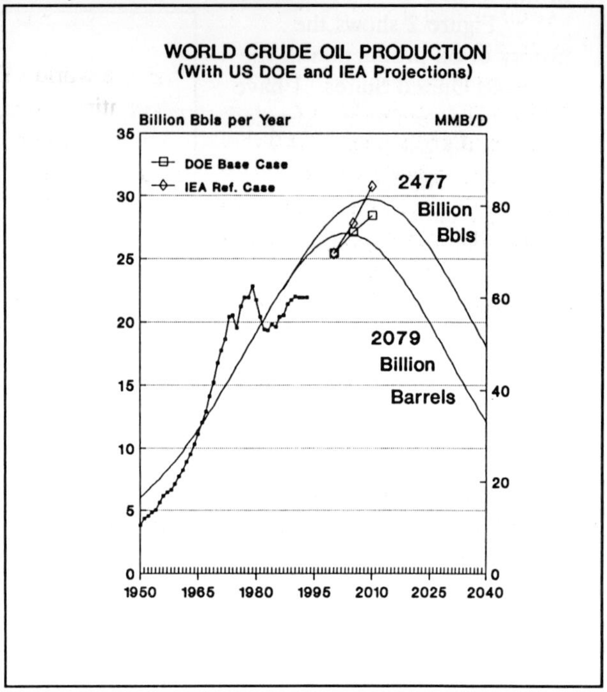

Figure 3

To provide a conservative estimate of the year of peaking, we choose the most optimistic assumptions: that the amount of ultimately-recoverable crude oil will be at the higher end (2477 billion barrels), and that growth in world demand will be at the lower end (USDOE). With these assumptions, one can calculate that the 519 billion barrels needed to consume the first half of the earth's recoverable crude oil would be consumed by the year 2012, *a mere 18 years into the future.*

---

[b] DOE and IEA projections are for total petroleum production, including natural gas liquids. I have assumed that crude oil production will account for 90 percent of their totals.

No one knows, of course, what will happen to world oil prices or global oil demand in the years ahead. Peaking in the early part of the next century can be expected if consumption rises rapidly. If consumption were to stay relatively flat -- through much higher prices or by introducing alternative energy sources -- a decline from resource depletion could be delayed by several decades more. The important point of this discussion is that the time when world crude-oil production will begin to decline is measured in decades, not centuries.

This analysis does not mean that the world will soon "run out" of oil or hydrocarbon fuels. Oil production will continue, though at a declining rate, for many decades after its peak. And there are enormous amounts of coal, heavy oils, and bitumen worldwide that could be used to produce liquid or gaseous substitutes for crude oil. But the facilities to make such synthetic fuels are extremely costly to build, are environmentally damaging to operate, and would lead to much larger carbon dioxide emissions than would occur from the consumption of products made from conventional crude oil. Given the risks that such emissions would entail, it would not seem to be prudent to build such plants.

5. GLOBAL WARMING

Air pollution and oil depletion are not the only issues of concern from the growing motor-vehicle fleet. Vehicles are major direct and indirect sources of greenhouse gases. Cars, trucks, and buses contribute to the atmospheric buildup of carbon dioxide, methane, nitrous oxide (N_2O), CFCs, and ozone. Of these, carbon dioxide is the largest contributor to increased warming and, worldwide, motor vehicles account for about 15 percent of global fossil-fuel emissions.[10]

Vehicle carbon-dioxide emissions have increased steadily with growth in the global fleet. Between 1971 and 1991, carbon-dioxide emissions grew by about 17 million tons per year, from 518 million metric tons of carbon to 900 million metric tons.[11] (See Figure 4.) Figure 5 shows that the highest growth rates in emissions are in the non-OECD countries. This increase in oil consumption occurred despite a reduction in the average fuel consumed and CO_2 released per motor vehicle of 1.5 percent per year. (See Figure 6.) Just as with attempts to control vehicle air-pollution, growth in vehicle use overwhelmed efforts to improve fleet fuel-efficiency. If the global fleet continues to grow according to recent trends, and if fuel-consumed-per-vehicle continues to decline at 1.5 percent per year, total vehicle carbon-dioxide emissions could increase by an additional 20 to 50 percent (relative to 1987) by the year 2010.[12]

Carbon-dioxide increases of this magnitude would make international efforts to control global warming exceedingly difficult. According to the UN's Intergovernmental Panel on Climate Change (IPCC), stabilizing atmospheric carbon-dioxide concentrations at present levels would require a reduction *in emissions* of 60 to 80 percent below today's levels.[13]

6. COPING WITH THE THREATS OF AIR POLLUTION, RESOURCE DEPLETION, AND CLIMATE CHANGE

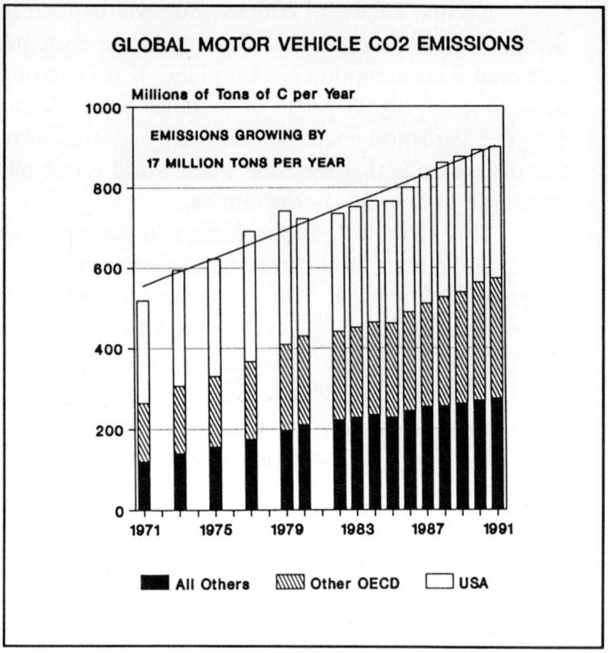

Figure 4

Given the projected long-term growth in global vehicle use in urban areas, how can we address the three closely linked problems of air pollution, global warming, and resource depletion? The brief review presented above suggests that a continuation of past trends -- that is, programs to improve new-vehicle performance (more efficient, less polluting) -- will not, by itself, solve these problems. History demonstrates all too clearly that building cleaner and more fuel efficient oil-powered vehicles, while certainly desirable and important as an interim goal, will not be sufficient to clear the air of pollution or stabilize -- much less significantly reduce -- carbon dioxide emissions as long as the global fleet continues to grow. And, of course, expanding the world fleet of oil-powered vehicles does not address at all the problem of eventual resource depletion.

7. SUSTAINABLE TRANSPORTATION TECHNOLOGIES

Today, motor vehicles depend totally on a depletable fossil fuel -- oil. Until economically and technologically attractive alternatives become widely available, it is unlikely that the worldwide pollution, security, climate, and resource problems posed by motor vehicles can be solved. People -- in both the developed and developing world -- will have little alternative but to continue buying oil-powered cars and trucks. As a result, air pollution will affect more and more people and oil consumption will rise as will carbon dioxide emissions.

The world desperately needs alternatives to oil-powered cars and trucks. And the burden for developing the technological alternatives rests squarely on the industrial countries that make the world's motor vehicles: 80 percent are made in the U.S., Japan, and Europe. There is no one else that can do it.

8. CLEAN AIR FROM NEW FUELS

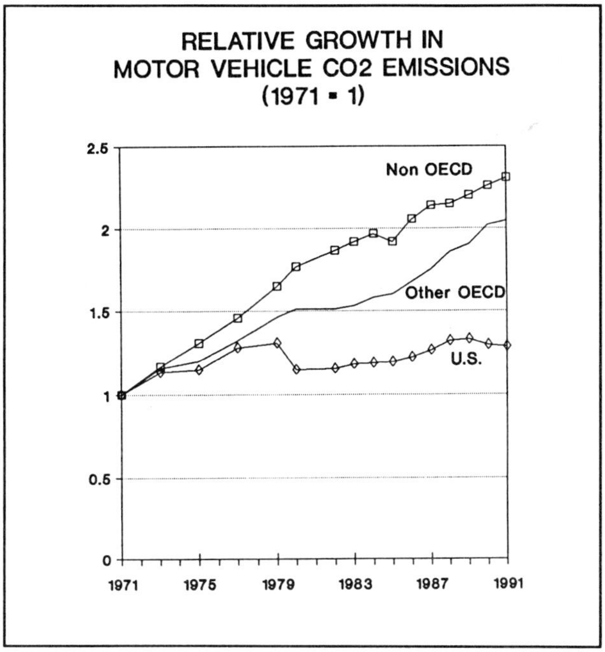

Figure 5

Of the various problems stemming from motor vehicle use, air pollution has received by far the greatest political attention. Clean-air legislation exists in most countries and the health and other effects of air pollution are more or less immediate and visible. Unlike global warming, there is little speculation about whether air pollution is real or threatening. In the United Sates, recent proposals to cut carbon monoxide and ozone concentrations have placed great emphasis on the introduction of so-called "alternative fuels" for vehicles. These fuels include methanol, compressed natural gas (CNG), ethanol, electricity, and hydrogen.

9. METHANOL

Methanol is currently made almost exclusively from natural gas. In the United States, methanol would be introduced in "flexible-fuel" vehicles that can burn methanol-gasoline blends ranging from pure gasoline to a mixture of 85 percent methanol and 15 percent gasoline (M85). In December 1991, the National Research Council (NRC) of the U.S. National Academy of Sciences released a major report on the American ozone problem and the potential benefits of methanol and other fuels in reducing ozone exposures.[14] According to the report, any ozone reductions from a switch to M85 would be modest and would be quite sensitive to the percentages of methanol and gasoline burned in the vehicles. The overall air quality impacts of using methanol blends and methanol oxygenates (MTBE) were also assessed by a group of university-base engineers in July 1993 in Science magazine.[15]

The alcohol fuels, [methanol and ethanol], especially when blended with gasoline and used in flexible fuel vehicles, provide little or no air quality advantages beyond the reduction of CO.

...the addition of oxygenates, such as methyl tertiary butyl ether (MTBE) and ethanol, has been found to have little or no effect on the problems of atmospheric reactivity and ozone formation.[16]

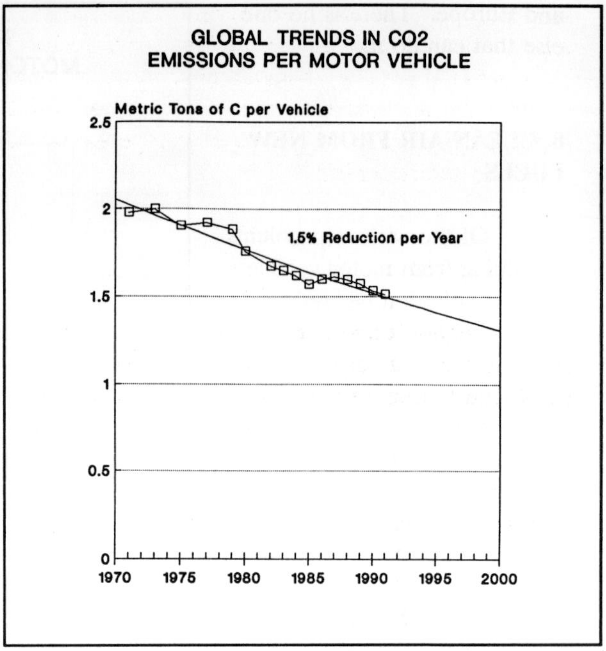

Figure 6

As for its global warming impacts, methanol, at least when made from fossil fuels, offers little greenhouse warming benefit. According to a Congressional Office of Technology Assessment (OTA) report, methanol "...use is expected to provide, at best, only a small greenhouse gas benefit over gasoline, and then only if the vehicles are significantly more efficient than gasoline vehicles."[17]

10. COMPRESSED NATURAL GAS (CNG)

Natural gas, primarily in the form of CNG, is another potential option to reduce air pollution and reliance on imported oil. CNG vehicles emit comparatively little carbon monoxide, reactive hydrocarbons, and particulates. According to the U.S. OTA and the National Research Council, a switch to CNG vehicles would definitely reduce ozone concentrations, at least in regions where reactive-hydrocarbon reductions are required.[18]

Per vehicle-kilometer traveled, CNG cars emit less carbon dioxide than gasoline-powered cars. But methane, the principal component of natural gas, is itself a potent greenhouse gas. On balance, using CNG vehicles could reduce overall greenhouse-gas emissions for light-duty vehicles by 15 percent and increase those of heavy-duty diesel-powered vehicles by about 5 percent.[19] Because CNG is a depletable fossil fuel, however, and its burning emits carbon dioxide, CNG vehicles aren't the answer to global long-term pollution and climate problems. But their limited use would reduce pollution emissions, especially from urban buses and trucks. In this limited use, CNG vehicles could make a positive, though ultimately transitional, contribution to reducing the risks of air pollution and heavy dependence on imported oil. Also, experience with CNG vehicles might make it easier to introduce hydrogen-powered vehicles later since the two technologies are similar.

11. ETHANOL

Ethanol is another alcohol fuel, commonly made from biomass feedstock. In a study of ethanol's impacts on air quality, Sierra Research -- a U.S. consulting firm -- concluded that gasohol (a mixture of 10 percent ethanol and 90 percent gasoline) would on average reduce CO concentrations by 25 percent, but would increase NOx emissions by 8 to 15 percent and evaporative emissions of VOCs by 50 percent.[20] Switching to gasohol, the researchers estimated, would increase ozone concentrations by at least a 6-percent. The academic study already cited concluded that the "addition of ethanol to gasoline is generally counterproductive with respect to ozone formation..[and that no] convincing argument based on combustion or atmospheric chemistry can be made for the addition of ethanol to gasoline."[21]

Estimates of the net greenhouse impacts of ethanol are subject to great uncertainty. All depend on which feedstock is used, what fuels are used in distillation, and how much energy is credited to the by-products (such as animal feeds, in the case of corn) of ethanol production. Calculating the greenhouse gas emissions for vehicles running on pure ethanol (E100), DeLuchi concludes that switching to motor vehicles running on E100 -- made from corn (as it is made in the United States) -- would *increase* greenhouse gas emissions by 20 percent (for light-duty vehicles) to 50 percent (for heavy-duty vehicles).[22]

12. ALCOHOL FUELS FROM OTHER BIOMASS SOURCES

Ethanol and methanol could be produced from other sources of biomass, including paper, municipal solid waste, agricultural and forestry residues and wastes, or low-grade "junk wood" stands.[23] But the removal of forest and agricultural residues would have to be carefully evaluated since it could reduce soil humus and increase erosion, water runoff, and nutrient loss.[24]

By far the largest potential source of alcohol fuels would be energy plantations, where crops or trees would be grown as a feedstock.[25] DeLuchi has estimated that under favorable conditions -- very energy-efficient processes using little or no fertilizer -- total emissions of all greenhouse gases from wood-based ethanol could decline by as much as 75 percent compared with those from gasoline.[26] He also points out, though, that

> ...if wood plantations require a lot of fertilizer, and if the manufacture of materials and equipment results in substantial emissions of greenhouse gases (and if other unfavorable conditions hold), then the biofuel cycles will provide no more than modest reductions -- and perhaps even no reduction at all -- in greenhouse gas emissions.[27]

Growing energy crops on a large scale entails some potentially significant environmental problems.[28] For example, growing short-rotation trees can lead to a loss of topsoil and nutrients and to soil compaction and increased water runoff. It can also require heavy applications of fertilizer and herbicide. Soil erosion where trees are harvested every five years has been estimated at two metric tons per hectare per year, while forest soils are formed at only 0.3 metric tons per hectare per year.[29] Detailed analyses of erosion rates for energy crops have yet to be made, and most existing data are on small field trials collected over short periods.[30]

Viewed solely as a means of cutting air pollution -- which is the focus of most advocates of these three fuels -- each of these energy sources has some merit, however small. All three would reduce carbon monoxide emissions relative to conventional gasoline. CNG would also reduce ozone concentrations. But when the impacts of the fuels on global warming are also taken into account, the overall attractiveness of these fuels declines significantly.

13. ZERO EMISSION VEHICLES: ELECTRIC AND HYDROGEN CARS

The only candidate vehicles that will curb air pollution, reduce oil consumption, and significantly cut greenhouse as emissions are electric vehicles (EVs) powered by batteries, flywheels, or hydrogen fuel cells.[31] California has adopted regulations requiring that, by the year 1998, 2 percent of new light-duty vehicles sold in the state must have zero emissions and that by 2003 ten percent, a requirement currently met only by electric vehicles.

14. AIR POLLUTION BENEFITS

Switching to EVs would significantly improve urban air quality, at least where relatively clean power plants are in service. The overall impacts on air quality, global warming, and oil use would be determined by the emissions from the power plants used to charge the EVs. Where clean power plants are used, there would be few carbon monoxide or hydrocarbons emissions compared with those from the cleanest gasoline vehicles. The California Air Resources Board -- taking into account the power plants used to generate electricity for the state -- estimates that an electric vehicle would have 97 percent lower emissions of carbon monoxide and reactive organic compounds compared with the cleanest gasoline vehicle available. Nitrogen oxides would be reduced 24 to 47 percent.[32] In a study of the effects of EVs on the northeastern United States, electric vehicles were estimated to reduce carbon monoxide and organic compound emissions by 99 percent and nitrogen oxide emissions by 70 percent.[33] Thus, if the power plants are reasonably clean, carbon monoxide and ozone problems would be greatly reduced by the introduction of electric vehicles. And -- unlike gasoline vehicles -- there would be no deterioration in electric vehicle emissions. Electric vehicles would also cut the emission of toxic substances such as benzene by about 99 percent. Offsetting these benefits, there could be increased emissions of sulfur dioxide and, hence, acid precipitation if high-sulfur emissions resulted from the generation of the electricity.

If EVs were recharged at night -- as they ought to be initially -- no new power plants would have to be built and ozone (which cannot form without sunlight) would thin out dramatically. If renewable energy sources were used to charge the batteries, pollution and greenhouse gas emissions would be very small indeed.

15. EVs AND GLOBAL WARMING

How much EVs can cut carbon dioxide emissions depends mostly on two factors: the electric efficiency of the vehicles and the emissions from the power plants that produce the electricity used to charge them. If EVs are charged by electricity from natural gas power plants, carbon dioxide emissions would fall by about 50 percent compared with similar gasoline powered vehicles. Charging electric vehicles with electricity made from coal, in contrast, would cut greenhouse gas emissions by only about 20 percent. In California, which derives much of its electricity from natural gas, hydro plants, wind machines, and geothermal energy, CO_2 emissions would be cut by about 70 percent compared with a conventional car. In the northeastern United States, where more coal is burned, these emissions would be cut by about 30 percent. In the longer term, emissions could be eliminated entirely by charging the batteries using either renewable or other non-fossil (perhaps nuclear) energy sources.

16. REDUCTION IN OIL USE

No matter how they are recharged, EVs will lead to reduced oil consumption. The largest reduction would occur where the electricity for recharging the EVs is generated without burning oil. But even where oil is used, there is a significant reduction. Taking into account refining (oil) and transmission (electricity) losses, we estimate that an EV recharged using an oil-fired steam electric plant would consume at least 30 percent less oil than a comparable gasoline-powered vehicle.

17. HYDROGEN-POWERED VEHICLES

Interest in hydrogen-powered vehicles has increased over the past decade as well, particularly in Japan and Europe. Production of hydrogen from water is the most likely long-term source of hydrogen with the primary energy supplied by non-fossil energy sources such as renewable or nuclear technologies. In this case virtually no carbon dioxide emissions would result.

Although hydrogen vehicles would have several advantages -- longer range and fast refueling -- their widespread commercial use is further down the road because of the lack of a supporting infrastructure. Still, like electric vehicles, they would form a natural link in developing a sustainable energy system, and their use would reduce oil imports, alleviate trade deficits, and cut air pollution and greenhouse gas emissions.

As a result of California's regulations, automakers around the world are developing electric- and hydrogen-powered vehicles. Electric-vehicle fleet tests are being conducted in France, Germany, Japan, and the United States. While electric-vehicle prices are still higher than those of conventional vehicles, as technology improves -- better batteries, more efficient heating and cooling systems for EVs, and so forth -- prices will drop and a global transition will begin to vehicles that will dramatically cut street-level pollution and -- as power plants become more efficient -- carbon dioxide emissions as well. These vehicles, re-charged eventually by emissionless power sources such as wind turbines, photovoltaic cells, or nuclear plants, will serve as one of the legs for a sustainable transportation system.

18. CONCLUSIONS AND RECOMMENDATIONS

For several important reasons -- reducing air pollution, cutting global CO_2 emissions, coping with limited world oil resources -- international efforts should be accelerated to develop and introduce electric and hydrogen vehicles. The industrialized countries should increase their research and development of better, cheaper batteries and fuel cells as well as renewable energy technologies to provide the needed electricity and hydrogen. It is also important that these sustainable transportation technologies be introduced as quickly as possible into the economies of industrializing countries where much of the growth in the global motor-vehicle fleet will be occurring.

19. REFERENCES

1. American Automobile Manufacturers Association. 1993. "Facts and Figures '93." Washington, D.C.
2. A. Fail, et al. April 1992. "Air Pollution from Motor Vehicles." World Bank.

3. World Resources Institute. 1992. World Resources 1992-93. Oxford University Press, N.Y. P. 196.
4. William Brown. 1994. "Dying From Too Much Dust." New Scientist. 12 March, 1994. pp 12-13.
5. William Brown, "Deaths Linked to London Smog," New Scientist, June 25, 1994.
6. UNEP and WHO. March, 1992. "Air Pollution in the World's Megacities." Environment.
7. J.J. MacKenzie and M.P. Walsh. 1990. "Driving Forces: Motor Vehicle Trends and Their Implications for Global Warming, Energy Strategies, and Transportation Planning." The World Resources Institute, Washington, D.C.
8. M. King Hubbert. "Techniques of Prediction As Applied to the Production of Oil and Gas." Oil and Gas Supply Modeling. National Bureau of Standards Special Publication 631, U.S. Department of Commerce, May 1982; M. King Hubbert. "Degree of Advancement of Petroleum Exploration in United States." The American Association of Petroleum Geologists Bulletin, Vol. 51, No. 11, November 1967.
9. C.D. Masters et al., "Resource Constraints in Petroleum Production Potential," Science Vol. 253, July 12, 1991, pp. 146-52.
10. See endnote 7.
11. Estimate based on data from the OECD.
12. See endnote 7.
13. Intergovernmental Panel on Climate Change (IPCC). 1990. "Climate Change, The IPCC Scientific Assessment." Cambridge University Press; 1992. "1992 IPCC Supplement, Scientific Assessment of Climate Change." Cambridge University Press.
14. National Research Council. Rethinking the Ozone Problem in Urban and Regional Air Pollution. December 1991, Washington, D.C.
15. J.G. Calvert et al. "Achieving Acceptable Air Quality: Some Reflections on Controlling Vehicle Emissions." Science, Vol. 261, 2 July 1993, pp. 37-45.
16. Ibid. p. 42.
17. U.S. Congress, Office of Technology Assessment. "Replacing Gasoline, Alternative Fuels for Light-Duty Vehicles." OTA-E-364, Sept. 1990, p. 71.
18. U.S. Congress, Office of Technology Assessment. "Replacing Gasoline, Alternative Fuels for Light-Duty Vehicles." OTA-E-364, September 1990, p. 98; National Research Council. "Rethinking the Ozone Problem in Urban and Regional Air Pollution." December, 1991, Washington, D.C., p. 402.
19. Mark A. DeLuchi. "Emissions of Greenhouse Gases from the Use of Transportation Fuels and Electricity, Vol. 1." ANL/ESD/TM-22, Vol.1, Center for Transportation Research, Energy Systems Division, Argonne National Laboratory, Argonne, Illinois, November 1991.
20. Sierra Research, Inc. "The Air Pollution Consequences of Using Ethanol-Gasoline Blends in Ozone Non-Attainment Areas." Sacramento, California May, 1990.
21. J.G. Calvert et al. "Achieving Acceptable Air Quality: Some Reflections on Controlling Vehicle Emissions." Science, Vol. 261, 2 July 1993, p. 42.

22. Mark A. DeLuchi. "Emissions of Greenhouse Gases from the Use of Transportation Fuels and Electricity, Vol. 1." ANL/ESD/TM-22, Vol.1, Center for Transportation Research, Energy Systems Division, Argonne National Laboratory, Argonne, Illinois, November 1991, p. 59.
23. C.E. Wyman et al. 1993. "Ethanol and Methanol From Cellulosic biomass," Renewable Energy, Sources of Fuels and Electricity, Ed. by T.B. Johansson, et al., Island Press, Washington, D.C.; "Meeting Transportation Needs Through Biofuels," Energy, February 1993, pp. 23-24; L.R. Lynd et al. 1991. "Fuel Ethanol from Cellulosic Biomass." Science, Vol 251, March 15, pp. 1318-1323.
24. U.S. Congress, Office of Technology Assessment. 1993. "Potential Environmental Impacts of Bioenergy Crop Production." Background Paper, OTA-BP-E-118, pp. 34.
25. Interlaboratory Report. 1990. "The Potential of Renewable Energy: An Interlaboratory White Paper," National Renewable Energy Laboratory, Golden, Colorado, p. B-5.
26. Mark A. DeLuchi. 1991. "Emissions of Greenhouse Gases from the Use of Transportation Fuels and Electricity, Vol. 1." ANL/ESD/TM-22, Vol.1, Center for Transportation Research, Energy Systems Division, Argonne National Laboratory, Argonne, Illinois, p. 126.
27. Ibid.
28. J.M. Ogden and M.A. DeLuchi. 1991. "Solar Hydrogen Transportation Fuels," presented at the Conference on Transportation and Global Climate Change: Long Run Options." Asilomar, CA, August, 1991; J.H. Cook et al. 1991. "Potential Impacts of Biomass Production in the United States on Biological Diversity." Annu. Rev. Energy Environ. pp. 401-31; D. Hall, D. et al. 1993. "Biomass For Energy: Supply Prospects." Renewable Energy, Sources of Fuels and Electricity, Ed. by T.B. Johansson et al., Island Press, Washington, D.C.; Interlaboratory Report. 1990. "The Potential of Renewable Energy: An Interlaboratory White Paper." National Renewable Energy Laboratory, Golden, Colorado.
29. D. Pimentel and J. Krummel. 1987. "Biomass Energy and Soil Erosion: Assessment of Resource Costs." Biomass, 14 pp. 15-38.
30. Interlaboratory Report. 1990. "The Potential of Renewable Energy: An Interlaboratory White Paper." National Renewable Energy Laboratory, Golden, Colorado, p. 23.
31. James J. MacKenzie. 1994. The Keys to the Car: Electric and Hydrogen Vehicles for the 21st Century. World Resources Institute.
32. California Air Resources Board. 1994. "Technical Support Document, Zero-Emission Vehicle Update."
33. Northeast States for Coordinated Air Use Management (NESCAUM). 1992. "Impact of Battery-Powered Electric Vehicles on Air Quality in the Northeast States, Final Report."

TOWARD GLOBAL PLANNING OF SUSTAINABLE USE OF THE EARTH
Development of Global Eco-Engineering
S. Murai (Editor)

It is too late to achieve sustainable development, now let us strive for survivable development

® Dennis L. Meadows, 1994

Institute for Policy and Social Science Research, University of New Hampshire, Durham, NH 03824, USA

Abstract

The popular term, "sustainable development" is mainly used by those who are searching for means and justifications to sustain economic growth long enough that there is some hope of stabilizing the global population through the influence of the demographic transition. But simulations with World3, our model of the causes and consequences of growth in the world's population and economy, show that it will take another century and productive capital stocks 60 times current levels to achieve this transition. And population would grow to 12-15 billion people. Since the global ecosystems are deteriorating already under the load of current population and current economies, it is a fantasy to think that sustainable development can be realized. Instead we face a future that will include overshoot and collapse in population and in important economic systems.

I summarize a recent study on the impending global food shortage to illustrate the proximity and the magnitude of the problems we face. Then I pose a number of questions that we need to begin addressing, if we are to structure a new international system that can carry us through the collapse with minimal damage to the long-term carrying capacity of the globe.

1. PREAMBLE

It is extremely important to note at the outset of this paper, that several questions I pose towards the end of this talk have not been the focus of my objective research. The issues dealt with here do arise directly out of my research, and I have reflected on them extensively. But many of the thoughts I offer here are *informed intuitions* rather than research findings. So the following is certainly incomplete and probably full of errors. My only defense is that I consider the issues to be at the same time profoundly important and almost totally ignored by serious scientists. So the criticisms which I may provoke here should serve the useful purpose of stimulating others to do a much better job.

My challenge to you is summarized nicely by a quote in the *Satires* of Quintus Horatius Flaccus, 35 B.C. "If you know better precepts than these, candidly tell me; if not, follow them as I do."

2. INTRODUCTION TO SUSTAINABLE DEVELOPMENT

The most widely-used definition of sustainable development was coined originally by the Brundtland Commission in their report, *Our Common Future*:

> to meet the needs of the present without compromising the ability of future generations to meet their own needs.

Everyone getting what they want, now and forever, is a wonderfully attractive notion, because it frees politicians from dealing very concretely with the problems associated with redistribution across space or time. As a consequence the phrase has become very popular; it has been the focus of hundreds of books and thousands of conferences over the past decade. Most nations have officially designated commissions or committees on sustainable development. The tenor of these can be judged by looking at the vision underlying the United States' President's Council on Sustainable Development.

> We are committed to the achievement of a dignified, peaceful, and equitable existence. We believe a sustainable United States will have an economy that equitably provides opportunities for satisfying livelihoods and a safe, healthy, high quality life for current and future generations.

Every politician and most industrial leaders enthusiastically support sustainable development, at least until someone finds a way of proving that its attainment would require them to give up any short-term advantage.

As an abstract goal, "sustainable development" is attractive. But until it incorporates explicit and well-founded notions of the globe's carrying capacity, it cannot serve as a concrete basis for planning. Worse, the enthusiasm for sustainable development diverts our attention from a set of issues that we need to address now.

3. MODES OF TRANSITION FROM GROWTH TO EQUILIBRIUM

As illustrated in Figures 1, 2, and 3, global population, global industry, and global use of materials, for example metals, are all growing exponentially at rates that have produced doublings every 2-3 decades. These three illustrations and all of those that follow in this report are taken from our recent book, *Beyond the Limits*.[1] These three are from (*BtL*, pp. 4,5, & 82).

There are only three possible futures for the exponential trends shown in the figures above. They are shown in Figure 4 (*BtL*, p. 108).

It was once assumed implicitly that physical growth on the planet could continue more or less indefinitely, that Figure 4a was our foreseeable future. Now the growing recognition of world-wide environmental stress has brought most analysts to acknowledge that the earth has a finite carrying capacity which we are already near. Thus sustainable development will require population growth

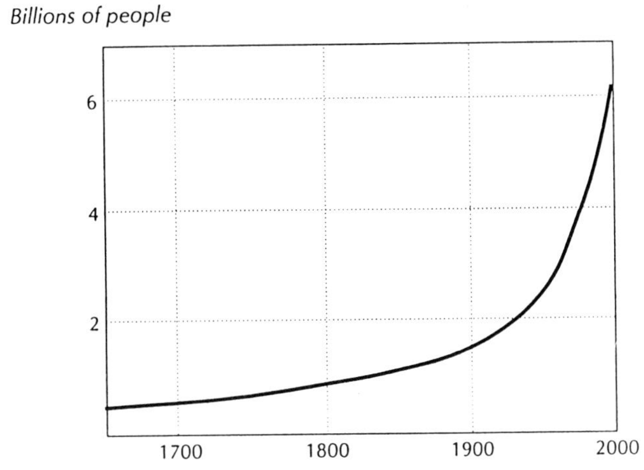

Figure 1. World population has been growing exponentially since the beginning of the Industrial Revolution. In 1991 the world population growth rate was estimated to be 1.7%, corresponding to a doubling time of 40 years.

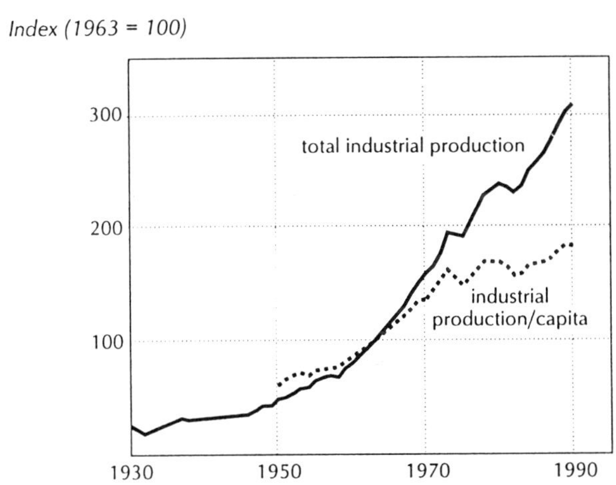

Figure 2. World industrial production, relative to the base year 1963, shows clear exponential increase, despite fluctuations due to oil price shocks. The 1970-1990 growth rate in total production has averaged 3.3% per year. The per capita growth rate has been 1.5% per year.

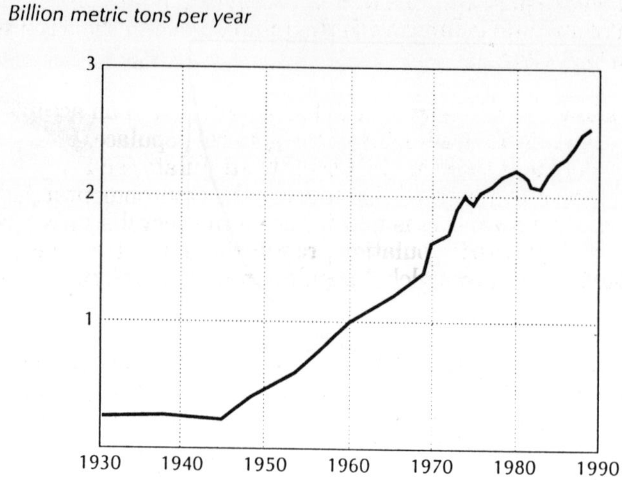

Figure 3. Total world metal consumption showed a slight downturn in the economic recession of the early 1980s, but then continued to rise.

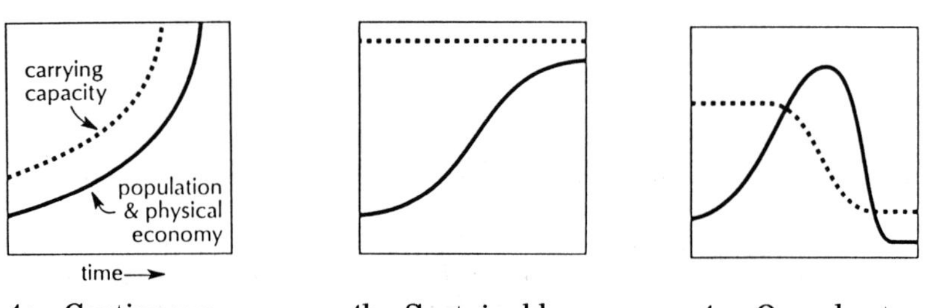

4a - Continuous growth

4b - Sustainable development

4c - Overshoot & collapse

Figure 4. There are three possible paths through which a population can approach its carrying capacity: 4a- continuous growth, 4b- sustainable development, and 4c-overshoot and collapse.

essentially to end. And it will require that the flow of energy and materials through the environment come into balance with the productive and the assimilative capacities of natural systems.

But history has only shown us one way to stabilize population in an acceptable fashion - by raising the material well-being of the general populace. So efforts at sustainable development today mainly seek to stimulate (and sometimes redistribute) economic growth while reducing the environmental damage it entails. The image of the future that is now inherent in every discussion of sustainable development is Figure 4b. Population growth slows as birth rates fall to equal the low death rates enjoyed by a global population with a uniformly long average lifetime.

There is some historical basis for expecting population to quit growing; that has happened already in many of the wealthy nations. Through a process called the *demographic transition* populations evolve through three stages. First, while the country is poor, the birth and death rates are both high and approximately equal; population growth is slow. Then economic growth occurs. It produces many social and economic changes that increase life expectancy, lowering the death rate. However, birth rates are still high and population growth is rapid. In the third stage, birth rates decline until they are approximately equal to the low death rates, and population growth is again slow or nonexistent. This process is illustrated in Figure 5 (*BtL*, p. 30) for three industrialized nations: Sweden, England & Wales, and Japan. Notice that it required well over 100 years to complete the transition in each of these lands.

Demographic statistics for the entire globe do show the first phase of the transition. As seen in Figure 6 (*BtL*, p. 25), the death rate (shown in persons/1000 people per year) did start to decline around 1850. Only after World War II did the birth rate start to show significant declines. However, even today, the death rate is falling about as fast as the birth rate. The results are shown in Figure 7 (*BtL*, p. 26), which indicates the number of people added to the global population (births-deaths) each year between 1900 and 2100. The current rate of increase is almost 100 million people per year. And conventional wisdom holds that the growth will stay near this level through the first several decades of the next century. United Nations experts expect that the global population will still be growing at the end of the next century, by which time global population will be 10-12 billion. And most of those added will be in the poorer nations.

Note that there is no comparable historical basis for expecting self-limitation in material consumption. We have a number of examples where important countries have achieved zero population growth through peaceful, internal means. I do not know of a single case where a major country has willingly stabilized its economic output and material standard of living.

Of course there are myriad examples of efforts to understand and to stabilize specific flows of energy, raw material, or pollution through the natural systems of a country. Some examples of effective efforts to reduce specific flows globally are even available - for example by the end of 1993 global production of chloro

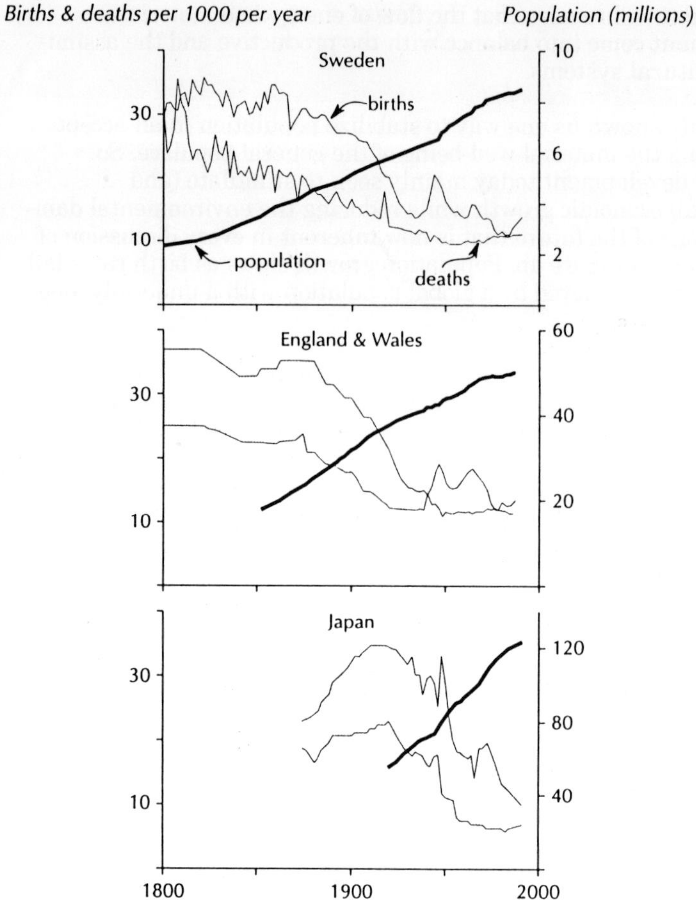

Figure 5. In the demographic transition a nation's death rate falls first, followed later by its birth rate. Sweden's demographic transition occurred over almost 200 years, with the birth rate remining rather close to the death rate; Japan took slightly more than a century.

fluorocarbons had declined to just 40 percent of their peak levels in 1988.[2] And the papers presented at this conference describe a very diverse variety of efforts to monitor and reduce environmentally-damaging flows. But there is no national effort to reduce all flows and no widespread to limit growth in material well-being for a population. Quite the contrary. Politicians everywhere are convinced that their constituents need much more growth in their standard of living.

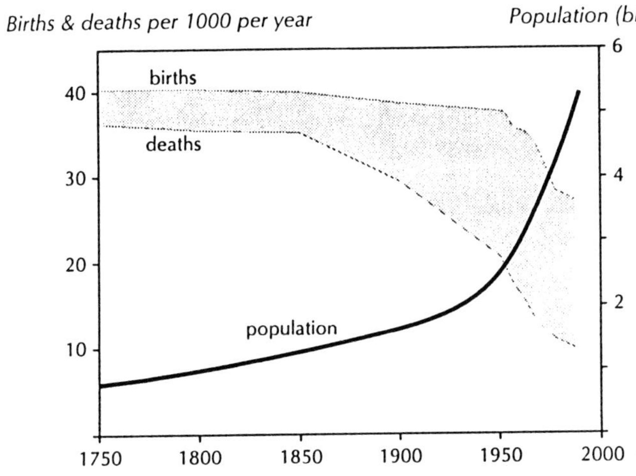

Figure 6. The shaded gap between births and deaths shows the rate at which the population grows. Until about 1970 the average human death rate was dropping faster than the birth rate, and the population growth rate was increasing. Since 1970 the average birth rate has dropped slightly faster than the death rate.

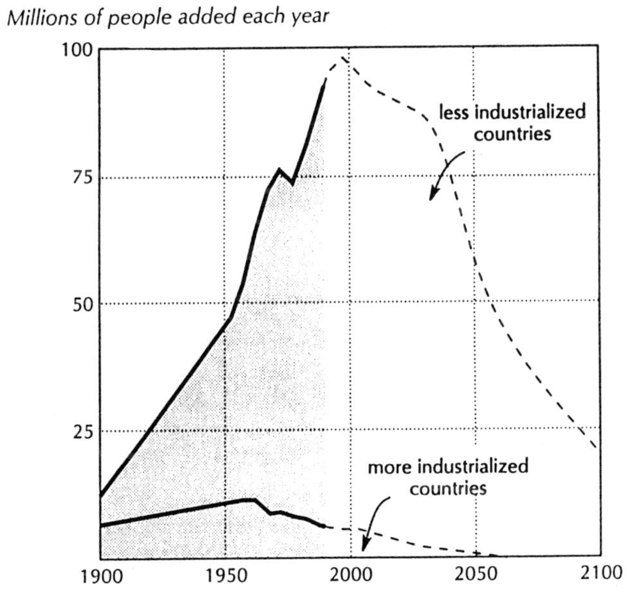

Figure 7. The number of people added to the world population each year has increased enormously and is projected to go on increasing for another decade under the World Bank's forcasts. These forecasts are very optimistic; they assume rapid drops in birth rates in the less industrialized countries.

Indeed most analysts urgently call for economic growth to continue long enough to give a decent living standard to the several billion people on the globe that live in poverty and to the new people who will be added to the global population. Sustainable development is therefore really used to mean sustainable growth - for an indefinitely long period during which we hope to continue getting all the goods from an expanding economy while somehow drastically reducing all the bads, meanwhile encouraging people to have lower birth rates.What would this really look like? What changes would we have to imagine, for this notion of sustainable development actually to be realized.

4. DIMENSIONS OF THE "SUSTAINABLE" TRANSITION

Over the past twenty years, I have worked intermittently with my colleagues on a global simulation model of population and economy, World3, that gives some answers to the above questions. By incorporating in the model a diverse, comprehensive, and profoundly optimistic set of assumptions about political, economic, cultural, and technological changes, World3 can give us a picture of the world in 2100 that has finally stabilized its population through pursuit of sustainable development with its reliance on sustained economic growth and the workings of the demographic transition. The outlines of that world are shown in Figure 8 (*BtL*, p. 121).

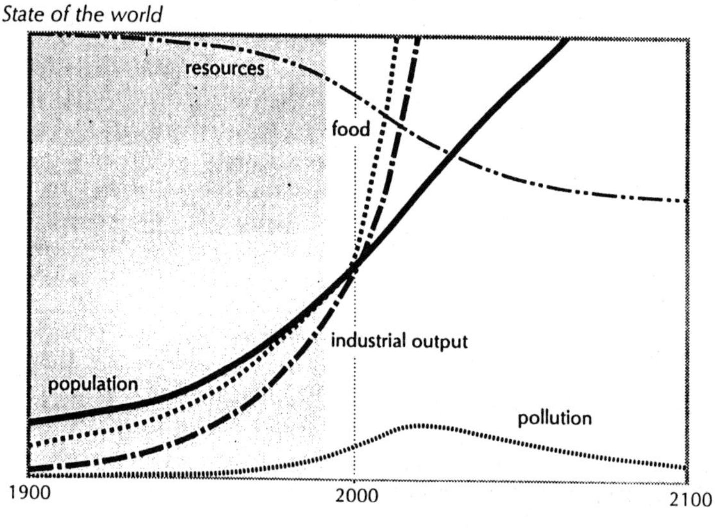

Figure 8. If all physical limits to the World3 system are removed, population grows to about 15 billion and levels off in a demographic transition. The economy grows until by the year 2100 it is producing 55 times the 1990 level of industrial output and 8 times the present level of food.

In Figure 8 curves for global aggregate population, unexploited nonrenewable resources, ambient levels of persistent pollution, food production, and industrial output trace out the gross physical dimensions of the future economy and society that are sought implicitly by the proponents of sustainable development.

World3 incorporates four different important physical limits on growth. Into the model we incorporated a limited stock of arable land, finite possibilities for yield from each hectare, a limited, though very large, stock of unexploited nonrenewable resources, and finite capacity of the global ecosystem to assimilate pollution.

When these physical limits in the World3 system are effectively removed, population grows to 15 billion and then levels off through the influence of a global demographic transition. The economy grows until in the year 2100 it is producing 55 times the 1990 level of industrial output and 8 times as much food as it did in 1990. This growth is achieved while using only 5 percent as many nonrenewable resources and producing only 15 percent as much pollution as today. In other words, the resource intensity of industrial growth is reduced by a factor over 1000. To achieve this outcome the world's economies would have to accumulate, site, and operate 60 times as much productive capital in the twenty-first century as they did during the whole of the twentieth century.

The current discussions about sustainable development are only useful to those who think the numbers and ratios cited above are feasible. I do not. As I will show, current food production levels are already stressing the ecosystem. And many of the papers presented at this meeting show that the current stock of capital is already deteriorating many important air, water, and soil resources. There is widespread evidence that we are already above the sustainable limits, that we have already overshot the carrying capacity.

5. SYMPTOMS OF OVERSHOOT

When society and economy are burdening the environment above sustainable levels you would expect to see a variety of symptoms. For example you would detect wide-spread evidence of:

- falling stocks of ground waters, forests, fish, soils.
- rising accumulations of wastes and pollutants.
- growing amounts of capital, energy, materials, and labor devoted to exploitation of more distant, deeper, or more dilute resources.
- more and more capital, energy, materials, and labor required to compensate for what were once free natural services (sewage treatment, flood control, air purification, pest control, restoring soil nutrients, preserving species).
- capital, energy, materials, and labor diverted to defend or gain access to

resources that are concentrated in a fewer and fewer places (such as oil in the Middle East).

- deterioration in physical capital, especially in long- lived infrastructure.
- reduced investment in human resources (education, health care, shelter) in order to meet consumption needs or to pay debts.
- increasing conflict over resources or pollution emission rights. Less social solidarity, more hoarding, greater gaps between haves and have-nots.

As the papers presented at this conference and the daily media make clear, all of those symptoms are evident today. If you accept this assessment, you are led to Figure 4c as a more appropriate vision of our future. Indeed, Figure 4c, which we label *overshoot and collapse* illustrates the dominant mode of behavior that has been observed in closed, natural systems when enough constraints are removed that one species can temporarily enter into a phase of uncontrolled exponential growth.

Overshoot and collapse is also the scenario that emerges from World3 when realistic assumptions about social delays, economic policies, and physical limits are reinserted in the model.

No single run of our model can be considered as a prediction of the future. Each simulation run illustrates general properties of a possible future. But a very typical example of what emerges in our model when society tries to sustain traditional policies related to economic and demographic growth is shown in Figure 9 (*BtL*, p. 135).

In this scenario the world society proceeds along its historical path as long as possible without major policy change. Technology advances in agriculture, industry, and social services according to established patterns.

The global population in this scenario rises from 1.6 billion in 1900 to over 6 billion in the year 2000. Total industrial output expands by a factor of 20 between 1900 and 1990, and it does so while using less than 15 percent of the earth's total stock of nonrenewable resources. In 1990 over 85 percent of these resources remain. Pollution in that year has just begun to rise significantly. Life expectancy is increasing, services and goods per capita are increasing, food production is increasing. But major changes are just ahead.

Just after the simulated year 2000 pollution rises high enough to begin to affect the fertility of the land. At the same time land erosion increases. Total food production quites rising after 2015. That causes the economy to shift more investment into the agriculture sector. But agriculture has to compete for investment with a resource sector that is also beginning to sense some limits.

As both food and nonrenewable resources become harder to obtain in this simulated world, capital is diverted to producing more of them. That leaves less

State of the world

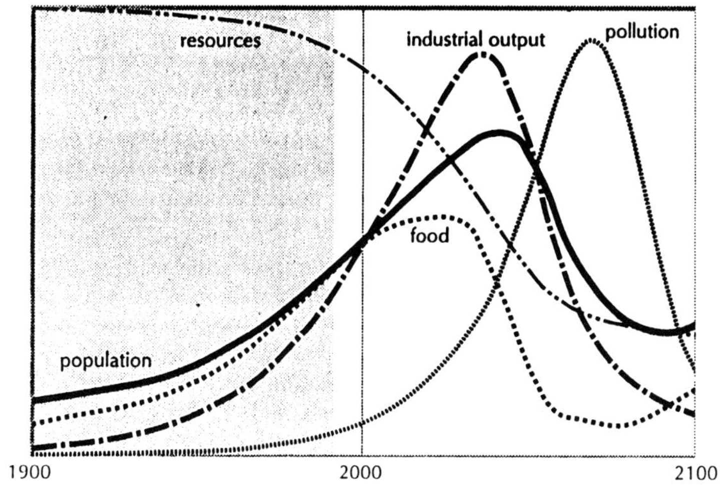

Figure 9. In this scenario population rises to more than 9 billion in 2040. These increased levels of demand generate much more pollution and soil erosion, which reduce land yield and forces much greater investment in agriculture. Eventually declining food raises the population death rate.

output to be invested in capital growth. Next the industrial capital plant begins to decline, taking with it the service and agricultural sectors. Finally population too begins to decrease, as the death rate is driven upward by lack of food and health services.

In this graph the resources line can be viewed usefully as a general indicator of stress or pressure exerted by the environment on the processes that sustain industrial and demographic growth.

Notice that for the entire century 1900- 2000, there is little deterioration in resources. Then "suddenly" they begin to decline rapidly. In 1990 the nonrenewable resources remaining in the ground would have lasted 110 years at the 1990 consumption rates. No serious resource limits would be obvious under those circumstances. But by 2020 the remaining resources constituted only a 30 year supply. Why did the shortage arise so fast? Because exponential growth rates increased the consumption while lowering the resources. Between 1990 and 2030 population increases by 50 percent and industrial output grows by 85 percent. The nonrenewable resource usage rate doubles. During the first 3 decades of the 21st century the rising population and industry use as many nonrenewable resources as the global economy used in the entire century before.

This kind of precipitous decline might seem unlikely after decades of steady progress. However, a recent study of the globe's food production potential illustrates this effect exactly.

6. OVERSHOOT OF GLOBAL FOOD DEMAND

In *Full House* Lester Brown and Hal Kane summarize the world's experience in raising cereal output from 1950 to 1990.[3] They use the medium-level projections of country populations by the Center for International Research of the U.S. Bureau of the Census. They incorporate in their analysis assumptions about technological advance, potential for irrigation, soil erosion, and other environmental and social factors. Their data are summarized in Table 1 below for the 4 largest nations and the next 9 - together these 13 countries account for about 70 percent of the world's population

Table 1
Grain production and consumption for the world's 13 largest countries, 1950-2030

United States, China, India, Former Soviet Union

year	grain production	grain consumption	net trade
1950	378	365	13
1990	959	926	33
2030	1099	1303	-204

Bangladesh, Indonesia, Iran, Pakistan, Egypt, Ethiopia & Etritrea, Nigeria, Brazil, Mexico

year	grain production	grain consumption	net trade
1950	57	59	-2
1990	167	199	-32
2030	273	420	-147

Compiled from data presented in *Full House*, pp.168, 178, 179.

After four decades of growth that generally matched production to consumption, limits on food output produce a rapidly growing deficit. A key factor in this precipitous decline is diversion of crop land to other economic uses. This happened in Japan, South Korea, and Taiwan, for example. The result was that for Japan, between 1960, when grain production peaked, and 1992 grain production dropped 33 percent. But of course as production declined, demand continued to climb steadily as a result of population growth and growing affluence. The net effect is that by 1993, Japan was importing 77 percent of its grain, South Korea 68 percent, and Taiwan 74 percent. At 27 million tons, Japan is today the world's leading importer of grain. China, the major cause of the deficits projected above, is entering this same transition.

The problems are not limited just to these 13 countries nor to the distant future. United Nations statistics reveal that food production per person declined in 94 countries over the period 1985-1989.[4]

7. IMPLICATIONS OF OVERSHOOT

Note that even after the decline in World3, our model still shows there to be more population, food, and industry than there were globally around 1950. So the scenario portrayed in Figure 9 does not imply the disappearance of our species - just an adjustment to the carrying capacity of the globe. We have witnessed and survived this adjustment in the past, but only on a national or regional scale. This time it will be global.

There have been various qualitative efforts to estimate what population could be supported globally from the income of the renewable energy and resource base. The sustainable population estimates that emerge from these efforts are generally around 2-3 billion. Our model results are consistent with that estimate.

Of course our model deals with global averages, and we all know that there will be great variation around the mean. Some countries might possibly pass through this period with little reduction in their standards of living. Others would suffer terribly.

It is possible that the end of growth will come through forces so sudden, so strong, and so pervasive that essentially all nations lose control of their future. But I believe it is much more likely that the countries that are clever, far-sighted, and nimble in their production patterns will retain substantial ability to control their own development and provide for their people.

The question here is in the timing and the identity of the the limiting factors. In this connection I find it useful to differentiate between Global and Universal Problems. Global problems affect everyone, and they can only be solved by concerted action of many nations. Examples are climate change, depletion of the ocean fisheries, and stratospheric ozone depletion. Universal problems affect everyone, but they can be solved locally through local initiative. Examples are soil erosion, ground water contamination, over-harvesting of the forests, and loss of local species. Japan can not save the ozone layer alone, but Japan has not had to enlist the help of India and other nations in its successful fight against soil erosion.

If growth is stopped by global problems, then most countries will be severely affected. If the principal pressure is exerted by universal problems, the richer countries can probably buy their way out, just as they have until now imported food to compensate for overshooting the capacity of their own agricultural soils.

8. NEW ETHICS AND MODES OF GOVERNANCE

So long as international relations can be based on the shared myth that all countries will pass through the demographic transition and achieve decent living standards, the ethical issues are relatively straightforward. But suppose it became generally accepted that many countries were going to experience population collapse through rising death rates? What then should be the stance of a responsible member of the international community? The question has many aspects.

I will pose a number of questions that occur to me. It is not a comprehensive list, nor do I know the correct answers. I just know that we will soon confront these questions either explicitly or by default.

What should be the policy related to immigration? There will be a greater supply of possible immigrants than can possibly be accommodated by any nation that wants to maintain something near to its current living standards for all its people. So some force will limit in-migration. What will it be; what should it be? Quotas? Military force? Something else?

The Red Cross has estimated that the number of refugees displaced by environmental problems is about 500 million. This number is bound to increase. What mechanisms of international assistance and governance can be implemented to deal with a large and more or less permanently stateless population of people. What should be the long-term goals for assisting these people?

Supporting a population through periods of famine makes sense, if there is a prospect of the people quickly returning to self sufficiency in a world where population and food needs are roughly in balance. But what should be the policy, when the world has entered into a period of chronic food deficit, and some nations have no prospect of regaining self-sufficiency?

The effects of overshoot and collapse will be imposed and moderated within each nation by the diverse flows which cross its borders. These days any modern country is linked to many lands through flows of information, money, pollution, energy, natural resources, manufactured goods, food, and people. As we enter the period of the transition, there will be interruptions in some of these flows - either unintentionally or by those who hope to alter the terms of trade by embargoes and sanctions. What goals and ethical norms should govern these flows; what mechanisms of governance will we need to attain the goals?

The transition will see drastic shifts in equity. As shortages grow in a region, there is a basic tendency for the powerful to manipulate the system so as to retain what they have. If they succeed, as they generally do, the result is rapidly escalating gaps between the rich and the poor. Should that be a concern for the international community? Should it be a basis for intervention?

When a population has overshot its carrying capacity, one priority is to protect the fertility of the most crucial resources, so as to minimize the decline in the number of people and the standard of living that will be possible after the transition. What and where are those resources? What rights and responsibilities does the international community have to protect them from short-term destruction?

9. THE CONCEPT OF SURVIVABLE DEVELOPMENT

The current conversations about sustainable development are suited to a world that is moving through a transition in which population growth slows through declining birth rates to an orderly accommodation with its limits and in which the earth can support real economic growth for many decades into the future to raise the living standards of the present and the impending poor. For the reasons cited above, I do not believe that we live on such a world. I believe we live on a planet where population is already above the carrying capacity. Thus I expect death rates to rise above the birth rates until they are sufficient to reduce population significantly below current levels. I believe this transition will be obvious during my life time and will be essentially complete by the end of the next century.

Fifty to one hundred years seems like a long time. But the costs and benefits of many actions discussed at this meeting will unfold over that time period. So the transition from growth to overshoot and collapse is well within the interval that should have important effects on our current efforts related to the theme of this conference, "Toward Global Planning of Sustainable Use of the Earth."

Almost every one of the papers presented at this meeting has dealt with some aspect of the threats to productivity of the globe's natural ecosystems, that is with some aspect of the threats to or the protection of the planet's carrying capacity. If the work presented here is to have real effects, the knowledge you are generating through your research will have to become embedded in the policies that govern the actions of corporations, governments, and citizens. The actions we should propose in order to achieve the best possible future for our species depend profoundly on assumptions we make about the modes of governance, about international standards of ethical conduct, about the efficacy of international agreements, about the amount and the allocation of discretionary capital investments within the global society over the coming century.

It would be serious mistake to assume that the actions we propose now will be implemented within a political and economic system like those we have today. Because both will change dramatically, as society moves further along the path illustrated by Figure 4c.

Path 4b requires policies of sustainable development; path 4c requires something else. I call it policies for *survivable development*. I chose the term because of the alliteration with "sustainable" and because of the stark contrast it offers with the term sustainable. The phrase is inappropriate in some ways. It implies the need for a kind of bunker mentality, and it paints the future in much more negative terms than I think are justified. Maybe *flexible, adaptive*, or *resil-*

ient development would be technically better. But *survivable development* is easier to associate with the special perspectives discussed here.

We need policies for our cities, nations, and international organizations that can address the real and pressing needs of people today while simultaneously providing ethics, technologies, capital, and governance that can survive the period of population collapse. They should be designed to minimize the potential for catastrophic discontinuities during the transition and leave us with the maximum possible set of options after balance has been reestablished.

We will not find those policies nor develop those technologies, if we remain preoccupied with the myths that underlie the ideas of sustainable development - if we continue to pretend that the whole world is going to move through the demographic transition that we have witnessed in a few rich countries. We will only find them, if we devote at least a portion of our resources to understanding implications of a future that is much less attractive and politically popular, one that explicitly acknowledges the probability of overshoot and collapse.

In our studies of alternative scenarios for population and industrial growth through the end of the next century, perception delays were shown to be a major factor governing the likelihood and the intensity of population and economic collapse. The shorter the delays in recognition and response, the less damage is done to the global ecosystem during the period of overshoot. Let us shorten the delays and quickly acknowledge our realistic options. It is too late now to strive for sustainable development; let us begin to understand the nature of a strategy geared instead to survivable development.

It is time to concede honestly that ecological, political, and economic realities make it impossible to meet the needs of either present or future generations except at population levels far below the current 5.6 billion. The level of global population that can be sustained at decent living standards without damage to the global ecosystem is not precisely calculable, but it is below current levels. Inevitably the population will decline until it is again below the support capacity of the globe. What mechanisms will produce the decline? What quality of political, cultural, and economic well-being will we have after? We have inadequate scientific tools for dealing with these questions, and they involve very troubling ethical issues. But our future as a species is likely to be more attractive if we start to face them honestly.

10. REFERENCES

1 Donella H. Meadows, Dennis L. Meadows, and Jørgen Randers, *Beyond the Limits*, Chelsea Green Publishing Co., P.O. Box 428, White River Jct., VT 05001, 1992.
2 Lester R. Brown, et. al., *Vital Signs 1994*, W. W. Norton & Co., NY 1994, p. 64.
3 Lester R. Brown and Hal Kane, *Full House*, W. W. Norton & Co., NY 1994.
4 United Nations Food and Agriculture Organization, *The State of Food and Agriculture 1990*, Rome, Italy, 1990.

TOWARD GLOBAL PLANNING OF SUSTAINABLE USE OF THE EARTH
Development of Global Eco-Engineering
S. Murai (Editor)

Global Mapping
—— Contribution of Mapping Community ——

Minoru Akiyama

Director for Environmental Geographic Information, Planning Division
Geographical Survey Institute
Kitasato-1, Tsukuba-shi, Ibaraki-ken 305 Japan

Abstract

Today, many scientists and government administrators are devoted to different studies to understand and solve global environmental problems. These studies need basic geographic information which describes the present conditions of the earth's environment and detects the changes. It is, therefore, the responsibility of surveying and mapping community to develop the Global Map, an integrated geographic data set which covers the whole land of the earth with consistent specifications and the spatial resolution of at least 1 km on the ground, and to provide it to the international community. In this regard, the Geographical Survey Institute of Japan and the Ministry of Construction of Japan have proposed to create global maps, a basic integrated geographic data set for global change researches under international cooperation. Last year we have studied on the specifications of the Global Map with scholars and specialists in the related fields. Here, the outline of the specification is shown.

1. Specifications of the Global Map

The Global Map is an integrated geographic data set which covers the whole earth with consistent specifications and the spatial resolution of at least 1 km on the ground. It needs to include the data of rivers, shorelines, roads, place names, terrain relief, vegetation, hydrological conditions, land use, soils, geology, climate, population, various economic indices, and so forth. More detail specifications of the Global Map is as follows.

(1) Data Types

Data Types developed as the Global Map are shown in the following table.

Grid Data	Graphic Data	Data currently available
Vegetation,	Boundaries of	Basic Geographic Data(USA)
Hydrological conditions	Plant Ecosystems	
Elevation	Watershed Boundaries	Elevation(USA)
Geomorphology	Boundaries of Landforms	
Land Use	Rivers, Shorelines	
	Administrative Boundaries	

(2) Forms of Descriptive Data

Descriptive Data is defined at each grid of Global Map Standard Grid System at the ground resolution of approximately one kilometer.

(3) Grid System

The Global Map Standard Grid System is defined by dividing the ITRF-92 geodetic coordinate system into 30–second sections for latitude and each 30 to 180–seconds sections for longitude, depending on latitude as shown below.

Latitudinal zone	Latitude	Spacing in the longitudinal direction
I	0 – 50 degrees	30 sec. (0.5 min.)
II	50 – 70 degrees	60 sec. (1 min.)
III	70 – 75 degrees	90 sec. (1.5 min.)
IV	75 – 80 degrees	120 sec. (2 min.)
V	80 – 90 degrees	180 sec. (3 min.)

(4) Forms of Boundary Data

Boundary Data is defined by dividing the lines of data into segments in which the necessary accuracy can be retained and by recording the coordinate values of the start and end points of each segment and its attributes.

(5) Coordinate System of Boundary Data

Coordinate values of the start and end points of a segment are recorded in longitude and latitude of the ITRF–92 geodetic coordinate system. The references for elevation should follow the national datum of each country. Coordinate values should have sufficient accuracy that the data points should be well defined for the Grid System, while considering the accuracy of the original data.

(6) Definition of Vegetation and Hydrological Condition data

Vegetation and Hydrological data are classified using the seasonal changes of plant activities (i.e., vegetation indices) derived from meteorological satellite images and using hydrological conditions for low plant activity periods as shown in the following table, and defined at each grid point of the Global Map Standard Grid System. The horizontal accuracy of vegetation and hydrological condition data must be better than 500m.

Class (Plant Activity)	Subclass (Low Plant Activity Periods)	Corresponding Vegetation and Climate Class
Constantly High		Tropical rain forest, evergreen broad–leaved forest
Changed with	Dry land	Coniferous forest, deciduous broad–leaved forest, deciduous coniferous forest, steppe, grassland, farmland, savanna
	Water(temporary dry land)	Paddy field
	Ice and snow	Tundra
Constantly Low	Dry land	Desert, urban areas
	Water	Lakes, rivers and oceans
	Ice and snow	Glaciers

(7) Definition of Geomorphology Data

Geomorphology data is a landform class defined at each Global Map Standard Grid point. Landforms are largely classified as higher, medium, lower, or lowest relief energy areas. Lowest relief energy areas are further classified as moderate slope, slight slope or flat surface. The horizontal accuracy of geomorphology data must be better than 50m.

Major Landforms (Relief Energy Area)	Micro Landforms	Example landforms in humid and temperate zone
Higher		Mountains
Medium		Hills
Lower		Plateaus
Lowest	Moderate slope	Alluvial fan
	Slight slope	Natural levee
	Flat surface	Delta

(8) Definition of Elevation Data

Elevation data is defined at each Global Map Standard Grid point and recorded in the unit of meter. The accuracy of elevation data must be better than 50m horizontally and 10m vertically, considering semi–automatic production of geomorphology and watershed boundary data, and consistency with existing global and national elevation data sets.

(9) Definition of Land Use Data

Land use is the state of the earth's surface defined at each Global Map Standard Grid point. The earth's surface is classified as water, ice and snow, barren lands, wetlands, forest lands, forest plantations, grasslands (including range lands), agricultural lands, or urban areas. If possible, it would be desirable to have subclasses shown in the table below. Subclasses may be added or eliminated within the range of the classes according to regional characteristics. The horizontal accuracy of land use data must be better than 50m.

Class	Subclass
Water	Man–made river and lake, natural river and lake
Ice and snow	Perennial snow, glacier, tundra vegetation
Barren land	Mining site, beach, unused land, barren land
Wetland	Forested wetland, non–forested wetland
Forest land	Thick forest, thin wood
Forest plantation	Forestry land, wood field
Grassland (including range land)	Range land, unused grassland
Agricultural land	Pasture, confined feeding operations, upland field, paddy field, other agricultural land
Urban or built–up land	Densely populated urban area, urban area of low density, transportation, park, transitional area

(10) Definition of Boundaries of Plant Ecosystem

Boundaries of plant ecosystems are lines enclosing sections of nearly homogeneous vegetation, hydrological and climate conditions. The class of each section should be determined by consulting experts when the boundary data is developed. Sections smaller than 100 x 100 km are ignored except for especially important ones. The horizontal accuracy of the boundary data must be better than 500m.

(11) Definition of Watershed Boundaries

Watershed boundary data consist of lines dividing catchment areas. Watersheds smaller than 10 x 10 km are ignored except for especially important ones. The horizontal accuracy of the boundary data must be better than 50m.

(12) Definition of Boundaries of Landforms

Boundary of landforms are defined as the lines enclosing sections having uniform geomorphological characteristics and proximity, and are drawn based on geomorphology data. The class of each section should be determined by consulting experts when the boundary data is developed. Sections smaller than 10 x 10 km are ignored except for especially important ones. The horizontal accuracy of the boundary data must be better than 200m.

(13) Definition of Rivers and Shorelines

Rivers and shorelines data include the properties of river banks and shores classified as natural or artificial for every 3 km. Riverbanks are merged into a line when the width of the corresponding river is less than 100m. The horizontal accuracy of the boundary data must be better than 50m.

(14) Definition of Administrative Boundaries

Administrative boundary data are boundaries dividing administrative sections and are drawn by considering their correspondence to statistical data. Administrative sections smaller than 10 x 10 km are ignored except for especially important ones. The horizontal accuracy of the boundary data must be better than 50m.

2. Toward the Development of the Global Map

The following two methods are considered practical to implement Global Mapping.

(1) Those data the necessary information of which can be provided mostly by satellite data are developed with satellite data. One or a limited number of organizations would be practical to implement this method.

(2) Those data which would be more efficiently developed with existing maps and other materials, or those which need to be provided by cooperative countries are developed with existing materials and satellite data. Each country can develop the data of its own land, or one or a limited number of organizations can do it with the provision of existing materials from each country.

By combining these two methods, the Global Map can be successfully developed with the presently available materials and technologies. At present, data items as vegitation, hydrology and land use are considered to be applicable to method (1). On the contrally, Elevation, geomorphology, watershed, landforms, and rivers and shorelines are considered to be applicable to method (2). However, there is possibility to develop the latter data items from satellite data with the method (1), if more advanced satellite data acquisition systems will be developed. In this regard, a stereo imaging satellite sensor of 5m ground resolution, multi–spectral image data of 20m ground resolution, multi–spectral image data of 250m spatial resolution with repeat cycle of less than a day, interferometric SAR images of 5m spatial resolution, real–time index data for plant photo–synthesis activity intensity of 1 km ground resolution should be necessary as future research target.

TOWARD GLOBAL PLANNING OF SUSTAINABLE USE OF THE EARTH
Development of Global Eco-Engineering
S. Murai (Editor)

Monitoring of Glacier Movement in Antarctica Using Satellite Images

K. Cho, H. Shimoda, T. Sakata[a], A. Takahashi[b], and F. Nishio[c]

[a]Tokai University Research & Information Center
2-28-4 Tomigaya, Shibuya-ku, Tokyo 151 Japan
kcho@keyaki.cc.u-tokai.ac.jp

[b]Communication Research Laboratory
4-2-1 Kitamachi, Nukii, Koganei-shi, Tokyo 184 Japan

[c]Hokkaido University of Education
1-15-55 Shiroyama, Kushiro-shi 085 Japan

Abstract

In this study, the movement of Shirase glacier and Kaya Glacier in Antarctica was analyzed using satellite images. The the velocity of the Shirase Glacier was about 2.9 km/year during 1988-1989 and was about 2.6 km/year during 1989-1991. While the velocity of the Kaya glacier was constantly 600 m/year for 1973-1991. The comparison of a SAR image with a near-IR image of this area suggested the use multi-sensor analysis for ice sheet monitoring.

1. INTRODUCTION

Global warming is widely recognized as one of the most important themes in the earth environmental problems. Discharge of the Antarctic ice sheet to the ocean is a "key" of the mass balance from the view points of global warming and sea level change. It is one of the basic approaches to assess the ice discharge comparing satellite images taken in different time.

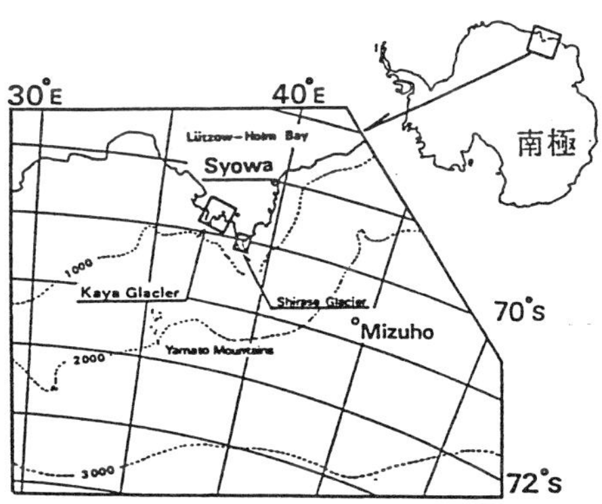

Figure 1. Map of study area.

In this study, the discharges of Shirase glacier and Kaya glacier were analyzed using satellite images. The both glaciers are main ice streams of Shirase and Prince Harald drainage basins respectively which are discharging ice to the Lutzow-Holm Bay (Figure 1). The analyzed area are the tips of the glaciers where the ice sheets are floating in the sea, and so called "floating ice tongue".

2. Shirase Glacier Movement

Through aerial photography surveys of Shirase glacier floating ice tongue performed in 1960th and 1970th, the flow velocity was estimated to be 2.5 km/year[1,2]. This was one of the fastest glacier velocities previously observed in Antarctica [2,3].

In this study, the recent flow velocity of the Shirase glacier was obtained from satellite images of 1988, 1989 and 1990. The same area of the images were extracted as shown in Figure 2. Table 1 shows the specs of the images. The edges and crevasse patterns of the floating ice tongue in the three images were traced manually, and movement vectors were obtained. The flow velocity of the glacier was about 2.9 km/year during 1988-1989 and was about 2.6 km/year during 1989-1991. This result shows that the velocity of 1988-1989 was rather faster than the other years. The authors have paid attention to the relationship of the velocity of the Shirase glacier and the sea ice extent in Lutzow-Holm Bay. The Bay was not covered with sea ice during 1988-1989, while the bay was covered with sea ice during 1988-1989. This suggests that the existence of sea ice may control the outflow velocity of the glacier.

Table 1. Specs of analyzed satellite images.

Acquisition date	Satellite	Sensor	Path/Row No.
1973.12.16	Landsat	MSS	158-109,110
1984.12.03	Landsat	MSS	150-109
1988.01.30	Landsat	TM	149-109
1989.02.23	MOS-1	MESSR	060-219W
1990.01.28	MOS-1	MESSR	059-219W
1991.01.15	MOS-1	MESSR	062-218E
1991.12.03	ERS-1	SAR	191-419

3. Kaya Glacier Movement

The flow velocity of the floating ice tongue of the Kaya glacier was obtained from satellite images of 1973, 1984, 1988, 1989 and 1991. The specs of the images are shown on Table 1. Figure 3 shows the four satellite images of a same part of the Kaya glacier. The images show that, beside the almost parallel movement of the ice sheet, the glacier is keeping the overall ice edge and crevasse patterns of this area for nearly 20 years. The velocity of the glacier was found to be almost 600 m/year constant for 1973-1991. This suggests the stability of sea ice condition in this area.

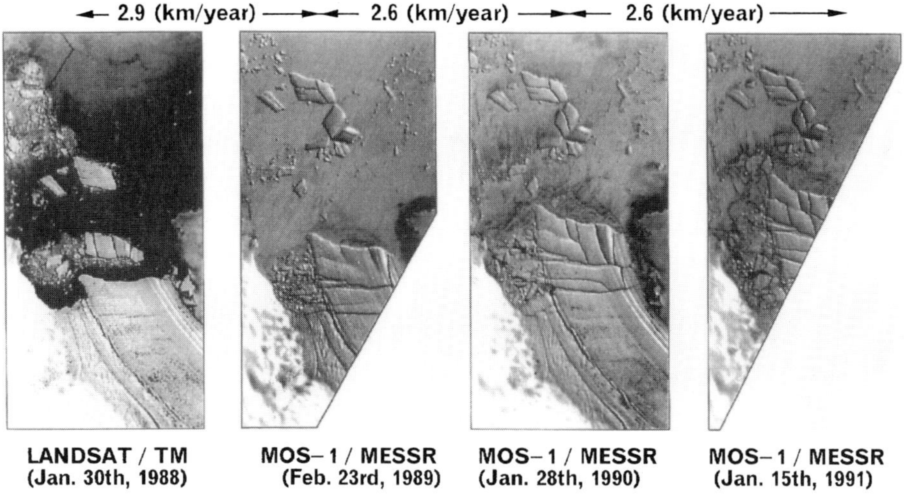

Figure 2. Movement of Shirase Glacier.

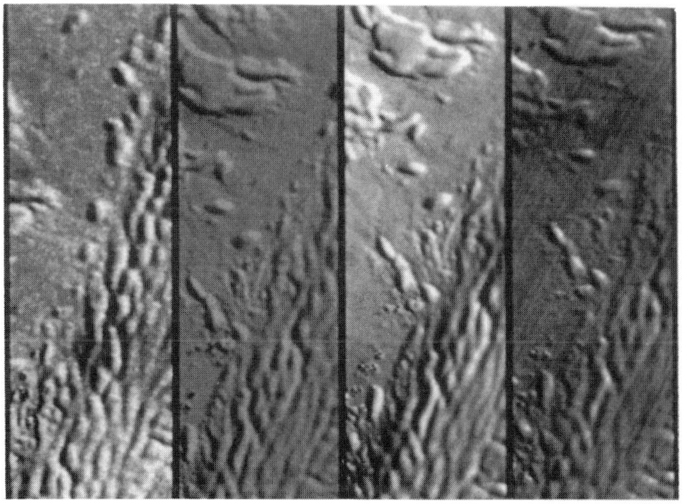

(a)1973 (b)1984 (c)1988 (d)1989

Figure 3. Movement of Kaya glacier.

4. Comparison of SAR image with Near-IR image

In this study, a ERS-1 SAR image of Kaya glacier was compared with a MOS-1 near-IR image as shown in Figure 4. The SAR image was acquired on December 3, 1991, and the near-IR image was acquired on January 15, 1992. The SAR image is a 3 look image (level 2.1), with 4x4 pixel average. Though the observation dates are nearly one year appear, there are no particular difference in ice sheet condition.

In the SAR image, sea ice and ice shelf could easily be distinguished, and floating ice tongue structure appeared much clearly than in the near-IR image. With it's all weather imaging ability, SAR is one of the most appropriate sensors to monitor glacier movement from space.

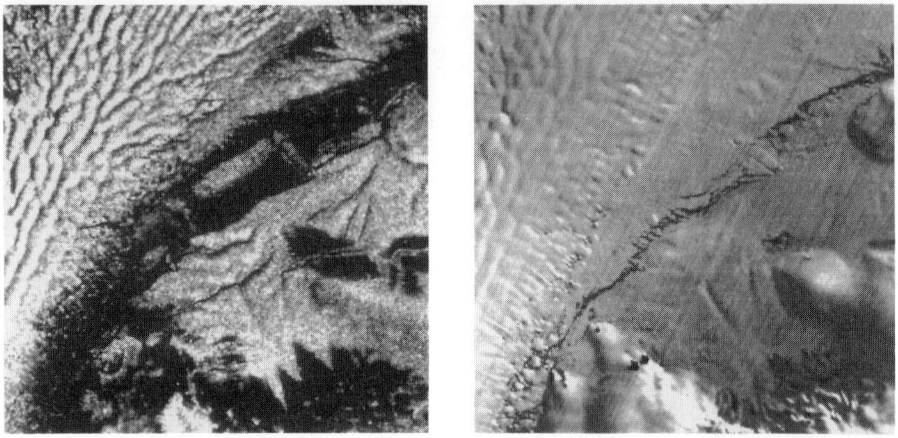

(a) ERS-1/SAR(1991) (b) MOS-1/MESSR(1991)
Figure 4. Satellite images of Kaya glacier

5. CONCLUSION

In this study, velocities of the Shirase glacier and the Kaya Glacier in Antarctica was obtained from satellite images. It was suggested that the velocities of glaciers may have close relationships with the sea ice extent in this area. The comparison of a SAR image with a near-IR image showed the use of multi-sensor analysis for ice sheet monitoring.

6. REFERENCES

1 Nakawo, M., Ageta, Y. and Yoshimura, A., Discharge of ice across the Soya Coast, Men. Natl Inst. Polar Res., Spec. Issue, 7 pp.235-244 (1978)

2 Fujii, Y., Aerophotographic interpretation of surface features and an estimation of discharge at the outlet of the Shirase drainage basin, Antarctica, The Antarctic. rec., No.72, pp.1-15 (1981)

3 Thomas R.H., Polar Research from Satellite, Joint Oceanographic Institutions, pp.37 (1990)

4 Takahashi, A., Cho, K., Simoda, H., Sakata, T, Nishio,F., An investigate of Antarctica ice sheet for visible and near-IR image and SAR image, Proc. of the 15th RSSJ Autumn Conference, pp.133-134 (1993)

TOWARD GLOBAL PLANNING OF SUSTAINABLE USE OF THE EARTH
Development of Global Eco-Engineering
S. Murai (Editor)

Prediction of Changes of Carbon Exchange between Atmosphere and Terrestrial Ecosystems by a Carbon Dynamics Model

Naohiro Goto[1], Yukio Yanagisawa[1], Akiyoshi Sakoda[2] and Motoyuki Suzuki[2]

1 Research Institute of Innovative Technology for the Earth, 9–2 Kizugawadai, Kizu–cho, Soraku–gun, Kyoto 619–02 Japan
2 Institute of Industrial Science, University of Tokyo, 7–22–1 Roppongi, Minato–ku, Tokyo 106 Japan

ABSTRACT

Terrestrial ecosystems play two important roles on global carbon cycle. One is a carbon source by deforestation, from which one to two GtC/yr is discharged. The second role is a carbon sink of about 100 GtC/yr by photosynthesis of plant. In natural forest, however, almost equal amount of the absorbed carbon is emitted by respiration of plant and decomposition of soil organic. The carbon emission and absorption could be divided into anthropogenic flux and non anthropogenic one. It should be recognized that the natural flux is one hundred times as much as the anthropogenic one. Then it is important to estimate the non anthropogenic fluxes between the atmosphere and the terrestrial ecosystems and to predict their changes by climate changes. A mathematical model was made to describe carbon dynamics between the atmosphere and the terrestrial ecosystems for these objectives. The model could represent global vegetation distribution and carbon storage in present climate condition. This model also could be applied for the prediction of changes of the vegetation distribution and the carbon storage in future climate conditions. In this paper we calculated (1) current carbon exchange between the atmosphere and the terrestrial ecosystems (2) the prediction of changes of carbon exchange by climate change.

1. INTRODUCTION

In order to prevent atmospheric CO_2 increase, two counterparts in terrestrial biosphere can be considered. One is to decrease carbon emission from biosphere by forest protection and the other is to increase carbon fixation by forestation. Simpson et al.[1] estimated Carbon flux in terrestrial ecosystems (net ecosystem production : NEP) as shown in Table 1. Though they didn't concluded which terrestrial biosphere is sink or source for CO_2, they concluded that land use is the most contributor to changes of carbon storage in terrestrial ecosystems. The terrestrial ecosystems fixes CO_2 through photosynthesis (net primary production : NPP) as shown in Table 2[2]. As shown in these tables, NPP is 10 ~ 100 times as much as NEP. If NPP fluctuates in 1 ~ 10 % due to climate change or atmospheric CO_2 fertilization, it corresponds NEP changes by one to two GtC, which is nearly equivalent of the carbon emission due to human activities. Simpson et al. also estimated total NEP in doubled CO_2 climate is −4.2 to +1.3 [GtC/yr]. These fluctuations will make serious impact on atmospheric CO_2 concentration because carbon emission by consumption fossil fuel is 5 [GtC/yr] and increase of atmospheric C storage is only 3 [GtC/yr]. This indicates importance of distinguishing anthropogenic flux and natural flux between the terrestrial ecosystems and the atmosphere. We constructed a mathematical model which describes carbon dynamics in terrestrial ecosystems[3]. In this paper we investigated natural carbon exchange and effects of climate changes on it.

Table 1 Current C fluxes in terrestrial biosphere (GtC/yr) from Simpson et al.(1993)[1]
Sink =(+), Source=(−)

Tundra/Boreal Forests	+0.5 to +0.7	Grasslands, Savannas and Deserts	0 to +0.6
Temperate Forests	+0.2 to +0.5	Agro−Ecosystems	−0.1 to +0.1
Tropical Forest	−2.2 to −1.2	Wetlands	+0.2
		TOTAL	−1.4 to +0.9

Table 2 NEP estimation for the biosphere (GtC/yr) from Whittaker and Likens (1973)[2]

Tropical rain forest	15.3	Temperate grassland	2.0
Tropical seasonal forest	5.1	Tundra and alpine meadow	0.5
Temperate evergreen forest	2.9	Desert scrub	0.6
Temperate deciduous forest	3.8	Rock, ice and sand	0.04
Boreal forest	4.3	Cultivated land	4.1
Woodland and shrubland	2.2	Swamp and marsh	2.2
Savanna	4.7	Lake and stream	0.6
		Total	48.3

2. MODEL RESULTS

2.1. Current Carbon Exchange

Seasonality and locality of carbon exchange between atmosphere and terrestrial ecosystems was calculated in order to validate this model. Fig.1 shows monthly global NPP variation, and Fig.2 shows monthly global carbon storage in terrestrial biomass and monthly carbon storage in atmosphere (CO_2 concentration was measured at Mauna Loa in 1992, carbon storage was calculated as 340 [ppm] = 730 [GtC]). As shown in Figs.1 and 2, this model can represent seasonal variation of NPP and carbon storage. There are differences in phase and amplitude between global carbon storage in biosphere and that in atmosphere at Mauna Loa as shown in Fig.2. Fig.3 shows monthly NPP variation in biosphere by every 10 latitudes zone. According to latitude zone, the phase and amplitude of NPP differ from each other. This variety of amplitude can be considered to be caused by area of vegetation. The latitude zones which have large amplitude are N60–50 and S0–10. Vegetation areas of these two latitude zone are 10.6 and 10.1 [10^6km^2] respectivity, while the areas of the latitude zones which have small amplitude, N80–70 and N20–30, are 1.2 and 2.5 [10^6km^2] respectivity. The parametric sensitivity analysis was calculated regarding environmental variables such as temperature and precipitation. Fig. 4 shows dependencies of the photosynthesis rates on (a) seasonal variation of temperature and precipitation (b) only seasonal variation of temperature (c) only seasonal variation of precipitation (c) neither variations (only solar radiation variation). In these cases other variables were kept at the optimal values for the photosynthesis rate. Cases (a) and (b) have a NPP peak in June and cases (c) and (d) have in March. This time gap is caused by photosynthesis rates in summer (north hemisphere). If the effect of temperature on photosynthesis rates is considered, the rates increase in summer because of temperature elevation. In cases (c) and (d) decreases of NPP in summer period come from respiration rate increases and/or photosynthesis decreases due to bad soil water condition (evapotranspiration from soil increase in summer). In Fig. 3 monthly NPP in S0–10 zone is similar to case (c) or (d) in Fig. 4. This means that this zone belongs to tropical zone and temperature is nearly constant through a year. Then only precipitation condition influences seasonality of NPP in this zone.

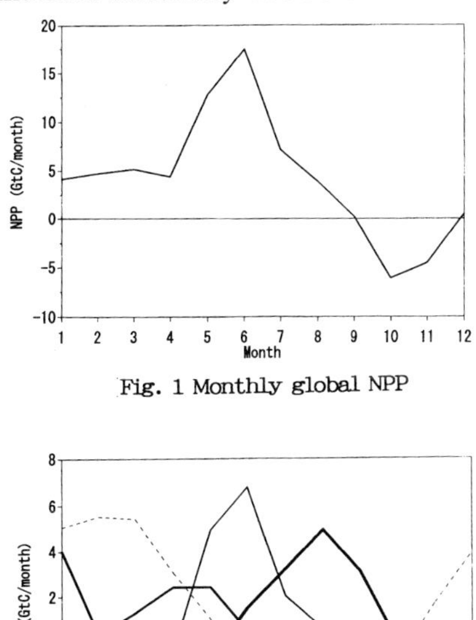

Fig. 1 Monthly global NPP

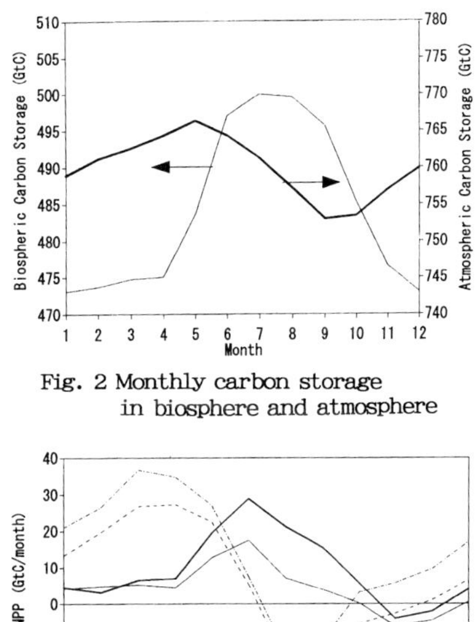

Fig. 2 Monthly carbon storage in biosphere and atmosphere

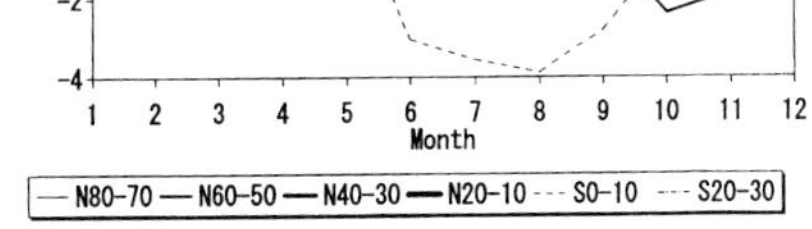

Fig. 3 Monthly NPP in six latitude zones

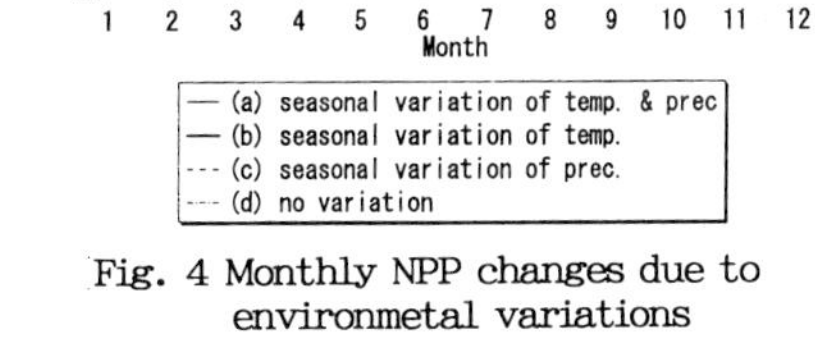

Fig. 4 Monthly NPP changes due to environmetal variations

Changes of monthly NPP in doubled CO_2 climate condition (climate change scenario is derived from GFHI GCM) are shown in Table 3. If it takes 100 years for the doubling, NPP decreases to 0.098 [GtC/yr]. This NPP decrease due to climate change is estimated to be as much as that due to deforestation. For example, global deforestation of closed forest is estimated at 0.14 [$10^6 km^2/yr$] in 1981–1990 by FAO[4]. Assumed that tropical forest is converted to grassland by this 0.14 [$10^6 km^2$] and there is no climate change, NPP decreases by 0.12 [GtC/yr].

Table 3 Change of global NPP (GtC/yr)

	Tropical forest	Temperate forest	Boreal forest	Grassland	No Vege- tation	Total
NPP (present)	30.2	4.6	3.9	11.0	0.2	49.8
(doubled CO_2)	22.8	1.9	1.6	13.6	0.3	40.0

3. CONCLUSION

We could indicate current carbon exchange between atmosphere and terrestrial ecosystems by the mathematical model. This model could be used to predicted future NPP change due to global warming. According to the model the changes of NPP is as much as that due to present human activities. There are plenty of trials to estimate carbon emission due to burning fossil fuel. However, there is few estimation of carbon emission from terrestrial biosphere including agriculture. Because it is difficult to classify anthropogenic carbon emission and natural one. In order to estimate the effect of various counterparts in terrestrial ecosystems, it is important to make such an emission inventory and further improvement of this model will be useful for it.

Acknowledgements:Study funded , in part, by New Energy Development Organization.

References

1 R.N.Simpson, M.Apps, S.Brown, C.V.Cole, J.Downing, L.S.Heath, D.S.Ojima, T.M.Smith, A.M.Solomon and J.Wisniewski, Water,Air,and Soil Pollution,70 (1993) 3.

2 R.H.Whittaker and G.E.Likens, Carbon and biosphere, U.S.Atomic Energy Commission, 281, 1973.

3 M.Suzuki and N.Goto, 1994, this volume

4 R.T.Watson, L.G.Meira, F.E.Sanhueza and A.Janetos, Climate Change 1992 (J.T.Houghton, B.A.Callander and S.K.Varney eds.), Cambridge Univ. Press, Cambridge, (1992) 29

Keyword: terrestrial ecosystems, carbon exchange, global warming

TOWARD GLOBAL PLANNING OF SUSTAINABLE USE OF THE EARTH
Development of Global Eco-Engineering
S. Murai (Editor)

Assimilating use of GIS for Socio-Economical Simulation for Global Change

Shintaro Goto

Kanazawa Institute of Technology
7-1 Ohgigaoka Nonoichi Ishikawa 921, Japan
Fax: +81-762-94-6711 E-mail: got@gaia.manage.kanazawa-it.ac.jp

Abstract

A model to find a balance of human impacts and the earth from the view point of food and energy is discussed. This model consists of satellite data and Edmonds & Reilly (E-R) model. The farmer combined with weather data and soil maps (ex. ISRIC data) can monitor the suitable place for cultivation and plantation and estimate the amount of CO2 fixation. And the latter can estimate the CO2 discharge. From the model, using GIS (Geographical Information System), proposed in this study, I can find the CO2 discharge when the population becomes the limit by E-R model, and can find the place for cultivation and forestation simultaneously.

1. INTRODUCTION

The influence between human impact and global environmental change is estimated from the view point of the human habitability by using the relationships between food production and food consumption such as S.Goto etc.1). This model is consists of satellite data and can analyze the distribution of vegetation. But this model can explain only one type of human activities such as food consumption.

On the other hand the Edmonds-Reilly model 2) is the model can estimate the energy consumption of each kind of energy and amount of CO2 discharge. But E-R model can't analyze the effect of the distribution of vegetation such as the local distribution of food production, because the spatial unit of E-R model is too large.

These study discuses about the model to analyze the balance the global environmental change and the human activities from the view point of food and energy consumption by using GIS coupling with E-R model.

2. METHODOLOGY

2.1 Estimation of Supportable Population Using Satellite Data

S.Goto etc.1) estimates the supportable population using potential arable land(Fig.1) made by using NOAA-AVHRR 3). This model is based on the assumption that humans consume only crops. In this study, I estimate the supportable population considering not only crop consumption but also meat consumption. The data used are as follows,

Crop production(1990): FAO 4)
Food Consumption(1987,1989): FAO 4)
Crop Import and Export(1980):FAO 5)
Livestock Import and Export(1990):FAO 6)
Slaughtered Livestock Number(1990):FAO 6)
Potential Arable Land Distribution(1990):S.Goto et al.1)

Fig.2 shows the calculation flowchart of supportable population.
(1) The grassland on the world vegetation map with the conditions being an attitude of no more than 1000m and a latitude of no more than 60° is defined as potential arable land. Deforestation is not considered.
(2) The following kinds of crops are selected as the principal food in this study: wheat, paddy rice, barley, maize, sorghum, potatoes, sweet potatoes, cassava, soybeans, oats, rye and millet. And following kind of livestock are selected as the principal meat: cow, pig, sheep and goat.
(3) Crop production per hectare is assumed to keep the highest level at present, and in potential arable land areas, it is assumed to decline in a linear fashion, as shown in Fig.3.
(4) Par capita food consumption is calculated by each country with a formula using net consumption divided by population.
(5) The base year is 1990 except food balance.

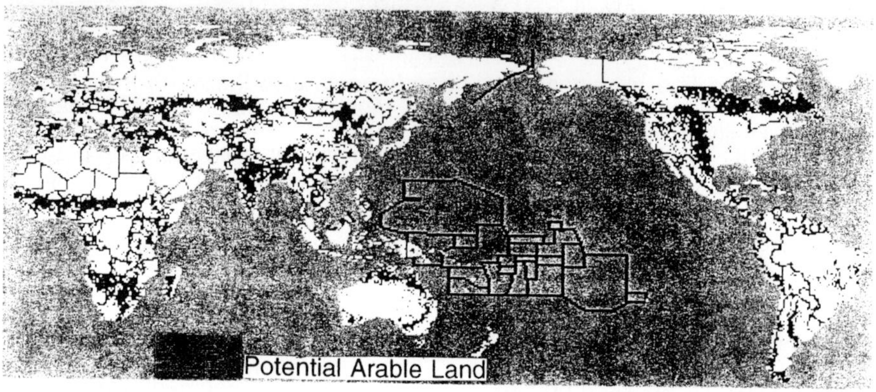

Fig.1 Potential arable land using NOAA-AVHRR

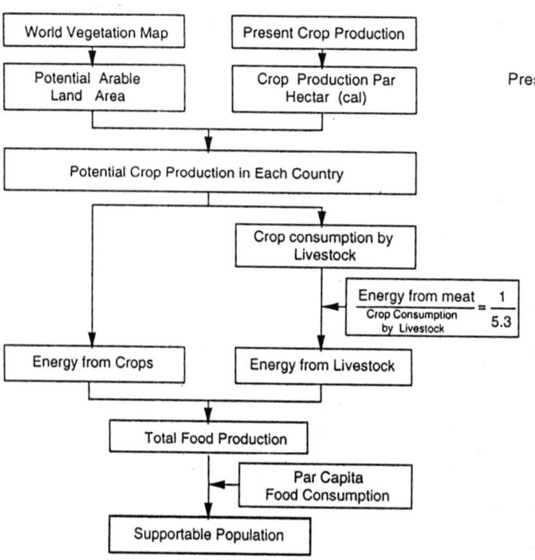

Fig.2 Calculation flowchart of supportable population

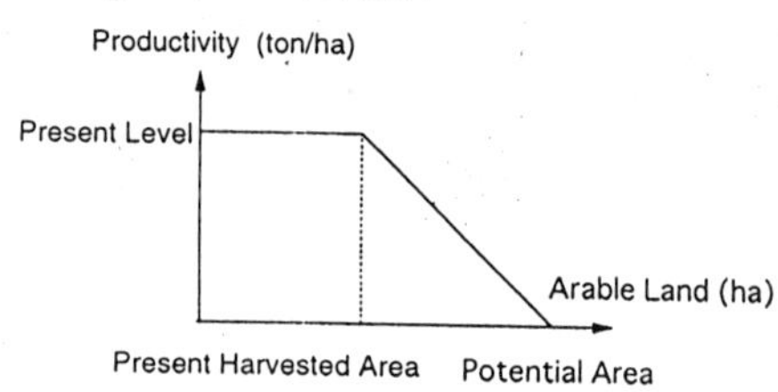

Fig.3 Assumed Productivity

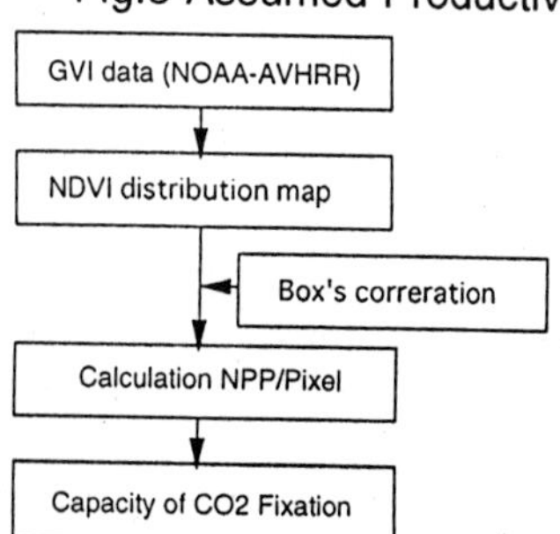

Fig.4 Calculation Flowchart of estimation of CO2 fixation

2.2 Estimation of CO2 Fixation Using Satellite Data

Fig.4 shows the calculation flowchart of estimation of CO2 fixation.

(1) Capacity of CO2 fixation on land is estimated by NPP (Net Primary Productivity) of the vegetation on land.

(2) The flux of CO2 from litter is not considered in detail. The amount of CO2 discharge from litter is equal as the amount of CO2 fixation in the vegetation.

(3) To estimate the capacity of CO2 fixation, the correlation between NDVI (Nomarized Difference Vegetation Index) and NPP is used. The correlation shown as Eq.1) was developed from correlation of 947 observed data by E.Box et al 7).

$$NDVI = 0.4 \ [\ 1\text{-}e\text{-}0.00068128\ NPP\ ] \quad \text{...} \ 1)$$

Where : NDVI = (NIR-VIS)/(NIR+VIS), NIR=a spectral measurement in the 0.7 to 1.3 and VIS=a spectral measurement in the 0.4 to 0.7.

Fig.5 shows the CO2 fixation distribution map using NOAA-AVHRR.

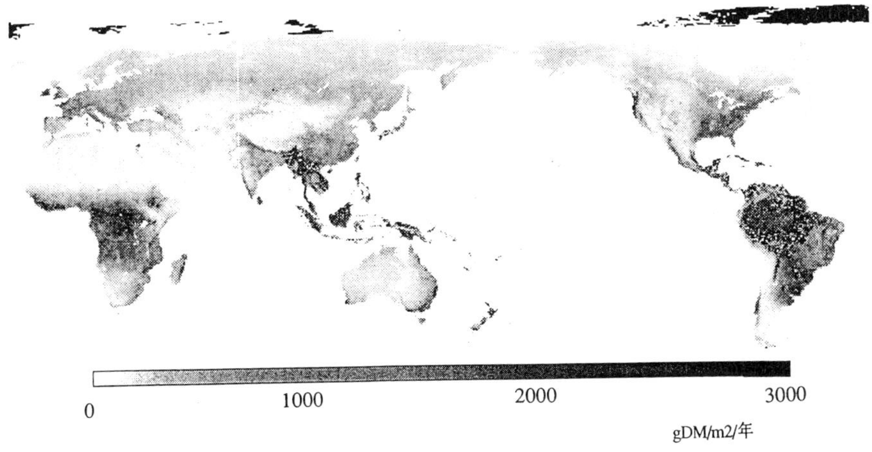

Fig.5 CO2 fixation distribution map using satellite data

2.3 E-R model

E-R model was developed by J.Edmond and J.Reilly. They divided the world into 9-district. And made it possible to estimate the CO2 discharge from 9-district from 1975 to 2100 every 25-years. The calculation flow is shown as follows,

(1) The demand and supply of each kind of energy is expressed as the function of each price of energy.

(2) The price is controlled by iteration till I find the status where the demand and supply are balanced.

(3) CO2 discharges from each region are calculated by multiplying the coefficient by fossil fuel at the status (2)

Original E-R model is fixed as the initial year is 1975. In this study this model is changed to
calculate from 1990.

3. RESULT

3.1 Supportable Population

The case, per capita food consumption is assumed to be the same level as the present in the respective countries, is set for the calculation of world's supportable population. The supportable populations for three cases are 7.4 billion.

3.2 The relationships between energy and supportable population

The following five cases are set for the calculation of CO2 discharge. And Fig.6 shows the capacity of CO2 discharge using E-R model. And Fig.7 shows the population growth used in this calculation based on UN 8).

Case 1: Par capita GNP is assumed to become the same level as the present in the countries where GNP is more than the average level, and is assumed to become the same level as the world's average in the countries where GNP is lower than the average level, in 2040.

Case 2: Par capita GNP is assumed ditto to Case 1, in 2065.

Case 3: Par capita GNP is assumed ditto to Case 1, in 2090.

Case 4: Par capita GNP is assumed ditto to Case 1, in 2115.

Case 5: No political control.

In Fig.7 supportable population 7.4 billion is shown by horizontal solid line.

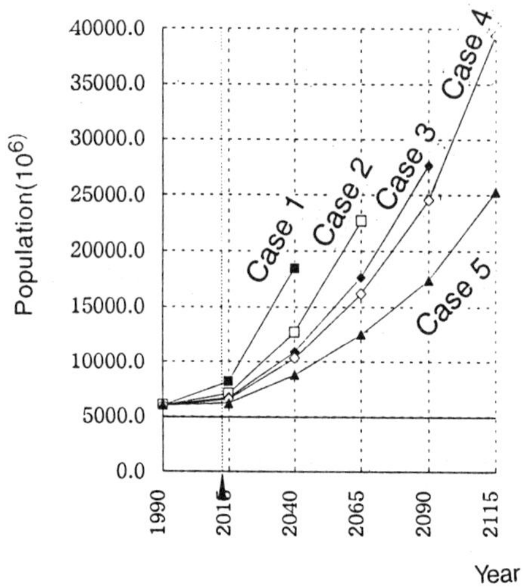

Fig.6 Capacity of CO2 discharge using E-R model

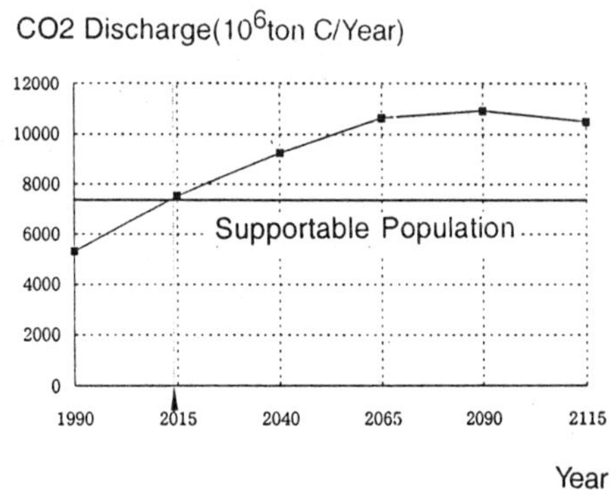

Fig.7 Population growth used in the calculation shown Fig.6.

4. THE MODEL TO ESTIMATE THE BALANCE OF GLOBAL ENVIRONMENTAL CHANGE AND HUMAN ACTIVITIES FROM THE VIEW POINT OF FOOD AND ENERGY

The Socio-Economical simulation model for Global Change, such as E-R model, can't analyze the effect of the distribution of each parameter such as the local distribution of food production, because the spatial unit is regional polygon. But if I find the place for cultivation and forestation, I need the distribution map of suitable place for the objects mentioned above.

Fig.8 shows the model to estimate the balance of global environmental change and human activities from the view point of food and energy. First, the supportable population is set up and estimate the capacity of CO2 discharge by E-R model. Secondly, the food production to support the population set up is estimated, as a result, by using GIS to process the satellite data, weather data and soil data, the area of cultivation for the food production is estimated, the land use map is planed to distribute the suitable place for cultivation and forestation, and the capacity of CO2 fixation is estimated. Finally if the amount of CO2 discharge and fixation are balanced the iteration is ended, and if not the iteration is repeated.

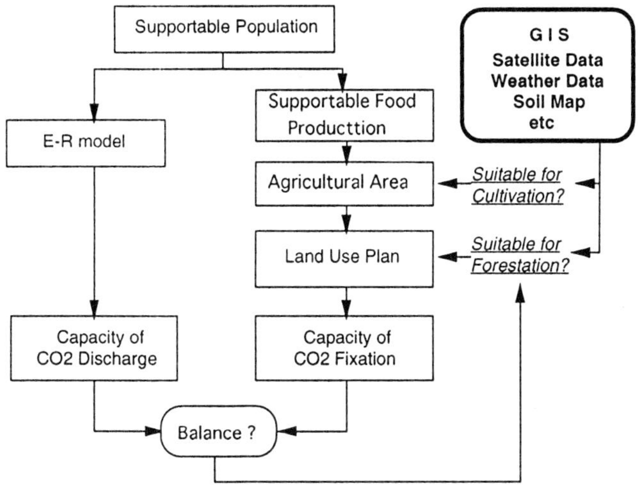

Fig.8 Estimation model of the balance of global environmental change and human activities from the view point of food and energy

5. CONCLUSION

The model combined GIS and E-R model can estimate the balance of global environmental change and human activities from the point of food and energy through the amount of CO2. Father studies are going to be paid attention to analyze the case study.

6. REFERENCES

1)S.Goto,S.Murai,Y.Honda and K.Asakura : A Study on the Relationships Between Human Activities and Biosphere Using Satellite Data, Proc. of the 17th International Society for Photogrammetry and Remote Sensing,vol.XXIX, PART B7 (1992)pp.439-445.
2)J.Edmonds and J.Reilly : The IEA/ORAU Long Term Global Energy-CO2 Model:Personal Computer Version A84PC,OAK LIDGE NATIONAL LABORATORY,1986.
3)Y.Honda and S.Murai:Vegetation Mapping Using Global Vegetation Index and Weather Data.,Proc. on the 10th Asian Conference on Remote Sensing,pp.A-2-4-1 ~pp.A-2-4-6, 1989.
4)FAO : Production Yearbook, vol.44,1990.
5)FAO : Food Balance Sheet 1987-87 Average,1985.
6)FAO : Agrostat-PC,1993.
7)E.O.Box, B.N.Holben and V, Kaib: Accuracy of the AVHRR Vegetation Index as a predictor of biomass,primary productivity and net CO2 flux, Vegetario, Vol.80,pp71-89,1989.
8)UN : Popuration Prospect, 1990.

TOWARD GLOBAL PLANNING OF SUSTAINABLE USE OF THE EARTH
Development of Global Eco-Engineering
S. Murai (Editor)

Urbanization, Motorization and the Environment

Yoshitsugu Hayashi, Hirokazu Kato and Krit Anurakamonkul
Department of Geotechnical and Environmental Engineering

1. Introduction

The energy consumption and pollutant emissions, affecting both local and global environment, generated by transport sector are rapidly increasing ahead of those by industrial and domestic sectors. In advanced countries where urban land uses are virtually saturated, the main causes of environmental deterioration may come from their motor-vehicle oriented transport systems and the utilization habit of their people. Whilst in developing countries the main causes may include the increasing trip length due to the rapid expansion of urban areas, the being worsened traffic congestion due to the mismatch between the delaying supply of transport infrastructure and the inundation of car ownership and so forth. Apparently these phenomena occur through the processes of urbanization and motorization.

In this study, the mechanism of the increase in energy consumption and environmental degradation is examined on the basis of stages of urbanization and motorization with reference to economic development. An international comparative study is subsequently conducted for four metropolises, namely London, Tokyo, Nagoya and Bangkok where they are located at different stages in terms of economic development, urbanization and motorization. The study also proposes a scenario analysis method to determine possible causes and results of environmental degradation in the area of land use and transport planning. The analogies of measures and outcomes previously applied for land-use and transport in the preceding countries are believed to provide useful information to future policy formulations, particularly in the developing countries where environmental problems are significantly becoming more serious.

2. Mechanism of Influences of Urban Development on Transport and the Environment

Evidently, it is found that transport sector plays the most important role to existing air pollution problems in the metropolises all over the world. This section aims to analyse the influences of economic growth and urban spatial development on travel patterns and energy consumption.

The mechanism of changes in energy consumption due to economic growth and spatial development can be illustrated as Fig. 1. The expansion of metropolitan economy generally results in immigration of people from the rural area. An increase in population and the higher spatial demand for property development and housing cause rapid spread of built-up area. As a consequence of this phenomenon, the average trip length becomes longer. An increase in income also contributes to higher car ownership, more car use and more trip frequency. On the other hand, the limited supply side, namely, the bottleneck of transport infrastructure development often occurs due to budget constraints and lower cost-effectiveness in sparse development areas. The lack of rail transit, especially in the suburban area encourages more private car use. The higher car use, higher trip rate, longer trip length and insufficient road surface all together cause worse congestion and more energy consumption thereby leading to air pollution problem along the roadsides.

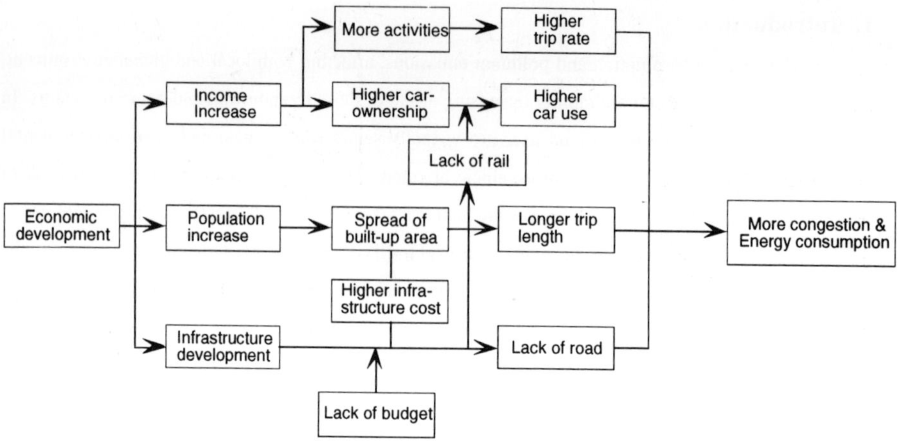

Fig. 1 Mechanism of influences of urban development on transport and the environment

3. A Scenario Analysis Method and its Phenomenal Interpretations

Through combining all the intuitive processes of analysis conducted in this study, a conceptual method for scenario analysis of environmental influences can be obtained as shown in Fig. 2.

In short, the process of scenario analysis method consists of six consecutive components described as follows:

(1) observe the past trends and consolidate them to get likely scenarios of economic growth

(2) examine urbanization scenarios

(3) examine motorization scenarios

(4) examine road improvement policy scenarios in comparison with the increase of traffic demand

(5) examine the relationship among urban radius, vehicle-drive distance and the consumed transport energy

(6) forecast the environmental impacts

At the moment it is still difficult to quantitatively forecast the future conditions of land use, transport and the consequent environment in a dynamic development stage for each metropolis by simple empirical examinations of the trend. One fundamental barrier is that there are many unforeseen and non-linearly changing factors emerging in the real-world economic framework. For instance, these could be unstable economic growth patterns, inconsistent increasing rate of car ownership and traffic demand which is generally higher than that of income in a growing period of developing countries, the inadequate financing to supply transport infrastructure or to support technological innovation for car emissions, and the shortage of funds to subsidize environmental improvement measures, etc. In other words, such shortcoming cases imply that it would be much useful to review the actual trends and to understand the

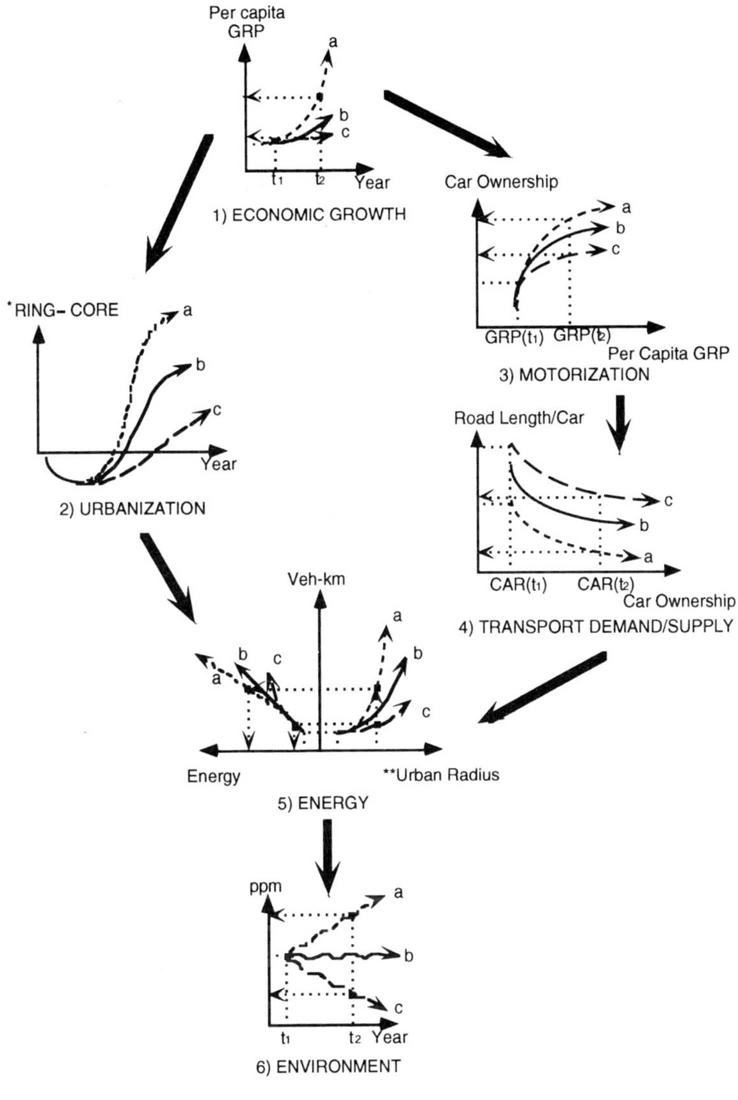

Fig. 2 Process of scenario analysis method

*Core: The agglomeration area or core city
Ring: The surrounding cities
Ring minus Core: A sign indicator of population change to identify the stage of urbanization cycle

**Urban radius: Equivalent distance index to represent magnitude of spatial development

development mechanism of elderly aged metropolises in the past, especially when comparing at the same economic development stage with another younger aged metropolis.

4. Conclusion

Fundamentally, the developing metropolis can incorporate the fact findings and results of analysis from the developed metropolises into its planning and evaluation process to accommodate future sustainable development and pleasant environment simultaneously. To a certain extent, this paper has explicitly pointed out the *common phenomena* under similar constraints that could take place again at the *same stages* of economic development, urbanization, motorization and the environment which are closely interrelated. Thus learning from the past experiences, we have tried to derive some guidelines for a late developing metropolis by appropriately shifting and/or accelerating the development cycles of preceding metropolises to depict the possible scenarios.

Although many attempts have been made mainly to compare and contrast the dynamic changes of urbanization, motorization and the environmental impact among the metropolises, one should keep in mind that our comparative study is typically based on an analogous estimates by *'cross-sectional analysis'* of the outcomes at various stages of development and at different points of time in order to characterize their common phenomena. It is recommended that further comprehensive research be also carried out through *'longitudinal analysis'* of the chronological entity and compatible development mechanism for each metropolis in order to gain an insight into interactions and actual outcomes thoroughly. As a result of this integrated analysis, it would be possible to realize a deeper and broader perspective towards sustainability of how and when such comparable dynamic development process keeps changing under diverse conditions. Likewise, it would enable us to depict a more reasonable and reliable portrait of our future environment.

References

1) Hayashi, Y., Tomita, Y., Doi, K. and Suparat, R. (1993): *An International Comparative Study on Land Use Transport Planning Policies as Control Measures of Urban Environment.* Selected Proceeding of the Sixth World Transport Conference on Transport Research, pp. 255-266

2) Newman, P. and Kenworthy, J. (1991): *Cities and Automobile Dependance:An International Source Book.* Gower

TOWARD GLOBAL PLANNING OF SUSTAINABLE USE OF THE EARTH
Development of Global Eco-Engineering
S. Murai (Editor)

A Comparative Analysis of Global Environmental Risks in the Japan Sea Rim :Ecological and Economic Vulnerability to Global Climate Change

Saburo Ikeda and Masaaki Kataoka

Institute of Socio-Economic Planning, University of Tsukuba, Tsukuba 305, Japan

Abstract

In this paper, we will present results of a comparative analysis on societal responses to degradation of environmental quality caused by global climate change and globalization of regional economies in the area of the Japan Sea Rim . First, we will address both ecological and socio-economic vulnerability in terms of resource management in this area. The interactions between economy and environment are measured through an environmental accounting system, focusing on pollution loads from land-based sources. Then, we will discuss societal responses to environmental risks in this marginal coastal areas taking account of the experiences of the Mediterranean and the Baltic Sea.

1. Introduction

Is it possible for us to take effective measures to the ongoing worldwide environmental changes for promoting sustainable growth of our society? That is the question which requires a wide variety of scientific investigations, ranging from scientific disputes on global warming in the fields of earth sciences, meteorology and ocean sciences up to socio-economic evaluation of its impacts on human dimensions of health, economy, culture, and so on.

The Sea of Japan is a semi-closed marginal sea which is linked with the Pacific and the East China Sea through narrow straits (Figure 1). So far the marginal sea has preserved a large amount of marine resources and efficient purification capacity. However, when some of the coastal countries on both the continent and islands take a development strategy which pays little considerations to the potential environmental hazards in future, the countries surrounding the marginal sea would suffer environmental degradation in the long run. First, polluted water from industrial and domestic wastes might dramatically increase at the continent side, and spread into the marginal sea due to the poor exchangeability of the sea water. Second, the oxidized air pollutants such as NOx and SOx generated on the continent would fly over to Japan and other islands countries with Asian monsoon and the westerly. Third, rise in sea level caused by global warming might degrade the quality of urban infrastructures, preventive ability of natural disaster, and environmental amenities in the coastal area.

The degree of such potential risks partially depends on the strategies that these continental countries take for their economic development. In such a case, various international environmental conflict would occur in the area and the conflict is expected to take the form of European style international environmental conflict such as in the Mediterranean and Baltic Seas. In this study[1], we focus our attention on the following environmental risk issues:

1) existing and future ecological vulnerability - risk profile in coastal resources
(fish stock, water and land resources, marine amenity, etc.)
2) societal responses to adapt or manage the potential risks
(policy implications of socio-economic, cultural and institutional responses)

2. Framework of Environmental Vulnerability Analysis

The Economic-Environmental Accounting System (EEAS) is used to give an overview on interactions between economy and environment [2]. In evaluating the resources, this accounting system has two accounting tables in the same manner as the double bookkeeping system: environmental balance sheet and material profit and loss table. Environmental balance sheet describes environmental resources, constraints on its utilization by the man and utilization potential of environment resources under the constraints. On the other hand, material profit and loss table shows resource input for maintenance of environment and benefits from utilizing quality of environmental capacity. The economy is divided into the following five sectors: public, agriculture, fishery, other industries and households. So far, we could make a cross-national comparison only in the case of China, Japan and South Korea in 1989 [1-2].

3. Scenarios of Climatic Impacts and Sea Level Rise

The following are such typical scenarios of regional climate impacts on the Japanese islands summarized by Yoshino and others [3]:

- In summer, Asian monsoons will become stronger, and rainfall is expected to decrease and become like tropical areas, sometimes torrential. When it pours, water will not be kept due to the steep terrain in mountain topology of Japanese river basins. That will cause not only loss of water resource but also erosion of moisture in soil of the Japanese islands.
- In winter, the atmospheric pressure pattern of the cold and high anticyclone over the northern continent is expected to weaken, accordingly, the snowfall will decrease in the coastal areas facing the Japan Sea. The rainfall in mountainous area will increase about 30 percent or more.
- The sea level is considered to have gone up by 10 to 20 cm over the last century because of the thermal expansion of sea water and possibly by melting ice bed. However, the monitoring data at the Japanese coast shows that the sea level in the coastlines along the Pacific Oceans was at its peak in 1950, and has been going down since then . The possible rise in the sea level depends not only on the degree of the global warming effects, but also the long-term coastal land movement in Japanese islands. By and large, it seems to be reasonable to take the assumption of the possible rise in the sea level from 80 cm to 120 cm by the year of 2100 along the Japan Sea, which the IPCC working group reached consensus. Then, it is obvious that a considerable portion of major Japanese metropolitan areas, such as Tokyo, Osaka and other municipal capitals along the Japan Sea might be under water.

4. Ecological Vulnerability

Based on some paleogeological and oceanographic studies [4] concerning oceanographic conditions of the sea, we can come up with the following hypothetical scenarios:

In the beginning, the temperature of surface water will rise. Then, thermal expansion of water body may result in rise in the water level. At the same time, the mixing and upwelling capacity between the surface water (8 - 17 °C) and cold bottom water (about 0.5 °C) may decline. The cold bottom water, which used to be adherently confined in the deep bottom layer (approximately 2,000 m to 3,000 m in depth) for many years, may have more difficulty in water exchange by getting less water supply from the upper layer as illustrated in Figure2.

In this context, this marginal sea has a high degree of vulnerability in potential increase of industrial wastes at the sea bottom due to the less exchangeability of the sea water. Other issue to be addressed is the dispersion of acid rain by crossing the Japan Sea. The recent study by the Central Research Institute of Electric Power Industry estimated the amount of sulfur dioxides generated by the Northeast Asian countries which amounted about 11 million ton per year by sulfur content [5] This is equivalent to the amount generated by the Northeastern part of the U.S. While we see a clear sign of increased acidity in rainfall and soils, so far, actual damages to the forest or fresh water have not clearly appeared in the Japanese coastal regions. Because it is said that Japanese soil is more resilient to sulfur acidity in terms of neutralizing capacity than continental soil. It is, however, a matter of time for Japanese trees and soil ecosystems to suffer the fatal damage.

5. Social and Economic Vulnerability

The direct impacts by the climatic changes and sea level rise include damages to urban river system, harbors, coastal low land and scenic views, setback of coastlines by erosion and high waves, etc.. These problems can cause a number of vulnerability to urban life lines, transportation and industrial and agricultural infrastructures by the increasing hazard potential. As other types of impacts on the regional economic system, we will have a various degree of impacts on industries, primarily on agriculture and climate-dependent industries such as brewing and fermentation, recreation and tourism, construction and transportation. However, in the regional economic sense, we should be aware the following indirect ones:

1) urban infrastructure of life lines: vulnerable to the increasing hazard potentials
2) industrial structure: vulnerable to lower energy consumption with higher recycle rate
3) trade structure: vulnerable to export and import of pollution loads which will be transferred through the various types of media such as atmosphere, river, sea, and ecological components

6. An Example of Critical Environmental Risk: Estimation of Water Pollution Loads by the Progress of Development Plans

We estimated the current level of water pollution loads and projected its future through a couple of alternative scenarios on possible economic growth paths of the Japan Sea Rim area. The method of estimation is the unit-per-product method. In the Japanese case, we adopted pollution load per million yen of a product. In the case of continental countries, we adopted the WHO method which had been applied to the case of Southeast Asian Seas for the UNEP water pollution control project [6].

The preliminary results of estimation and simulation show that China, which has currently 60 million people in Heilongjiang and Jilin Province, is critical in controlling total water pollution loads to the Japan Sea. Water pollution loads from Japanese and Korean industrial and domestic sources might expect to decrease their discharges into the Sea. However, China and Russia now have plans to promote agriculture and food and raw materials industries in this basin areas. Such development plans of acceleration of economic growth through free market economy and huge size of population in China will bring about significantly higher water pollution loads to the Japan Sea through Songhua and Amur river, which is already nicknamed as the 'Death River' by Russian mass media. Amur river flows into sea at the Tatar strait, which connects the Japan Sea and the Sea of Okhotsk.

Development plans in the Northeast Asian Area become increasingly realistic through favorable international political climate of that area. The United Nations Development Program took up a large scale development project of Tumen River basin which runs between China, Russia and North Korea into the Japan Sea. The project gives China an opportunity to have direct ship access to the Japan Sea through the new ship route. In February 1993, China and Russia agreed that China will have a lease on a Russian port named Zarbini, which lies nearby the mouth of Tumen river. The regional governments in this area now have ever closer international economic linkages after the resolution of the east and west conflicts in spite of uncertainties on the future of Russian and North Korean economy.

7. Concluding Remarks: Toward Transnational Environmental Cooperations in North East Asia - Lessons from the Mediterranean and the Baltic Sea -

The cases of Mediterranean and the Baltic Sea illustrates how fears of oil pollution and dumped wastes were seriously perceived in most European countries in the 1970s. and finally formed transnational organizations for marine environmental management through multilateral conventions in the mid-1970s. The Helsinki convention for the Baltic and the Barcelona one for the Mediterranean became a landmark of multilateral environmental cooperations as the first attempt in the world on comprehensive transnational marine environment protection. Political conflict of interests between the north and south, and the east and west was finally solved by a redistribute scheme of financial burden sharing and technological transfer to the South on environmental management and monitoring [7-8]. In both cases, the convention facilitated scientific cooperations and mutual assistance to combat with marine pollution. Monitoring stations were installed and environmental scientists and technicians in developing countries were trained. These factors contributed not only to expansion of scientific knowledge on marine environment but also to common understanding on the situation and problems.

In the case of Japan Sea, the political complexity such as the east-west, the north-south and ideology-territory issues has still facilitated conflicts on marine resources. At the same time, it prevented joint research and information exchange on marine environment. At most, marine fishery resources were protected only through bilateral fishing agreements. In addition, the Japan Sea area has been a front of military confrontations and therefore, was left economically underdeveloped in each countries. These factors prevented international disputes on marine pollution in the Japan Sea.

In October 1992, UNEP sponsored the Conference of North East Asian Regional Environmental Planning Specialists at Beijing, taking a first step to transnational environmental cooperation in that area. Around the same time, the Northeast Asian Conference on Environmental Cooperation was held in Niigata, and policy executants of China, Japan, Mongolia, Russia and South Korea jointly met for the first time. All Japanese prefectures on the Japan Sea coast organized the Conference on International Exchange on Fishery in the Sea of Japan, and began transnational exchange on fishery techniques and joint research on fishery resource [9]. With the dissolution of political conflicts and the progress in economic development projects through international cooperations, we can expect degradation of environment in the Japan Sea area. We must learn lessons from the cases of the Baltic and the Mediterranean and take a step for fruitful transnational environmental cooperation in the Japan Sea area.

References

1. S. Ikeda (1993): Ecological and Economic Vulnerability and Resilience to Global Climate Change - A Comparative Analysis of Regional Environmental Risks in the Japan Sea Rim, Research Report to the Min. of Sci. & Education Japan.
2. Y. Kitabatake (1990): An Environment-Economic Accounting System - An Extension of Double Entry Bookkeeping System to the Man's Utilization of Environmental Resources -, Insti. of Socio-Economic Planning Discussion Paper No.429, The University of Tsukuba.
3. M.Yoshino (1991): Impact of Climate Change on Agricaulture from the Viewpoint of East Asia, in "The Global Environment (ed. by K.Takeuchi)", Springer.
4. Y. Kasano (1991): The Origin of the Japan Sea, Aoki Pub., Tokyo.
5. S.Fujita et al (1991): Preliminary Inventory of Sulphur Oxide Emission in East Asia, Atmospheric Environment, Vol.25A, p.1409-1411.
6 WHO Western Pacific Regional Centre (1981): Preliminary Assessment of Land-based Sources of Pollution in East Asian Seas.
7. P.Haas(1990): Saving the Mediterranean, The Politics of International Environmental Cooperation, Columbia University Press.
8. R. Hjorth(1992): Building International Institutions for Environmental Protection, The Case of Baltic Sea Environmental Cooperation, Linkoeping University.
9. Environmental Agency and Niigata Prefecture (1992): Northeast Asian Conference on Environmental Cooperations: reports and presentations by the delegates.

Figure 1. Geographic Location of the Regions Surrounding the Japan Sea

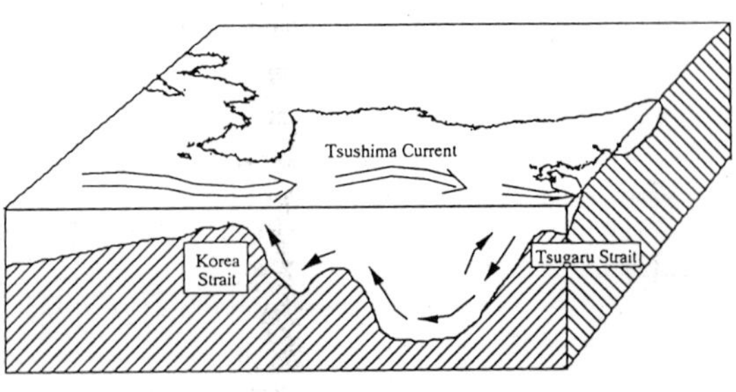

Figure 2. Water Circulation in the Sea of Japan : a Schematic Illustration

TOWARD GLOBAL PLANNING OF SUSTAINABLE USE OF THE EARTH
Development of Global Eco-Engineering
S. Murai (Editor)

Land use, chemical weathering and atmospheric CO_2

Kikuo Kato and Kaori Komaki

Institute for Hydrospheric–Atmospheric Sciences, Nagoya University, Chikusa, Nagoya 464–01, Japan

Abstract

Chemical weathering of rocks is well known to had affected on the CO_2 level of the atmosphere in the geologic time. However, the following facts are considered to suggest that chemical weathering caused by human land use affects on the budget of anthropogenic CO_2. Both the concentration and $\delta^{13}C$ of the atmospheric CO_2 show little variations in the 13–18th centuries, when human land use was spreading, and also show smaller variations than expected from CO_2 uptake by the ocean, even in the budget of anthropogenic CO_2 after the Industrial Revolution.

The calculated amount of CO_2 uptake by chemical weathering is around 0.5×10^{15} g–C/yr. This value is too small to affect on the atmospheric CO_2 level but may become larger according as studies on runoff and especially runout progress worldwidely. Therefore, we assume that variation of the atmospheric CO_2 concentration caused by human land use should be examined in connection with variation in chemical weathering and propose a international program for worldwide investigations on runoff and especially on runout.

Key words: Land use, global change, chemical weathering, anthropogenic CO_2, global carbon budget, oceanic CO_2 uptake.

1. INTRODUCTION

Chemical weathering of rocks had obviously affected on the CO_2 level of the atmosphere in the geologic time [1–2]. However, it has generally been considered that chemical weathering could have affect on atmospheric CO_2 in the time scales over million years. Recently, from the studies on ice cores from both Antarctic and Greenland ice sheets, the CO_2 concentration of air trapped in the last glacial ice is found to be lower level than at present–day [3–4]. The CO_2 concentration of air increased from 200 ppmv in the last glacial maximum, 17,000–20,000yrBP, to 280 ppmv at 10,000yrBP. In connection with variation of the atmospheric CO_2 concentration in the glacial to interglacial, variation in terrestrial chemical weathering in the time scale of thousand years was found by using Sr isotope ratios [5–6]. Here we report some facts which are considered to suggest that chemical weathering caused by human land use affects on the budget of anthropogenic CO_2 in the shorter time scale.

2. CONCENTRATION-$\delta^{13}C$ RELATION OF ATMOSPHERIC CO_2

Both the concentration and $\delta^{13}C$ of the atmospheric CO_2 after the 13th century were determined from the studies on Antarctic ice cores [7–8]. In the 13–18th centuries, when human land use and hence deforestation was spreading, the atmospheric CO_2 concentration shows a very little variation but its $\delta^{13}C$ seems to lack in variation. This indicates the existence of some causes to perturb both an increase in the concentration (280 ppmv) and a 1decrease in the $\delta^{13}C$ (–6.4 per mil) of the atmospheric CO_2, which was caused by the addition of terrestrial biomass carbon with $\delta^{13}C$ of –25 per mil.

After the Industrial Revolution, the addition of anthropogenic CO_2 to the atmosphere caused a significant increase in the atmospheric CO_2 and was accompanied by a decrease in $\delta^{13}C$ of the atmospheric CO_2. Figure 1 shows the correlation be–tween the concentration and $\delta^{13}C$ of CO_2 extracted from air trapped in a polar ice core. While this is approximately linear, the slope deviates largely from that for the anthropogenic CO_2 with $\delta^{13}C$ of –25 per mil, as pointed out by Siegenthaler [10]. This supports the existence of some causes to perturb both the concentration and $\delta^{13}C$ of the atmospheric CO_2.

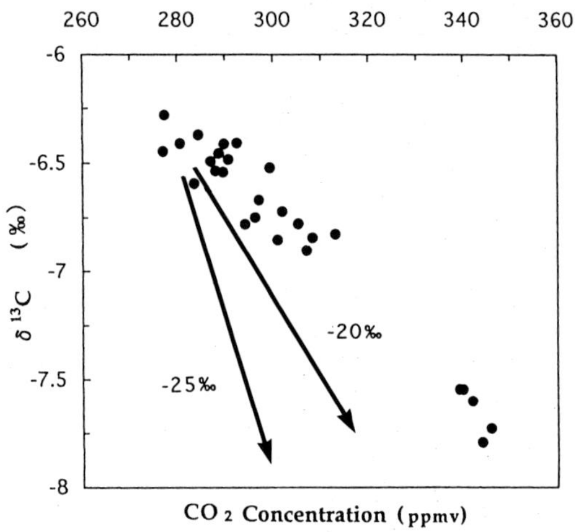

Figure 1 The correlation between concentration and $\delta^{13}C$ of CO_2 extracted from air trapped in an ice core from Siple Station, Antarctica. The arrows indicate the effect of the addition of CO_2 with $\delta^{13}C$ values of –25 and –20 per mil.

3. CARBON ISOTOPE BUDGET FOR ANTHROPOGENIC CO_2

Even in the present–day budget of anthropogenic CO_2, both the concentration and $\delta^{13}C$ of the atmospheric CO_2 show smaller variations than expected.

Emissions estimated of anthropogenic CO_2 are both 5.4×10^{15} g–C/yr from fossil fuels and 0.6×10^{15} g–C/yr from land use and deforestation [11]. Observed accumulation in the atmosphere is 3.4×10^{15} g–C/yr and that in "so–called missing sinks" is 3.6×10^{15} g–C/yr. Here, the $\delta^{13}C$ values of the anthropogenic CO_2 of both fossil fuels origins and land use and forestation origins are around –27 per mil, and that (around –12 per mil) of the CO_2

fraction added to the atmosphere can be calculated from the observed concentration and $\delta^{13}C$ of the atmospheric CO_2. Accordingly, the $\delta^{13}C$ value of "so–called missing sinks" can also be calculated. The $\delta^{13}C$ value calculated of –40 per mil is greatly smaller than –25 per mil of the anthropogenic CO_2.

Quay et al. [12] observed a decrease by 0.4 per mil between 1970–1990 in the $\delta^{13}C$ value of the dissolved inorganic carbon (DIC) in the surface waters of the Pacific Ocean. This means that the $\delta^{13}C$ value of the fraction of DIC added to the ocean in the same period is –55 per mil. This value is unbelievably small and suggest the existence of some important missing sinks.

Figure 2 shows the global carbon cycle through major reservoirs [10–11]. Calculation based both on the decrease by 0.4 per mil between 1970–1990 in the $\delta^{13}C$ value and on annual increase of DIC in surface ocean water gives – 19 and –55 per mill for the $\delta^{13}C$ value of the fraction of DIC added to the ocean given by IPCC 1990 [11] and Siegenthaler [10], respectively. These results indicate that the latter is better estimation than the former. This means a decrease by half of oceanic uptake of anthropogenic CO_2 and supports the existence of some important missing sinks, as above mentioned.

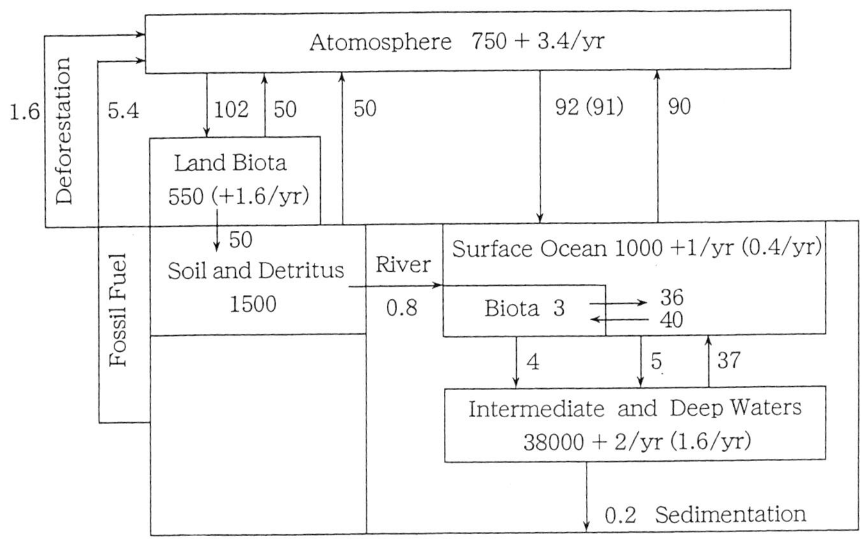

Figure 2 Global carbon cycle through major carbon reservoirs and fluxes after IPCC 1990 [11] and after Siegenthaler [10] given in parentheses. Numbers are in GtC for reservoirs and GtC/yr for fluxes.

4. CO_2 UPTAKE AMOUNT BY CHEMICAL WEATHERING

We calculated the amount of CO_2 uptake by terrestrial chemical weathering in the world by using world average chemical composition of river water [13] and chemical weathering reactions proposed by Mackenzie and Garrels [14].

Consequently, the CO_2 uptake amount of chemical weathering in the world is around 0.5×10^{15} g–C/yr, which is much the same that many studies have estimated [15–16].

On the other hand, we determined the chemical composition of Midorikawa River, Kumamoto Prefecture, Japan, and calculated the CO_2 uptake amount of terrestrial chemical weathering in its drainage basin. The amount calculated is 6×10^{10} g–C/yr. This value is considered unexpectedly large from its drainage area of 1,100 km^2. This indicates strongly that the estimated amount in the world is probably underestimated.

No other causes but chemical weathering are assumed as significant ones to perturb both an increase in the concentration and decrease in the $\delta^{13}C$ of the atmospheric CO_2 caused by the addition of the anthropogenic CO_2 of both fossil fuels origins and land use and forestation origins. We consider the CO_2 uptake amount of chemical weathering in the world become larger according as studies on runoff progress worldwidely. Especially, there are very few studies on runout, which is very important on the meaningful estimation of its CO_2 uptake amount. Therefore, we believe that the variation of the atmospheric CO_2 concentration caused by deforestation due to land use should be examined in connection with variation in terrestrial chemical weathering, and propose a international program for worldwide investigations on runoff and especially on runout "Toward global planning of sustainable use of the Earth".

REFERENCES

1 R.A. Berner, Amer. J. Sci., 291 (1991) 339.
2 L. Francois and C. Walker, Amer. J. Sci., 292 (1992) 81.
3 A. Neftel, H. Oeschger, J. Schwander, B. Stauffer and R. Zumbrunn, Nature, 295 (1982) 220.
4 J.M. Barnola, D. Raymond, Y.S. Korotkevitch, and C. Lorius, Nature, 329 (1987) 408.
5 A.N. Dia, A.S. Cohen, R.K. O'Nions and N,J, Schackleton, Nature, 356 (1992) 786.
6 G. Munhoven and L.M. Francois, NATO ASI Series I, 17 (1994) 39.
7 H. Friedli, E. Moor, H. Oeschger, U. Siegenthaler and B. Stauffer, Geophys. Res. Lett., 11 (1984) 1145.
8 A. Neftel, E. Moor, H. Oeschger and B. Stauffer, Nature, 315 (1985) 45.
9 H. Friedli, E. Lotscher, H. Oeschger, U. Siegenthaler and B. Stauffer, Nature, 324 (1986) 237.
10 U. Siegenthaler, NATO ASI Series I, 15 (1992) 367.
11 J.T. Houghton, G.J. Jenkins and J.J. Ephraums (eds), Climate Change: The IPCC Scientific Assessment, Cambridge University Press, Cambridge, 1990.
12 P.D. Quay, B. Tilnrock and C.S. Wong, Science, 256 (1992) 74.
13 D.A. Livingstone, USGS Prof. Paper No. 440 Chap. G (1963).
14 F.T. Mackenzie and R.B. Garrels, Amer. J. Sci., 264 (1967) 507.
15 E.T. Sundquist, AGU Geophys. Monogr. Ser., 32 (1985) 5.
16 M. Meybeck, Amer. J. Sci., 287 (1987) 401.

TOWARD GLOBAL PLANNING OF SUSTAINABLE USE OF THE EARTH
Development of Global Eco-Engineering
S. Murai (Editor)

Effects of Harvesting Impacts on Forest Ecosystem and Global Environment

Shigeo Kobayashi[a]

[a]Forestry & Forest Products Research Institute, Tsukuba, Ibaraki, 305 Japan

Abstract

Forest harvesting impacts on forest ecosystem were surveyed at three different site. Results of the forest harvesting vary with tree species, time of cutting, tree yarding methods, local precipitation patterns, topography, soil characteristics, and the methods of reforestation. Severe variations of changes in soil properties are caused by the compaction or the removal and the deposition of topsoil by a tractor. Soil physical properties were severely changed. Fine porosity increased and coarse porosity decreased as a result of compaction and decomposition of soil organic matter. According changes of porosity composition, the saturated hydraulic conductivity rates decreased notably and bulk density increased. Although natural forest stores a lot of minerals, forest harvesting causes these mineral contents to decrease. But these minerals accumulate again in the forest ecosystem following the process of forest recovery.

1. INTRODUCTION

Forest is decreasing at a rate of 16.9 million hectares per year in tropics and forest land is converted to agricultural land, pasture and plantation. Decrease and degradation of the forest affects not only the production of timber but also forest ecosystem and global environment in various ways. Concerns have been raised about harvest effects on site productivity, water supply from the forest land, soil and nutrient losses, natural disaster, biodiversity and green house gas emission (EPA et al. 1990). Environmental changes must be initiated by forest harvesting.

Recently, scientific interest has focussed on effects of forest harvesting in a variety of soil factors, such as the deterioration of soil physical properties, nutrient loss, organic matter decomposition and land degradation. However, results of the studies vary with tree species, time of cutting, tree cutting and yarding methods, local climate patterns, topography, geology, soil characteristics and methods of reforestation (Kobayashi et al. 1987). Therefore, many case studies in different site will be effective to clarify the mechanism of soil degradation and aggradation. In this study, I selected three site such as planted coniferous forest in Japan, natural coniferous forest in U.S.A. and natural mixed *Dipterocarp* forest in Brunei where different cutting and yarding methods were applied. Variations in forest structure, soil and microclimate were surveyed before and after forest harvesting.

The purpose of this study are (1)to clarify the soil changes at degradation phase after forest harvest, (2)to clarify the losses of carbon and nitrogen after forest harvest, (3)to clarify the aggradation process of soil for ten years after forest harvest, and (4)to discuss the forest rehabilitation methods for the conservation of environment and sustainable development of forest resources.

2. SITE DESCRIPTION and METHODS

Three study site were selected such as Takahara mountain in Japan, Andrews experimental forest in U.S.A. and Labi hill in Brunei (Table 1, Tabl 2). Other three site of Mikouchi in Japan, Madan in Papua New Guinea and Thon Pha Phoon in Thailand, are under survey.

Takahara mountain is located in central Japan. The topography of this site extends from interfluve down to alluvial toeslope and ranges from 550 m to 470 m above sea levels. The climate near this site (in Yaita city) is mild with wet summer and warm, dry winter. Annual precipitation normally exceeds 1385 mm and is concentrated in summer. Mean temperature in August is 24.8 ℃, and January is 2.1 ℃. This site is covered with Kanto loam underlained several igneous rocks and is classified in brown forest soil groups. This 47 years old Hinoki (*Chamaecyparis obtsusa*) plantation was clearcut harvested using cable skyline yarding systems in April, 1979.

Andrews experimental forest is located in the Cascade mountains about 80 km east of Eugene, Oregon U.S.A.. The topography of this site shows from a high terrace down a colluvial slope and ranges from 733 m to 683 m above sea levels. The parent material of this site is glacial deposits and classified in Cryumbrept soil groups. The climate is also mild. Annual precipitation exceeds 2500 mm and is concentrated in the winter. Little or no rain falls during Jun, July, and August. Mean temperature in January is 2 ℃, and July 20 ℃. Before harvest, the forest consisted of Douglas-fir (*Pseudotsuga menziesii*), western hemlock (*Tsuga heterophylla*) and western red ceder (*Thuja plicata*). Douglas-fir was dominant with 43 stems/ha in density, averaging 56 m in height, 1.2 m in D.B.H. and 490 years in age. This study site was clearcut harvested using tractor and high leed yarding systems in Jun, 1985.

The study site is situated about 45 km from Sungai Liang, in Brunei Darussalam at about 100 m above sea level. An experimental plot was set 1ha from ridge to stream. Parent material is sand and mud stones. Soil in this study site is classified into red-yellow podzolic soil which showed thin A0 and A horizons, and pipestone at a depth of 150 cm. This site consisted of mixed *Dipterocarp* forest such as *Dryobalanops aromatica*, *Dipterocarpus globosus*, *Shorea laxa*, *Shorea exlliptica*, *Dipterocarpus exalatus*, *Dipterocarpus geniculatus*, *Shorea macrophylla*, *Shorea parvifolia*, *and Upuna borneensis*. *Dryobalanops aromatica* was the dominant species. A large amount of *Dipterocarp* seedlings were observed on the forest floor before harvesting. This study site was applied the selective cutting using tractor yarding system in October, 1987.

Topsoils were surveyed in 54 pits of Takahara mountains, 54pits of Andrews experimental forest and 100 pits of Labi hill before and after harvesting. Morphological properties of the soil profiles were examined and soils were sampled from surface to 10 cm in depth for chemical and physical analysis. Organic carbon and total nitrogen were analyzed by the dry combustion method. Soil physical properties were determined for samples taken at the site using 400 cc cylinder by Mashimo's method.

Table 1. Environmental Condition of Study Site

Study Site	Climatic condition	Elevation	Soil	Parent material	Forest type
Takahara mt. Tochigi Pref. JAPAN	Temperate	470m - 550m	Forest Brown Soil	Volcanic Ash	Chamaecyparis obtusa Cryptomeria japonica Plantatio (Height:12.5m, D.B.H.:0.15m)
Andrews Experiment Forest Oregon, U.S.A.	Cool Temperate	683m - 733m	Cryumbrept	Glacia Deposit	Pesudotsuga menziesii Natural Forest (Height:56m, D.B.H.:1.2m)
Labi Hill BRUNEI	Moist Tropics	90m - 142m	Red-Yellow Podzolic soil	Sand and Mud Stones	Mixed Dipterocarp Natural Forest (Height:50m, D.B.H.:0.5m)

Table 2. Forest Logging Methods

Study Site	Cutting Method	Yarding Mehod	Site Preparation	Cutting Date	Survey Date
Takahara Mt. Tochigi Pref. JAPAN	Clear-cutting	Sky line yarding (Tayler method) Stem only	Slash accumulation	1979 April	before:1978 Oct. after :1979 Oct.
Andrews Experimental Forest Oregon U.S. A.	Clear-cutting	High-Lead method Tractor Stem only	Slash burning	1985 Jun	before:1985 May after :1986 Aug.
Labi Hill BRUNEI	Sellective cutting	Tractor Stem only	Slash scatter	1987 November	before:1987 July after :1988 July

3. RESULTS and DISCUSSION

3.1. Soil properties changes after forest harvest

Results of the forest harvesting vary with tree species, time of cutting, tree yarding methods, local precipitation patterns, topography, soil characteristics, and the methods of reforestation (Table 3). Tree felling and yarding directory affect changes in soil physical properties and soil chemical properties are changed indirectly due to climate condition following these forestry activities. The accumulation of litter and cull logs is apparently caused by stem harvesting. Therefore, severe variations of changes in soil properties are caused by the removal or the deposition of topsoil by a tractor, thickness of A horizon decreases from 11.2 cm, 4.6 cm, 9.8 cm to 10.3 cm,

Table 3. Changes of Soil Properties before and after Forest Harvesting

Soil Property	Takahara Mt. Tochigi Pref. JAPAN	Andrews Exp. Forest Oregon U.S.A.	Labi Hill BRUNEI	(unit)
A0 thickness				
before	3.7±1.9	4.6±4.2	4.8±2.1	cm
after	2.1±2.1	8.1±9.0	4.5±4.5	cm
rate	(56.8%)	(176.1%)	(93.8%)	
A thickness				
before	11.2±3.8	4.6±2.0	9.8±3.9	cm
after	10.3±5.4	2.5±2.4	6.3±5.0	cm
rate	(92.0%)	(54.3%)	(64.3%)	
Bulk density				
before	0.379±0.075	0.524±0.084	1.068±0.153	g/cc
after	0.440±0.081	0.591±0.114	1.228±0.290	g/cc
rate	(116.1%)	(112.8%)	(115.0%)	
Saturated hydraulic conductivity				
before	446±196	254±187	61±75	cc/min.
after	252±129	53±79	46±75	cc/min.
rate	(56.5%)	(20.9%)	(75.4%)	
Non capilary porosity				
before	47.7±4.0	48.5±4.9	35.5±4.9	%
after	40.5±4.8	45.3±6.3	24.7±9.0	%
rate	(84.9%)	(93.4%)	(69.6%)	
Capilary porosity				
before	35.4±3.3	24.0±3.4	21.4±2.6	%
after	39.8±4.0	25.4±4.7	27.2±5.2	%
rate	(112.4%)	(105.8%)	(127.1%)	
Total carbon				
before	15.41±3.31	7.77±7.32	3.9±1.4	%
after	12.38±2.96	9.23±3.54	2.8±1.7	%
rate	(80.3%)	(118.8%)	(71.8%)	
Total nitrogen				
before	0.75±0.14	0.23±0.05	0.21±0.05	%
after	0.65±0.14	0.25±0.06	0.16±0.08	%
rate	(86.7%)	(108.7%)	(76.2%)	

2.5 cm, 6.3 cm at Takahara mountains, Andrews experimental forest and Labi hill with organic matter decomposition in topsoil. This compaction decreases non capillary porosity of topsoil from 47.7 %, 48.5 %, 35.5 % to 40.5 %, 45.3 %, 24.7% respectively. But capillary porosity increases from 35.4 %, 24.0 %, 21.4 % to 39.8 %, 25.4 %, 27.2 % respectively. Bulk density also increases from 0.379 g/cc, 0.524 g/cc, 1.068 g/cc to 0.44 g/cc, 0.591 g/cc, 1.228 g/cc. As a result of these changes, the hydraulic conductivity of topsoil decreases from 446 cc/min., 254 cc/min., 56 cc/min. to 252 cc/min., 53 cc/min., 46 cc/min. before and after harvesting respectively. Typical decrease of saturated hydraulic conductivity rate is shown in Andrews experimental forest resulted from repellency revelation. Soil organic matter is accelerated its decomposition because of removal of vegetation cover. Contents of organic carbon and total nitrogen decrease at Takahara mountains and Labi hill after harvesting. But soil organic matter in Andrews experimental forest increases caused by tillage.

Severe variations of changes in soil properties are caused by the compaction or the removal and the deposition of topsoil by a tractor. Therefore, direct impacts of harvest to the soil are evaluate as compaction, removal of soil, tillage and accumulation and/or mixing of organic matter. Indirect impacts of harvest resulted in mineralization of topsoil, decomposition of organic matter, repellency revelation and erosion. Soil physical properties were severely changed. Fine porosity increased and coarse porosity decrease as a result of compaction and decomposition of soil organic matter. According changes of porosity composition, the saturated hydraulic conductivity rates decrease notably and bulk density increases.

Many studies have examined forest harvesting impacts on soil physical properties (Dyrness 1967, Wert et al. 1981, Froehlich et al. 1983, Gent et al. 1984). While the studies of saturated hydraulic conductivity, bulk density and porosity composition are valuable indicators of harvesting effects on the forest soils and useful in predicting the deterioration of soil, they may be insensitive to change in soil water conditions. I observe these soil changes and interpret them as resulting from both direct and indirect effects.

3.2. Carbon and nitrogen masslosses after forest harvest

Organic matter in the soil is accelerated its decomposition rate and total carbon and total nitrogen contents decrease because organic matter mineralizes the inorganic matter which is lost from the forest ecosystem. Kobayashi (1982) found that organic matter in the topsoil decomposed rapidly in the first six months after harvest. Binkley (1984) observed rapid decomposition at the interface of the humid layer and mineral soil in a clearcut site, using cellulose filled litterbags.

Carbon masslosses from topsoil in Brunei, Japan and U.S.A. are estimated about 19.2, 9.3 and 5.1 ton/ha respectively (Table 4). Nitrogen masslosses from topsoil in Brunei, Japan and U.S.A. are about 0.96, 0.24 and 0.19 ton/ha. Carbon and nitrogen masslosses from topsoil increase according to cool temperate forest, temperate forest and tropical moist forest after harvesting, although storage of carbon and nitrogen in soils indicated reverse trend (Tsutsumi 1989). It is also said that temperature and water condition effect on litter and soil organic matter decomposition, and high temperature and enough water condition accelerate litter and soil organic matter decomposition (Jansson et al. 1985). Decomposition rate of soil organic matter is higher in Labi hill than in Andrews experimental forest caused by climate. Therefore, tropical rain forest plays a key role in maintaining the global carbon balance.

It remains to be proved that relationship between soil water condition and decomposition ratio will be discussed more explicitly and accelerated soil organic matter decomposition will be clarified its effect on soil recovery from degraded log over forest site.

Table 4. Carbon and Nitrogen Loss/Emission after Forest Harvesting

	Takahara Mt. Tochigi Pref. JAPAN	Andrews Experimenta Forest Oregon U.S.A.	Labi Hill BRUNEI	(unit)
Total Carbon				
before harvest	65.41	18.73	40.82	Ct/ha
after harvest	56.11	13.64	21.66	Ct/ha
(loss)	(- 9.31)	(- 5.09)	(- 19.16)	Ct/ha
Total Nitrogen				
before harvest	3.18	0.55	2.20	Nt/ha
after harvest	2.95	0.37	1.24	Nt/ha
(loss)	(-0.24)	(-0.19)	(-0.96)	Nt/ha

3.3. Aggradation process in soils after forest harvest

Although natural forest stores a lot of nutrients and minerals, forest harvesting causes these nutrient and mineral contents to decrease. But these nutrients and minerals accumulate again in the forest ecosystem following the process of forest recovery (Jordan 1985). Figure 1 shows aggradation process through degradation phase in Takahara mountains. A0 and A thickness, saturated hydraulic conductivity rate, contents of total carbon and nitrogen had decreased for two years after forest harvesting and then slowly recover. These recovery rate is depending on landform unit. Concerning on saturated hydraulic conductivity, it is very clear that recovery rate reach only less than half of initial rate at seepageslope, convex creepslope and alluvial toeslope. If these saturated hydraulic conductivity do not recover until next harvest time, land becomes to degrade which will be trigger of surface erosion, nutrient losses and landslide.

Changing soil properties as a result of forest harvest also depend on characteristics of landforms, so landform units are classified to related with soil changes in this aggradation process. Dyrness (1967) discussed slope steepness, Sollins et al. (1981) also pointed out the importance of slope position and Binkley (1984) suggested the elevation influenced harvest impacts and soil changes. Changing soil patterns could be associated with landform units effectively. Seepage slope, convex creepslope, and transportational midslope are considered to be the most damaged and difficult recovered units so that the forestry practice will be paid attention to these units.

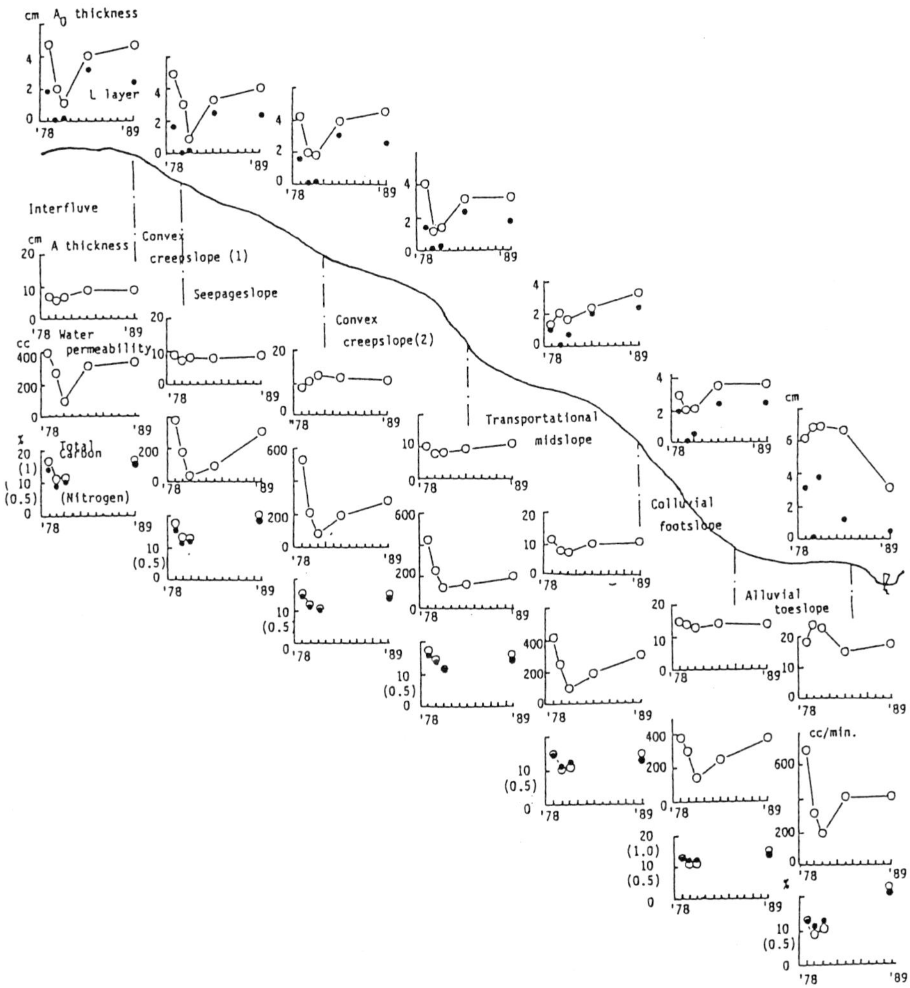

Figure 1. Aggradation process of forest soil properties after 10 years clear-cutting in Takahara mountains

Land degradation will be advanced, if second harvest apply the recovered forest not to recover soil physical properties at initial stage before harvest. In the same way, global warming will be accelerated, if contents of organic carbon and nitrogen in topsoil do not recover at the second forest harvest time. Therefore, sustainable use of forest resources must be considered its rotation and capacity of ecosystem. Rehabilitation of logged over forest or afforestation of degraded land must be applied using adequate silvicultural treatments.

4. CONCLUSION

1) Changes of soil properties are caused by direct influence of forestry activities after forest harvest. Following forest harvest, soil properties also change by indirect influence of soil mineralization and organic matter decomposition and erosion.

2) These impacts decrease the thickness of A0 and A horizons, organic carbon, total nitrogen. Changes soil physical properties are mainly represented by decreased saturated hydraulic conductivity rate, coarse porosity and increased bulk density.

3) Carbon and nitrogen losses from topsoil increase from cool temperate forest to tropical moist forest after harvesting.

4) During aggradation process, soil properties such as thickness of A0 and A horizons, the saturated hydraulic conductivity rate, organic carbon and total nitrogen increase again.

5) Land degradation will be advanced, if second harvest apply the recovered forest not to recover soil physical properties at initial stage before harvest. In the same way, global warming will be accelerated, if contents of organic carbon and nitrogen in topsoil do not recover at the second forest harvest time. Therefore, sustainable use of forest resources must be considered its rotation and capacity of ecosystem.

6. REFERENCES

1 Binkley, D. 1984. Dose forest removal increase rates of decomposition and nitrogen release? For. Eco. Manag. 8, 229-233.

2 Dyrness, C.T. 1967. Soil surface conditions following skyline logging. USDA Forest Service, Research Note, PNW-55, 1-8.

3 EPA, IBAMA, USP(1990):Tropical forestry response options to global climate change, IPCC, AFOS, pp.531.

4 Froehlich, H.A. and McNabb, D.H. 1983. Minimizing soil compaction in Pacific Northwest forests. Forest soils and treatment impacts (ed. Stone, E.L.). Proc. of Sixth North Ame. For. Soils Coff, 159-192.

5 Gent, J.A., Ballard, R., Hassan, A.E. and Cassel, D.K. 1984. Impact of harvesting and site preparation on physical properties of Piedomont forest soils. Soil Sci. Soc. of Amer. J, 48, 173-177.

6 Jansson, P.-E. and Berg, B. 1985. Temporal variation of litter decomposition in relation to simulated soil climate. Long-term decomposition in a Scots pine forest V. Can. J. Bot. 63, 1008-1016.

7 Jordan, C.F. 1985. Nutrient cycling in tropical forest ecosystems. John Wiley & Sons, pp190.

8 Kobayashi, S. 1982. Changes in soil caused by forest clear-cutting. Pedologist, 26, 150-163. (in Japanese)

9 Sollins, P. and McCorison, F.M. 1981. Nitrogen and carbon solution chemistry of an old growth coniferous forest watershed before and after cutting. Water Reso. Res., 17, 1409-1418.

10 Tsutsumi,T.ed.(1989):Forest Ecology, Asakura, Tokyo, pp.166. (in Japanese)

11 Wert, S. and Thomas, B.R. 1981. Effects of skid roads on diameter, height, and volume growth in Douglas-fir. Soil Sci. Soc. of Ame. J., 45, 629-632.

TOWARD GLOBAL PLANNING OF SUSTAINABLE USE OF THE EARTH
Development of Global Eco-Engineering
S. Murai (Editor)

Lakes and Reservoirs; Reflecting Waters of Sustainable Use

S. Matsui[a], S. Ide[b] and M. Ando[b]

[a]Laboratory for Control of Environmental Micropollutants, Kyoto University, 1-2 Yumihama, Otsu City, Shiga 520, Japan

[b]International Lake Environment Committee Foundation, 3-13-25 Kusatsu, Kusatsu, Shiga 525, Japan

Abstract

The International Lake Environment Committee (ILEC) has been conducting, in cooperation with the United Nations Environment Programme (UNEP), a project entitled the "Survey of the State of World Lakes" since 1986 with the aim of collecting and compiling environmental data on as many important lakes and reservoirs of the world as possible to be used as the basis for environmentally sound management of lakes. The results so far obtained from the survey have been published in five volumes as the "Data Book of World Lake Environments", which contain sets of detailed data on 217 lakes from 73 countries all over the world.

The survey carried out by ILEC and UNEP illustrated that the following six major environmental problems are now widespread among lakes and reservoirs around the world, leading to a crisis of water resources and aquatic ecosystems: 1) *Lowering of water level*, 2) *Rapid siltation*, 3) *Acidification*, 4) *Toxic contamination*, 5) *Eutrophication*, and 6) *Extermination of ecosystems and biota.*

A preliminary analysis on surveyed lakes and reservoirs also demonstrated some interesting correlations between 1) the volume of lake water and transparency, 2) suspended solids concentrations and the ratio of forest area to the whole catchment area, 3) suspended solids concentrations and the area of farmlands in the catchment area per unit volume of lake water, and 4) total nitrogen and total phosphorus concentrations in lake water, indicators of the degree of eutrophication, and the catchment area population per unit volume of lake water.

1. LAKES AS PRECIOUS RESOURCES

The earth is called the blue planet. So does it look from space due to plenty of water covering its surface, and it is also called the planet of water. The water that can be of direct use to human beings is, however, very little. Of the world's water, more than 97% is salt water and only the remaining 3% fresh water. Most of the fresh water is frozen as ice caps, or found deep underground. Fresh water most easily available to us exists in rivers and lakes, but it amounts to only 0.26% of all the fresh water on the earth. On this small amount of fresh water, most of the world's population depend for the supply of domestic, agricultural and industrial waters.

Because of the importance of freshwater to life, people have always chosen to live, if possible, close to it since ancient times. Ironically this is creating a crisis in rivers and lakes, which are now threatened with many types of pollutions and ecological disturbance induced by mankind's intensive activities. Lakes and reservoirs are the most important water resources for the human, which hold a 100 times greater amount of freshwater than rivers. In this paper we will try to overview the state of world lakes in the hope of attaining sustainable use of lake water as a most precious resource.

2. SURVEY OF THE STATE OF WORLD LAKES

ILEC has been collaborating with UNEP in a project called the "Survey of the State of World Lakes" for more than eight years. This project was launched to provide the basis for environmentally sound management of lakes and reservoirs in the world through gathering and disseminating limnological and socio-economic data and information on as many important lakes and reservoirs as possible. The findings from the project have been published by ILEC in five volumes entitled the "Data Book of World Lake Environments". These volumes contain sets of detailed data on 217 lakes from all over the world: 64 from Asia, 4 from Oceania, 56 from Europe, 20 from Africa, 61 from North America, and 12 from South America. The lakes are from 73 countries and 43 of them are reservoirs, while the rest (174) are natural lakes.

3. SIX MAJOR ENVIRONMENTAL PROBLEMS

The data collected by ILEC indicates that various environmental problems are now widespread among lakes and reservoirs around the world, leading to a crisis of water resources and aquatic ecosystems. There are some common features in the disruptions which may be classified into the following six categories: 1) *Lowering of water level,* 2) *Rapid siltation*, 3) *Acidification,* 4) *Toxic contamination*, 5) *Eutrophication,* and 6) *Extermination of Ecosystems and Biota* . Figure 1 illustrates the interactions of those six major problems, in which they are related to one another.

The survey revealed, for instance, that many lakes and reservoirs are suffering from eutrophication: 54% of studied lakes in Asia and Oceania, 53% in Europe, 28% in Africa, 48% in North America, and 41% in South America, and some of them are damaged by accelerated siltation: 14% of them in Asia and Oceania, 5% in Europe, 19% in Africa, 13% in North America, and 16% in South America. The deterioration of water quality and aquatic ecosystems, resulting from the above mentioned environmental problems, as well as the decreasing amount of available freshwater is a global problem in the sense that it is taking place simultaneously over much of the world.

4. PRELIMINARY ANALYSIS

We have not yet undertaken comprehensive analyses of the data collected as this survey will continue for another few more years. Nevertheless a preliminary analysis demonstrated some interesting relationships and correlations as follows:

1) Transparency is one of the most easily measurable index of water quality and a large number of observations are available. The volume of lake water was most closely correlated with transparency as shown in Figure 2. Logarithmic scales were used for both axes;

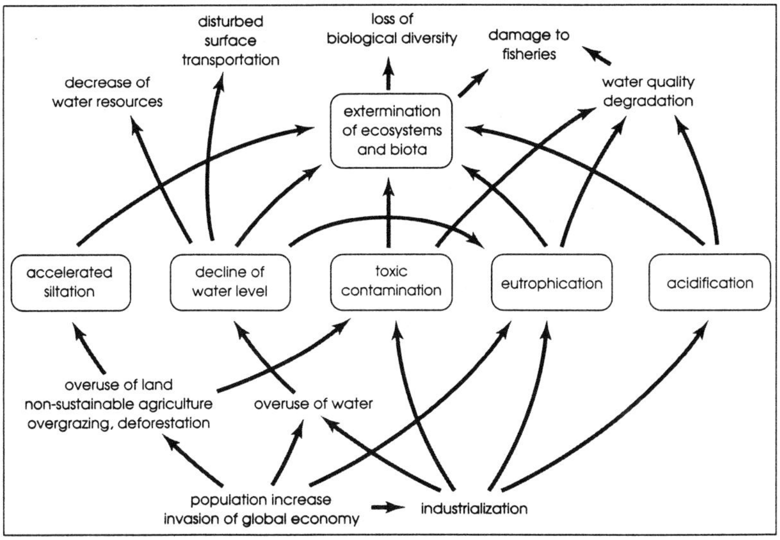

Figure 1. The six major environmental problems in world lakes and reservoirs [3].

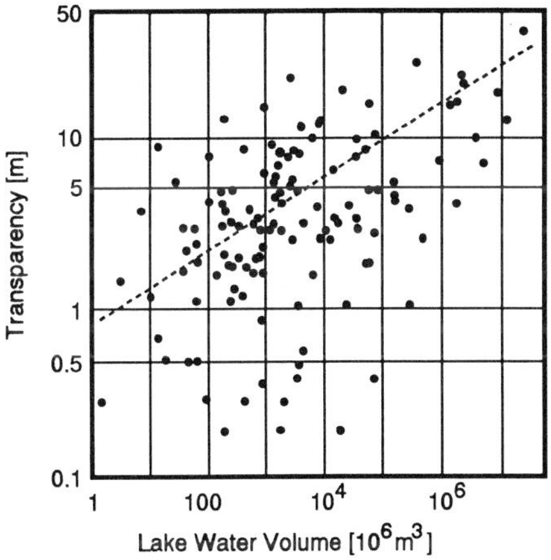

Figure 2. Relationship between Transparency and Lake Water Volume [2].

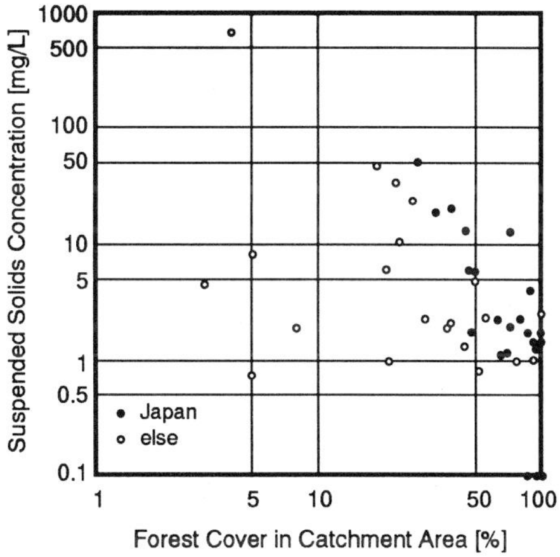

Figure 3. Relationship between SS and Percent Forest Cover in Catchment Area [2].

2) As far as lakes in Japan were concerned, suspended solids concentration (SS) of lake water was found to increase significantly if the area ratio of forest in the catchment area came to be less than 50% (See Figure 3);
3) The progress of farmland development seems to be inevitably associated with the acceleration of soil erosion. As illustrated in Figure 4, the area of farmlands--including both crop fields and pastures in this case--in the catchment area per unit volume of lake water was the most influential factor for the concentration of suspended solids; and
4) Figure 5 shows the relationships between total nitrogen (Total-N) and total phosphorus (Total-P) concentrations in lake water, indicators of the degree of eutrophication, and the catchment area population per unit volume of lake water. The population increase in the catchment area is the leading factor for accelerating eutrophication in world lakes.

More comprehensive and detailed analyses are expected to be made after the completion of this survey to reaffirm the factors responsible for various parameters of water quality.

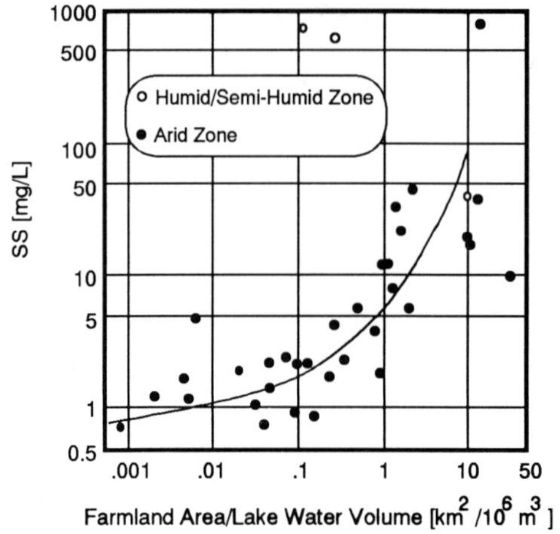

Figure 4. Relationship between SS and Farmland Area in Catchment Area [2].

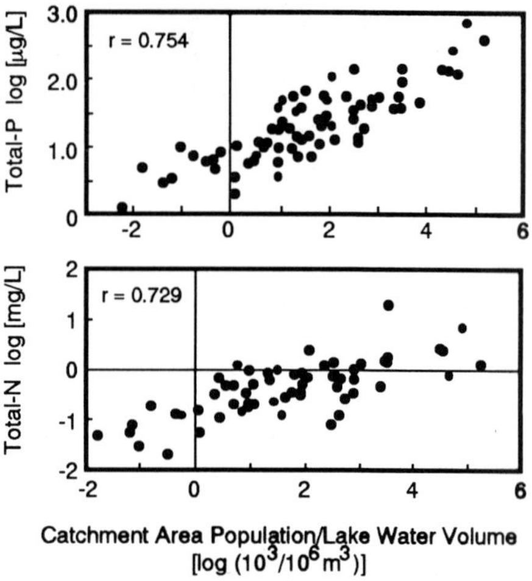

Figure 5. Relationship between Total -P and Total-N Concentration and Population in Catchment Area [2].

5. REFERENCES

1 ILEC and Lake Biwa Research Institute (eds.), Survey of the State of World Lakes, Vols I-V, Kusatsu/Nairobi, ILEC/UNEP, 1988-1993.
2 T. Kira, State of the environments of world lakes: from the survey by ILEC/UNEP, in Proceedings of symposium on water resources management with the views of global and regional scales, Nairobi/Otsu, UNEP/ILEC, pp 48-54, 1991.
3 The pollution of Lakes and Reservoirs (UNEP Environment Library), Nairobi, UNEP, 1994.

TOWARD GLOBAL PLANNING OF SUSTAINABLE USE OF THE EARTH
Development of Global Eco-Engineering
S. Murai (Editor)

Rainfall climatology over Asian monsoon region

Jun Matsumoto

Department of Geography, University of Tokyo
7-3-1, Hongo, Bunkyo-ku, Tokyo 113, Japan

Abstract

Rainfall climatology is performed over Asian monsoon region using 5-day mean raingauge station data. By utilizing time-series of 5-day mean precipitation, a map of onset and retreat of summertime wet season is shown over the whole Asian monsoon region. The onset and withdrawal pattern in South Asia is almost similar to the previous studies, while those in Southeast and East Asian monsoon regions is found to be not a simple north-south migration. Early onset of Assam region, the Inland Indochina Peninsula, northwestern Philippines, eastern Tibet and early retreat of central South China are presented.

1. INTRODUCTION

Rainfall climatology is important in order to understand the climatic environment and its variation in both global and local scales. It is also necessary for the use and planning of water resources over the specific region. Most of the past rainfall climatologies have been based on monthly mean data. However, the seasonal changes and intraseasonal variations of rainfall are very pronounced over Asian monsoon region. Sometimes we experience abrupt seasonal changes in relation to onset and retreat of wet monsoon season. Therefore, monthly data are not sufficient for more minute recognition of climatic environment over the monsoon region.

The maps showing onset and retreat of summertime wet season have been shown over India using mean rainfall data [1]. Recently some studies present similar maps using satellite derived cloud data [2, 3, 4]. The latter can show the large-scale situation. But since they are based on cloud data, it is desirable that the results be checked by rainfall data. Although the onset pattern of the summertime wet monsoon season over the whole monsoon region is shown by compositing South and East Asian situations [5], the definition of wet season is not the same over the whole region.

In the present study, it is attempted to examine the rainfall climatology over Asian monsoon region using 5-day mean precipitation data. 5-day mean rainfall data are computed from daily rainfall data, which are provided from meteorological offices of each country [6]. The data at about 480 stations are utilized within the studied domain (Equator - 45°N, 60 - 160°E). The data period is twelve years (1975-77, 79-87) so as to match the out going longwave radiation (OLR) data [6]. The stations with less than 5 years data missing are utilized.

2. DEFINITION OF ONSET AND WITHDRAWAL OF WET SEASON

It is rather difficult to define the onset and withdrawal of the wet season by pure objective method. However, in order to compare the large-scale situation, some simple definition is needed. In the present study, the basic definition of the wet season onset (withdrawal) is that the first pentad when the pentad precipitation exceeds (lowers) annual mean pentad precipitation (Pm = (Annual precipitation) ÷ 73) in at least three consecutive pentads after lowering (exceeding) it in more than three consecutive pentads. This definition is applied to the stations with annual precipitation of more than 500mm. Winter precipitation maximum is also evident in some stations, but it is not discussed here because its location is limited.

The onset and withdrawal of wet season can easily be determined at most stations of South and Southeast Asia (e.g., Fig. 1A). In some cases, however, there are some 'breaks' of wet season when pentad precipitation lowers Pm in more than three consecutive pentads (e.g., marked 'B' in Figs. 1B and 1C). These breaks are neglected subjectively by considering the situation of neighboring stations.

They cannot simply be determined at many stations over East Asia, because of spring high precipitation (e.g., marked 'S' in Figs. 1D and 1E). However, it generally occurs that the precipitation diminishes substantially in one or two pentads between spring rain and Mei-yu or Baiu pre-summer rainy season. The onset of summer wet season can be defined subjectively by this rainfall diminution. It is noted that this spring rain can only be recognized over East Asian monsoon region. The boundary of this phenomenon provides the regional division between tropical Asian monsoon region and East Asian (Fig. 2A).

3. DISTRIBUTION OF ONSET AND WITHDRAWAL OF WET SEASON

Figure 2 shows the mean onset and withdrawal of summer-time wet season over the whole monsoon Asia derived from the definition mentioned in Chapter 2.

The onset and withdrawal pattern over India is almost similar to the former studies [1, 4], although early beginning of wet season over Assam region and the Himalayan foothill region is more prominent in this study. It is noted the early beginning of wet season in the inland region of Indochina Peninsula and the northwestern part of Philippines. Also early beginning of wet season is observed over eastern Tibetan Plateau. The situation over China is different from the preceding study over northern part [5]. It is because precipitation increases substantially over whole East China at the beginning of Mei-yu season. The separation of pre-summer and mid-summer rainfall as well as that and autumn rainfall is not conducted in the present study. In case of withdrawal of wet season, the situation is again similar to the preceding study over India [1]. However, it isnoted that the wet season ends earliest over central South China.

It should be noted that over east coast region of South and Southeast Asia, pre-winter or winter rainfall is conspicuous (e.g., Fig. 1F). In the preceding map by Indian meteorologists [1] also shows the termination of this winter rain. However, since this is limited over east coast region only, it is not considered in the preceding study. The termination of winter rain is in early December over Southeast India and mid-December in coastal Vietnam and eastern part of Philippines.

In the present study present the mean condition only. Year to year variations should be investigated in the near future.

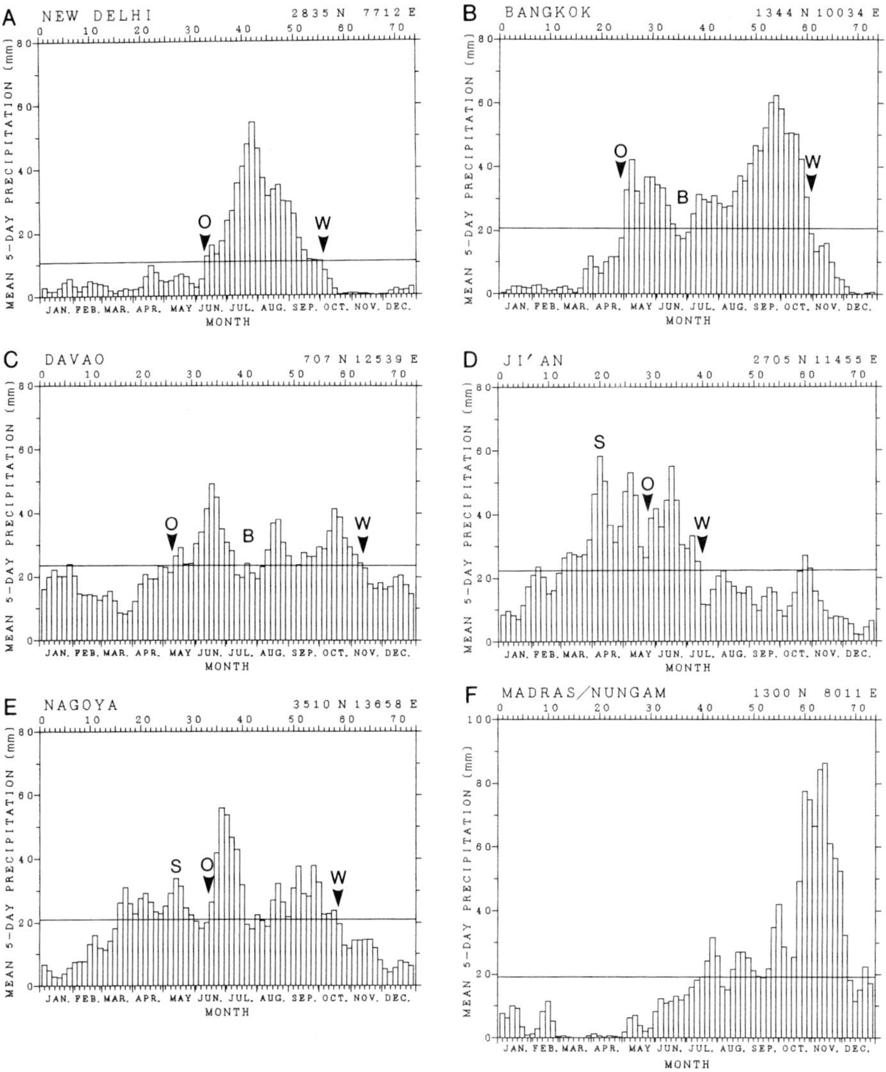

Figure 1. Time-series of mean pentad precipitation at New Delhi (A), Bangkok (B), Davao (C), Ji'an (D), Nagoya (E) and Madras (F).
Horizontal line is annual mean pentad precipitation (Pm). 'O' and 'W' indicate the onset and withdrawal of summer-time wet season, respectively. 'B' and 'S' denotes the period of outstanding break of the summer wet season and spring rainfall, respectively.

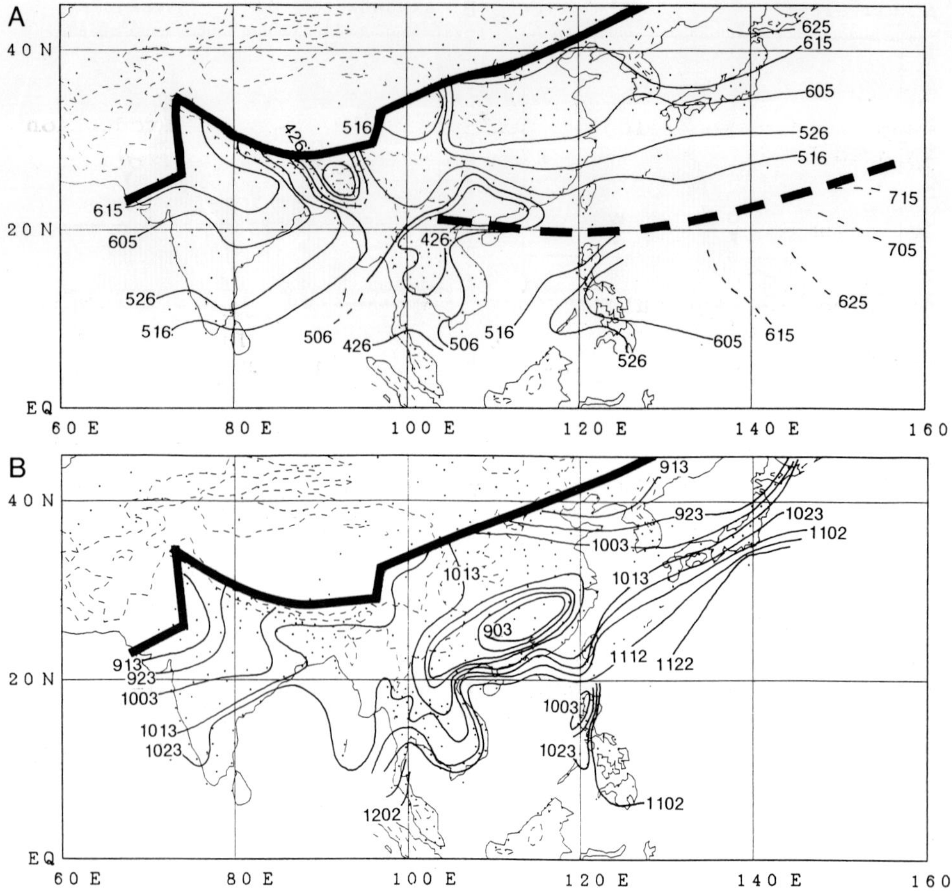

Figure 2. Mean onset (A) and withdrawal (B) of summer wet season over the whole monsoon Asia.
Numeral indicates the beginning date of onset or withdrawal pentad of wet season. Thick line is the 500mm isohyet of annual precipitation. Thick broken line indicates the boundary between South and Southeast Asian region and East Asian region.

4. REFERENCES

1 India Meteorological Department, Climatological Atlas for Airmen, India Meteorological Department, New Delhi, 1943.
2 M. Tanaka, J. Met Soc. Japan, 70 (1992) 613.
3 M. Tanaka, J. Met Soc. Japan, 72 (1994) 255.
4 T. Murakami and J. Matsumoto, J. Met Soc. Japan, 72 (1994).
5 S. Tao and L. Chen, Monsoon Meteorology, Oxford University Press, New York, 1987.
6 J. Matsumoto, Climate over East and Australian Monsoon Regions. Part II: 5-day mean precipitation and OLR. Dept. of Geography, Univ. of Tokyo, 1992.

TOWARD GLOBAL PLANNING OF SUSTAINABLE USE OF THE EARTH
Development of Global Eco-Engineering
S. Murai (Editor)

Proposal for Technologies for Sustainable Biomass Production in Arid Lands

S. Matsumoto

Division of Agriculture and Agricultural Life Sciences, University of Tokyo, Yayoi 1-1-1, Bunkyo-ku, 113, Tokyo, Japan

Abstract

The polyacrylic acid solution formed a sol with sodium salts, permeated the soil, and, by using calcium salts, underwent gelation to form an aquiclude. Furthermore, by altering it was possible to vary the position of the aquiclude within the soil, an indispensable requirement for the practical application of this method. The effectiveness of this new method for economical irrigation showed that the areas produced roughly equal harvest, but that the experimental area used only 1/10 of the amount of irrigation water required by the normally irrigated area, clearly indicating the superiority of the new method for water saving. The new-developed desalinization system features with transparent panels constructed of Iupilon fine could collect pure water from sea water by 10L/m^2 on a fine summer day in Tokyo.

1. Formation and problems with past methods

In the past, various methods for forming an aquiclude have been proposed. For example, a method has been advocated using vinyl sheets buried 40 - 50cm beneath the soil surface, in an attempt to develop paddy-like conditions in sandy soils. However, this method requires the removal and replacement of large quantities of soil aroud the rhizosphere, inorder to lay down the vinyl. The extensive labor required has prevented this method from gaining widespread use. Later, a method was developed involving the injection of molten asphalt into the ground at a depth of 40 - 50cm, creating an impermeable moisture barrier (asphalt moisture barrier, AMB). The effectiveness of this method was recognized after testing in Abu Dhabi in the United Arab Emirates. However, in addition to the large size of the asphalt-injection equipment, the asphalt layer is completely impermeable to water, frequently resulting in the formation of stagnant salt water where the surface soil layer is thin. Further, there is a tendency for salt to accumulate near the surface. For these reasons and others, this method, too, has never seen widespread use.

2. DEVELOPMENT DETAILS AND MERITS OF THE NEW METHOD

It was surmised that if, simply by sprinkling liquid over the surface of the ground, an aquiclude membrane that is not entirely impermeable could be formed within the soil, this would be an ideal method for creating a water-saving layer for irrigation use. A laboratory experiment was therefore conducted on the use of a poly sodium acrylate solution showed in Table 1. It became clear that the polyacrylic acid solution formed a sol with the sodium salt,

Table 1
Characteristics of the polymers used for aqiclude formation

Polyacrylic acid	Polymerization Sodium saturation rate	2,000 0.52
Polyacrylic acid solution	Concentration pH Viscosity	0.2% 6.5 17 cp

permeated the soil, and, by using calcium salts, underwent gelation to form an aquiclude. Furthermore, by altering the sodium saturation with calcium, it was possible to vary the position of the aquiclude within the soil profile, an indispensable requirement for the practical application of this method [2-4].

3. EFFECTS OF THIS NEW METHOD OF AQUICLUDE FORMATION

In order to verify the effectiveness of this new method for economical irrigation and the control of salt accumulation, a field experiment was conducted with barely as a prototype crop. We prepared an experimental area with an aquiclude at a depth of 35cm beneath the soil surfac by using the poly sodium acrylate solution, together with a control area without an aquiclude. Fig. 1 shows that the two areas produced roughly equal harvest, but that the experiment area used only 1/10 of the amount of irrigation water required by the normally irrigated area, clearly indicating the superiority of the new method for water saving.

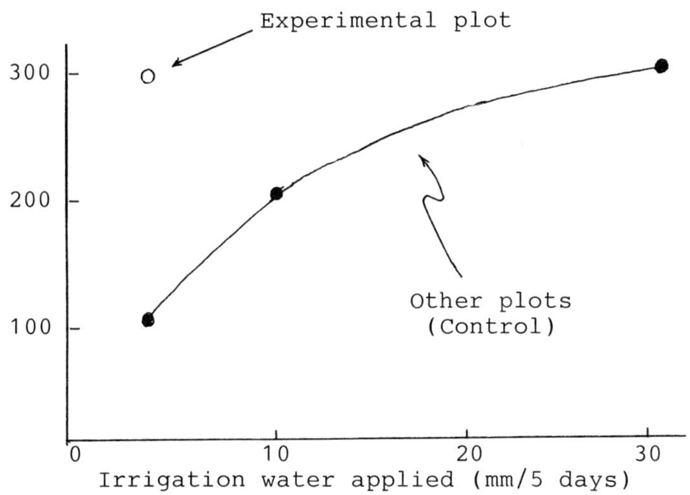

Figure 1. Water saving by using an aquiclude evaliated by barely yield.

4. AN EFFICIENT DESALINIZATION SYSTEM THAT USED SOLAR ENERGY

4.1 Conventional Systems

Of conventional systems, a typical one is the basin-type distillation plant. Such plants started to appear around the world in the 1960s; the largest, with an evaporation area of almost 10,000 square meters, is in Pakistan [5]. In this type of system, the basic structure is the transparent roof panels through which sunlight passes are generally constructed of glass, acrylic, or polycarbonate. However, roof panels made of such materials must be installed at a steep incline, and evaporated water that condenses on the underside of the roof panels flows down sluggishly and sometimes falls off the condensation surface, resulting in inefficient water collection.

4.2 The new system

This newly-developed system features transparent panels

constructed of Iupilon Fine, which is laminated on the underside with a macromolecular film comprised of long-lasting natural materials that allow condensation to immediately flow down the angled roof. This eliminates the falling of droplets of condensation and makes possible roof panel inclines of only four to ten degrees. Fig. 2 shows the mechanism by which droplets of condensation flow down the roof panels in this system.

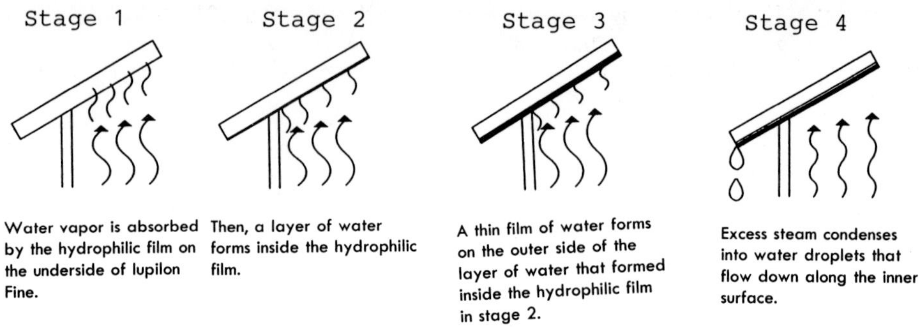

Figure 2. The mechanism by which droplets of condensation flow down.

Results of the experiment using the new system showed that 8.5L/m^2 of pure water was collected from sea water from 10 a.m. to 4 p.m. under the condition of air temperature 24°C and sea water temperature 60°C, while 5.0L/m^2 of pure water was collected using polycarbonate film under the same condition.

5. REFERENCES

1 Komatsu Ltd., Interim Report on Asphalt Moisture Barrier Construction in Arid Region (1978) 70.
2 S. Matsumoto and Y. Owa, Transactions of 14th ICSS, 7(1990) 210.
3 Y. Owa, US Patent No. 4 637 759 (1975)
4 Y. Owa, US patent No. 4 690 589 (1987)
5 Mitsubishi Gas Chemical Co., Green Age, 2(1994) 14

TOWARD GLOBAL PLANNING OF SUSTAINABLE USE OF THE EARTH
Development of Global Eco-Engineering
S. Murai (Editor)

An evaluation method of environmental quality for establishing SEEA

H. Morisugi[a] and E. Ohno[b]

[a]Department of Civil Engineering, Gifu University, 1-1 Yanagido, Gifu 501-11, Japan

[b]Institute of Socio-Economic Planning, University of Tsukuba, 1-1-1 Tennodai, Tsukuba, Ibaraki 305, Japan

Abstract

The SEEA (System of integrated Environmental and Economic Accounting) is presented as a new index aimed at both economic development and environment preservation, which is the SNA (System of National Accounting) joined with the environmental quality change. Because of the necessity for evaluation of environmental quality, this paper presents a benefit measurement method of the environmental quality improvement, which can measure not only its direct impacts but also its repercussion effects by using the general equilibrium model. By applying the concept of equivalent variation for the benefit definition, it shows theory to express the defined benefits in the form of consumer's surplus.

1. INTRODUCTION

Study on benefit measurement of environmental quality improvement had began started in the 1960s [1]. Because most of the environmental quality is not traded in the market and do not have the market price, many economists employ indirect non-market evaluation methods which include the averting behavior approach [for example, 2], the weak complementarity approach [for example, 3], and the hedonic price approach [for example, 4]. While the averting behavior approach exploits the substitutability between environmental quality and other inputs into production, the weak complementarity approach values environmental quality change by using the complementarity of environmental quality. The hedonic price approach has been used primarily to value environmental disamenity, which are reflected both in housing prices and in wages. When the environmental quality has no appropriate averting or mitigating behavior, the indirect method cannot be used to estimate benefits such kinds of environmental quality.

The contingent valuation method is the direct non-market evaluation method, which basically asks people what they are willing to pay for a benefit and/or what they are willing to receive by way of compensation to tolerate a cost. Examples of environmental quality which have been valued by using the contingent valuation method include improvements in water quality for fishing or swimming [5], improvements in air pollution for viewing [6], and so on.

Because those previous studies did not deal with the repercussion effects of the environmental quality improvement, this paper presents a benefit measurement method which can measure not only its direct impacts but also its repercussion effects by using the general equilibrium model. By applying the concept of equivalent variation for the benefit definition, it shows theory to express the defined benefits in the form of consumer's surplus.

2. SOCIO-ECONOMIC MODEL

2.1. Assumptions

In order to simplify the socio-economic model, which expresses the benefit incidence from any urban development project, we will make the following assumptions: (1) the society is composed of four agents; household, private firm, landowner, and government; (2) three kinds of markets exist in the society; the composite goods market, land market, and labor market; (3) when households and private firms demand land service, they may only rent the land from landowner; and (4) the society is in a long run equilibrium.

2.2. Each Agent's Behavior

<Household's Behavior> Suppose that the household behaves by maximizing its utility U. The utility is expressed by the consumption level of composite goods (x_1) and land service (x_2), and the environmental quality level (q) which may consist of some socio-economic indices and cannot be controlled, where it is assumed that the utility function does not have the labor supply (x_3). The household maximizes its utility by controlling x_1, x_2, and x_3 under its budget constrains. Its behavior is formulated by

$$V = \max_{x_1, x_2, x_3} U[x_1, x_2, q], \tag{1.a}$$

$$\text{s.t.} \quad c\, x_1 + r\, x_2 - w\, x_3 = \pi + \sigma - g \, \{= \Omega\}, \tag{1.b}$$

where V is the indirect utility; c is price of composite goods; r is land rent for the household; w is wage rate; π and σ are property incomes distributed from the firms' profit and the landowners' income, respectively; and g is lump sum tax charged for the household. Solving equations (1.a) and (1.b), the functions (x_1, x_2, and x_3) and the indirect utility function (V) can be derived;

$$x_1 = x_1[c, r, w, q, \Omega], \quad x_2 = x_2[c, r, w, q, \Omega], \quad x_3 = x_3[c, r, w, q, \Omega], \tag{2.a, b, c}$$

$$V = V[c, r, w, q, \Omega]. \tag{2.d}$$

<Private Firm's Behavior> Suppose that the private firm maximizes its profit expressed by the production level of composite goods (y_1), and the consumption level of land service (y_2) and labor (y_3). The private firm maximizes its profit by controlling y_1, y_2, and y_3 under its constraint of production function, and its behavior is formulated by

$$\Pi = \max_{y_1, y_2, y_3} c\, y_1 - R\, y_2 - w\, y_3 - G, \tag{3.a}$$

$$\text{s.t.} \quad y_1 = f\,[y_2, y_3, q], \tag{3.b}$$

where Π is the maximum profit; p is price of composite goods; R is land rent for the private firm; w is wage rate; G is lump sum tax charged for the private firm; and the production function (f) is a function of y_2, y_3, and the environmental quality level (q). Solving equations (3.a) and (3.b), the functions (y_1, y_2, and y_3) and the profit function (Π) can be derived;

$$y_1 = y_1[c, R, w, q, G], \quad y_2 = y_2[c, R, w, q, G], \quad y_3 = y_3[c, R, w, q, G], \tag{4.a, b, c}$$

$$\Pi = \Pi[c, R, w, q, G], \tag{4.d}$$

where this profit (Π) is distributed among the households.

<Landowner's Behavior> Suppose that only one landowner exists and rents the land (X_2) to the households and the land (Y_2) to the private firms, and that the landowner's income

(rX_2+RY_2) is distributed among the households. Its behavior is formulated by

$$X_2 = N x_2, \quad Y_2 = M y_2, \tag{5.a, b}$$

where N is the number of households; M is the number of private firms; and assuming that there is no competition between X_2 and Y_2.
<Government's Behavior> Suppose that the government finances I for the environmental quality development by using its revenue from taxes. Then its budget is formulated by

$$I = N g + M G. \tag{6}$$

2.3. Market Equilibrium Conditions

The socio-economic model in this paper will be constructed within the framework of the general equilibrium theory, so the equilibrium conditions on the market are to be considered. The following conditions are held;

- composite goods: $M y_1 = N x_1 + I/c,$ (7.a)
- land : $X_2 = N x_2, \quad Y_2 = M y_2,$ (7.b, c)
- labor : $M y_3 = N x_3,$ (7.d)
- private firm's profit : $M \Pi = N \pi,$ (7.e)
- landowner's income : $r X_2 + R Y_2 = N \sigma,$ (7.f)
- tax : $N g + M G = I.$ (7.g)

3. BENEFIT OF ENVIRONMENTAL QUALITY IMPROVEMENT

3.1. Benefit Definition

In the previous chapter we constructed a socio-economic model so that each agent's response towards the environmental quality improvement project can be grasped. When the environmental quality (q) is changed by the project, the price of land service, composite goods, the property income, the wage rate and so on will change according to the market equilibrium. As a result, the households' utility, the private firms' profit, and the landowner's income will change. An increase of the households' utility level evaluated in monetary terms is called the net benefit from the project. In this paper the social net benefit from the project is defined by using the concept of Equivalent Variation (EV). The benefit is defined as the value of EV satisfying the following equation;

$$V^B = V[c^A, r^A, w^A, q^A, \Omega^A + EV], \tag{8}$$

where the superscript (A) and the superscript (B) indicate the condition without the project and with the project, respectively. Equation (8) shows that the benefit is equal to the minimum compensation amount which the household needs in order to give up the project while maintaining the welfare level achieved in the case with the project.

3.2. Benefit Evaluation Model: Generalized Price Approach

By using the expenditure function (e), the benefit (EV) of equation (8) can be derived as

$$EV = e[c^A, r^A, w^A, q^A, V^B] - e[c^A, r^A, w^A, q^A, V^A]. \tag{9}$$

By assuming the utility function to be quasi-linear function with respect to income, and applying the equilibrium conditions (7.a)-(7.g), the social net benefit ($SNB = N \times EV$) can be derived from equation (9) as

$$SNB \doteqdot (1/2)\{N(\Omega^A_q + \Omega^B_q) + M(\Pi^A_q + \Pi^B_q)\}(q^B - q^A) - I, \tag{10}$$

where Ω_q and Π_q are the marginal value of Ω and Π on the marginal change of q, respectively; and equation (10) is derived by applying the Taylor expansion (the second order expansion) on the transformation of equation (9). The first term $(1/2)\{***\}(q^B-q^A)$ is the benefit and the second term (I) is the cost of the project.

Suppose that the complemental market goods of the environmental quality (q) is only x which is one of them (x_1, x_2, x_3, y_1, y_2, and y_3). The first term of equation (10) can be transformed as follows.

$$(N\Omega^A_q + M\Pi^A_q)(q^B - q^A) = (1/2)(X[p[q^B], q^A, \Omega^A] + X^A)(p^A - p[q^B]), \tag{11}$$

$$(N\Omega^B_q + M\Pi^B_q)(q^B - q^A) = (1/2)(X^B + X[p[q^A], q^B, \Omega^B])(p[q^A] - p^B), \tag{12}$$

where X is the total of x and assumed to be $X[p, q, \Omega]$; p is price of X; $p[q^B]$ is the price level which stands for the minimum price level needed by the household in order to give up the change ($q^A \rightarrow q^B$) and remain at q^A while maintaining the welfare level that would have been achieved by the level of q^B; and $p[q^A]$ is the price level which stands for the minimum price level needed by the household in order to accept the change ($q^A \rightarrow q^B$) and remain at q^B while maintaining the welfare level that would have been achieved by the level of q^A. Then the benefit of the project is the average value of equations (11) and (12), which are illustrated as the shaded part in figure 1.

As a result, the benefit of the environmental quality improvement can be measure in the same way as the benefit of price change in associated private goods such as transport improvement if we can find the price change of a private good equivalent to the environmental quality change.

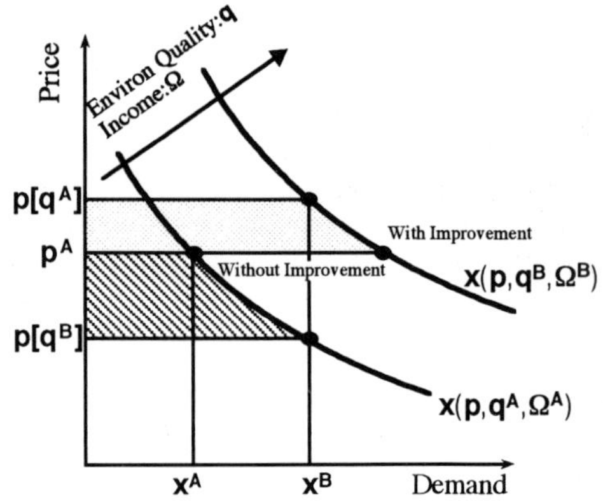

Figure 1. Benefit of Environmental Quality Improvement

4. REFERENCES

[1] M.L. Cropper and W.E. Oates, Environmental Economics: A survey. Journal of Economic Literature 30: 675-740, 1992.

[2] T.J. Bartik, Evaluating the benefits of nonmarginal reductions in pollution using information on defensive expenditures. Journal of Environmental Economic Management 15: 111-27, 1988.

[3] D. Feenberg and E.S. Mills, Measuring the benefits of water pollution abatement. Academic Press, 1980.

[4] Y. Kanemoto, Hedonic prices and the benefits of public projects. Econometrica 56: 981-9, 1988.

[5] R.T. Carson and R.C. Mitchell, The value of clean water: The public's willingness to pay for boatable, fishable, and swimmable quality water. Discussion Paper 88-13. University of California at San Diego, 1988.

[6] Decision Focus Incorporated, Development and design of a contingent value survey for measuring the public's value for visibility improvements at the Grand Canyon National Park. Revised Draft Report, 1990.

TOWARD GLOBAL PLANNING OF SUSTAINABLE USE OF THE EARTH
Development of Global Eco-Engineering
S. Murai (Editor)

Global Water Resources Assessment from Multiple Geo-Information

T. Oki[a], K. Musiake[a], H. Matsuyama[b] and K. Masuda[b]

[a]Institute of Industrial Science, University of Tokyo, 7-22-1, Roppongi, Minato-ku, Tokyo 106, Japan

[b]Department of Geography, Tokyo Metropolitan University, 1-1-1, Minami-Osawa, Hachioji, Tokyo 192-03, Japan

Abstract

Global hydrological cycle is one of the key issues in the global environmental problems. The change of precipitation, evapotranspiration and runoff will have serious effect for the societal activities, and quantitative estimation of current and future hydrological cycle is crucial to build up a master plan against the possible global climate change. In this study, the global distribution of water balance and hydrological cycle is estimated using atmospheric and river basin water balance method. The results show the latest estimation of global water balance, and they will be used for the validation of present general circulation model (GCM)s. They can also be used for the model refinement of land-surface hydrological processes and regional water balance assessments.

1 Introduction

In the research issues addressing the problems of global change, the atmospheric water balance can play a significant role: it can help in the estimation of global water circulation and balance, in the validation of the GCM grid scale hydrological models, and in the interpretation of the model forecasts for water resources assessment.

The basic concept of using atmospheric data to estimate terrestrial water balance was firstly presented by Starr and Peixòto [1]. The application of this concept to regional studies or the region and period with special observations has been troublesome because there were only a few scattered observational stations of upper air soundings. Currently, "Objective Analyses Data" are prepared for prescribing initial values for daily numerical weather forecasting, using GCM and various observational data. Now, such data sets are indeed available, and their spatial resolution seems much improved than the operational rawinsonde network. Therefore it is worth to apply such latest atmospheric data for the estimation of the global water balance.

Table 1: Annual water balance (mm year^{-1}) in each continent. Estimated from ECMWF vapor flux convergence mean from 1989 to 1992 and Legates and Willmott [2] precipitation.

	All Lands	Asia	Eur.	Afr.	N.A.	S.A.	Au.	Ant.
$-\nabla_H \cdot \vec{Q}$	238	244	195	−7	314	758	17	129
P	852	762	746	780	709	1,600	915	620
E	614	518	551	787	395	842	898	491

2 Atmospheric and river basin water balance

Some applications of the atmospheric and river basin water balance method for the estimation of global hydrological cycles are introduced in this section.

Firstly, the large-scale mean evapotranspiration can be estimated if large-scale mean precipitation is available. The equation

$$E = \frac{\partial W}{\partial t} + \nabla_H \cdot \vec{Q} + P \tag{1}$$

is obtained from the atmospheric water balance, and W, $\vec{Q}$, E, P represent precipitable water, vertically-integrated two-dimensional vapor flux, evapotranspiration, and precipitation, respectively.

Secondary, from combined atmospheric and river basin water balance, we obtain

$$\frac{\partial S}{\partial t} = -\frac{\partial W}{\partial t} - \nabla_H \cdot \vec{Q} - \nabla_H \cdot \vec{R}_o, \tag{2}$$

where S and $\vec{R}_o$ represent the storage in the basin and surface runoff, respectively. S includes snow accumulation in addition to soil moisture and ground water storage. Even though an initial value is required to obtain the absolute value of storage, the atmospheric water balance can be useful in estimating the seasonal change of total water storage in a river basin.

Next, with the assumptions of the annual water balance, simple equations of

$$-\nabla_H \cdot \vec{Q} = (P - E) = \nabla_H \cdot \vec{R}_o \tag{3}$$

are obtained. In this equation, the water vapor convergence (precipitation−evaporation) and runoff are equal over the annual period.

The annual water transport in the north-south direction over continents $R^L(\phi_0)$ and oceans $R^O(\phi_0)$ can be estimated from the equations of :

$$R^L(\phi_0) = -\int_{-\frac{\pi}{2}}^{\phi_0} \oint_{\text{Land}} R_e^2 \cos\phi \, \nabla_H \cdot \vec{Q} \, d\lambda d\phi - \int_{-\frac{\pi}{2}}^{\phi_0} D(\phi)\, d\phi, \tag{4}$$

$$R^O(\phi_0) = -\int_{-\frac{\pi}{2}}^{\phi_0} \oint_{\text{Ocean}} R_e^2 \cos\phi \, \nabla_H \cdot \vec{Q} \, d\lambda d\phi + \int_{-\frac{\pi}{2}}^{\phi_0} D(\phi)\, d\phi, \tag{5}$$

where $D(\phi)$ represents the total discharge from continents to the oceans at latitude ϕ, and R_e and λ represent the radius of the earth and longitude, respectively. $\oint_{\text{Land}} d\lambda$ and $\oint_{\text{Ocean}} d\lambda$ mean the zonal integration only over land and ocean respectively.

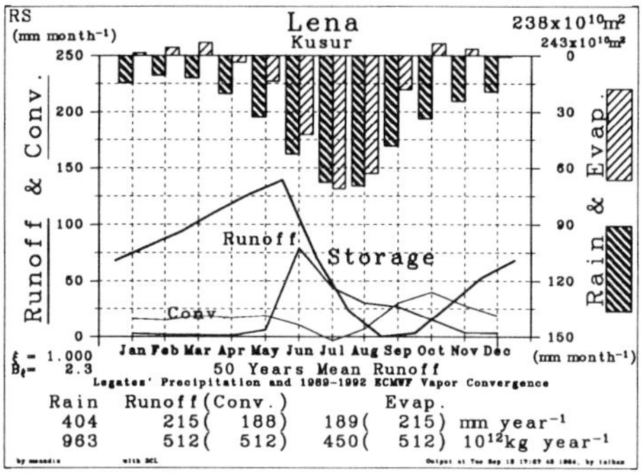

Figure 1: Estimated mean water balance in Lena river basin.

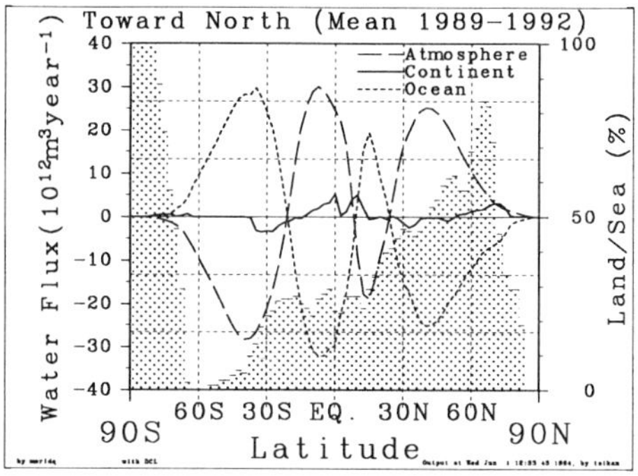

Figure 2: The annual fresh water transport by atmosphere, ocean and rivers.

3 Data and Results

Atmospheric vapor flux convergence is calculated from the the objective analysis dataset by the European Centre for Medium Range Weather Forecasts (ECMWF). The algorithm to estimate the $-\nabla_H \cdot \vec{Q}$ from ECMWF data set and detailed discussion on the data quality compared with observed runoff are described in the Oki et al. [3]. The data period from 1989 to 1992 is used in this study.

Global water balance is calculated from the estimated $-\nabla_H \cdot \vec{Q}$ and digital precipitation data estimated by Legates and Willmott [2]. Both data are compiled in 2.5°mesh grids, and the summary is shown in Table 1. Evapotranspiration is calculated from equation (1). The continents including large dry areas, such as Africa and Australia, have comparatively large evapotranspiration. In the case of South America, $-\nabla_H \cdot \vec{Q}$ is larger than the other continents but precipitation is extremely large, and consequently the evapotranspiration is estimated as large as the dry continents.

The application of the atmospheric water balance method to a river basin is shown in Fig. 1 for an example. Discharge data is from Global Runoff Data Centre (GRDC) and long term mean is used. In order to meet the mean annual $-\nabla_H \cdot \vec{Q}$ with the mean annual discharge, the correction value of $B_\xi = 2.3$ (mm month^{-1}) was uniformly added to mean monthly $-\nabla_H \cdot \vec{Q}$ in the calculation of equations (1) and (2). Sudden decrease of total storage from June in Fig. 1 corresponds to the snow melt in the Lena river basin.

4 Global Annual Fresh Water Transportation

The fresh water supply to the ocean has an important effect for the thermohaline circulations of the ocean, because it changes the salinity. In this study, zonal mean discharge $D(\phi)$ is estimated from $-\nabla_H \cdot \vec{Q}$ by the atmospheric water balance equation (3) at the location of river outlet of each basin. Annual fresh water transport from continents to each ocean was estimated and shown in Table 2. It should be noted that the total amount of fresh water transport from the surrounding continents has the same order of magnitude as the fresh water supply that comes directly from the atmosphere expressed by $-\nabla_H \cdot \vec{Q}$.

Table 2: Annual fresh water transport from continents to each ocean (10^{15} kg year^{-1}). 'Inner' indicates the runoff to the inner basin. $-\nabla_H \cdot \vec{Q}$ indicates the direct fresh water supply from the atmosphere to the ocean.

	N.Paci.	S.Paci.	N.Atlan.	S.Atlan.	Indian	Arctic	Inner	Total
Asia	4.7	0.4	0.2		3.4	2.7	0.0	11.4
Europe			1.7		0.0	0.6		2.3
Africa			−0.3	0.9	−0.5		−0.2	−0.1
N.America	2.9		4.7			1.1		8.7
S.America	0.4	0.4	5.7	8.1				14.6
Australia		0.1			0.1			0.2
Antarctica		1.0		0.1	0.8			1.9
Total	8.0	1.9	12.0	9.1	3.8	4.4	−0.2	39.0
$-\nabla_H \cdot \vec{Q}$	9.7	−10.8	−12.7	−13.5	−13.9	2.2		−39.0
Grand Total	17.7	−8.9	−0.7	−4.4	−10.1	6.6	−0.2	0.0

From equations (4) and (5), annual water transport in the latitudinal direction was estimated (Fig. 2). Transportation by the atmosphere and by the ocean have almost the same absolute value in each latitude but with different sign, and the water transport by rivers is about 10 % of these other fluxes globally.

This number may be underestimated because $-\nabla_H \cdot \vec{Q}$ tends to be smaller than river discharge observed at the ground surface. The peak at 30°S is mainly carried by the Parana River in South America, and the peak around 10°N is also carried by rivers in south America. Russian large rivers, such as Ob, Yenisey and Lena, carry the freshwater towards north and these transports can be seen around 70°N in Fig. 2. These results indicate that rivers play not a negligible role in the climate system not only by the exchange of energy and water within the basin, but also through the fresh water transport by runoff, affecting the hydrological circulation among atmosphere, oceans and continents.

Acknowledgments

The authors wish to thank the European Centre for Medium-Range Weather Forecasts and Global Runoff Data Centre for providing data sets. The ECMWF data set used here is from the archive in the Center for Climate System Research, University of Tokyo.

References

[1] V. P. Starr and J. Peixòto. On the global balance of water vapor and the hydrology of deserts. *Tellus*, 10:189–194, 1958.

[2] D. R. Legates and C.J. Willmott. Mean seasonal and spatial variability in gauge-corrected global precipitation. *Intl. J. Climatol.*, 10:111–127, 1990.

[3] T. Oki, K. Musiake, H. Matsuyama, and K. Masuda. Global atmospheric water balance and runoff from large river basins. *Hydrol. Proces.*, 1994. in print.

TOWARD GLOBAL PLANNING OF SUSTAINABLE USE OF THE EARTH
Development of Global Eco-Engineering
S. Murai (Editor)

Agriculture Monitoring Using Remote Sensing Data

Genya SAITO and Nobuyuki MINO

Remote Sensing Laboratory, National Institute of Agro-Environmental Sciences
3-1-1 Kannondai, Tsukuba, Ibaraki Japan

Abstract

Using NOAA/AVHRR Local Area Coverage(LAC) data received in Japan, we made NDVI Maximum Value Composites(MVCs) and got 12 Japan Vegetation Index (JVI) for one year. Cropping pattern was analyzed in agricultural areas in Japan using JVI.

1. Introduction

Many studies have shown that satellite remote sensing technique is a powerful tool for agriculture monitoring. However LANDSAT data can only be acquired once in 16 days and are highly restricted by clouds. This makes LANDSAT data not very reliable data source for temporal variation analyses in countries like Japan. NOAA/AVHRR data on the other hand, can be acquired daily and comparatively free from cloud restriction.

National Oceanic and Atmospheric Administration(NOAA) has prepared weekly Global Vegetation Index(GVI) images using NOAA/AVHRR data since April,1982. A Very high correlation is reported between vegetation Index and the biomass production[1)].

Spatial resolution of GVI images are more than 10km. It is far too large a value to apply to monitoring agricultural crop pattern in Japan. So we prepared monthly Japanese Vegetation Index(JVI) images which have spatial resolution of 1km using NOAA/AVHRR data.

2. Methods

2.1. Sensor and data

We used NOAA-11/AVHRR data to prepare JVI images. Scanning width of NOAA/AVHRR is 2800km. Spatial resolution is 1.1km at nadir. The JVI images were created using AVHRR CH1(visible red:580 - 680nm) and CH2(near-infrared: 725 - 1100nm) measurements.

2.2. NDVI

Some remote sensing data are related to the activity of the vegetation. Often visible and near-infrared reflectances are combined to compute vegetation indices in order to bring out the difference of radiometric response between the active green canopy and bare soils. Normalized Differential Vegetation Index(NDVI)[2] is most commonly used to estimate the amount of vegetation. NDVI is defined as follows:

$$NDVI=(CH2-CH1)/(CH2+CH1) \quad \text{..................} \quad (1)$$

where CH1, CH2 are observation data of CH1 and CH2

The value of the NDVI takes theoretically between -1 and 1. The higher the vegetation activity is the closer the NDVI gets to 0.5. Where there is no vegetation cover(bare soil), the NDVI is nearly 0. The values of the NDVI become negative in cloud, snow and water areas.

Figure 1 Agricultural Areas of Cropping Pattern Analysis

2.3. Japan Vegetation Index (JVI)

Using NOAA/AVHRR Local Area Coverage(LAC) data received in Japan, daily LAC datasets were created. We then compiled monthly NDVI images using NDVI Maximum Value Composites(MVCs) and got 12 JVI for one year.

Cropping pattern was analyzed in agricultural areas in Japan using JVI. These areas were shown figure 1.

3. Results

3.1. Temporal variations of NDVI in agricultural area

JVI images from April, 1991 to March, 1992 were compiled. In April, the almost values of the NDVI to the north of Kanto district are less than 0.2 while those to the south of Kanto district are higher than 0.2. In June, the values of the NDVI throughout Japan become higher than 0.2 except in the agricultural areas(paddy fields and upland farming fields) where the values of NDVI are about 0.1. This is because crops in those areas are at the beginning stage of their growth. In August, the values of the NDVI higher than 0.25 are found everywhere except urban areas like Tokyo and Osaka. At this time of the year, the agricultural areas are widely covered with grown crops. In October, after the harvest time, the values of the NDVI fall steeply in the agricultural areas.

3.1.1. Paddy fields

As for paddy fields, we selected the following 10 sites; Ishikari Plain, Hachiro-gata, Sendai Plain, Niigata Plain, Aizu Basin, Kanto Plain(near Utsunomiya), Noubi Plain, Okayama Plain, Tokushima Plain and Saga Plain. Shown in figure 2 are monthly NDVI variations from April, 1991 to March, 1992. Figure 2(a) represents northern part of Japan, (b) central part and (c) south-western part respectively. NDVI variations of Niigata Plain are shown both in (a) and (b) as a standard used for purposes of comparison.

In northern part of Japan shown in figure 2(a), the NDVIs reach their maximum values either in July or in August and keep values less than 0.05 from December to March everywhere. Only in May, a clear difference in the value of the NDVI can be found. The kind of crops planted in paddy fields not used for rice and the area of these fields make this difference. There are few paddy fields for winter wheat in Niigata while there are 20% of all paddy field in Sendai, 40% in Ishikari and Hachiro-gata. In Hachiro-gata, winter wheat grows fast resulting in an apparent peak in May while there is no such peak in Ishikari. Less snowfall in Sendai makes wheat and vegetable grow in wintertime resulting in the NDVIs about 0.05.

In central part of Japan, as shown in figure 2(b), the general shape of the two curves of Niigata and Aizu is the same. The NDVIs of both Utsunomiya and Noubi reach their maximum value in August and maintain a certain amount of value even in wintertime. This is partly because their winter wheat and vegetables grow in winter and also because their rice yield late in comparison with other areas. Average NDVIs are low in Noubi because of the urbanization.

In south-western part of Japan, two-crop system is possible which results in an apparent double-peak shape of NDVI time profiles in figure 2(c). These peaks represent the harvest time of rice and winter crops mainly winter wheat.

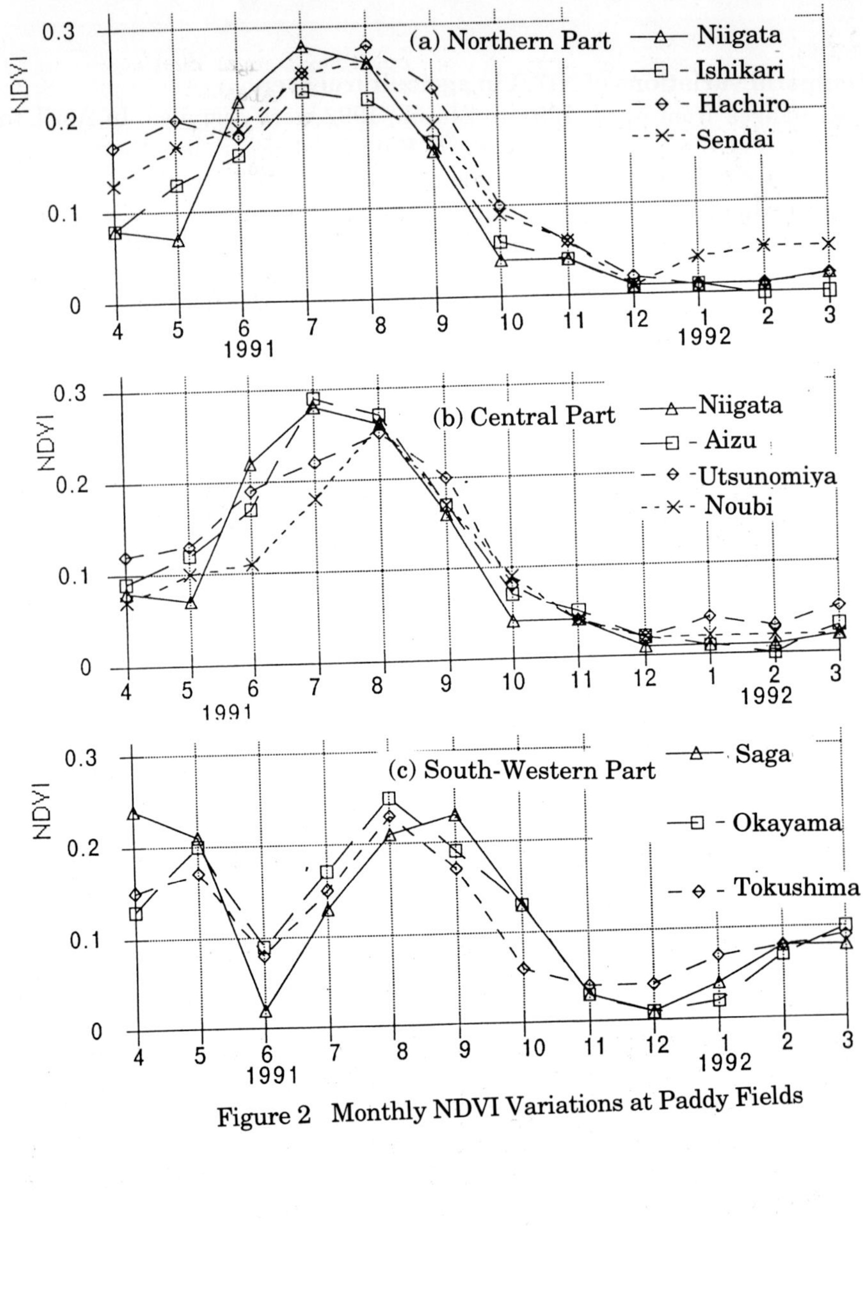

Figure 2 Monthly NDVI Variations at Paddy Fields

3.1.2. Upland farming fields

As for upland farming fields, we selected the following 6 sites; Tokachi Plain, Abashiri Plain, Hokuso Plateau in Chiba prefecture, Tsumakoi Plateau in Gumma prefecture, Nansatsu district in Kagoshima prefecture and Miyakonojyo Basin in Miyazaki prefecture. Monthly NDVI variations are shown in figure 3.

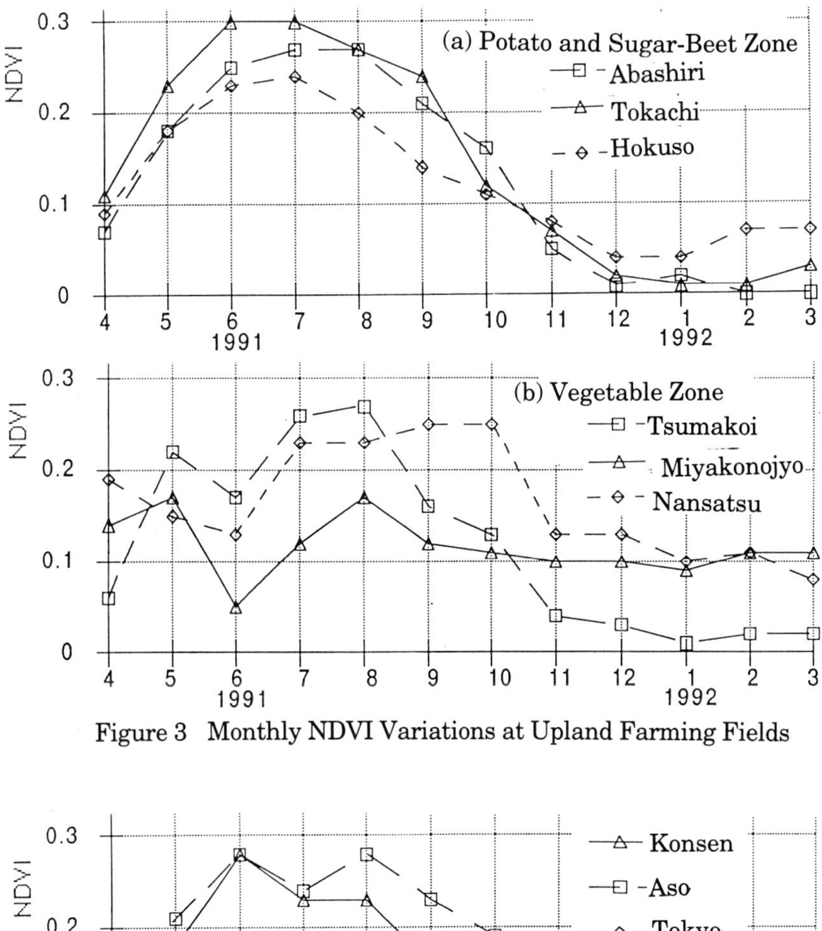

Figure 3 Monthly NDVI Variations at Upland Farming Fields

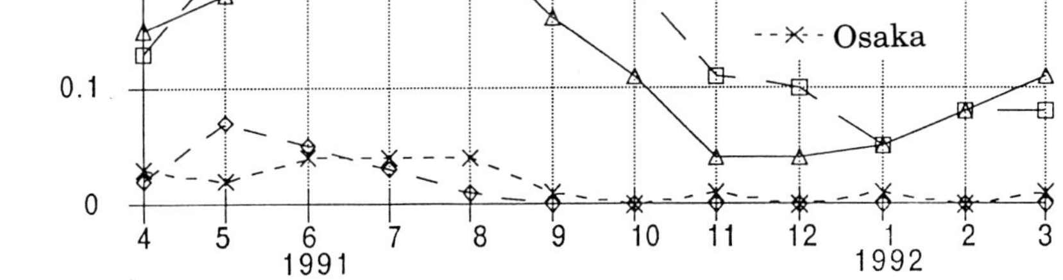

Figure 4 Monthly NDVI Variations at Grassland and Urban Areas

Shown in figure 3(a) are Tokachi, Abashiri and Hokuso that are located in potato and sugar-beet zone. The NDVI time profiles show an apparent peak in summertime. Snowfall in Tokachi and Abashiri keeps NDVIs almost zero from December to March.

Shown in figure 3(b) are Tsumakoi, Miyakonojyo and Nansatsu that are located in vegetable zone. The NDVI curves in these areas have multiple peaks according to the dominant vegetables in each area. Miyakonojyo and Nansatsu located in south-western part of Japan maintain a certain value of NDVI even in wintertime.

3.1.3. Grassland

As for grassland, we selected Konsen Plain in Hokkaido and Aso district in Kyusyu. In both areas the NDVIs in figure 4 increase as pasture plant grows from Spring to Summer. In Summer, the NDVIs drop because of grazing.

3.1.4. Urban areas

As for urban areas, we selected Tokyo and Osaka. Figure 4 shows the NDVI time profiles of the two cities. The NDVIs in Tokyo show a kind of seasonal variations while the NDVIs in Osaka keep low values all through the year.

Acknowledgments

This work was performed as JVI collaborated project. We wish to thank this project members, Dr. Yoshizumi YASUDA Chiba University, and Azumi GOTOH Japan Weather Association, etc.

References

1) C. O.Justice, J. R. G. Townshend, B. N. Holben and C. J. Tucker (1985) : Analysis of the phenology of global vegetation using meteorological satellite data, Int. J. Remote Sensing 6, 1271-1318

2) H. W. Yates, J. D. Tarpley, S. R. Schneider, D. F. Mcginnis and R. A. Scofield (1984) : The role of meteorological satellite in agricultural remote sensing, Remote Sensing of Environ., 14, 219-233

TOWARD GLOBAL PLANNING OF SUSTAINABLE USE OF THE EARTH
Development of Global Eco-Engineering
S. Murai (Editor)
1995 Elsevier Science B.V.

Vegetation Mapping using Remote Sensing Data in Indo-Chinese Peninsula

Masataka TAKAGI, Shunji MURAI and Yoshiaki HONDA
Institute of Industrial Science,
The University of Tokyo.
7-22, Roppongi, Minato-ku, Tokyo 106, JAPAN
Tel. +81-3-3402-6231
Fax. +81-3-3479-2762

1. Introduction

A land cover of Indo-Chinese Peninsula is changing by commercial logging and sifting cultivation, so that current condition of land cover must be monitored. Therefore, we attempt to generate wide coverage and high resolution vegetation map using remote sensing data.

By the way, almost satellite sensors can be classified into two categories. One of them is wide coverage, low resolution and short recurrent period sensor. One another is narrow coverage, high resolution and long recurrent period sensor. In this situation, generation of wide coverage and high resolution data set required multi sensor data fusion technology with both categorized satellite sensor.

405 scenes of MOS-1 MESSR data were used for high resolution data in order to cover Indo-Chinese Peninsula. These images were synthesized to generate mosaic image. Then we developed a special radiometric correction system for the mosaic image by using a cloud free composite image of NOAA AVHRR data which are wide coverage and frequent data. This system was very helpful to apply high resolution satellite data to wide area.

2. Data Used

In this study, MOS-1 MESSR data and NOAA AVHRR data were used. MOS-1 MESSR data has two advantages: the resolution is higher than NOAA; and MOS-1 data is cheaper than other satellite data (LANDSAT TM or SPOT). 405 scenes of MOS-1 data were used to cover the whole test area. A notice was paid careful attention to select cloud free scenes. It was impossible to acquire those data at the same season because of changeable weather. So, those data were received from 1988 to 1990 in different season.

NOAA AVHRR data also has two advantages: it is widely coverage data; and it is frequent data enough to generate a cloud free composite image in about a week. In this study, two scenes of NOAA data taken on Feb. 10, 1991 and Feb. 21, 1991 were used as a reference for radiometric corrections.

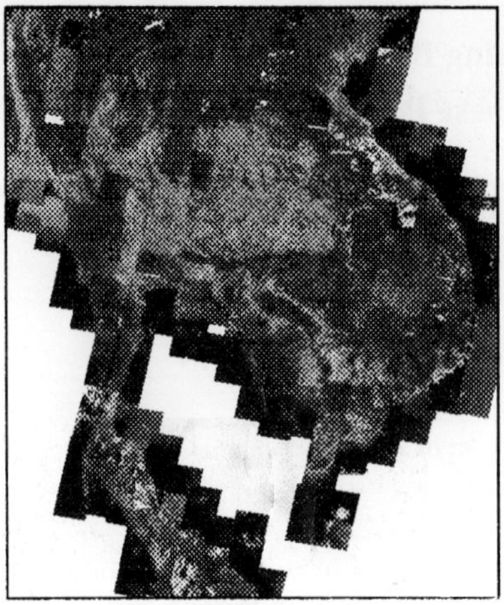

Figure 1 MOS-1 Mosaic Image

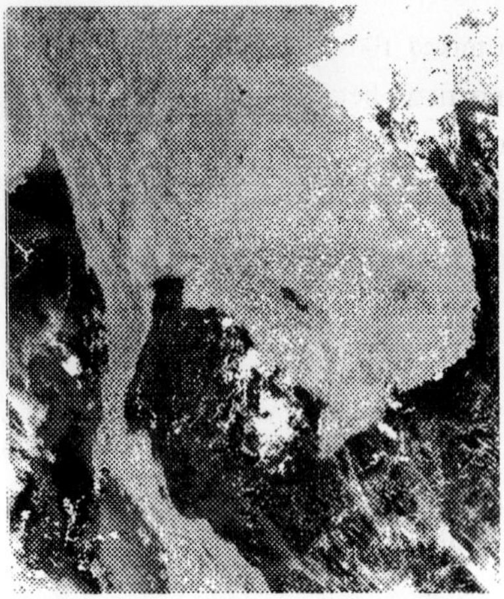

Figure 2 NOAA AVHRR Image

3. Methodology

At first, geometric correction should be carried out for 405 scenes of MOS-1 original data. Ground control points were selected from WDBII (World Data Bank II) which is vector data of coast line, river network and international boundary. And then each data were resampled at 10" grid interval in equal division of longitude and latitude. Figure 1 shows geometrically corrected mosaic image. In this mosaiced image, there are much different color balance in each scene, because each MESSR data are selected on different date with regardless of rainy season or dry season. Such color difference came from various air condition. Therefore, it disturbs land cover classification. In case of visual interpretation, land cover classification can carry out from such mosaiced image. However, visual interpretation needs much skill and much labor force, so that we tried to develop radiometric correction for automated vegetation mapping.

In this study, NOAA AVHRR data were used for radiometric correction as reference. Concerning wave length characteristics of MOS-1 MESSR and NOAA AVHRR, the second channel and the third one of MOS-1 data roughly correspond to the first one and the second one of NOAA data respectively. Therefore, those channels of mosaiced MOS-1 data can be corrected radiometrically by using NOAA data. Figure 2 shows used NOAA AVHRR data which were synthesized from two scenes of NOAA AVHRR data in order to generate cloud free composite image. The cloud free image of NOAA data were resampled at 100" grid interval.

Radiometric correction is based on statistical method. Grid interval of resampled MOS-1 image corresponds 10" while one of resampled NOAA image corresponds 100". Therefore, 100 pixels of MOS-1 data correspond to one pixel of NOAA data. Assuming that a pixel of NOAA data is a mixed 100 pixels of MOS-1 data, a gain value and an offset value can be determined by comparison between the average of MOS-1 data and NOAA data. The gain value can be determined by the ratio of NOAA data to MOS-1 data. The offset value can be led from difference between NOAA data and MOS-1 data as follows:

$$\text{Gain} = \frac{X_{noaa}}{\overline{X}_{mos}} \quad \text{------------------ (Eq. 1)}$$

$$\text{Offset} = X_{noaa} - \overline{X}_{mos} \quad \text{---------- (Eq. 2)}$$

X_{noaa} : Pixel Value of NOAA AVHRR Data

$\overline{X}_{mos}$: Average of MOS-1 MESSR Data

4. Results and Vegetation Mapping

A radiometrically corrected mosaic image was made by previous method with a high resolution of 10" pixel interval (Figure 3). From the corrected image, forest watch monitoring was concluded by using NDVI (Normalized Differential Vegetation Index). An accumulated histogram of NDVI value in each country (Thailand, Cambodia, Laos and Vietnam) is shown in Figure 4. The histogram was compared with land use data made by FAO (Figure 5). In Laos or Vietnam, forest area was more extensive than other country from our results. On the other hand, Cambodia was the most extensive forest area from FAO's land use data. It means forest area in Cambodia were decreased recently. FAO estimated from statistical method, which would result in a big gap from the real status because of a rapid change by commercial logging or shifting

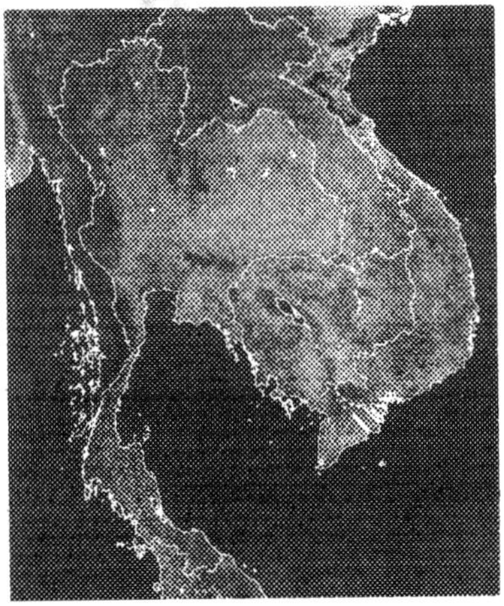

Figure 3 Corrected MOS-1 Mosaic Image

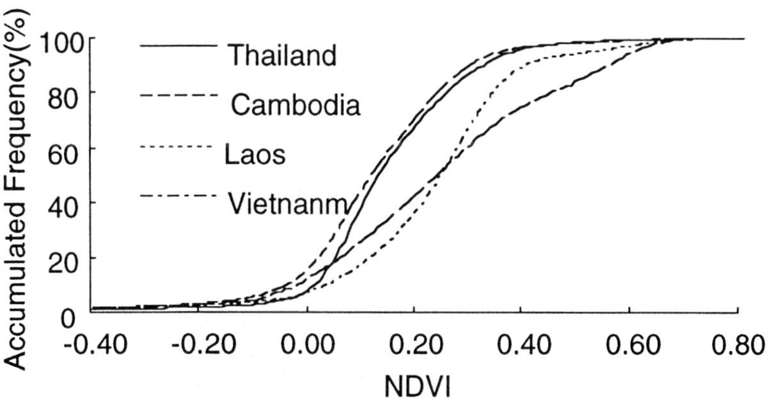

Figure 4 Accumulated Histogram of NDVI

cultivation, especially in Cambodia. Therefore, the land use data of Thailand or Vietnam is reliable. An vegetation classification of NDVI image was executed by using the land use data of Thailand. A threshold value for classification can be led from comparison between accumulated histogram and the land use data. For example, the forest area ratio of 28% correspond to 0.22 of NDVI. The vegetation map was generated by using previous threshold value. Figure 6 shows classification image of vegetation. From this image, forest area in Thailand and Cambodia is decreased very much.

5. Conclusions

(1) Radiometric correction for 405 MOS-1 MESSR images taken in different seasons was made with a success using NOAA AVHRR data as a reference. It resulted in generation of cloud free mosaic with high resolution.

(2) Vegetation mapping for Indo-Chinese Peninsula can be made using NDVI calculated from radiometrically corrected mosaic.

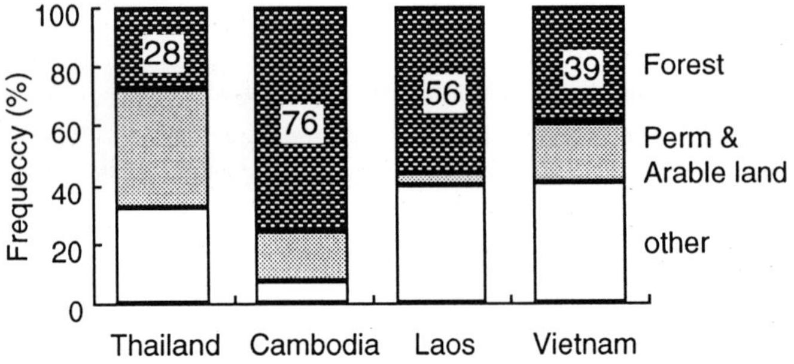

Figure 5 Histogram of Landuse made by FAO

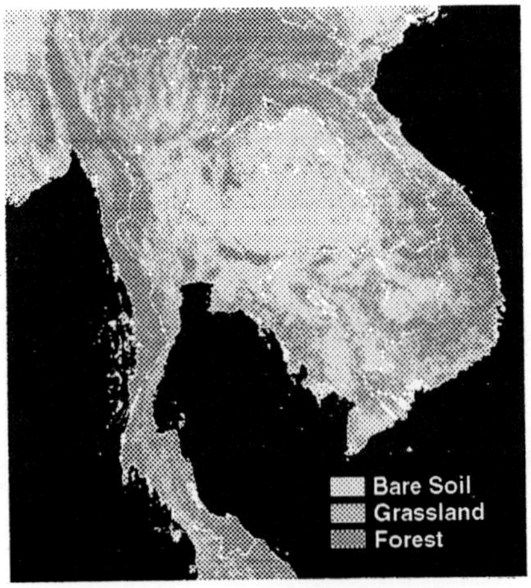

Figure 6 Vegetation Map

TOWARD GLOBAL PLANNING OF SUSTAINABLE USE OF THE EARTH
Development of Global Eco-Engineering
S. Murai (Editor)

Climatic Classification of the World by Cluster Analysis

Hiroshi Tani[a], Ikuo Horiguchi[b], Seiji Hayakawa[a] and Haruhiko Yamamoto[a]

[a]Department of Biological Science, Faculty of Agriculture, Yamaguchi University, Yoshida 1677-1, Yamaguchi-shi 753, Japan

[b]Department of Agricultural Engineering, Faculty of Agriculture, Hokkaido University, Kita-9 Nishi-9, Kita-ku Sapporo-shi 060, Japan

Abstract

In order to investigate the change of climate zone in the last sixty years, climatic classification of the world was performed by cluster analysis using monthly normal values of air temperature and precipitation. Variables used in the analysis were monthly mean temperature and precipitation of about eight hundred stations in the world. Eleven types were obtained by clustering the stations using twelve monthly mean temperatures as variables. The result of this clustering gave the climatic types of the first order. The second order types were given by clustering the stations in each type of the first order by using precipitation as the variable. Two classification maps are compared from a viewpoint of detecting the spread or narrowing of climate zones.

1. INTRODUCTION

There is concern that global warming caused by an increase of greenhouse gases might increase the global average temperature by one degree by the year 2025 and three degrees by the end of the next century. It has been predicted that global warming would change climate zones in spatial and cause drastic changes in vegetation. It is interesting to investigate how the climate zone is changing, or has changed in the last half century.

Many researchers have made attempts at climatic classification of the world up to the present. They used proper indices or external factors such as vegetation related to climate for the classification. A classification making reference only to climatic data and with no subjective judgement in fixing boundaries between different climatic types is required. In this paper, we present climate maps made by cluster analysis based on only climatic data, and detect the change of climate zone.

2. MATERIALS AND METHOD

The world climatic data which include temperature and precipitation of monthly mean values for about eight hundred stations were obtained from the Japan Meteorological Agency. Three data sets which have a difference in observation period, from 1931 to 1960 (abbreviated to period I), from 1951 to 1980 (period II) and from 1951 to 1980 (period

III), are used in this study. Since not all stations had the mean period of thirty years, data whose total period is less than fifteen years were omitted. The total stations of each data set decreased to 836, 803 and 812 in chronological order, as a result of this omission. For preprocessing of the classification, the climatic data in the southern hemisphere were shifted six months because the season in the southern hemisphere has a six months' difference from the northern hemisphere.

The clustering method applied in the analysis was Ward's minimum variance method in hierarchical clustering. The number of clusters was determined by considering pseudo F statistic measuring and pseudo t^2 statistic measuring. At first, the stations were divided into eleven clusters by using the monthly mean temperature from 1951 to 1980 as variables. The result of this clustering gave the climatic types of the first order. Next, the second order types were given by clustering the stations in each type of the first order by using precipitation as the variable.

It is necessary to classify the data sets of period I and period III on the same basis for comparing each other. The authors classified them by using the eleven annual temperature variations of climate types derived from clustering data set from 1951 to 1980 as reference data. That is, in a twelve dimensional space, a station was classified into the type with the minimum Euclidean distance.

3. RESULTS AND DISCUSSION

A climatic map was made by plotting the climatic type of each station on the world map and drawing boundary lines manually for the period from 1951 to 1980 (Figure 1). Both the first order type derived from temperature data (represented by number) and the

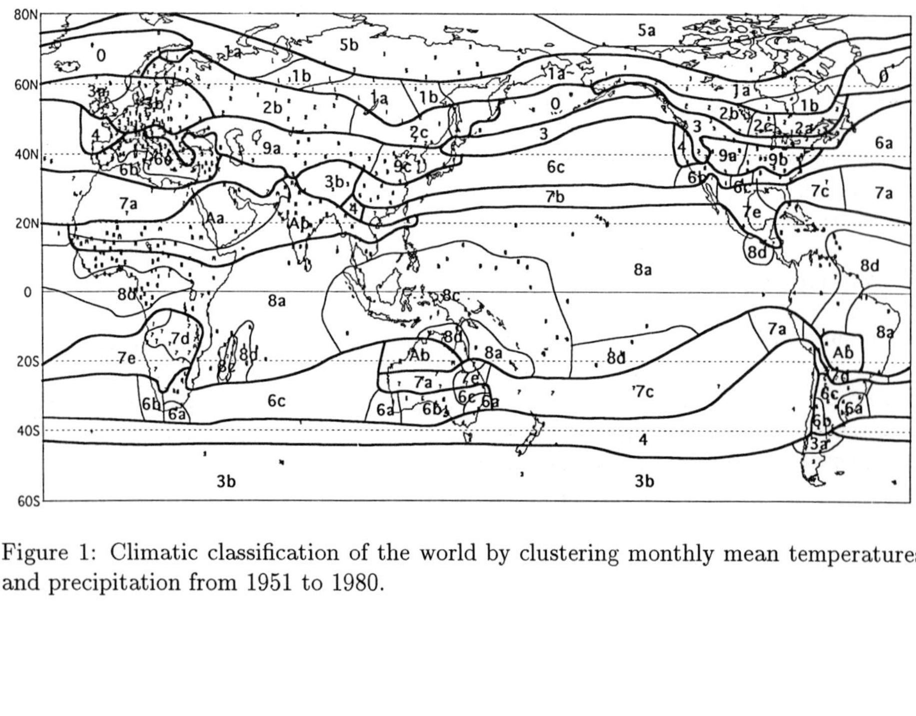

Figure 1: Climatic classification of the world by clustering monthly mean temperatures and precipitation from 1951 to 1980.

second order type by precipitation data (alphabet) are shown in the figure. This climatic map differs from another major climatic maps, which is characteristic of the result by cluster analysis.

Figure 2 shows annual temperature variations of each climatic type. These were obtained by averaging monthly mean temperatures for the stations of each climate type.

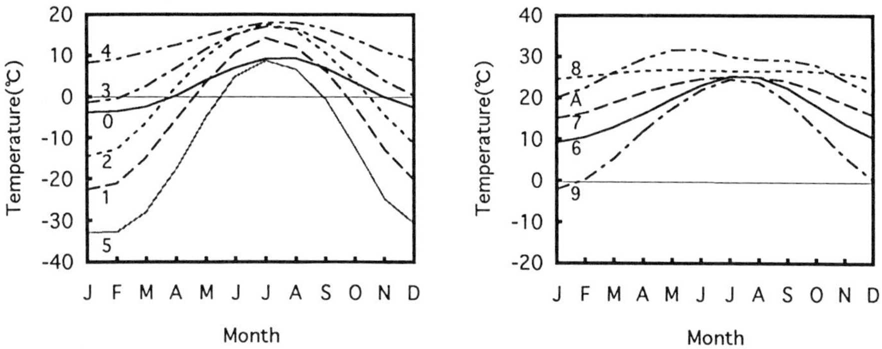

Figure 2: Annual temperature variation of climatic type derived by clustering monthly mean temperatures from 1951 to 1980.

Climatic classification was made for the data set from 1931 to 1960 by the minimum distance method using the annual temperature variations above-mentioned (Figure 3).

Figure 4 shows climatic classification for the data set from 1961 to 1990. Comparing Figures 3 and 4, the most remarkable change between the two periods is that the boundaries between different climatic types shifted southward in the southern hemisphere. Furthermore, the climatic type (6) of period I was included in type (7) of period III in South Africa and Australia. In the northern hemisphere, a slight shift to the northward was seen in East Europe and Central Asia. Another shift was not judged clearly because of the sparse distribution of meteorological stations.

3. CONCLUSIONS

The shifts of climatic zone in sixty years were detected in several places by the climatic classification of the world with cluster analysis and classification by the minimum distance method on the basis of climatic data. Since it was difficult to determine the boundaries between different climatic types where meteorological stations were distributed sparsely, further collection of climatic data is necessary.

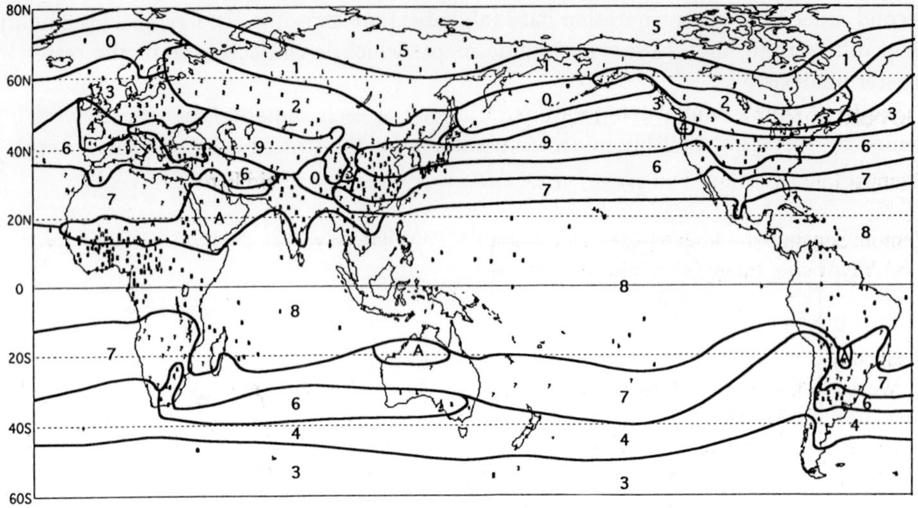

Figure 3: Climatic classification of the world by using monthly mean temperatures from 1931 to 1960.

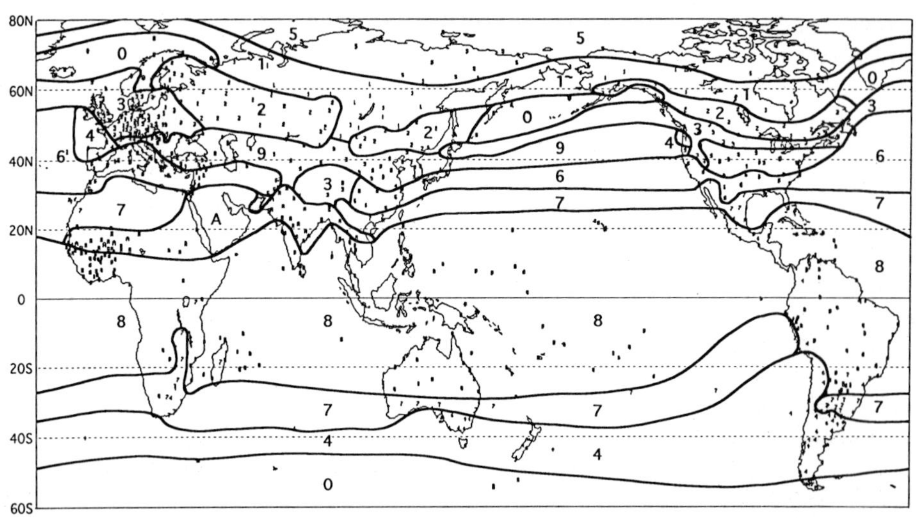

Figure 4: Climatic classification of the world by using monthly mean temperatures from 1931 to 1960.

TOWARD GLOBAL PLANNING OF SUSTAINABLE USE OF THE EARTH
Development of Global Eco-Engineering
S. Murai (Editor)

Development of Global Database

Ryutaro Tateishi, Ahn Chun-Hyun, Wen Cheng-Gang and L. Kithsiri Perera

Remote Sensing and Image Research Center, Chiba University
1-33 Yayoi-cho, Inage-ku, Chiba 263 Japan

Abstract

We are in the age of full recognition of the importance of global environmental research. The first step of the global environmental research is the production of global datasets of key environmental variables in order to know the present situation and change of the earth environment. The attempts to produce global datasets by the authors are reviewed in this paper. These include land cover dataset of Asia, global land evapotranspiration and water surplus/deficit. Furthermore some considerations are given for the development of a truly useful global database.

1. LAND COVER OF ASIA

The Asian Association on Remote Sensing(AARS) established the working group on "1-km Land Cover Database of Asia" in October 1993 [1] . The working group (Chairman: Ryutaro Tateishi) consists of 44 members from 27 Asian and Oceanian countries as of December 1994. It aims to produce land cover dataset of Asia and Oceania using 1-km AVHRR data. This chapter reviews the initial attempt to produce land cover dataset of the region using 16-km Global Vegetation Index(GVI) data and also describes the plan of future activities.

Land cover is one of the most important variables in global environmental research. Many projects are going on to develop large scale land cover datasets: for example, global area by IGBP, the territory of USA and global area by EROS Data Center(USA), the territory of Canada by Manitoba Remote Sensing Center(Canada), South America and the territory of Russia by Woods Hole Research Center(USA). However there has been no project for the whole Asia. This is why the working group was established in the AARS.

1.1. Land cover classes

Determination of land cover classes depends on applications of the land cover data. The main purpose of the development of land cover data by the working group is global or regional environmental research.

Based on this understanding, land cover classes were determined by communication between the chairman and WG members through their first-year activity. The determined classes have the characteristics as follows.

- Hierarchical
- The 1st level has only three classes which are Vegetation, Non-vegetation and Water.
- The 2nd level are (1) Forest and shrub, Grassland, Mixed vegetation, Agriculture, and

Wetland, for Vegetation, (2) Wetland without vegetation, Bare soils and rocks, Tundra, Perennial snow or ice, and Built-up area, for Non-vegetation, and (3) Water, Water with seasonal change, and Tidal flat, for Water.
- Number of classes is 70 but additional classes are possible to include.

1.2. Used data

The main source data for the project is Scaled Calibrated Vegetation Index(SCVI) data in GVI products by NOAA/NESDIS [2] . The used SCVI data is weekly data in 1992. It has 8 bit and based on Plate Carree projection with 2500 by 904 pixels for the region from 75 degree North to 55 degree South. The nominal resolution is 16km.

The other used data are temperature by Legates and Willmott [3] , precipitation data by Legates and Willmott [3] , evapotranspiration data by Ahn and Tateishi [4] [7] , and DEM (ETOPO5) [3] . All these datasets cover global area and the former three datasets consist of monthly datasets.

1.3. Classification

K-means cluster analysis was applied to monthly SCVI data which were derived from weekly data by the authors. 113 clusters in global area were obtained. The mean and standard deviation of temperature, precipitation, evapotranspiration, and elevation for each cluster were calculated. The datasets and results of the processing were distributed to WG members in order to identify each cluster's land cover.

1.4. Future activity

(1) The first land cover dataset using nominal 16-km resolution SCVI data will be developed by the end of 1994. 8-km Global Land AVHRR Data Set (Pathfinder product) will be used in the next step. Finally we plan to use 1-km AVHRR data set.
(2) Modification of land cover class system is necessary. Hierarchical land cover system will be kept. There are many land cover class systems in different global or regional projects. The working group decided that the highest (1st) level of class system should be common within various projects, and that the 2nd and lower level may have regional characteristics, for example dominant land cover type only in Asia.
(3) The working group is always open to cooperate with other land cover projects. The contribution to the development of global land cover dataset is one of the objectives of the working group.

2. EVAPOTRANSPIRATION AND WATER SURPLUS/DEFICIT

Monthly global land 30-minute latitude/longitude grid data for (1) potential evapotranspiration, (2) evapotranspiration, and (3) water surplus/deficit were developed by the authors.

2.1. Potential evapotranspiration [4]

Potential evapotranspiration has been defined as the evapotranspiration rate of a "short green crop, completely shading the ground of uniform height and never short of water" [5] . The Priestley-Tailor method [6] was used to estimate potential evapotranspiration in this study .

The used data are temperature by Legates and Willmott, albedo by Matthews, cloudiness by

Leemans and Cramer, ETOPO5, and precipitation by Legates and Willmott. All these data are provided on CD-ROM by NOAA-EPA [3] .

2.2. Evapotranspiration [7]

Monthly evapotranspiration was estimated based on a simple water balance model using monthly precipitation, monthly potential evapotranspiration and monthly soil moisture. Precipitation data were obtained from CD-ROM by NOAA-EPA [3] , and potential evapotranspiration data were developed by the authors. Soil moisture data were derived by the by Donker's method [8] using Water Holding Capacity data from UNEP/GRID-Geneva.

2.3. Water surplus/deficit [7]

The water surplus is defined as the excess of moisture beyond the plant water needs when the soil capillaries are recharged with water, that is the moisture which ultimately percolates to the water table and is lost as stream flow. On the other hand, the water deficit is defined as the difference between climatic water demand (potential evapotranspiration) and the actual water loss, that is the amount of water which must be supplied by irrigation to keep vegetation growing at an optimum rate.

According to the above definition, monthly water surplus/deficit were derived from monthly precipitation, monthly potential evapotranspiration, monthly evapotranspiration, and monthly soil moisture.

2.4. Availability

The above described monthly global potential evapotranspiration, evapotranspiration, and water surplus/deficit datasets are or will be available through UNEP/GRID-Tsukuba, NOAA/NGDC(USA), Rutgers University (USA).

3. CONSIDERATIONS

From the experience of the development of previously described datasets, some considerations are made as follows.

3.1. Technological problems

(1) Harmonized land cover class system should be established with the international agreement. The existing national and regional land cover maps/data and also even global land cover data developed by different researchers have different class system and different definition of land cover classes. It is neither necessary nor possible to allow only one class system because there are different applications using land cover dataset. And also it is almost impossible to change land cover classes of existing land cover maps to different new standard classes. A new land cover class system should be established to meet scientific needs for global environmental research and to clarify its relationships with other existing land cover class systems.

The efforts to establish harmonized land use class system have started by UNEP and FAO [9] . This efforts should be extended to land cover class system with land cover specialists from as many countries as possible.

(2) Accuracy of the developed datasets should be estimated and recorded. When developed data are derived using a model and input data, as in the case of potential evapotranspiration described above, the accuracy of output data are affected by input error and model error. If

input error is unknown, output error cannot be estimated correctly. Therefore accuracy of the data must be attached whenever a new dataset is developed. Another difficult part of the accuracy assessment is model error. In order to evaluate model error, estimated values by the model must be validated by real value(ground truth or measurement).

3.2. Cultural bias

Global datasets have been developed by researchers of Western and developed countries(Europe and North America). Very limited contribution has been made by researchers of Asia, South America and Africa. South America is environmentally critical because of forests in Amazon, and Africa is also critical because of its desert and population growth. This is why Western researchers are paying more attention on these continents.

Concerning dataset development of continental scale, Asia does not have a high priority and limited researchers of Asia have contributed to the development of dataset of the whole Asia. The activity by the Asian Association on Remote Sensing(AARS) to develop land cover dataset of Asia and Oceania is one of the growing contribution from Asia.

3.3. International cooperation

There are many international projects and individual researchers which have been and are developing global datasets. Some organizations such as UNEP/GRID, NOAA/NGDC, US corps of Engineers and Rutgers University collect, manage and distribute global datasets. However there are limited communication among different dataset producers, and among different datasets distributors.

Communication among dataset producers and distributors are necessary to solve common problems in order to develop truly useful global datasets for environmental research. The following subjects should be discussed.

(1) Clear definition of necessary data for global environmental research.

NASA did a valuable work to summarize "Physical Quantities Needed for Global Change Studies" [10] . Since this is just a simple summary, we have to define more detail about the data necessary for global environmental research.

(2) Identification of technological or methodological problems.

There are various problems to be solved. For example:

- How to estimate and record accuracy of data is an important question.
- Most data are based on latitude/longitude system. Is it the best map projection to describe global variables?
- What is the suitable data format to describe various kinds of global environmental variables in land, ocean and atmosphere?

(3) Strategy to build global database

Many projects and researchers are trying to develop global datasets. How can we integrate those datasets? At present, some organizations are collecting datasets and building database based on their own philosophy. What is the best way to facilitate useful global database.

(4) Use of national or regional datasets to develop global datasets.

There is a possibility to develop global dataset by integrating national or regional datasets. In order to do this, a survey of the existence of national or regional datasets and their availability will be the first action to be carried out.

The International Society for Photogrammetry and Remote Sensing(ISPRS) Working Group IV/6 is promoting communications among global dataset producers and distributors through workshops in order to discuss the above problems.

4. CONCLUSIONS

- The AARS has started the efforts to develop land cover dataset of Asia and Oceania.
- Monthly global land 30-minute potential evapotranspiration, evapotranspiration and water surplus/deficit datasets were developed by the authors. They are or will be available through UNEP/GRID, NOAA/NGDC and Rutgers University.
- The ISPRS WG IV/6 is promoting communications among global dataset producers and distributors for the development of better global database.

5. REFERENCES

1 Ryutaro Tateishi, Wen Cheng-Gang, and L. Kithsiri Perera, 1994, Working Group Report and Land Cover Database of Asia, Proceedings of the 15th Asian Conference on Remote Sensing, M-3-1 ~ M-3-6
2 Kidwell,K.B, 1990, Global Vegetation Index User's Guide, NOAA/NESDIS, Washington, D.C.
3 NOAA-EPA Global Ecosystems Database Project, 1992, Global Ecosystems Database Version 1.0, NOAA/ National Geophysical Data Center, Boulder, CO, USA
4 Ahn,C.H. and R.Tateishi, 1994a, Development of Global 30-minute Grid Potential Evapotranspiration Data Set, Journal of Japan Society of Photogrammetry and Remote Sensing, Vol.33, No.2, 12-21
5 Penman,H.L., 1948, Natural evapotranspiration from open water, bare soil and grass, Proc. Roy. Soc. London A193, 120-146
6 Priestley,C.H.B. and R.J. Tailor, 1972, On the assessment of surface heat flux and evaporation using large-scale parameters. Monthly Weather Review, 100, 81-92
7 Ahn,C.H. and R.Tateishi, 1994b, Development of Global Land Surface Evapotranspiration and Water Balance Data Sets, Journal of Japan Society of Photogrammetry and Remote Sensing, Vol.33, No.5, 48-61
8 Donker,N.H.W., 1987, BTRBLN: A Computer Program to Calculate Water Balance, Computer and Geoscience, Vol.13, No.2, 95-122
9 UNEP/FAO, 1994, Report of the UNEP/FAO Expert Meeting on Harmonizing Land Cover Land Use Classifications, GEMS Report Series No.25
10 NASA, 1988, Earth System Science: A Closer View, Earth System Science Committee, NASA Advisory Council, Washington, D.C.

TOWARD GLOBAL PLANNING OF SUSTAINABLE USE OF THE EARTH
Development of Global Eco-Engineering
S. Murai (Editor)

Climatological Estimation of Dry Matter Production and Carbon-Absorption of Natural Vegetation in the Indonesian Archipelago

Z. Uchijima[a], S. Ohta[b] and Y. Obuchi[b]

[a] Faculty of Science, Ochanomizu University, Otsuka 2-1-1, Bunkyo-ku, Tokyo 112, Japan

[b] Department of Basic Human Sciences, School of Human Sciences, Waseda University, Mikajima 2-579-15, Tokorozawa, Saitama 359, Japan.

Abstract

The tropical, closed rain forests of the Indonesian archipelago are a home to the rich wildlife and tribal people. This district is also a center of the rapid population growth and intensive industrialization in the world. It is therefore expected that the massive deforestation of the tropical forests affects the regional environment of this district as well as the global carbon balance. The Chikugo model, which can evaluate the net primary productivity of natural vegetation from climatic data, was used to evaluate the potential and actual net primary production. The net primary production data were converted into the carbon sink strength of this district.

1. INTRODUCTION

The Indonesian archipelago, which lies between the south-eastern tip of Asia and Australia, is a center of the tropical, closed canopy forests of the Asia–Pacific region. These islands are mostly clothed in tropical rain forests. The rich wildlife in these islands is well supported by their highest plant production in the world. Furthermore, the tropical rain forests are also home to millions of tribal people. They rely entirely on hunted and gathered food for their subsistence. However the rapid population growth and the acute economic development in this region are causing the large-scale deforestation of the productive tropical rain forests. Such large-scale deforestation should give severe influences on the global carbon balance as well as the rich wildlife and tribal people in the forest. In this report, the Chikugo model, which enables us to calculate the net primary productivity of natural vegetation from climate data, was used to estimate the total net production and the total carbon absorption with and without the consideration of effects of human land use on forests. The calculation was made under four climate conditions such as the current climate and warmed climates due to doubled CO_2 concentration.

2. METHODS and DATA

The following Chikugo model (Uchijima & Seino, 1985, 1988; Ohta *et al.*, 1993) was used to calculate the net primary productivity:

$$NPP = 0.29\,[\exp(-0.216\,RDI^2)]\,R_n\,, \tag{1}$$

where NPP is the net primary productivity (tonne dry matter $ha^{-1}\,yr^{-1}$), RDI ($=R_n\,/\,lr$) is the annual radiative dryness index (>0), R_n is the annual net radiation (kcal cm^{-2}), r is the annual precipitation (cm), *l* is the latent heat of water evaporation (kcal g^{-1}). The total net production (TNP, tonne dry matter yr^{-1}) was simply calculated from

$$TNP_0 = \sum_{i=1}^{n} A_i\,\overline{NPP}_i\quad , \tag{2}$$

$$TNP_a = \sum_{i=1}^{n}\sum_{j=1}^{m} E_j\,A_{ji}\,\overline{NPP}_i\quad , \tag{3}$$

where TNP_0 and TNP_a are the potential and the actual total net production, respectively. A_i and $\overline{NPP}_i$ are the land area (ha) and mean NPP of the *i* - th land unit, respectively, and E_j and A_{ji} denote, respectively, the production efficiency (0.0–1.0) for the *j* -th land use and the land area for *j* - th land use in the *i* - th land unit.

The climatic data of the Indonesian archipelago were converted in digital form on a uniform 1° latitude by 1° longitude meshes. The current climate data were collected and compiled from: (1) Agroclimatological Data for Asia (FAO, 1987); (2) Climatic Table for the World (JMA, 1994). In order to generate the warmed climate scenarios for the doubled CO_2-concentration, the climate deviations projected by the three GCMs (GISS, GFDL and UKMO) for $1\times CO_2$ and $2\times CO_2$ conditions were used together with current climate data.

3. RESULTS and DISCUSSION

Figure 1 shows the distribution pattern of carbon sink strength for current climate and warmed climates. The NPP-data obtained by the Chikugo model were converted into the carbon sink strength using the conversion constant of 0.45 tonne C $tonne^{-1}$ dry matter. Under the current climate conditions, the carbon sink strength is over a range from 13.8 to 3.6 with the average of 11.9 tonne C $ha^{-1}\,yr^{-1}$. Under the warmed climate conditions obtained by the three GCMs, the carbon sink strength is expected to increase by 8% for the UKMO, 16% for the GFDL, and 20% for the GISS compared with that for the current climate. The increase in the carbon sink strength was mainly due to the increase in solar radiation under the warmed climate conditions.

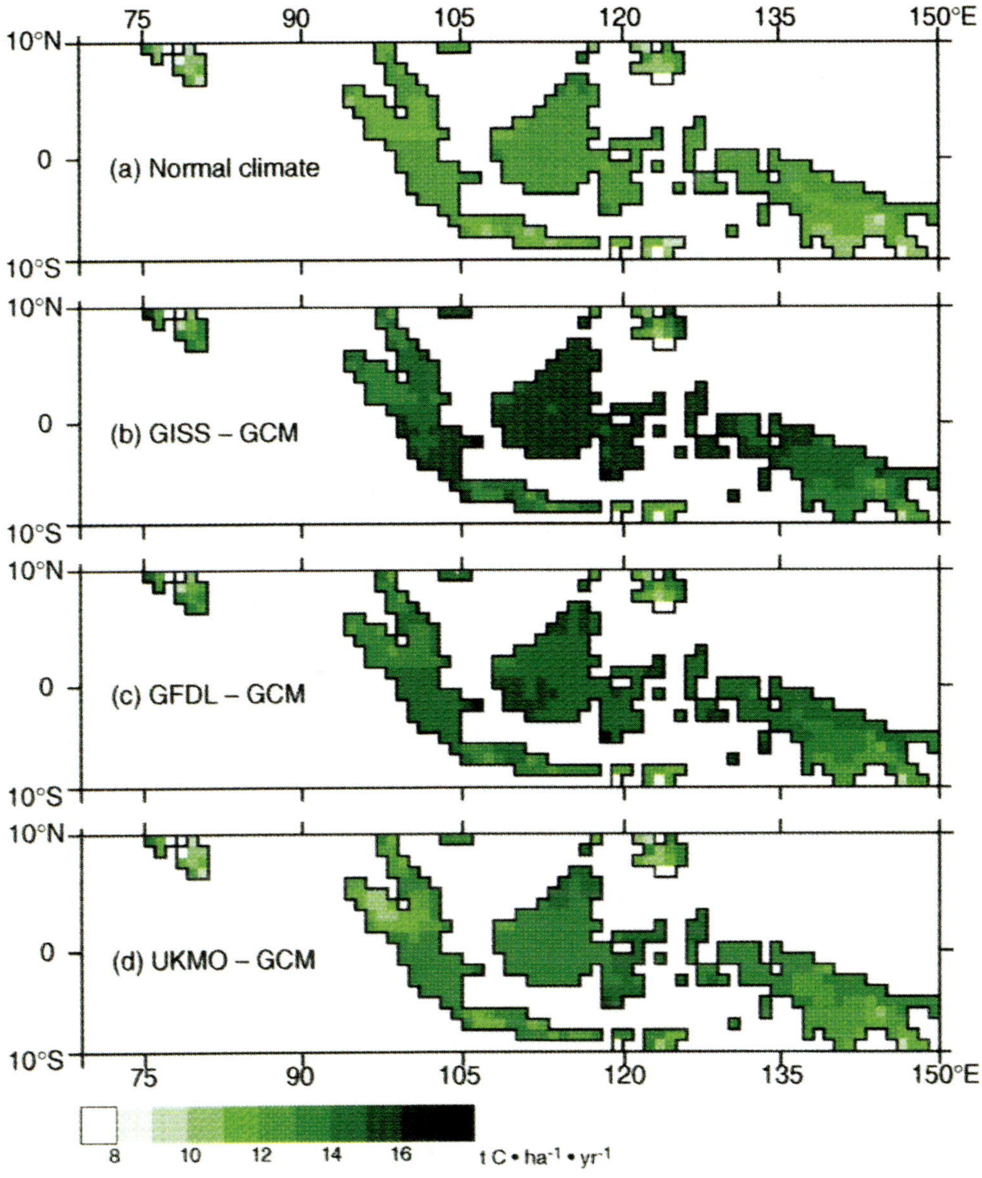

Figure 1. Geographical distribution of potential carbon sink strength (tonne C ha^{-1} yr^{-1}).

To evaluate influences of land use on vegetation activity, using a relatively rough data of the land use in the Indonesian islands, the land area of the each island was divided into two sub-areas such as the forest-sub-area with $E_1 = 1.0$ and the non forest-sub-area with the $E_2 = 0.7$. The results so obtained are summarized in Table 1. Table 1 indicates that the ratio TC_a / TC_0 that reflects the influence of human activity on vegetation ranges from 72.2% of Java island densely populated from 94.5% of Irianjaya island sparsely populated. Under the conditions that the carbon loss due to soil respiration is ignored, the Indonesian islands are expected to be a strong carbon sink to the atmosphere.

Table 1 Total carbon absorption of Indonesian islands under current climate

Island	Forest Area (%)	Non forest Area (%)	Carbon absorption ($\times 10^8$ t C yr^{-1})		TC_a / TC_0
			TC_0	TC_a	
Sumatra	43	57	5.45	4.50	0.829
Kalimantan	63	37	7.16	6.34	0.889
Sulawesi	52	48	2.05	1.82	0.889
Maluku Islands	70	30	0.32	0.29	0.910
Irianjaya	82	18	5.72	5.40	0.946
Bali	23	77	0.05	0.04	0.769
Java	7	93	0.70	0.62	0.721
Total			20.88	18.53	0.850

TC_0 and TC_a denote the potential and actual total carbon absorption, respectively.

The data of soil respiration of tropical soils (e.g., Kobak, 1988) were used to evaluate roughly the ecosystem net carbon absorption (tonne C ha^{-1} yr^{-1}) of the Indonesian islands (Sakurai *et al.*, 1994). The calculation yielded that the ecosystem net carbon absorption becomes about 36.8% of TC_a indicated in Table 1. This implies that the Indonesian islands with dense tropical rain forests are acting as an intensive carbon sink to the atmospheric carbon balance.

4. REFERENCES

1. K. I. Kobak, Carbon Cycle and Biological Element, USSR, Leningrad, 1988.
2. S. Ohta, Z. Uchijima and Y. Oshima, Ecological Research, 8 (1993) 199-213.
3. Y. Sakurai, M. Yokozeki, H. Iihama and Z. Uchijima (in preparation).
4. Z. Uchijima and H. Seino, J. Agr. Met., 40 (1985) 343-352.
5. Z. Uchijima and H. Seino, JARQ 21 (1988) 244-250.
6. Z. Uchijima, S. Ohta and Y. Obuchi, Annual Report of Carbon Sink Project-JIFPRO (1994) 18-55.

TOWARD GLOBAL PLANNING OF SUSTAINABLE USE OF THE EARTH
Development of Global Eco-Engineering
S. Murai (Editor)

Sustaining Global Life-Support Systems: The UNU Approach

J.I. Uitto

Academic Officer, The United Nations University
53-70, Jingumae 5-chome, Shibuya-ku, Tokyo 150, Japan

Abstract

The UNU programme on environmentally sustainable development is designed to reflect the priorities identified in the Agenda 21 emanating from the 1992 Earth Summit. The central tools of UNU's programme implementation consist of research, advanced training, and dissemination of knowledge, with specific emphasis placed on capacity building in developing countries. The programme area Sustaining Global Life-Support Systems approaches environmentally sound development from the three broad and mutually complementary entry points: (i) eco-restructuring; (ii) the resilience and absorptive capacity of ecosystems/regions; and (iii) governance. A significant strain of research focuses on integrated social and natural science studies on sustainable land management.

1. BACKGROUND

The aim of this paper is to outline the approach of the United Nations University (UNU) towards addressing global environmental problems identified in the United Nations *Agenda 21* (UN 1992). Here focus is given to integrated studies of ecosystems, encompassing research both into natural as well as social and policy sciences. The present paper highlights research carried out in the Asian region.

The central tools of UNU's programme implementation consist of research, advanced training, and dissemination of knowledge. The University works through global networks of scholars and institutions. A central objective of UNU's work is to promote institution and capacity building in developing countries. The collaboration of academic institutions and scholars from both the North and the South within UNU's research programmes as well as the University's post-graduate education and training activities are geared towards reaching these goals.

The University's present programme is aimed at generating policy contributions that could be of relevance to the international community in addressing some of the most crucial elements of the human condition in the current decade. The academic programme is clustered around five areas: (i) Universal Human Values and Global Responsibilities; (ii) New Directions for the World Economy; (iii) Sustaining Global Life-Support Systems; (iv) Advances in Science and Technology; and (v) Population Dynamics and Human Welfare.

2. SUSTAINING GLOBAL LIFE-SUPPORT SYSTEMS

In global environmental change, generally two different levels can be identified. First, there is global environmental change in the narrow sense, encompassing those changes that are truly global and affect the Earth as a whole. Global climate change and atmospheric ozone depletion typify this. Secondly, there are those changes in the environment, such as deforestation and soil erosion, that occur locally, but which relate to global phenomena through their cumulative effects. The two basic types of problems call for varying responses and approaches of study.

UNU's programmes within the area of 'Global Life-Support Systems' reflect the above perspectives in responding to *Agenda 21* (UNU 1993). The work has been organized according to five complementary programmes: (i) Eco-Restructuring for Sustainable Development; (ii) Integrated Studies of Ecosystems; (iii) Information Systems for Environmental Management; (iv) Natural Resources in Africa; and (v) Environmental Law and Governance.

The programme approach is designed following three major entry points or perspectives. Eco-restructuring refers to the profound economic, social and technological transition at macro- and micro-levels towards sustainability, and the implications to population and energy policy, consumption patterns, and the use of natural resources.

The second entry point aggregates issues of environmental sustainability from the point of view of the capacity of ecosystems to support, resist or recuperate from the long-term impact of major transformations, both human-induced and natural. Particular emphasis is placed on fragile ecosystems and vulnerable places, critical resources, and the methodology for environmental management and monitoring.

The third entry point is governance, meaning the ways by which sustainable development is managed at the local, national and global levels, including the legal and institutional frameworks.

3. INTEGRATED STUDIES OF ECOSYSTEMS

In order to ascertain maximal impact of the activities, the underlying principles of focus, integration and complementarity of the various research and capacity building efforts are applied. Building upon earlier work, UNU's research is multidisciplinary in nature, combining natural sciences with social and policy science perspectives.

Environmental studies, with a particular focus on tropical ecology and the human interactions, have been in the centre of the University's programme since its inception (Manshard and Uitto 1993). A central issue for the research efforts has been that of agricultural expansion, sustainable land management and agroforestry, with geographical emphasis on the humid tropics (*e.g.* Clarke and Thaman 1993; Kunstadter, Bird and Sabhasri 1986; Manshard and Morgan 1988).

The programme on Integrated Studies of Ecosystems emphasizes the importance of local and regional sustainability through appropriate environmental and resource management. It aggregates issues of environmentally sustainable development from the entry point of the capacity of ecosystems and their ability to support, resist or recuperate from the long-term impact of human activities. The approach is to focus on integrated studies of geographically delineated fragile ecosystems and vulnerable regions.

One strain of activities emphasizes regional studies on sustainable environmental futures intended to identify the driving forces of environmental transformation, analyze environmental issues related to global change, as well as specific environmental issues facing selected region. The aim is to consider constraints and possible solutions toward achieving improved environmental management, and to examine the policy options available for planners and development practitioners.

The study on Southeast Asia concluded that the present rates of population and economic growth based on natural resource exploitation and degradation are clearly unsustainable. Specific areas of unsustainability and increasing vulnerability, as well as needs for future research and action were identified (Brookfield and Byron 1993). These included land use issues, especially in the upland areas, and intensification of production, having both social and environmental consequences.

Similarly, UNU's work in the field of Mountain Ecology and Sustainable Development has produced a notable amount of research results related to sustainable development of mountain and highland areas in the world, and is providing vital contacts for mountain scholars and resource managers through extensive networking for research and capacity building. The programme played a crucial role in the preparation of Chapter 13 on Managing Fragile Ecosystems: Sustainable Mountain Development in *Agenda 21*, and continues to take an active role in its implementation.

The research has challenged the conventional notion that deforestation in the Himalaya has resulted in serious damage downstream, by demonstrating the pivotal role of subsistence farming in the hills in land management (Ives and Messerli 1989). The work continues with detailed field research in Bangladesh aimed at broadening the understanding of seasonal flooding and its causes.

4. THE POPULATION-ENVIRONMENT NEXUS

A major effort is the collaborative research on Population, Land Management and Environmental Change (PLEC). PLEC addresses the question of the interrelationships between population dynamics and environmental change. The greater part of the tropical and sub-tropical developing countries is occupied by farming and pastoral people, most of them small farmers. It is understood that only limited areas of natural habitat can be conserved from all interference; most biodiversity conservation must take place in a context of human use, increasing population densities and commercialization.

At the core is the dynamic nature of the population-environment nexus which affects land management by small farmers at global, regional and local scales. PLEC seeks to examine and disaggregate the processes of adaptation of indigenous resource management systems and land use through a network of field-based research projects in key agro-ecological zones of tropical and sub-tropical environments (Brookfield and Padoch 1994).

In the Asian region, field research has been established in the hill tribe areas of Northern Thailand with a corresponding cluster in the Yunnan province of China. This dual cluster enables a comparative approach to studying the sustainability of smallholder agricultural systems in two areas with similar natural environments, but widely differing social, political and environmental conditions. Another cluster conducts research in Papua

New Guinea investigating both the overall changes in land use, as well as carrying out detailed village-level studies.

The ultimate goal is to provide researched options for the better management of land and resources for small-scale producers. The programme addresses a highly topical issue affecting the sustainability of the world's food production: the ecological changes caused by and affecting small- and medium-scale farming operating under conditions of rapid population growth, socio-economic change, and environmental stress. The key concepts - production pressure and agrodiversity - contain both the problems as well as the potential solutions.

5. CONCLUDING REMARKS

UNU's Global Life-Support System programme area seeks to contribute to the process towards sustainable development through research into new and innovative solutions to global and regional environmental and resource management. Similarly, the University's training activities are aimed at promoting knowledge transfer and capacity building in developing countries. The University's activities are also geared towards greater awareness building and dissemination of knowledge amongst decision-makers and the general public alike in an effort to further environmentally sound development.

REFERENCES

Brookfield, H. and Y. Byron, eds. 1993. *South-East Asia's Environmental Future: The Search for Sustainability*. Tokyo-New York-Paris: United Nations University Press and Kuala Lumpur-Oxford-Singapore: Oxford University Press.

Brookfield, H. and C. Padoch. 1994. Appreciating Agrodiversity: A Look at the Dynamism and Diversity of Indigenous Farming Practices. *Environment* 36 (5): 6-11, 37-45.

Clarke, W.C. and R.R. Thaman, eds. 1993. *Agroforestry in the Pacific Islands: Systems for Sustainability*. Tokyo-New York-Paris: United Nations University Press.

Ives, J.D. and B. Messerli. 1989. *The Himalayan Dilemma: Reconciling Development and Conservation*. London-New York: Routledge.

Kunstadter, P., E.C.F. Bird and S. Sabhasri, eds. 1986. *Man in the Mangroves: The Socio-economic Situation of Human Settlements in Mangrove Forests*. Tokyo: The United Nations University.

Manshard, W. and W.B. Morgan, eds. 1988. *Agricultural Expansion and Pioneer Settlements in the Humid Tropics*. Tokyo: The United Nations University.

Manshard, W. and J.I. Uitto. 1993. Environmental Studies in the Humid Tropics: UNU Programmes and Future Perspectives. In *Environmentally Sound Socio-Economic Development in the Humid Tropics: Perspectives from Asia and Africa*, ed. J.I. Uitto and M. Clüsener-Godt, pp. 7-29. Tokyo: The United Nations University.

UN. 1992. *Agenda 21*. New York: United Nations.

UNU. 1993. *Programme on Environmentally Sustainable Development: UNU Agenda 21*, Advisory Team Report. Tokyo: The United Nations University.

AUTHOR INDEX

SUBJECT INDEX